Herbal Drugs: Ethnomedicine to Modern Medicine

K.G. Ramawat (Ed.)

Herbal Drugs: Ethnomedicine to Modern Medicine

Springer

Editor
Prof. Dr. K.G. Ramawat
M.L. Sukhadia University
Botany Dept.
Udaipur-313002
India

ISBN: 978-3-540-79115-7 e-ISBN: 978-3-540-79116-4

Library of Congress Control Number: 2008935113

Cover design: WMXDesign GmbH, Heidelberg

Printed on acid-free paper

9 8 7 6 5 4 3 2 1

springer.com

About the editor

Professor K.G. Ramawat (born in 1952) received his M.Sc. (1974) and Ph.D. (1978, Plant Biotechnology) from the University of Jodhpur, Jodhpur, India and became a faculty member in January of 1979. He joined M.L. Sukhadia University as an Associate Professor in 1991 and became a Professor in 2001. He served as Head of the Department of Botany (2001–2004), was in charge of the Department of Biotechnology (2003–2004), was a member of the task force on medicinal and aromatic plants at the Department of Biotechnology (Government of India, New Delhi; 2002–2005), and was a coordinator of the UGC-DRS and DST-FIST programs (2002–2007). He did his postdoctoral study at the University of Tours, France (1983–85) and subsequently worked as visiting professor at the University of Tours (1991) and University of Bordeaux 2, France (1995, 1999, 2003, 2006). He visited Poland under the auspices of an INSA-PAN academic exchange program (2005). He has published more than 100 research papers and review articles in reputed journals and books. He has edited two books on the biotechnology of secondary metabolites and of medicinal plants (Scientific Publishers, Enfield, USA and Springer verlag, Heidelberg, Germany). Professor Ramawat has completed several major research projects from UGC, CSIR, ICAR, DBT, and DST, and has supervised the doctoral theses of 16 students. He has been a member of the Plant Tissue Culture Association of India since 1991.

Preface

Considerable progress has been made in our healthcare system, in particular with respect to sensitive diagnostic tools, reagents and very effective and precise drugs. On the other hand, high-throughput screening technology can screen vast numbers of compounds against an array of targets in a very short time, and leads thus obtained can be further explored. In developing countries, the exploding population exerts pressure not only on natural resources but also on the human population itself, whose members strive to become successful and advance in society. This leads to increased blood pressure, anxiety, obesity-associated lipid disorders, cardiovascular diseases and diabetes. Most of these diseases result in disturbed family life, including sexual behaviour.

Despite technological developments, herbal drugs still occupy a preferential place in a majority of the population in the Third World and terminal patients in the West. Herbal drugs, in addition to being cost effective and easily accessible, have been used since time immemorial and have passed the test of time without having any side effects. The multitarget effects of herbs (holistic approaches) are the fundamental basis of their utilization. This approach is already used in traditional systems of medicine like Ayurveda, which has become more popular in the West in recent years. However, the integration of modern science with traditional uses of herbal drugs is of the utmost importance if ones wishes to use ancient knowledge for the betterment of humanity. This book will try to bridge this gap and will be a valuable source for herbalists, traditional and modern medical practitioners, and researchers in botany, ethnobotany, pharmacy, phytochemistry and agriculture. Contributions on herbs used for beneficial effects on memory, sexual behaviour, neurodegeneration, erectile dysfunction, inflammation, cardiovascular diseases, cancer prevention, stroke and central nervous system disorders will provide vital information to readers.

Finally, I would like to acknowledge my contributors, who have gone to great lengths to ensure the high scientific quality of the book. I would also like to thank my colleagues at Springer.

July 2008 *K.G. Ramawat*

Contents

List of Contributors

Bharat B. Aggarwal Cytokine Research Laboratory, Departments of Experimental Therapeutics, University of Texas, M.D. Anderson Cancer Center, 1515 Holcombe Blvd., Houston, TX 77030, USA, e-mail: aggarwal@mdanderson.org

Amina A. Aly Atomic Energy Authority, National Centre for Radiation Research and Technology, Natural Products Department, 110 Nile Street, Dokki, Cairo, Egypt, e-mail: liugs@ibcas.ac.cn

Jan Barciszewski Institute of Bioorganic Chemistry, Polish Academy of Sciences, Noskowskiego 12/14, 61-704 Poznan, Poland, e-mail: Jan.Barciszewski@ibch.poznan.pl

Lisa A. Beltz Department of Biology, University of Northern Iowa, Cedar Falls, IA 50614, USA, e-mail: lisa.beltz@uni.edu

Giuseppina Benoni Department of Medicine & Public Health, Section of Pharmacology, University of Verona, Policlinico G.B. Rossi, Piazzale L.A. Scuro, 37134 Verona, Italy, e-mail: Giuseppina.benoni@univr.it

Jochen Beyer Victorian Institute of Forensic Medicine, Department of Forensic Medicine, Monash University, 57-83 Kavanagh St, Southbank, Victoria 3006, Australia, e-mail: jochenb@vifm.org

Maheep Bhatnagar Animal Biotechnology and Neuroscience Laboratory, Department of Zoology, M.L. Sukhadia University, Udaipur-313001, India, e-mail: mbhatnagar@yahoo.com, m.maheep@gmail.com

Brian F.C. Clark Department of Molecular and Structural Biology, Aarhus University, Gustav Wieds Vej 10C, 8000 Aarhus C, Denmark, e-mail: bfcc@mb.au.dk

Laura Cuzzolin Department of Medicine & Public Health, Section of Pharmacology, University of Verona, Policlinico G.B., Rossi, Piazzale L.A. Scuro, 37134 Verona, Italy, e-mail: laura.cuzzolin@univr.it

S. Dass Laboratory of Biomolecular Technology, Department of Botany, M.L. Sukhadia University, Udaipur-313001, India

Olaf H. Drummer Victorian Institute of Forensic Medicine, Department of Forensic Medicine, Monash University, 57-83 Kavanagh St, Southbank, Victoria 3006, Australia

Kalina Duszka Institute for Genomics and Bioinformatics, Graz University of Technology, Petersgasse 14, 8010 Graz, Austria, e-mail: kalina.duszka@tugraz.at

Thomas Efferth German Cancer Research Center, Pharmaceutical Biology (C015), Im Neuenheimer Fedl 280, 69120 Heidelberg, Germany, e-mail: t.efferth@dkfz.de

James S. Ferraro Department of Physiology, Southern Illinois, University School of Medicine, Carbondale, IL 62901-6512, USA

Shaily Goyal Laboratory of Biomolecular Technology, Department of Botany, M.L. Sukhadia University, Udaipur-313001, India

Peter J. Houghton Department of Pharmacy, King's College London, Franklin-Wilkins Building, London, UK

Melanie-Jayne R. Howes Royal Botanic Gardens, Jodrell Laboratory, Kew, Richmond, TW9 3AB, Surrey, UK, e-mail: m.howes@kew.org

Guo Juan Key Laboratory of Photosynthesis and Environmental Molecular Physiology, Institute of Botany, Chinese Academy of Sciences, 20 Nanxincun, Xiangshan, Haidian District, 100093 Beijing, P.R. China

Peter Kaplan Comenius University, Jessenius Faculty of Medicine, Department of Medical Biochemistry, Malá Hora 4, 037 54 Martin, Slovak Republic

S.S. Katewa Laboratory of Ethnobotany and Agrostology, Department of Botany, University College of Science, M.L. Sukhadia University, Udaipur-313001, India, e-mail: sskatewa@yahoo.com

Hocheol Kim Department of Herbal Pharmacology, College of Oriental Medicine, Kyung Hee University, Seoul 130–701, Korea, e-mail: hckim@khu.ac.kr

Ján Lehotský Comenius University, Jessenius Faculty of Medicine, Department of Medical Biochemistry, Malá Hora 4, 037 54 Martin, Slovak Republic, e-mail: Lehotsky@jfmed.uniba.sk

Gongshe Liu Key Laboratory of Photosynthesis and Environmental Molecular Physiology, Institute of Botany, Chinese Academy of Sciences, 20 Nanxincun, Xiangshan, Haidian District, 100093 Beijing, P.R. China

Frank Massino Senetek PLC, 851 Latour Court, Suite A, Napa, CA 94588, USA, e-mail: FMSenetek@aol.com

Meeta Mathur Laboratory of Biomolecular Technology, Department of Botany, M.L. Sukhadia University, Udaipur-313001, India

H.C. Hans H. Maurer Department of Experimental and Clinical Toxicology, Saarland University, 66421 Homburg/Saar, Germany

G.P. McGregor Philipps-University of Marburg, Faculty of Medicine, Deutschhausstraße 2, 35037 Marburg, Germany, e-mail: mcgregor@staff.uni-marburg.de

Jörg Melzer Institute of Complementary Medicine, Department of Internal Medicine, University Hospital Zurich, Raemistr. 100, 8091 Zurich, Switzerland, e-mail: joerg.melzer@usz.ch

Anil Mittal Cancer Biology Laboratory, School of Life Sciences, Jawaharlal Nehru University, New Delhi, India

Laura L. Murphy Department of Physiology, Southern Illinois University School of Medicine, Carbondale, IL 62901–6512, USA, e-mail: lmurphy@siumed.edu

Martina Pavlikova Comenius University, Jessenius Faculty of Medicine, Department of Medical Biochemistry, Malá Hora 4, 037 54 Martin, Slovak Republic

K.G. Ramawat Laboratory of Biomolecular Technology, Department of Botany, M.L. Sukhadia University, Udaipur-313001, India, e-mail: kg_ramawat@yahoo.com

Reinhard Saller Institute of Complementary Medicine, Department of Internal Medicine, University Hospital Zurich, Raemistr. 100, 8091 Zurich, Switzerland

Beata Saniova Clinic of Anesthesiology and Intensive Medicine, Kollarova 2, SK-03601 Martin, Slovak Republic

Gautam Sethi Cytokine Research Laboratory, Departments of Experimental Therapeutics, University of Texas, M.D. Anderson Cancer Center, 1515 Holcombe Blvd., Houston, TX 77030, USA

Jyoti Shah Department of Urology, St. George's Hospital, Blackshaw Road, London SW17 OQT, UK, e-mail: jyoti.shah@imperial.ac.uk

Chen Shuangyan Key Laboratory of Photosynthesis and Environmental Molecular Physiology, Institute of Botany, Chinese Academy of Sciences, 20 Nanxincun, Xiangshan, Haidian District, 100093 Beijing, P.R. China

Rana P. Singh Cancer Biology Laboratory, School of Life Sciences, Jawaharlal Nehru University, New Delhi, India, e-mail: rana_singh@mail.jnu.ac.in

Bokyung Sung Cytokine Research Laboratory, Departments of Experimental Therapeutics, University of Texas, M.D. Anderson Cancer Center, 1515 Holcombe Blvd., Houston, TX 77030, USA

Peter Urban Comenius University, Jessenius Faculty of Medicine, Department of Medical Biochemistry, Malá Hora 4, 037 54 Martin, Slovak Republic

R. Verpoorte Department of Pharmacognosy, Section Metabolomics, Institute of Biology Leiden, PO Box 9502, 2300RA Leiden, The Netherlands, e-mail: VERPOORT@LACDR.LeidenUniv.NL

Chapter 1
Medicinal Plants: A Renewable Resource for Novel Leads and Drugs

R. Verpoorte

Abstract Present-day drug development is strongly focused on finding active compounds on well-defined targets using high throughput screening approaches. Unfortunately it seems that this approach is becoming less and less successful, as in most cases already good compounds are on the market, and the rapidly rising costs of drug development will make it increasingly difficult to make an economically competitive novel drug for any major disease. In other words, the reductionist approach presently used is becoming less successful. The time has come to rethink drug development. Many Western medicines are based on traditional knowledge from Europe and the Mediteranean region. This is why interest is rapidly increasing in Indian and Chinese medicine, both of which represent a very long tradition of apparently safe use. However, these healthcare systems are different from Western medicine, so novel methods are required to verify the efficacy and safety of the therapies. As it often concerns personalized medication with complex mixtures, a reductionist approach of screening for a single active compound on a known target will in many cases not be successful, as more than one target may be involved; in addition, and complicating the situation even more, synergism and prodrugs may be involved. Systems biology as a novel holistic way of dealing with biological problems seems here an interesting option. Systems biology means proceeding without a hypothesis, just observing, measuring as many parameters as possible in a biological system and afterwards using chemometrics to reveal any meaning in the data. This approach has already proven successful in studying medicinal plants and, in combination with the classical natural-product-based drug lead finding, is expected to be a major issue in the coming years. As present-day patent laws require innovative and unexpected findings, the development of old knowledge does not fit this requirement. Therefore, to support the development of evidence-based traditional medicines, it would be of great interest if some sort of protection could be obtained for companies developing such medicines so that they could earn back their huge R&D investments.

R. Verpoorte (✉)
Department of Pharmacognosy, Section Metabolomics, Institute of Biology Leiden, PO Box 9502, 2300RA Leiden, The Netherlands
e-mail: VERPOORT@LACDR.LeidenUniv.NL

K.G. Ramawat (ed.), *Herbal Drugs: Ethnomedicine to Modern Medicine*, DOI 10.1007/978-3-540-79116-4_1, © Springer-Verlag Berlin Heidelberg 2009

Keywords Herbal medicines · Traditional medicine · Systems biology

1.1 Introduction

Since ancient times humans have explored their environment for plants that could be used to cover all their basic needs: food, shelter, fuel and health. This has resulted in the use of a large number of plants; in particular, food plants' extensive breeding has resulted in high-yield crops. In the case of medicinal plants, such breeding has largely not yet taken place as nature could provide a sufficient supply. The number of medicinal plants has been estimated to be on the order of 40,000 to 70,000 [1], which means that almost 25% of all plant species have some sort of medicinal use somewhere in the world. This heritage from our ancestors has continued to develop in Western medicine and has resulted in the isolation and production of pure active compounds (e.g. morphine, atropine and digoxin) and later in the development of novel synthetic compounds based on this knowledge (e.g. local anaesthetics based on cocaine, analgesics based on morphine). Some of these synthetics based on natural products have been very successful, e.g. acetylsalicylate, which development was based on the use of Salix bark as analgesic. In other cases the result has not been so successful, e.g. the acetyl derivative of morphine (heroin). This illustrates that many medicines in the West have originated in phytotherapy, as occurred in European/Mediterranean region.

In addition, the statistics on novel drugs developed in recent decades show that natural products are a major source of inspiration for drug development [2], with only 30% of all novel molecules (of the 1184 so-called novel chemical entities or NCEs) introduced into the market in the period 1981–2006 being pure synthetic and all others being natural products or natural product related. These statistics also show that the number of novel chemical compounds reaching the market is decreasing every year. The high costs (approx. 1000 million euros) and long duration (more than 10 years), as well as the fact that for most major ailments good medicines that are already available hampers the development of novel drugs by the pharmaceutical industry. Recently problems with serious side effects caused that several novel medicines had to be taken of the market shortly after their introduction. This does not also help to increase efforts at novel drug development.

At the same time the strong emerging economies of countries like India and China have led to greater interest in local healthcare systems, which are even considered an important (cheap) alternative to expensive treatments using Western drugs (see Chapters written by Pandey et al., Melzer and Saller, and McGregor (this volume)). Moreover, after thousands of years of extensive and widespread use of traditional medicines, the question arises as to why we should not consider these medicines again using all the tools of modern science [3, 4]. Further studies may lead to the discovery of novel modes of action, novel biologically active compounds, confirmation of traditional use, or, in the worst case, the fact that no activity is present and even that a given medicine's use can carry risks of toxicity (see Chapters written by Cuzzolin and Benoni, and Benoni and Cuzzolin (this volume)). With 80% of the world's population using such traditional medicine, it makes sense to devote much more resources to such studies. The discovery of the antimalarial compound artemisinin in traditional Chinese medicine some 30 years ago has led to an efficient novel medicine used to treat malaria. But it has also led to totally new potential applications, e.g. in treating cancer (see Chapter written by Efferth (this volume)). Many more hidden gems may be found through studies of traditional medicine (Please specify the title and author you are referring to.).

One of the problems in studying traditional medicines is the totally different healthcare systems they are embedded in, e.g. different ways of classifying diseases, personalized medicines, and the complex mixtures of ingredients in traditional medicines. Current approaches to drug development may pick up some interesting compounds with high activity, but high throughput screening (HTS) will only detect compounds with strong affinity to a target enzyme or receptor; it will miss prodrugs (such as salicin the compound in Salix bark that in the human body is converted via glucolysis and oxidation into salicylate). Also, the synergy between compounds will not be observed in HTS, as one may, for example, envisage that artemisinin may have synergy with other antitumor compounds in a plant. The study by Stermitz et al. [5] showing the synergy between berberine and 5′-methoxyhydnocarpin is now a classical example of synergy between two compounds from one plant. The way traditional Chinese medicines are made and the different roles that each plant traditionally was thought to play in fact point to the possible importance of synergy between ingredients [6]. A recent study on the effects of ginkgo on peripheral blood flow is a beautiful example of the totally different concept of activity of such a traditional medicine and present-day Western pharmacology. Boelsma et al. [7] showed in a placebo-controlled, double-blind clinical trial that a standardized ginkgo preparation caused different effects in different subjects, which would be unacceptable to the Western way of thinking. However, their systems-biology-type of approach showed that in fact the preparation lowered peripheral blood flow in those people who had an above-average peripheral blood flow level, increased it in those who had a below average level, and in the case of the average level did not produce any effect. In other words ginkgo normalizes peripheral blood flow, a concept that does not match the reductionist approach of drug development, using the single-target, single-compound paradigm.

On the other hand, HTS may pick up well-known compounds such as adenosine and GABA in their respective receptor binding assays, thus masking possible other active compounds, but it would confirm the rationale behind the traditional use of a traditional medicine used to treat hypertension [8].

In fact the holistic ideas of traditional healthcare systems demand a holistic approach to studying their activity [4]. First, instead of trying to find an active compound, clinical trials could be considered as a way to confirm activity before trying to understand the activity. In well-established ancient healthcare systems such as in Asia, such experiments could be done in relation to current treatments. The fact that these medicines have been used for several thousand years and are still used extensively means that acute toxicity is unlikely to occur, though long-term toxicity might be a point for some further research (see also Chapters written by Cuzzolin and Benoni, and Benoni and Cuzzolin (this volume)).

In an approach using clinical studies, systems biology enters the picture. Systems biology aims at studying an organism under different conditions without a working hypothesis. Instead one tries to measure as many parameters as possible and use multivariate analysis or other related statistical tools to asses all the data and draw conclusions from this, i.e. the hypothesis comes afterwards. These data may include physiological parameters (e.g. blood pressure, pulse), chemical parameters (using metabolomics to measure e.g. metabolites in body fluids, metabolites in a medicinal plant), the proteome and the transcriptome. Using such a holistic approach prodrugs and synergy may be found. Also new modes of action can be revealed in this way. In any case I think that the different medical systems could learn from each other and in that way make some major steps forwards and become the source of novel ideas and concepts. Combining the best of all approaches would be to the great benefit of all people's healthcare the world over.

That said, one may also wonder why the pharmaceutical industry shows such little interest in traditional medicine. Besides the fact that the above-described problems of prodrugs and synergism do not fit their present expertise for drug development, the major reason might be that of patents. It is not impossible that the activity of a traditional medicine is due to a well-known compound, e.g. GABA or adenosine, which would thus not lead to a novel active and patentable compound. Moreover, patenting of a traditional medicine might be difficult, as a patent requires some sort of innovation, something unexpected [9]. Finding antidiabetic activity in a traditional antidiabetes medicine would thus not be accepted as an innovation, and even a compound isolated for such a plant might be difficult to patent. It would be of great value to all of humanity if any industry developing a traditional medicine with a view towards an evidence-based medicine would also be given some years of protection to be able to earn back the enormous investment needed to develop an evidence-based traditional medicine.

Ginkgo may again serve as an example. There is one ginkgo preparation (see also Chapters written by Howes and Houghton, Bhatnagar, Shah, Lehotsky et al., Melzer and Seller (this volume)) that has been studied extensively in clinical trials and shown to be active. An analysis of six different preparations for sale as an over-the-counter drug on the Dutch market, one of them being an evidence-based preparation, showed that the other five had lower, and some even very low, levels of the compounds thought to be involved in the activity, but the health claims were the same as for the proven one [10, 11]. One problem facing a country such as the Netherlands that has no clear legislation regarding phytotherapy, as the government is in general unfavourably disposed towards phytomedicines, is that a de facto laissez-faire policy is established that leads to the suboptimal use of herbal medicine.

1.2 Conclusion

There is an urgent need to convince Western pharmacologists that traditional medicines can be a major source of novel medicines, as well as novel concepts, but that a different approach to studying these medicines is required.

References

1. Schippmann U, Leaman D, Cunningham AB (2006) Plants as source of medicines: new perspectives. In: Bogers RJ, Craker LE, Lange D (eds) Medicinal and Aromatic Plants – Agricultural, Commercial, Ecological, Legal, Pharmacological and Social Aspects. Springer, Dordrecht, p 75
2. Newmann DJ, Cragg GM (2007) J Nat Prod 70:461
3. Verpoorte R, Kim HK, Choi YH (2005) J Ethnopharmacol 100:53
4. Verpoorte R, Kim HK, Choi YH (2006) Plants as source of medicines: new perspectives. In: Bogers RJ, Craker LE, Lange D (eds) Medicinal and Aromatic Plants – Agricultural, Commercial, Ecological, Legal, Pharmacological and Social Aspects. Springer, Dordrecht, p 261
5. Stermitz FR, Lorenz P, Tawara JN, Zenewicz LA, Lewis K (2000) Proc Natl Acad Sci USA 97:1433
6. Wang M, Lamers RJAN, Korthout HAAJ, van Nesselrooij JHJ, Witkamp RF, van der Heijden R, Verpoorte R, van der Greef J (2005) Phytother Res 19:173
7. Boelsma E, Lamers RJAN, Hendriks HFJ, van Nesselrooij JHJ, Roza L (2004) Planta Med 70:1052
8. Schripsema J, Fung SY, Verpoorte R (1996) Screening of plant cell cultures for new industrially interesting compounds. In: DiCosmo F, Misawa M (eds) Plant Cell Culture Secondary Metabolism. Toward Industrial Application. CRC, Boca Raton, FL, p 1
9. Koopman J (2006) Plants as source of medicines: new perspectives. In: Bogers RJ, Craker LE, Lange D (eds) Medicinal and Aromatic Plants – Agricultural, Commercial, Ecological, Legal, Pharmacological and Social Aspects. Springer, Dordrecht, p 129
10. Choi YH, Choi HK, Hazekamp A, Bermejo P, Schilder YDC, Erkelens C, Verpoorte R (2003) Chem Pharmaceut Bull 51:158
11. Choi YH, Choi HK, Peltenburg-Looman AMG, Lefeber AWM, Verpoorte R (2004) Phytochem Anal 15:325

Chapter 2
The Chemical Diversity of Bioactive Molecules and Therapeutic Potential of Medicinal Plants

K.G. Ramawat, S. Dass and Meeta Mathur

Abstract The therapeutic use of herbs is as old as human civilization and has evolved along with it. The vast majority of people on this planet still rely on their indigenous system of medicine and use herbal drugs. The Indian and Chinese systems of medicine are well established with written records going back around 3000 years. Medicinal plant drug discovery continues to provide new and important leads against various pharmacological targets including cancer, malaria, cardiovascular diseases and neurological disorders. Interest in herbal drugs and natural medicine is undergoing a renaissance at the present time. The medicinal properties of plants are due to the presence of active principles. These bioactive secondary metabolites are synthesized by two principal pathways: shikimic acid or aromatic amino acid, and mevalonic acid. Alkaloids, phenolics and terpenoids constitute many pharmacologically active compounds. Several natural-product drugs of plant origin have either recently been approved by the US Food and Drug Administration (FDA), including arteether, galanthamine and triotopium, or are in clinical trials. Although drug discovery from medicinal plants continues to provide an important source of new drug leads, this work is constrained by the unavailability of sufficient plant material, selection and implementation of appropriate high-throughput screening bioassay and the production of bioactive compounds in large quantities. This article reviews the use of herbs in traditional systems and bioactive therapeutic molecules responsible for this activity.

Keywords Herbal drugs · Traditional medicine · Secondary metabolites · Alkaloids · Terpenes · Polyphenolics

2.1 Introduction

The world's population will be more than 7.5 billion in the next 10 to 15 years. This increase in population will occur mostly in the southern hemisphere, where 80%

K.G. Ramawat (✉)
Laboratory of Biomolecular Technology, Department of Botany, M.L. Sukhadia University, Udaipur-313001, India,
e-mail: kg_ramawat@yahoo.com

K.G. Ramawat (ed.), *Herbal Drugs: Ethnomedicine to Modern Medicine*,
DOI 10.1007/978-3-540-79116-4_2,

of the population still relies on a traditional system of medicine based on herbal drugs [1]. As civilizations grew from 3000 BCE onwards in Egypt, the Middle East, India and China, the uses of herbs became more sophisticated and written records were prepared. The specific plants to be used and the methods of application for particular ailments were passed down through oral history. Later on, information regarding medicinal plants was recorded in herbals [2]. Historically, herbal drugs were used as tinctures, poultices, powders and teas followed by formulations, and lastly as pure compounds. Medicinal plants or their extracts have been used by humans since time immemorial for different ailments and have provided valuable drugs such as analgesics (morphine), antitussives (codeine), antihypertensives (reserpine), cardiotonics (digoxin), antineoplastics (vinblastine and taxol) and antimalarials (quinine and artemisinin). Some of the plants which continue to be used from Mesopotamian civilization to this day are *Cedrus* spp., *Cupressus sempervirens, Glycirrhiza glabra, Commiphora wightii* and *Papaver somniferum* [1, 3, 4]. About two dozen new drugs derived from natural sources were approved by the FDA and introduced to the market during the period 2000–2005 and include drugs for cancer, neurological, cardiovascular, metabolic and immunological diseases, and genetic disorders [5]. Seven plant-derived drugs currently used clinically for various types of cancers are taxol from *Taxus* species, vinblastine and vincristine from *Catharanthus roseus*, topotecan and irinotecan from *Camptotheca accuminata*, and etoposide and teniposide from *Podophyllum peltatum* [6]. It is estimated that the worldwide market potential for herbal drugs is around US$40 billion [6]. A similar situation also exists for plant-based food additives, fragrances and biopesticides. Mostly, herbal drugs are collected from the wild, and relatively few species are cultivated. Overexploitation of plants, particularly when roots, tubers and bark are used for commercial purposes, has endangered 4,000 to 10,000 species of medicinal plants [7]. To counter overexploitation of natural resources and the consequent threats to biodiversity, alternative biotechnological methods and sustainable practices have been recommended. Several world organizations and governments have established guidelines for the collection and utilization of medicinal plants [8, 9].

2.2 Traditional Use of Medicinal Plants

Traditional medicine is the sum total of the knowledge, skills and practices based on the theories, beliefs and experiences indigenous to different cultures used in the maintenance of health, prevention of diseases and improvement of physical and mental illness. In practice, traditional medicine refers to the following components: acupuncture (China), Ayurveda (India), Unani (Arabic countries), traditional birth attendant's medicine, mental healer's medicine, herbal medicine, and various forms of indigenous medicine. Complementary or alternative medicine refers to a broad set of healthcare practices that are not part of a country's own tradition and are not integrated into the dominant healthcare system. Traditional medicine has maintained its popularity in all regions of the developing world, and its use is rapidly spreading in industrialized countries [1]. Knowledge of plants and of healing have been closely linked from the time of human beings' earliest social and cultural groupings. The medicine man was usually an accomplished botanist. Even in historical times, botany and medicine continued to be virtually one and the same discipline until about 1500 CE, when they began to separate from their close association, to the advantage of both sciences.

Knowledge of the medicinal plants used in the drugs of traditional systems of medicine (TSM) has been of great significance, especially as a lead for the discovery of new single-molecule medicines for modern system of medicine. To determine the chemical nature of such compounds, isolation of a substance in pure form using various separation techniques, chemical properties and spectral characteristics are a prerequisite for establishing its correct structure. Thus, medicinal plants are used in crude or purified form in the preparation of drugs in different systems. In countries like India, China and others with well-founded traditional systems of medicine, plant-based formulations occupy an important place in health management [1–10]. However, the recent broadening of the horizons of drug discovery, due to advances in instrumentation and bioinformatics (computational methods), has opened up new avenues for use of this knowledge in drug development research [2–5]. Structural novelty and new modes of action are common features of plant drugs. This has been shown by anticancer agents like vinblastine, vincristine and paclitaxel, cardiovascular agents like forskolin, anti-HIV agents like calanoid, and antihyperlipidemic agents like guggulsterones.

2.3 Ancient Systems of Medicine

2.3.1 Traditional Indian Medicine

The word Ayurveda is derived from 'Ayur', meaning life, and 'veda', meaning knowledge. Ayurveda means the science of life. It is an ancient system of health care and longevity. Ayurveda takes a holistic view of human beings, their health and illness. It aims at positive health, which has been defined as a well-balanced metabolism coupled with a healthy state of being. Disease, according to Ayurveda, can arise from the body and/or mind due to external factors or intrinsic causes. Ayurvedic treatment is aimed at the patient as an organic whole and treatment consists of the salubrious use of drugs, diet and certain practices. This doctrine was conceived when science was not developed enough to understand even the human body, let alone drug molecules [6–11].

Ayurveda, perhaps the most ancient of all medicine traditions, is probably older than traditional Chinese medicine. The origin of Ayurveda is lost in prehistoric antiquity, but its characteristic concepts appear to have matured between 2500 and 500 BCE in ancient India. The earliest references to drugs and diseases can be found in the Rigveda and Atharvaveda, dating back to 2000 BCE. Atharvaveda, comprised of 6599 hymns and 700 prose lines, is considered as the forerunner of Ayurveda.

The 'Samhitas,' or encyclopedia of medicine, were written during the postvedic era, and include 'Charak Samhita' (900 BCE), 'Sushruta Samhita' (600 BCE) and 'Ashtang Hridaya'(1000 CE). Later on, many more treatises were prepared and the use of medicinal plants is described in 'Nighantu Granthas' between the 7th and 16th centuries. The most basic concept of Ayurveda is that all living beings derive their subsistence from three essential factors (three doshas), namely vaata, pitta and kapha, which operate in unison. It believes that the human body is composed of living and non-living environments including earth, water, fire, air and space. Illness is the consequence of imbalance between the various elements, and it is the goal of treatment to restore this balance [11, 12].

Ayurvedic drugs are also attracting much attention for diseases for which there are no or inadequate drugs for treatment in modern medicine, such as metabolic and degenerative disorders. Most of these diseases have multifactorial causation, and there is a growing realization that in such conditions, a combination of drugs, acting at a number of targets simultaneously, is likely to be more effective than drugs acting at one target. Ayurvedic drugs, which are often multicomponent, have a special relevance for such conditions [12]. For various reasons, Ayurveda has not incorporated much of modern science/scientific tools. Investigation of the biological activity of multicomponent Ayurvedic drugs will bring Ayurveda into the mainstream of scientific investigations.

2.3.2 *Traditional Chinese Medicine*

Traditional Chinese medicine (TCM) has been in practice for more than 200 years and includes acupuncture, massage (tuina), breathing exercise (qi gong) and dietary therapy. TCM has been an integral part of China's healthcare system along with conventional Western medicine. TCM products were safe and effective for the treatment of many human diseases before Western medicine was introduced in China. Famous texts in TCM include the *Yellow Emperor's Inner Classic* (Hung Di Nei Jing; ~200 BCE to 100 CE), *Divine Husband-man's Classic of Materia Medica* (Shen Nong Ben Cao Jing; 25-220 AD) and cold-induced disorders (Shang han Lun; 220 AD). The most complete reference to Chinese herbal prescriptions is *Chinese Materia Medica*, published in 1977. It lists nearly 6000 drugs, of which 480 are of plant origin. This ancient system of medicine, believed to be more than 5000 years old, is based on two separate theories about the natural laws that govern good health and longevity, namely 'Yin and Yang', which are in opposition to each other, and the five elements (wu xing). The five-element theory is similar to the four humours and elements of the Greeks or the three humours of Ayurveda. The five elements are earth, metal, water, wood and fire, each of which is linked to the main organ systems of the body—spleen, lungs, kidney, liver and heart, respectively. It considers that an unbalanced diet, lifestyle or environment will disrupt the body balance, which in turn manifests as symptoms of diseases. The aim of the practitioner of TCM is to restore health by removing the cause, correcting abnormal functioning, opposing the imbalance and normalizing the flow of energy. *Angelica polymorpha* var. *sinensis, Artemesia annua, Ephedra sinica, Paeonia lactiflora, Panax ginseng, Rheum palmatum* and *Peuraria lobata* constitute the important medicinal plants of TCM [3, 13, 14].

2.3.3 Traditional Egyptian Medicine

Although Egyptian medicine dates from at least 3000 BCE, the last known and most important pharmaceutical record is the *Papyrus Ebers* (1500 BCE). Use of *Ricinus communis* seeds, *Citrilus colocynthes, Senna alexandrina* and *Prunica granatum* roots in large quantities is mentioned in the ancient Egyptian literature. These uses were later documented by the Greek physician Dioscorides (100 CE). Writings of the Greeks, such as Hippocrates (460–377 BCE) and Galen (130–200 CE), also used parts of the *Papyrus Ebers*. Therefore Greek, and ultimately modern, medicine has its origin in Egyptian or Nile Valley civilization [15].

2.3.4 Traditional Arabic Medicine

The Babylonians, Assyrians and Sumerians comprise one of the oldest civilizations, and several plants were domesticated during this early period. Several medicinal plants are mentioned in civil laws carved on stone and commissioned by the King of Babylon (1700 BCE). The Arabs established drugstores in the eighth century, and the Persian pharmacist Avicenna described all Greco-Roman medicine in his book *Canon of Medicine*. This text forms the basis of distinct Islamic healing system known today as Unani-Tibb. *Papaver somniferum* was known to the Sumerians in 4000 BCE as *hul gil* (joy plant). The most frequently used medicinal plants in the Middle East are: *Allium cepa, Astracantha gummifera, Carthamus tinctorius, Carum carvi, Ferula asafoetida, Lawsonia inermis, Papaver somniferum, Peganum harmala, Prunus dulcis, Prunica granatum, Salvadora persica, Senna alexandrina, Sesamum indicum, Trachyspermum ammi, Trigonella foenum-graecum* and *Vitis vinifera* [3, 16].

2.3.5 African, European and Other Traditional Systems of Medicine

Africa is considered the cradle of *Homo sapiens*' emergence. Though traditional African medicine is the oldest and perhaps the most diverse of all healthcare systems [3], detailed documentation on the use of medicinal plants in Africa is lacking. With rapid urbanization, traditional oral knowledge is dwindling fast, e.g. knowledge of traditional oral knowledge of the Khoisan, the Nguni and the Sotho-speaking peoples [17]. Traditional African medicine is holistic, involving both body and mind. Famous African medicinal plants include *Acaccia senegal* (source of gum Arabic), *Aloe ferox, Aloe vera, Artemisia afra, Asplanthus linearis, Boswellia sacra, Catha edulis, Commiphora myrrha, Harpagophytum procumbens, Catharanthus roseus*, etc.

Like Africa, South American countries are also rich in biodiversity and diverse healing cultures, but information on the use of medicinal plants is sparse. The famous medicinal plants from this region are *Cinchona pubescens, Erythroxylum coca, Ilex paraguariensis, Paullinia cupana, Spilanthes acmella* and *Uncarina tomentosa*. The European healing system is believed to have originated with Hippocrates (460–377 BCE) and Aristotle (384–322 BCE). Subsequent naturalists like Theophrastus (~300 BCE), Dioscorides (100 CE) and the pharmacist Galen (130–200 CE) recorded the use of medicinal plants. The philosophy was based on the belief that the body is composed of earth, wind, fire and water, similar to the Indian system [14]. The famous book *De Materia Medica* by the Greek physician Dioscorides was the standard reference in Europe for more than 1000 years. The use of herbal teas and decoctions is still very popular in Europe, e.g. teas prepared from *Humulus lupulus, Rosmarinum officinalis, Hypericum perforum* and *Valeriana officinalis* [14].

Though traditional and alternative medicine and its practitioners exist in Europe, it is not officially recognized and is punishable under the law in France, Italy, Spain and other countries, while it is unregulated in UK. This requires provisions in pharmacopoeias to include herbal drugs. Allopathic medicine is practiced predominantly in developed countries, and herbal drugs are categorized as food supplements and are not reimbursed by the social security system.

2.4 Exploration of Medicinal Plants

Plants are a great source of therapeutic molecules. In thc carly 20th ccntury, taxonomic surveys established the identity of plants, followed by ethnomedical surveys documenting the use of plants as medicine and other uses. The identification of active principles of medicinal plants leads to the use, misuse and abuse of substances of vegetal origin. The use may be curative (e.g. vincristine and vinblastine, reserpine, ephedrine, aspirin, morphine, digoxin) or narcotic abuse (cocaine, morphine and cannabis), and misuse has made several plants endangered species, e.g. *Podophyllum hexandrum, Taxus baccata, Coptis teeta, Picrorhiza kurroa* and *Nardostachys jatamansi* [18]. This overexploitation has resulted in depletion in germplasm resources, particularly in Third World countries, and urgently warrants the development of alternative biotechnological methods for micropropagation, the study of seed and reproductive biology, and, last but not least, social awareness [11]. It is estimated that approximately 1500 plant species in India are threatened including 124 endangered species [19]. About 250,000 species of higher plants are yet to be investigated for pharmacological activity. Plants can be a source of effective remedies for Alzheimer's, Parkinson's, epilepsy, migraine, arthritis and schizophrenia. Increased demand for natural drugs has led to the domestication of several plants such as

Catharanthus roseus and *Taxus baccata*, and several others (*Psoralea corylifolia, Carthamus tinctorius*) are being evaluated for agronomic traits [9, 11, 20].

Secondary metabolites have evolved in nature in response to needs and challenges of the plant environment. Nature has been carrying out its own combinatorial chemistry for over three billion years [21]. The total number of natural products produced by plants has been estimated at over 500,000 [22]. Ethnobotanical and traditional usage of medicinal plants serves as a source of information for the isolation of active compounds, e.g. as direct therapeutic agents (D-tubocurarine from *Chondrodendron tomentosum*), as the starting drug for semisynthesis (diosgenin from *Dioscorea floribunda*), the model drug for new synthetic drugs (cocaine from *Erythroxylum coca*), for the synthesis of local anaesthetics and, lastly, as taxonomic markers for identification [2, 3, 23]. Drug discovery from plants requires the combined efforts of botanists, pharmacognosists, phytochemists and other scientists to screen the products. Improvements in isolation techniques to meet the demand for pharmacology, the generation of large numbers of samples from correctly identified plants from the tropics for high-throughput screening, elaborate arrangements for preclinical (pharmacology, toxicology, pharmacokinectics and drug delivery) and, lastly, for clinical trials are required for drug development. This process may take more than 10 years [24].

2.5 Approaches to Drug Discovery

Isolation of compounds in their pure form and evaluating them for their pharmacological properties leading to drug discovery is a long, tedious, time-consuming and expensive path. Drug discovery from natural sources has become very expensive (currently estimated at US$500 million) and time consuming (5 to 6 years in the 1980s to 15 to 22 years in the 21st century). Two-thirds of the cost goes to leads that fail during the clinical trials [25]. Fifty percent of all potential drugs fail because of adsorption, distribution, metabolism, excretion or toxicity problems [26]. Due to the long periods involved in discovery and preclinical and clinical trials, which require large numbers of volunteers (1000 to 5000 for long-term effects), many companies have abandoned the search for natural products and turned their attention towards combinatorial chemistry and modification/analogue synthesis of existing drugs. Therefore, the new approaches like metabolomics and high-throughput screening (HTS) are used to screen and evaluate several metabolites in a short time.

In the postgenomic era, pharmaceutical researchers are evaluating vast numbers of protein sequences to formulate novel strategies for identifying valid targets and discovering leads against them [27]. Modern drug discovery often involves screening small molecules for their ability to bind to a preselected protein target. Drug discovery can also involve screening small molecules for their ability to modulate biological pathways in cells or organisms, without regard to any particular protein target. Thus, the establishment of various techniques of the genomic sciences, such as rapid DNA sequencing, together with combinatorial chemistry, cell-based assays and automated HTS, has led to a new concept of drug discovery [28]. In this concept, interaction between biologists and chemists, as well as scientific reasoning, has been replaced by a very high number of samples processed. With rapid industrialization, an HTS system has been developed to screen not just a few hundred but hundreds of thousands of chemical compounds in a short amount of time. HTS was created in the early 1990s for the rapid screening of large numbers of extracts/compounds. This requires the identification of disease-specific targets by basic research or by a genomic approach, which is used to design/develop a bioassay used in the HTS system [29]. Under an HTS setup, large numbers of hypothetical targets are incorporated into cell-based assays and exposed to large numbers of compounds representing numerous variations on a few chemical themes. It is assumed that this experimental design would be suitable to identify many substances, which can modify the target in question. Such molecules are then isolated in greater quantity and evaluated on more complex models (cells, animals) to a certain efficacy. About 50 million screening tests have been conducted so far using different molecules, different concentrations and different bioassays [28]. These technologies generated vast amounts of information on natural products obtained from plants and microorganisms and have had a historic impact on modern medicine.

New bioactive molecules are desirable to pharmaceutical companies for their growth and economic viability. Previously, the search for new bioactive molecules was carried out either by activity-based separation and purification or by pure compounds obtained through slow and tedious chemical methods. These methods were time consuming, yielded a small number of products, and resulted in the failure of the product/process using critical evaluation parameters at various stages of clinical/toxicological tests. The history of gene expression analysis began when laboratory methods were developed to examine the expression of individual known genes. The northern blot technique, developed in the late 1970s, hybridizes labelled DNA or RNA probes of known genes to RNA blots. The resulting expression patterns of mRNA transcripts can then be read [30]. Now technology has reached its pinnacle where large numbers of genes can be sequenced or gene products can be analyzed. The ability to rapidly survey and compare gene expression levels between reference and test samples using new technology such as differential display, READS (improved differential display), expressed sequence tag (EST), serial analysis of gene expression (SAGE), and DNA microarray has made drug discovery genome oriented. By comparing gene products of two samples it is possible to identify gene products or disease targets. To find new drugs in the postgenomic era, pharmaceutical researchers must evaluate vast numbers of protein sequences and formulate novel intelligent strategies to identify valid targets and discover promising molecules against them. This is helpful in the identification of small molecules that selectively target proteins. By identifying protein function first, efficacy is gained that makes it possible later on to focus resources on protein families of interest [27]. Although the sequencing of the human and other genomes is a result of the synergy between innovations in automation and bioinformatics, it is in principle a breakthrough in synchronous use and process management. New technologies offer rapid and simultaneous analysis of the sequence and function of all genes in the genome.

Mass sequencing is done today in factory-like setups, with many to several hundred sequencers running 24 h, 7 d a week [31]. The companies involved in such work are Incyte Pharmaceuticals, Human Genome Sciences, Millennium, and PE Celara. These known sequences will be useful in the rapid drug discovery for target diseases and in the identification of lead molecules [29].

2.6 Bioactive Molecules of Medicinal Plants

Plant cells produce two types of metabolites. Primary metabolites are involved directly in growth and metabolism, viz. carbohydrates, lipids and proteins. Primary metabolites are produced as a result of photosynthesis and are additionally involved in cell component synthesis. Most natural products are compounds derived from primary metabolites such as amino acids, carbohydrates and fatty acids and are generally categorized as secondary metabolites. Secondary metabolites are considered products of primary metabolism and are generally not involved in metabolic activity viz. alkaloids, phenolics, essential oils and terpenes, sterols, flavonoids, lignins, tannins, etc. These secondary metabolites are the major source of pharmaceuticals, food additives, fragrances and pesticides [4, 6, 32, 33].

In general, primary metabolites obtained from higher plants for commercial use are high-volume, low-value bulk chemicals. They are primarily used as industrial raw materials, foods or food additives such as vegetable oils, carbohydrates (sucrose, starch, pectin and cellulose) and proteins. It is the medicinal plants that are rich in secondary plant products, and it is because of these compounds that these are termed 'medicinal' or 'officinal' plants. These secondary metabolites exert a profound physiological effect on mammalian systems; thus they are known as the active principle of plants. With the discovery of the physiological effect of a particular plant, efforts are being made to know the exact chemical nature of these drugs (called active principle) and, subsequently, to obtain these compounds by chemical synthesis [4].

Here we present a brief account of physiologically active primary and secondary metabolites, and more prominent compounds belonging to the alkaloid, terpenoid and phenolic groups. This will provide readers an overview of the biosynthesis, diversity and distribution of such compounds.

Besides secondary plant products, several primary metabolites exert strong physiological effects. In this category, proteins are the principal compounds having such diverse functions as blood agglutinants from Fabaceae, hormones (e.g. insulin), various snake venom poisons, ricin from *Ricinus communis*, and abrine and precatorine from *Abrus precatorius*. Other examples of primary metabolites exerting a strong physiological effect include certain antibiotics, vaccines and several polysaccharides acting as hormones or elicitors [34, 35].

The carbon skeleton of all the compounds are derived from carbohydrates synthesized by photosynthesis. The synthesis of various classes of secondary metabolites from primary metabolites is presented in schematic form in Fig. 2.1. The majority

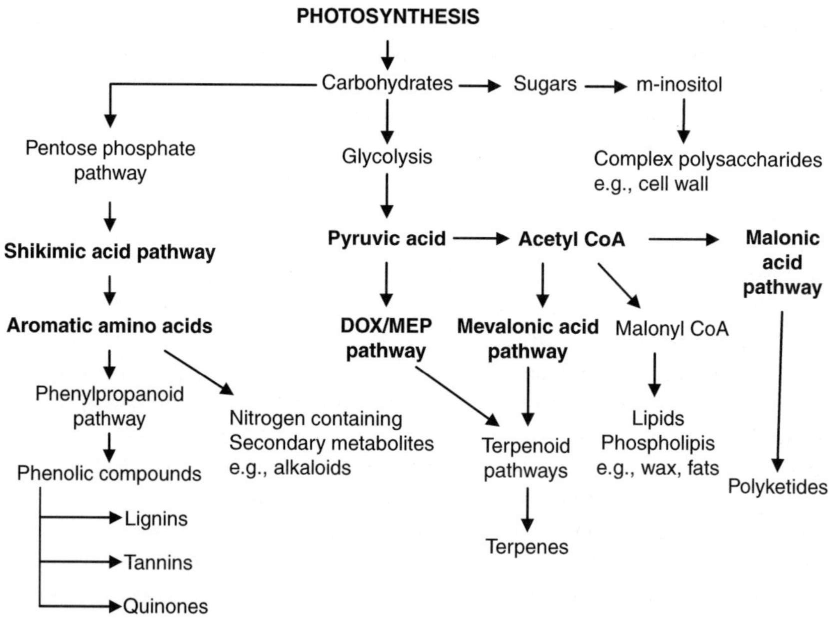

Fig. 2.1 Principal biosynthetic pathways leading to synthesis of secondary metabolites

of secondary metabolites are synthesized via two principal biosynthetic pathways: (1) shikimic acid pathway producing a pool of aromatic amino acids, which in turn are converted into diverse compounds such as phenolics (lignins, tannins, quinones) and alkaloids [36], and (2) acetyl-CoA mevalonic acid pathway, leading to a vast array of terpenoids [37].

2.6.1 Alkaloids

Around 12,000 alkaloids of various types have been known to occur in all land plants, including more than 150 families. The important families are Apocynaceae, Papaveraceae, Fabaceae, Ranunculaceae, Rubiaceae, Rutaceae, Solanaceae, and less common lower plants and fungi (ergot alkaloids). In plants, alkaloids generally exist as salts of organic acids like acetic, oxalic, citric, malic, lactic, tartaric, tannic and other acids. Some weak basic alkaloids (such as nicotine) occur freely in nature. A few alkaloids also occur as glycosides of sugar such as glucose, rhamnose and galactose, e.g. alkaloids of the solanum group (solanine), as amides (piperine), and as esters (atropine, cocaine) of organic acids [4, 38].

In plants, alkaloids may be present systematically in whole plants, or they may be accumulated in large amounts in specific organs like roots (aconite, belladonna), stem bark (cinchona, pomegranate) and seeds (nux vomica, *Areca*). In angiosperms, alkaloids are more common in dicots than in monocots.

Alkaloids have known to humans for several centuries. They are a diverse group of low-molecular-weight, nitrogen-containing compounds found in about 20% of plant species (Fig. 2.2). Morphine, an alkaloid from latex of the opium poppy, was isolated by F.W. Serturner in 1806, whose structure could be confirmed in 1952 due to the stereochemical complexity of the molecule [39]. Later on, other alkali-like active principles were isolated and identified, e.g. narcotine in 1817 by Robiquet, emetine in 1817 by Pelletier and Magendie, and so on. The term 'alkaloid' was coined by W. Meibner, a German pharmacist, meaning 'alkali like'. Later it was demonstrated that the alkalinity was due to the presence of a basic nitrogen atom. The first alkaloid synthesized was coniine in 1886 by Ladenburg, which had already been isolated in 1827 [4].

According to Pelletier [40] "an alkaloid is a cyclic organic compound containing nitrogen in a negative oxidation state which is of limited distribution among living organisms". Sometimes it is not possible to draw a clear line between true alkaloids and certain plant bases. Simple bases, such as methylamine, trimethylamine and other straight-chain alkylamines, are not considered alkaloids. Other compounds such as betaines, choline and muscarine (present in fig agaric, *Amanita muscarea*) are also excluded from alkaloids by some experts. These compounds are synthesized from amino acids and categorized as biological amines or protoalkaloids because their nitrogen is not involved in a heterocycle system. Similarly, polyamines (putrescine, spermine, spermidine) are also excluded. Some authorities even exclude the phenylalkylamines, such as β-phenylethylamine (mistletoe, *Viscum album*; barley, *Hordeum vulgare*), dopamine (banana, *Musa sapientum*), ephedrine (Ma huang, *Ephedra sinica*), mescaline (peyote, *Lophora williamsii*) and tryptamine (*Acacia* species). Widely distributed vitamin B_1 (thiamine) is not categorized as an alkaloid even though it contains a nitrogen in heterocycle and has physiological activity. Similarly, purine-based compounds (caffeine, theophilline, theobromine) are also excluded by some workers, as they are not derived from amino acids [4]. However, a neutral compound such as colchicine from autumn crocus (*Colchicum autumnale*), in which the nitrogen is present in an amide group, is an alkaloid because of other traits like medicinal properties and restricted distribution in plants. Other examples of neutral compounds such as alkaloids are piperine from black pepper (*Piper nigrum*), indicine-n-oxide (*Heliotropium indicum*), di-n-oxide trilupine (*Lupinus barbiger*, L. *laxus*), betaines e.g. stachydrine (*Medicago sativa*) and trigonelline (in fenugreek, garden peas, oats, potatoes, coffee, hemp) [4].

The potent physiological activity of many alkaloids has also led to their use as pharmaceuticals, stimulants, narcotics and poisons. Alkaloids currently in clinical use include the analgesics morphine and codeine, the anticancer agent vinblastine, the gout suppressant colchicine, the muscle relaxant (+) tubocurarine, the antiarrythemic ajmalicine, the antibiotic sanguinarine and the sedative scopolamine. The plant alkaloids like caffeine in tea and coffee and nicotine in all preparations (smoking, chewing) of tobacco are widely consumed daily [4].

Piperidine alkaloids such as coniceine, coniine and *N*-methyl coniine are present in *Conium maculatum*. Apart from tobacco alkaloids, nicotinic acid and its derivatives are a major pyridine alkaloid present in plants. The most commonly occurring

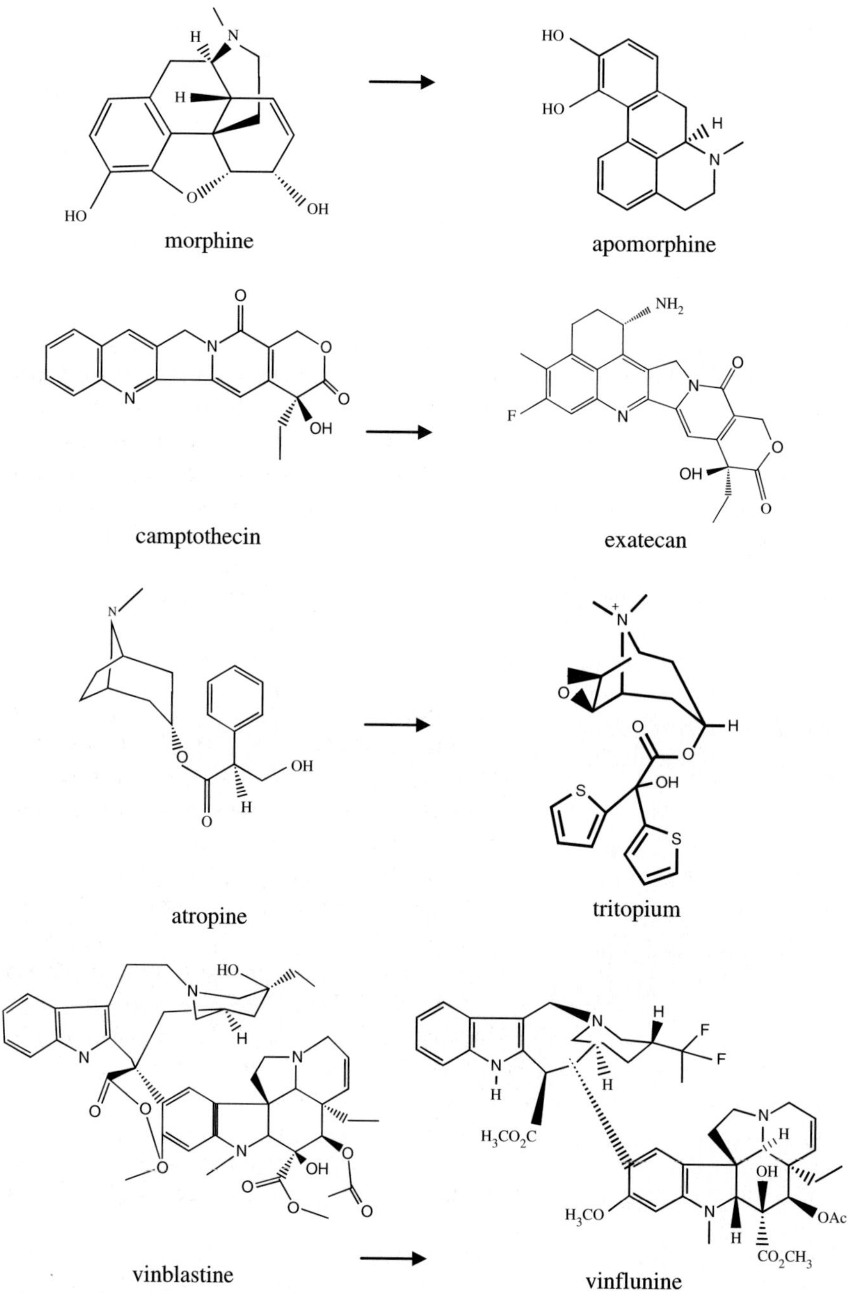

Fig. 2.2 Alkaloids

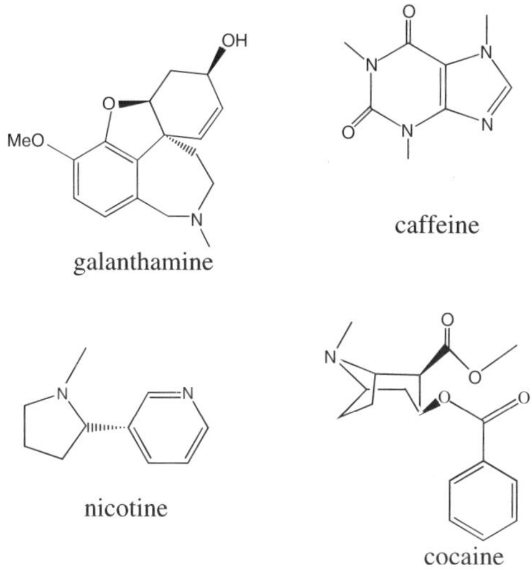

Fig. 2.2 (continued)

compound is trigonelline (*N*-methyl-nicotinic acid), which is present in *Trigonella foenum-graecum*. Anticholinergic alkaloids hyoscyamine, atropine and hyoscine (scopolamine) are found principally in plants of the family Solanaceae. *Datura* species contain more than 30 alkaloids. *D. stramonium* and *D. innoxia* are the main sources of hyoscyamine and scopolamine respectively. Other species known to contain tropane alkaloids are *Atropa belladonna* (Deadly nightshade), *Hyoscyamus niger* (Henbane), *H. muticus, Dubosia* hybrids, *D. myoporoides* and *D. leichhardtii* [41]. Nicotine and tropane alkaloids are formed in the roots and transported to the aerial parts of the plant [42]. The tropane alkaloids possess an 8-azabicyclo octane nucleus and are found in plants of three families, Solanaceae, Erythroxylaceae and Convolvulaceae.The attractive berries of these plants contain these alkaloids, which are toxic. Fewer than three berries of Henbane (*Hyoscyamus niger*) or deadly nightshade (*Atropa belladonna*), both of which contain scopolamine and hyoscyamine, can cause death in infants. Both the tropane ring moiety of the tropane alkaloids and the pyrrolidine ring of nicotine are derived from putrescine by way of *N*-methyl putrescine (Fig. 2.3). Because putrescine is metabolized to polyamines such as spermidiene and spermine, the *N*-methylation of putrescine catalysed by putrescine *N*-methyltransferase is the first committed step in the biosynthesis of these alkaloids [43, 44].

Galanthamine hydrobromide is an Amaryllidaceae alkaloid obtained from *Galanthus* and *Crinum* species that has been used traditionally in Russia, Bulgaria and Turkey for neurological conditions. It is currently obtained by synthesis and used as an acetyl cholinesterase inhibitor in the treatment of Alzheimer's disease [45].

Cocaine in Coca (*Erythroxylum coca*) was the first local anaesthetic to be discovered. Leaves of a few species of *Erythroxylum*, indigenous to Peru and Bolivia,

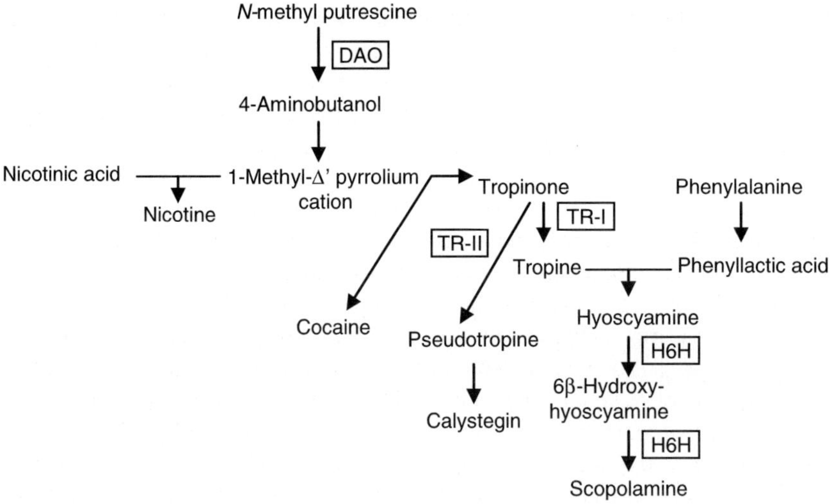

Fig. 2.3 Biosynthesis of tropane alkaloids from N-methyl putrescine

contain 0.6 to 1.8% cocaine. The leaves of the plant have been used for centuries by natives to increase endurance and to promote a sense of well-being. Cocaine was isolated in 1859 by A. Niemann for its central-nervous-system-stimulating activity, which can lead to dependence liability; cocaine has been a drug of abuse. These alkaloids are synthesized from tropic acid. Outside the Solanaceae, tropane alkaloids occur in two other plant families. Within the Erythroxylaceae, the genus *Erythroxylon* comprises about 200 widely distributed, tropical species found mainly in S. America and Madagascar. *E. coca* is the only plant currently cultivated for cocaine production.

About 2500 benzyl isoquinoline alkaloids have been isolated from diverse taxa, including well-established drugs like morphine, sanguinarine, berberine and palmatine. The biosynthesis of these alkaloids begins with the conversion of tyrosine into dopamine and 4-hydroxyacetaldehyde [39]. Isoquinoline-type alkaloids show strong pharmacological activities like those of morphinan-, protoberberine- and benzophenenthridine-type alkaloids, and they are widely distributed in the plant kingdom, mainly in Papaveraceae, Berbidaceae, Ranunculaceae and Menispermaceae.

The opium poppy (*Papaver somniferum*) is one of the oldest cultivated plants. The therapeutic use of latex obtained from unripe capsules of poppy was recorded by Theophrastus in the third century BCE. Dioscorides (100 CE) described the curative properties of the opium poppy and presented the different uses for both latex and extracts of whole plants. Laticifers are found associated with the vascular bundle in plant parts. The morphinan alkaloids, morphine, codeine and thebaine, are found both in roots and in aerial plant parts and specifically accumulate in vesicles within laticifers. Details of opium alkaloid biosynthesis, the enzymes involved and localization are given elsewhere [39], [46–48]. These alkaloids are used in modern medicine for the treatment of pain, cough and diarrhoea. Two new semisynthetic derivatives of age-old drugs, morphine and atropine, have been developed and are being used clinically for Parkinson's disease and for chronic obstructive pulmonary disease, respectively. Apomorphine hydrochloride [49] is a dopamine receptor agonist, while triotropium bromide is derived from atropine [50].

The antimalarial activity of the alkaloid quinine obtained from the bark of *Cinchona* species (*C. ledgeriana, C. pubescens* and *C. officinalis*) has been known for many centuries. Besides their pharmaceutical use, *Cinchona* alkaloids are also used in the food and soft drink industry because of their bitter taste [4].

Terpenoid indole alkaloids comprise a group of about 3000 compounds with well-known compounds such as antineoplastic agents vinblastine and camptothecin, the antimalarial drug quinine and the rat poison strychinine. Terpenoid indole alkaloids consist of an indole component provided by tryptamine and a terpenoid moiety derived from secologanin. Tryptophan, a shikimic acid biosynthetic pathway product, is converted to tryptamine by tryptophan decarboxylase (TDC). The biosynthesis of these alkaloids is discussed elsewhere in detail [39, 42], [51–53]. New derivatives of vinblastine and camptothecin are currently in clinical trials for their anticancer properties, such as vinflunine and exatecan, respectively [2]. Besides vincristine and vinblastine, *Catharanthus roseus* is famous for its more than 100 secologanin-derived monoterpene indole alkaloids (MIAs). The monomeric MIA ajmalicine is used in the treatment of circulatory disorders, whereas the heterodimeric MIAs vinblastine and vincristine are powerful antitumour drugs [54].

Purine alkaloids are widely distributed within the plant kingdom (Fig. 2.2) and have been detected in at least 90 species belonging to 30 genera. Caffeine and threobromine, methylated derivatives of xanthine, are generally the main purine alkaloids and are regularly accompanied in low concentrations by the two methylxanthines theophylline and paraxanthine, as well as by methylated uric acids such as theacrine, methylliberine and liberine. Purine alkaloids, being present in tea and coffee, are widely consumed in the human diet across the continents. Plant species from different families are made into a pleasant stimulant, e.g. coffee (*Coffea arabica, C. robusta*), tea (*Camellia sinenesis*), cocoa (*Theobroma cacao*), mate (*Ilex pguariensis)* and cola (*Cola nitida*) [55].

2.6.2 Phenolics

Simple phenolics are termed compounds and have at least one hydroxyl group attached to an aromatic ring, e.g. catechol. Most compounds having a C_6C_1 carbon skeleton, usually with a carbonyl group attached to the aromatic ring, are phenolic compounds [56].

Simple phenylpropanoids are defined as secondary metabolites derived from phenylalanine, having a C_6C_3 carbon skeleton, and most of them are phenolic acids, e.g. cinnamic acid, *o*-coumaric acid, *p*-coumaric acid, caffeic acid and ferulic acid [57]. A simple phenylpropanoid can conjugate with an intermediate from the shikimic acid pathway, such as quinic acid, to form compounds like chlorogenic acid. Phenolic compounds having a $C_6C_3C_6$ carbon skeleton include flavonoids (including anthocyanins) and isoflavonoids.

Phenolic compounds are generally synthesized via the shikimate pathway (Fig. 2.4), but the polyketide pathway can also provide some phenolics, such as orcinols and quinones. Phenolic compounds derived from both pathways are quite common, e.g. flavonoids, stilbenes, pyrones and xanthones [58].

The shikimate pathway, a major biosynthetic route for both primary and secondary metabolism, starts with phosphoenol pyruvate and erythrose-4-phosphate and ends with chorismate [59]. Chorismate is an important branching point since it is the substrate for all subsequent products (Fig. 2.4) [60]. A great diversity of phenolic compounds are synthesized through these intermediate products, e.g. phenylalanine is a common precursor for C_6C_3 and $C_6C_3C_6$ compounds, and their polymers such as tannins and lignins [38].

Resveratrol (3, 5, 4′-trihydroxystilbene) is an oligomeric polyphenol found as dimer, trimer and tetramer in the families Vitaceae, Dipterocarpaceae, Cyperaceae, Gnetaceae and Leguminosae. Resveratrol is synthesized from phenylalanine, mediated by the enzyme stilbenes synthase, while chalcone synthase converts phenylalanine into flavonoids. Resveratrol is implicated in the prevention of cancer and

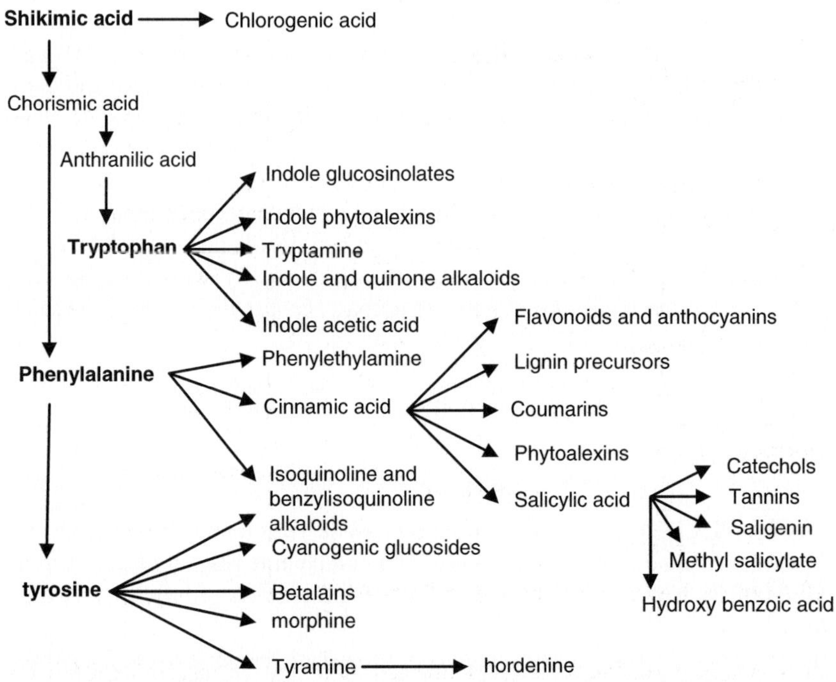

Fig. 2.4 Biosynthesis of diverse phenolic compounds from pool of amino acids formed by shikimic acid pathway

cardiovascular diseases in vasoprotection and neuroprotection [61–63]. The details about the mode of action of resveratrol are discussed in reviews on resveratrol [62–64].

The phenolic group includes metabolites derived from the condensation of acetate units (e.g. terpenoids), those produced by the modification of aromatic amino acids (e.g. phenylpropanoids, cinnamic acids, lignin precursors, hydroxybenzoic acids, catechols and coumarins), flavonoids, isoflavonoids and tannins. The phenolics derived from aromatic amino acids, and their precursors, are just some of the very wide range of compounds (Fig. 2.5) derived from shikimic acid [57]. A phenyl group having three carbon side chains is known as a phenylpropanoid, such as hydroxycoumarins, phenylpropenes and lignans. The phenylpropenes are important components of many essential oils, e.g. eugenol in clove oil (*Syzygium aromaticum*) and anethole and myristicin in nutmeg (*Myristica fragrans*) [65].

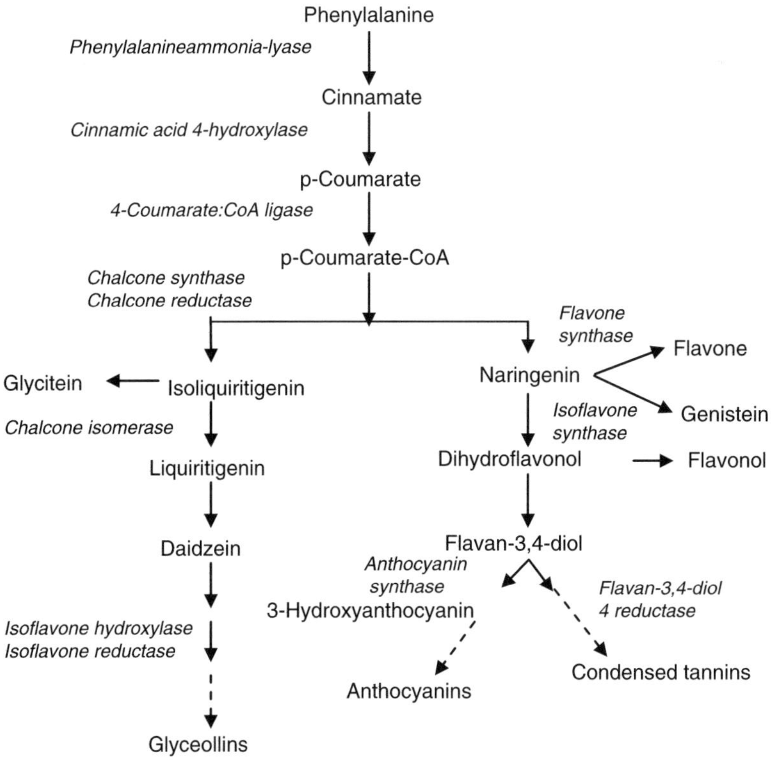

Fig. 2.5 Biosynthetic pathway of major flavonoids and anthocyanins

Flavonoids have two benzene rings attached by a propane unit and are derived from flavones. They are found throughout the plant kingdom, whereas isoflavonoids are more restricted in distribution, and are present in the family Fabaceae, in which they are widely distributed and function as antimicrobial, anti-insect compounds, as an inducer of nodulation genes of symbiotic *Rhizobium* bacteria, or as allelopathic agents [66]. Flavonoids are brightly coloured compounds generally present in plants as their glycosides. Different classes within this group differ by additional oxygen-containing heterocyclic rings and hydroxyl groups and include the chalcones, flavones, flavonols, flavanones, anthocyanins and isoflavones. Anthocyanins impart red and blue pigment to flowers and fruits and can make up as much as 30% of the dry weight of some flowers. Flavanones, flavonols and anthocyanins normally exist as their glycosides. The isoflavonoids are rearranged flavonoids, in which this rearrangement is brought about by a cytochrome P-450-dependent enzyme which transforms the flavanones' liquiritigenin or naringenin into the isoflavones daidzein or genistein, respectively. Simple isoflavones such as daidzein, and coumestans such as coumestrol, have sufficient estrogenic activity to seriously affect the reproduction of grazing animals and are known as phytoestrogens [56]. Isoflavones exhibit estrogenic, antiangeogenic, antioxidant and anticancer properties [66, 67].

Major sources of isoflavones for humans are pulses, particularly soybeans and chick peas [68]. Epidemiological studies suggest a link between consumption of soy isoflavones and reduced risks of breast and prostate cancers [66–68]. Isoflavones also possess other health-promoting activities, such as chemoprevention of osteoporosis, and prevention of postmenopausal disorders and cardiovascular diseases [69, 70]. Phenoxodiol, a synthetic analogue of diadzein, is being developed as a therapy for cervical, ovarian, prostate, renal and vaginal cancers and induces apoptosis through inhibition of antiapoptopic proteins [71]. The other important compounds (Fig. 2.6) in this group include quercitin (flavonoid), silybin (flavonolignan) and genistein (isoflavone).

2.6.3 Terpenes

The functional diversity of chemicals within plants is best demonstrated by terpenoids. More than 30,000 terpenoids [72] have been identified. The terpenes have a simple unifying feature by which they are defined and by which they may be easily classified. This generality, referred to as the isoprene rule, was postulated by Otto Wallach in 1887. This rule describes all terpenes as having fundamental repeating 5-carbon isoprene units [73]. Thus, terpenes are defined as a unique group of hydrocarbon-based natural products that possess a structure that may be hypothetically derived from isoprene, giving rise to structures that may be divided into isopentane (2-methylbutane) units.

The actual biosynthesis route to terpenes is not so simple. Two different biosynthetic pathways produce the main terpene building block, isopentenyl diphosphate (IPP). The first classical biosynthetic route is known as the MVA (mevalonic acid) pathway. This takes place in the cytosol, producing sesquiterpenes [54, 76]. It is now known that the actual 5-carbon building blocks in vivo are the interconvertible isomers isopentenyl pyrophosphate (IPP) and dimethylallyl pyrophosphate (DMAPP). These two building blocks are condensed together in a sequential fashion by the

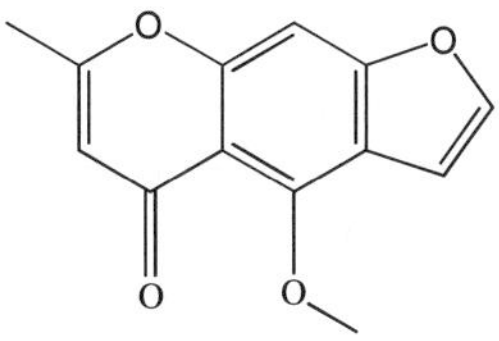

visnagin

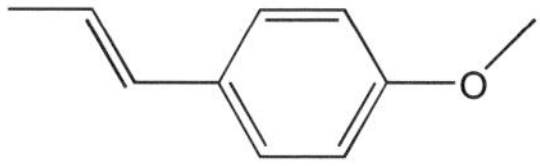

trans- anethole

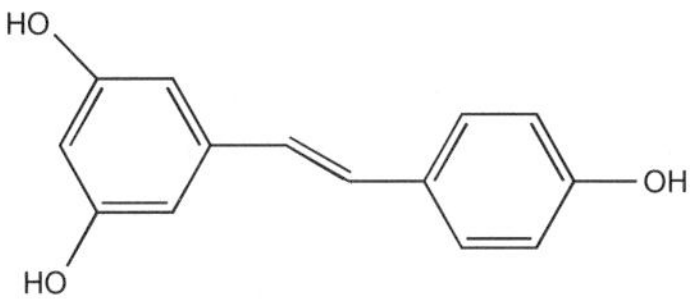

resveratrol

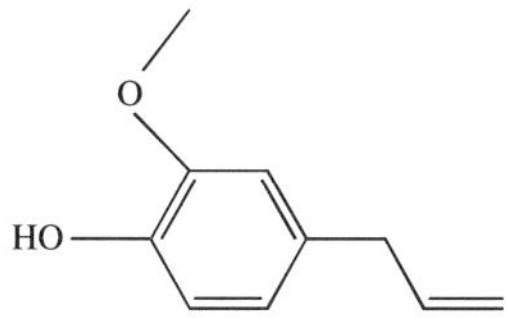

eugenol

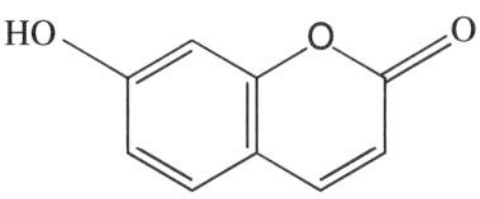

umbelliferone

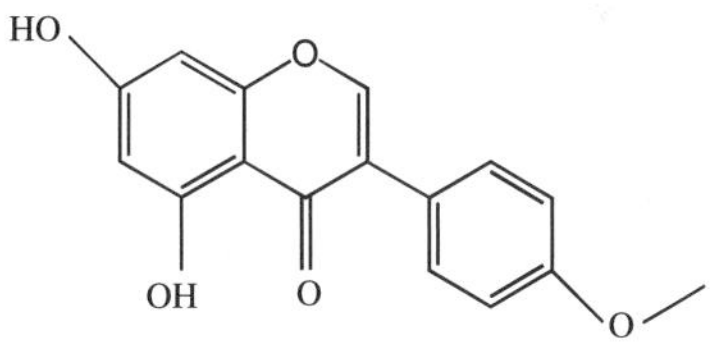

biochanin-A

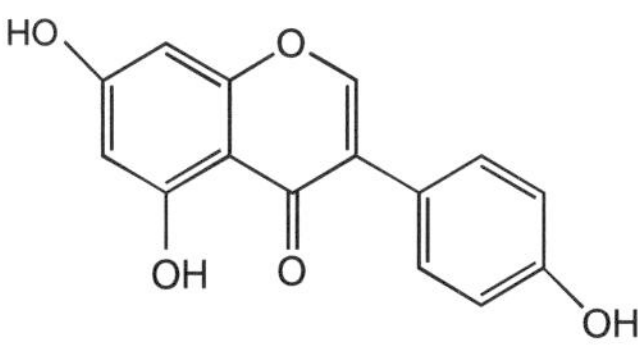

genistein

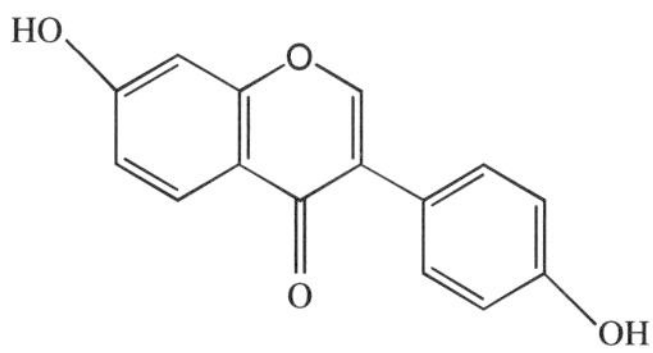

daidzein

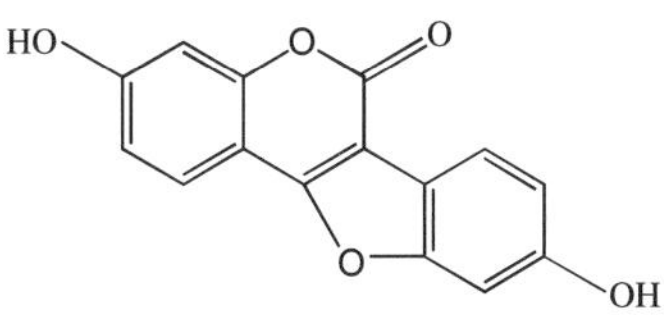

coumestrol

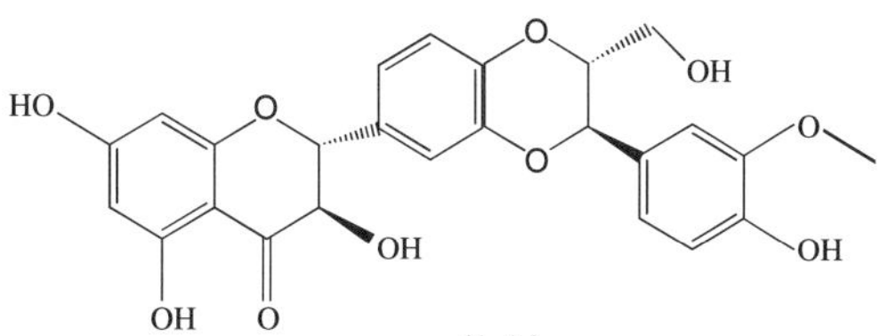

silybin

Fig. 2.6 Phenols

action of enzymes called prenyltransferases. The products include geranyl, farnesyl and geranyl geranyl pyrophosphate, squalene and phytoene, which are the direct precursors of the major families of terpenes. The key intermediate in the process was mevalonic acid (MVA), a 6-carbon compound. MVA is formed by the enzymatic reduction of the 3-hydroxy-3-methylglutaryl coenzyme A (HMG-CoA), which in turn is formed by the head-to-tail condensation of three molecules of acetate. MVA is enzymatically converted to IPP with the loss of carbon dioxide, and subsequently IPP and DMAPP are incorporated directly into cholesterol.

The second biosynthesis route to terpenes is referred to as either the MEP (methylerythriol-4-phosphate) or DOX (1-deoxy-D-xylulose) pathway [74]. When first discovered, this new plastid-bound pathway was distinct biochemically and was identical to that found in bacteria and probably is a legacy of prokaryotic endosymbiotic ancestors [75]. In this case, IPP is derived, not from MVA, but from 1-deoxyxylulose 5-phosphate (1-DXP), formed from the glycolytic intermediates glyceraldehyde 3-phosphate and pyruvate. The key step in the biosynthesis is the skeletal rearrangement and reduction of 1-DXP to form 2C-methylerythritol 4-phosphate (MEP) using the biological reducing agent NADPH as cofactor. MEP is converted to IPP via a chemical sequence involving the removal of three molecules of water. Thus, in higher plants, there are two pathways for generating terpenes. Here, IPP is formed in the chloroplast, mainly for the synthesis of more volatile mono- and diterpenes (Fig. 2.7). The evidence indicates that there may be sharing of intermediates across these pathways, a sort of biosynthetic crosstalk [77]. Various classes of terpenes classified by the number of 5-carbon units are given in Fig. 2.8. In plants, the MEP pathway leads to monoterpenes, diterpenes, the prenyl side chains of chlorophylls and carotenoids as well as to the phytohormones such as abscisic acid, gibberellins and *trans*-cytokinins. The first of the seven enzymatic steps of the MEP pathway is catalysed by the enzyme 1-deoxy-D-xylulose 5-phosphate synthase [54]. The monoterpenoids are major components of many essential oils and are economically important as fragrances and perfumes. Common acyclic compounds include myrcene, geraniol and linalool. Cyclic structures include menthol, camphor, pinene and limonene (Fig. 2.8).

Sesquiterpenes, C_{15} or compounds having 3-isoprene units, exist in aliphatic bicyclic and tricyclic frameworks. A member of this series, farnesol, is a key intermediate in terpenoid biosynthesis. Arteether is derived from artemisinin, a sesquiterpene lactone isolated from *Artemisia annua*, and currently used as an antimalarial drug (Fig. 2.8). Several derivatives of artemisinin are in various stages of clinical trials as antimalarial drugs in Europe [78] and as antineoplastic agents (see Chap.11).

The diterpenes are not considered essential oils and constitute a component of plant resins because of their higher boiling point. These are composed of four isoprene units. Gibberellic acid, a plant growth regulator, and taxol are diterpenes.

Triterpenes, C_{30} compounds, are composed of six isoprene units and are biosynthetically derived from squalene. These are high-melting-point colourless solids and constitute a component of resins, cork and cutin. Triterpenoids produce several

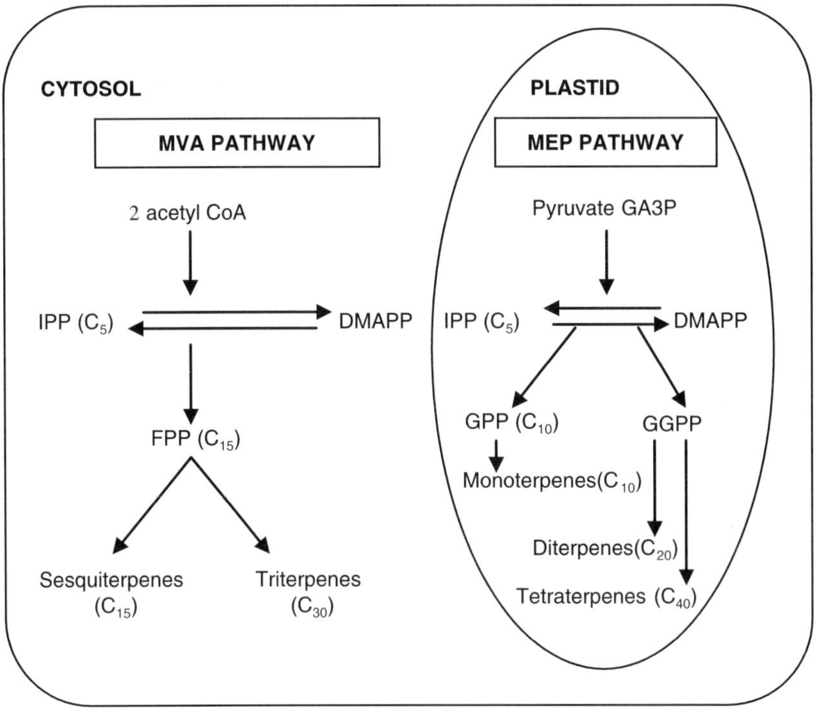

Fig. 2.7 Cytosolic MVA and plastidial MEP pathways for the synthesis of Terpenoids

pharmacologically active groups such as steroids, saponins and cardiac glycosides. Azadirachtin, a powerful insect antifeedent, is obtained from seeds of *Azadirachta indica*. Other triterpenes include the limonins and the cucurbitacins, which are potent insect steroid hormone antagonists [79].

All plant steroids hydroxylated at C_3 are sterols. Steroids are modified triterpenes and have profound importance as hormones (androgens such as testosterone and estrogens such as progesterone), coenzymes and provitamins in animals. Many progesterones are derived semisynthetically from diosgenin. Saponins are C_{27} steroids widely distributed in monocot families like Liliaceae, Amaryllidaceae and Dioscoreaceae, and in dicot families, e.g. Scrophulariaceace and Solanaceace. Saponins are composed of two parts: the glycone (sugar) and the aglycone or genin (triterpene). Commercially important preparations based on saponins include sarsaparilla root (*Sarsaparilla*), licorice (*Glycerrhiza glabra*), ivy leaves (*Hedera*), primula root (*Primula*) and ginseng (*Panax ginseng*). The ammonium and calcium salt of glycyrrhizic acid are referred to as glycyrrhizins. They are 50 to 100 times sweeter than sucrose [80].

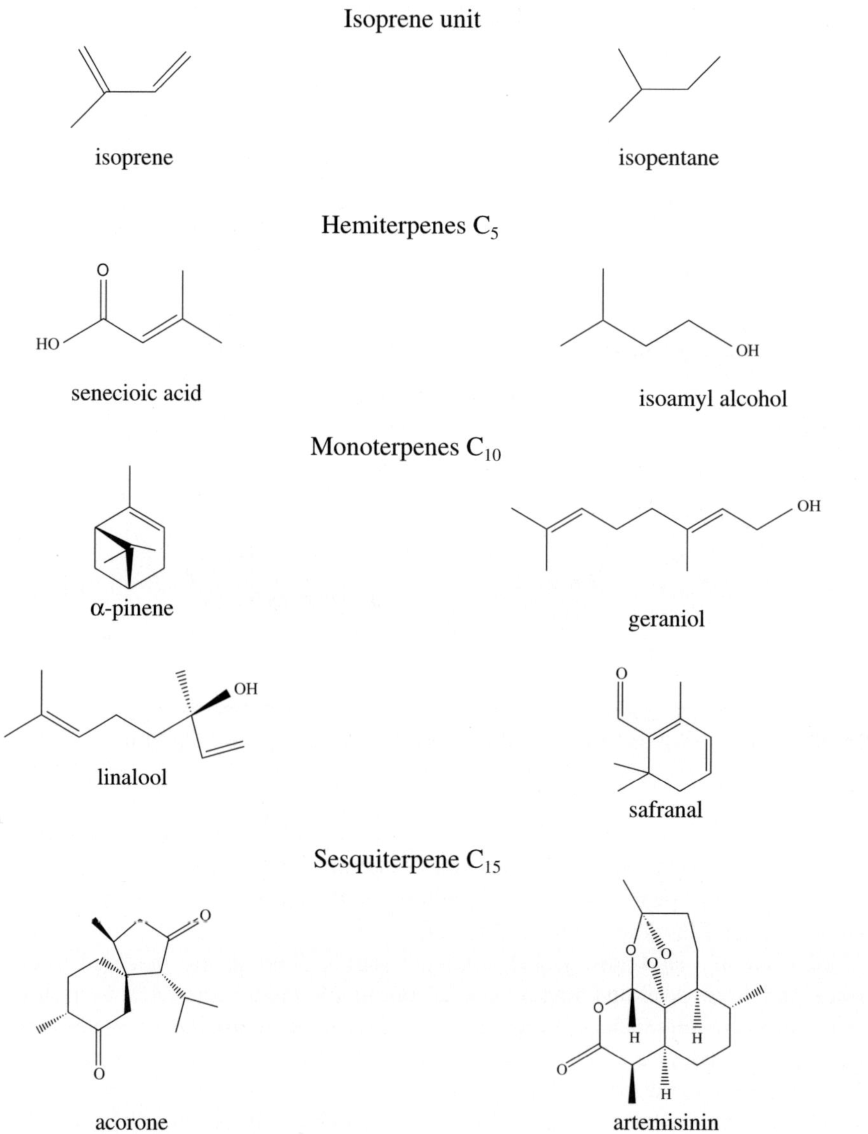

Fig. 2.8 Basic building unit and various classes of terpenes

Diterpene C_{20}

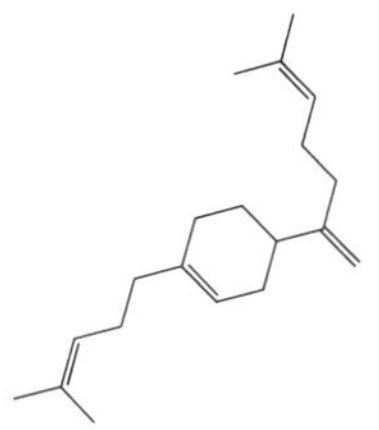

p-camphorene

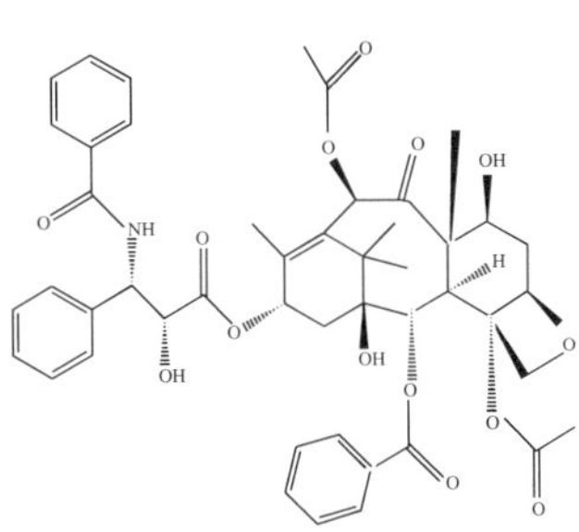

taxol

Triterpene C_{30}

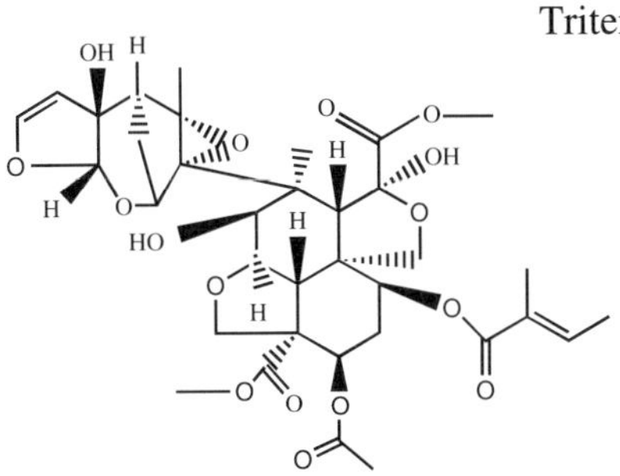

azadirachtin

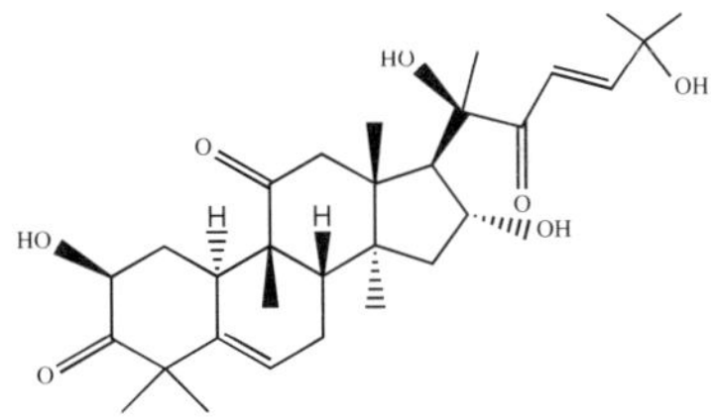

cucurbitacin-D

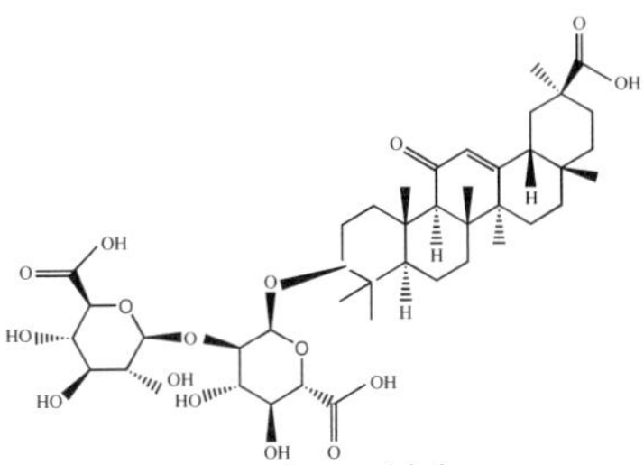

glycyrrhizin

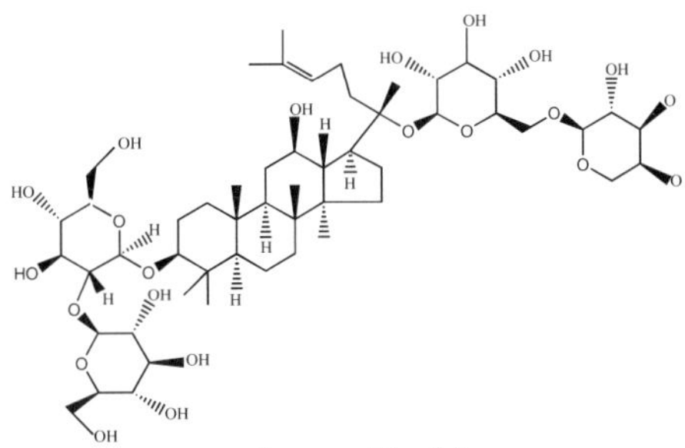

ginsenoside rb2

Fig. 2.8 (continued)

2.7 Conclusion

The therapeutic potential of herbs has been well recognized by various indigenous systems of medicine. Besides their therapeutic use, herbs are disease preventors and also used as cosmetics, dietary supplements and for reducing obesity. The priority of developed countries is different from that of developing countries in relation to medicinal plants. Developed countries are looking for leads to develop drugs from medicinal plants, while developing countries would like to have cheap herbal formulations as these countries cannot afford the long path of drug discovery using pure compounds. The indigenous system of medicine is officially recognised in India. The government funds education on this system, reimburses the cost of treatment by this system, and approves drugs based on Ayurvedic formulations.

About one-fourth of the drugs approved during the period 1981–2002 was either natural products or based on natural products [81]. Despite the ups and downs in funding for drug discovery from medicinal plants, new drugs that are in the pipeline for approval by the US FDA include morphine-6-glucoronide (a derivative of morphine with fewer side effects than morphine), vinflunine (a modification of vinblastine for cancer), exatecan (an analogue of camptothecin for cancer) and calanolide A (a dipyranocoumarin from *Calophyllum lanigerarum* var *anstrocoriaceum*, an anti-HIV drug).

Drug discovery from medicinal plants has traditionally been lengthier and more complicated than other drug discovery methods. As a result, many pharmaceutical companies have reduced their efforts and funding for natural-product research. Overexploitation of medicinal plants in developing countries needs proper attention to conserve biodiversity while at the same time catering to the needs of herbal drugs by sustainable utilization and production. Current research in drug discovery from medicinal plants involves a multidisciplinary approach combining botanical, phytochemical, biological and molecular techniques. There is a need to improve technology for the rapid isolation of active compounds in large quantities for evaluation with the scientific collection of plant material and maintenance of biodiversity. It is also desirable to have collaborative work with profit-sharing agreements between leading institutes, pharmaceutical companies of developed countries, and organizations in developing countries where most medicinal plants are still unexplored.

Acknowledgements Research on medicinal plants in the laboratory is supported by funds from the UGC-DRS programme and the DST-FIST programme.

References

1. Ramawat KG, Goyal S (2008) The Indian herbal drugs scenario in global perspectives. In: Ramawat KG, Merillon JM (eds) Bioactive Molecules and Medicinal Plants. Springer, Berlin Heidelberg New York, p 323
2. Balunas MJ, Kinghorn AD (2005) Life Sci 78:431
3. Gurib-Fakim A (2006) J Mol Asp Med 27(1):1
4. Ramawat KG (2007) Secondary metabolites in nature. In: Ramawat KG, Merillon JM (eds) Biotechnology: Secondary Metabolites. Science Publishers, Enfield, CT, p 21
5. Chin YW, Balunas MJ, Chai HB, Kinghorn AD (2006) AAPS J 8:E239
6. Patwardhan B, Vaidhya ADB, Chorghade M (2004) Curr Sci 86:789
7. Canter PH, Thomas H, Ernest E (2005) Trends Biotechnol 23:180
8. Kathe W (2006) Revision of the "Guidelines on the conservation of medicinal plants" by WHO, IUCN,WWF and TRAFFIC: process and scope. In: Bogers RJ, Craker LE, Lange D (eds) Medicinal and Aromatic Plants. Springer, Dordrecht, The Netherlands, p 109
9. Merillon JM, Ramawat KG (2007) Biotechnology for medicinal plants: research needs. In: Ramawat KG, Merillon JM (eds) Secondary Metabolites: Plants and Microbes, 2nd edn. Science Publishers, Enfield, CT, p 1

10. Jia W, Zhang L (2005) Challenges and opportunities in the Chinese herbal drug industry. In: Demain AL, Zhang L (eds) Natural Products: Drug Discovery and Therapeutic Medicine. Humana, Totowa, NJ, p 229
11. Ramawat KG, Sonie KC, Sharma MC (2004) Therapeutic potential of medicinal plants: an introduction. In: Ramawat KG (ed) Biotechnology of Medicinal Plants: Vitalizer and Therapeutic. Science Publishers, Enfield, CT, p 1
12. Dev S (2006) Selection of Prime Ayurvedic Plant Drugs Ancient-Modern Concordance. Anamaya, New Delhi
13. Zhang L, Zia W (2007) Chinese herbal drug industry: past, present and future. In: Ramawat KG, Merillon JM (eds) Biotechnology: Secondary Metabolites: Plants and Microbes, 2nd edn. Science Publishers, Enfield, CT, p 417
14. Chevallier A (1996) The Encyclopaedia of Medicinal Plants. Dorling Kindersley, London, p 17
15. Heinrich M, Barnes J, Gibbons S, Williamson EM (2004) Fundamentals of Pharmacognosy and Phytotherapy. Churchill Livingstone, Elsevier, UK
16. Theis N, Lerdau M (2003) Int J Plant Sci 163:S93
17. Taylor JLS, Rabe T, McGaw LJ, Jager AK, Staden JV (2001) Plant Growth Regul 34:23
18. Gupta R, Chadha KL (1995) Medicinal and aromatic plants research in India. In: Chadha KL, Gupta R (eds) Advances in Horticulture: Medicinal and Aromatic Plants, vol 11. Malhotra, New Delhi, p 429
19. Nayar MP, Sastry ARK (1987) Red Data Book of Indian Plants, vol II. In: Nayar MP, Sastry ARK (eds) Botanical Survey of India. Calcutta, p 142
20. Croteau R, Ketchum EB, Long RM, Kaspera R, Wildung MR (2006) Phytochem Rev 5:75
21. Holland HD (1998) Science 275:38
22. Mendelson R, Balick MJ (1995) Econ Bot 49:223
23. Wagle A, Kelkar GD, Heble MR (2007) Production of steroids and saponins. In: Ramawat KG, Merillon JM (eds) Biotechnology: Secondary Metabolites: Plants and Microbes, 2nd edn. Science Publishers, Enfield, CT, p 209
24. Reichert JM (2003) Nat Rev Drug Discovery 2:1
25. Wilson EK (2002) Chem Eng News 80:35
26. Demain AL, Zhang L (2005) Natural products and drug discovery. In: Demain AL, Zhang L (eds) Natural Products: Drug Discovery and Therapeutic Medicine. Humana, Totowa, NJ, p 3
27. Betz SF, Baxter SM, Fetrow JS (2002) DDT 7:865
28. Drews J (2000) Science 287:1960
29. Kessmann H, Schnurr B (2002) Industrial strategies for the discovery of bioactive compounds from plants. In: Oksman-Caldentey KM, Barz WH (eds) Plant Biotechnology and Transgenic Plants. Dekker, New York, p 45
30. Lennon GG (2000) DDT 5:59
31. Bertels S, Formann S, Jas G, Bindseil KU (2000) Synergistic use of combinatorial and natural product chemistry. In: Grabley S, Thierike R (eds) Drug Discovery from Nature. Springer, Berlin Heidelberg New York, p 72
32. Ravishankar GA, Bhagyalakshmi N, Ramachandra Rao S (2007) Production of food additives. In: Ramawat KG, Merillon JM (eds) Biotechnology: Secondary Metabolites: Plants and Microbes, 2nd edn. Science Publishers, Enfield, CT, p 103
33. Pal A (2007) Production of insecticides. In: Ramawat KG, Merillon JM (eds) Biotechnology: Secondary Metabolites: Plants and Microbes, 2nd edn. Science Publishers, Enfield, CT, p 131
34. Audi J, Belson M, Patel M, Schier J, Osterloh J (2005) J Am Med Assoc 294:2342
35. Ramos MV, Sampaio AH, Cavada BS, Calvete JJ, Grangeiro TB, Debray H (2001) Glycoconjugate J 18: 391
36. Mustafa NR, Verpoorte R (2007) Phytochem Rev 6:243
37. Eisenreich W, Bacher A, Arigoni D, Rohdich F (2004) Cell Mol Life Sci 61:1401
38. Wink M (2000) Biochemistry of Plant Secondary Metabolism: Annual Plant Review, vol 2. Academic, Sheffield, UK, p 151
39. Facchini PJ (2001) Annu Rev Plant Physiol Plant Mol Biol 52:29

40. Pelletier SW (1982) The nature and definition of an alkaloid. In: Pelletier SW (ed) Alkaloids: Chemical and Biological Perspectives. Wiley, New York
41. Oksman-Caldentey KM (2007) Curr Pharma Biotech 8:203
42. Rothe G, Hachiya A, Yamada Y, Hashimoto T, Drager B (2003) J Exp Bot 54:309
43. Gomez-Galera S, Peiacho AM, Gene A, Capell T, Christou P (2007) Plant Cell Rep 26:1689
44. Sato F, Hashimoto T, Hachiya A, Tamura K, Choi KB, Morishige T, Fujimoto H, Yamada Y (2001) PNAS 98:367
45. Howes MJR, Houghton PJ (2003) Phytother Res 17:1
46. Facchini PJ, Hagel M, Liscombe DK, Loukantana N, MucLeod BP, Samamani N, Zulak KG (2007) Phytochem Rev 6:97
47. Weid M, Ziegier J, Kutchan TM (2004) PNAS 38:13957
48. Bird DA, Franceschi VR, Facchini PJ (2003) Plant Cell 15:2626
49. Deleu D, Hanssens Y, Northway MG (2004) Drugs Ag 21:377
50. Koumis T, Samuel S (2005) Clin Ther 27:377
51. Heijden RVD, Verpoorte R, Ten Hoopen HJG (1989) Plant Cell Tiss Org Cult 18:231
52. Verpoorte R, Contin A, Memelink J (2002) Phytochem Rev 1:13
53. Memelink J, Kijne JW, Van der Heijden R, Verpoorte R (2001) Adv Biochem Eng/Biotech 72:105
54. Oudin A, Courtois M, Rideau M, Clastre M (2007) Phytochem Rev 6:259
55. Eteng MU, Eyong EU, Akpanyung EO, Agiang MA, Aremu CY (1997) Plant Foods Hum Nutr 51:231
56. Dewick PM (2002) Medicinal Natural Products. Wiley, West Sussex, UK
57. Bennett RN, Wallsgrove RM (1994) New Phytol 127:617
58. Bruneton J (1999) Pharmacognosy Phytochemistry Medicinal Plants, 2nd edn. Lavoisier, Paris
59. Herrmann KM, Weaver LM (1999) Annu Rev Plant Physiol Mol Biol 50:473
60. Mustafa NR, Verpoorte R (2005) Planta 222:1
61. Ates O, Cayli S, Altinoz E, Gurses I, Yucel N, Sener M, Kocak A, Yologlu S (2007) Mol Cell Biochem 294:137
62. Delmas D, Jannin B, Latruffe N (2006) Mol Nutr Food Res 49:377
63. Vitrac X, Krissa S, Decendit A, Deffieux G, Merillon JM (2004) Grapevine polyphenols and their biological effects. In: KG Ramawat (ed) Biotechnology of Medicinal Plants. Science Publishers, Enfield, CT, p 33
64. Baur JA, Sinclair DA (2006) Nat Rev 5:493
65. Archer AW (1988) J Chromatogr 438:117
66. Dixon RA, Sumner LW (2003) Plant Physiol 131:878
67. Dixon RA, Ferreira D (2002) Phytochem 60:205
68. Lamartinier CA (2000) Am J Clin Nutr 71:1705S
69. Alekel DL, St Germain A, Pererson CT, Hanson KB, Stewart JW, Toda T (2000) Am J Clin Nutr 72:844
70. Uesugi T, Toda T, Tsuji K, Ishida H (2001) Biol Pharm Bull 24:368
71. Kamsteeg M, Rutherford T, Sapi E et al (2003) Oncogene 22:2611
72. Buckingham J (1998) Dictionary of Natural Products. Chapman & Hall, London
73. Croteau R (1998) Discoveries Plant Biol 1:329
74. Lichtenthaler HK, Rohmer J, Schwender J (1997) Physiol Plant 101:643
75. Cvejic JH, Rohmer M (2000) Phytochem 53:21
76. Janses BJM, de Groot A (2004) Nat Prod Rep 21:449
77. Jux A, Gleixner W, Boland W (2001) Angew Chem Int Ed Engl 40:2091
78. van Agtmael MA, Eggelte TA, van Boxtel CJ (1999) Trends Pharmacol Sci 20:199
79. Miro M (1995) Phytother Res 9:159
80. Brielmann HL, Setzer WN, Kaufman PB, Kirakosyan A, Cseke LJ (2005) Phytochemicals: the chemical components of plants. In: Cseke LJ, Kirakosyan A, Kaufman PB, Warber SL, Duke JA, Brielmann HL (eds) Natural Products from Plants. Taylor Francis, Boca Raton, FL, p 1
81. Newmann DJ, Cragg GM, Snader KM (2003) J Nat Prod 66:1022

Chapter 3
Indigenous People and Forests: Perspectives of an Ethnobotanical Study from Rajasthan (India)

S.S. Katewa

Abstract The state of Rajasthan has a sizable tribal population existing in the still surviving deciduous forests of the Aravalli and Vindhyan ranges. Though having undergone varying degrees of change, a substantial population even today can be seen thriving in primitive conditions with preserved traditions. Of such groups, the Bhils are the largest, followed by Garasias and Damors. The Kathodias and Sahariyas are still sociologically not very well known. Living close to nature, the tribals have acquired unique knowledge about the properties and uses of wild plants, most of which are not known to the outside world. Until a decade ago, nothing was known about the ethnobotany of the tribes of Rajasthan. Exhaustive field work in tribal villages with a macro-level perspective brought forth interesting revelations from the panorama of their lives. The present work highlights useful ethnobotanical information about the uses of wild plants by the tribals of Rajasthan as food, medicine, veterinary medicine, material culture, etc. This folk wisdom, if subjected to scientific scrutiny, could benefit humankind in many ways.

Keywords Ethnobotany · Rajasthan · Tribals · Indigenous knowledge

3.1 Introduction

Indigenous people throughout the world possess knowledge of their surrounding flora and fauna. People's knowledge and perceptions of the environment and their relationship with it are often important elements of cultural identity. In India, the traditional folklore healthcare system has a long history and is very deeply rooted in rural and tribal populations. It was practiced long before the beginning of the Christian era and, perhaps, in the "pre-Vedic" periods to which belong the Mohanjodaro and Harappan civilizations. The familiarity with plant species producing medicines,

S.S. Katewa (✉)
Laboratory of Ethnobotany and Agrostology, Department of Botany, University College of Science, M.L. Sukhadia University, Udaipur-313001, India,
e-mail: sskatewa@yahoo.com

K.G. Ramawat (ed.), *Herbal Drugs: Ethnomedicine to Modern Medicine*,
DOI 10.1007/978-3-540-79116-4_3, © Springer-Verlag Berlin Heidelberg 2009

essential oils and insecticides dates back to the beginning of civilization. The Aryans and Dravidians had good knowledge about the rich heritage of herbal drugs in India. The traditional healthcare practice of indigenous people pertaining to human health is known as ethnomedicine. Indian civilization has played a pioneer role from time immemorial in utilizing plants such as indigenous drugs. Ethnomedicinal practices are the mother of all other traditional systems such as Ayurveda, Siddha, Unani and even today's most popular modern therapeutic system or allopathic system [1]. Herbal medicine is not just a poor man's substitute for conventional medicine but a valuable form of treatment in its own right.

It has been estimated that approx. 80% of the population of developing countries is partially or fully dependent upon herbal drugs for primary healthcare, and higher plants are known to be the main source of drug therapy in traditional medicine [2–4]. The World Health Organization (WHO) estimated that 80 to 90% of the world's population relies mainly on local herbal practitioners [2]. The WHO in its 29th and 30th assemblies (1976–1977) gave formal recognition to traditional medicine and recommended the inclusion of traditional healers in the national healthcare programme [5]. The major classical systems of medicine used on the Indian subcontinent like Ayurveda, Siddha and Unani together use about 1200 plant species to treat human ailments, but the tribals of India use more than 7500 plant species [6]. India is one of the world's 12 hotspots, having the largest plant biodiversity, and it has almost 45,000 plant species of which 15,000 to 20,000 are used for medicinal purposes [7].

Demographically, the Aravalli hills are the homeland of people belonging to tribes, subtribes and different ethnic and religious groups such as the Bhil, Meena, Garasia, Kathodi, Saharia, Bhagora, Damor, etc. A major portion of its population, irrespective of ethnicity, is of rural background, most of them living in isolated pockets and remote areas maintaining a primitive state of economic life and sometimes completely cut off from modern amenities. However, the Aravalli hills are very rich in herbal medicinal flora of angiosperm and lower plants. The inhabitants of this region are still completely dependent upon natural plant resources for their daily needs. Likewise, these tribal people possess perfect knowledge of plant use, which they have acquired over the course of their centuries-old experiences with herbal plants [1].

Ethnomedicinal therapy plays a vital role in the primary healthcare of tribals and rural populations of the Aravalli hills and hold great potential in the discovery of new drugs of herbal origin which are easily metabolized in the human body and have no adverse side effects. The search for new potential therapeutic compounds like reserpine, quinine, ephedrine, cocaine, emitin, khallin, colchicines, digoxin, taxol, vinblastine, gugulipid, vincristine, artemisinine, etc. from medicinal plants, with a rich ethnobotanical lore, gave impetus to ethnobotanical research throughout the world [3, 4].

Because of the various ethnic groups scattered throughout the state of Rajasthan, extensive ethnomedicinal studies of plants in Rajasthan have been carried out [1], [8, 17]. The phytochemistry of several species has been carefully examined to validate the role of the active principle, viz. *Ephedra* species [18], *Aegle marmelos* [19], *Cocculus pendulus* [20], *Curculigo orchioides* [21], *Commiphora wightii* [22, 23], *Cayratia trifolia* [24] and *Pueraria tuberosa* [25].

Unfortunately, much of this knowledge and many of the cultures are fast disappearing in most regions. It is, therefore, essential to rescue the disappearing knowledge and to revitalize and return it to the local communities. With this objective in mind, some ethnobotanical investigations were conducted by the author and his associates in the tribal-dominated areas of Rajasthan [17], [27–30].

3.2 Study Site and People

Rajasthan is the largest state of India and is located in the north-western part of the country. Geographically it lies between $23°3'$ to $30°12'$ longitude and $69°30'$ to $78°17'$ latitude (Fig. 3.1). The vegetation of Rajasthan has considerable diversity due to a variety of ecosystems found in different parts of the state. Apart from the predominant desert ecosystem, there are grasslands, scrub jungles, wetlands and deciduous forests, each supporting a characteristic assemblage of plant species. Many tribes are scattered in different parts of Rajasthan, but the southern part of the state is the homeland of several tribes which have their own separate identities.

Living close to nature, these tribes have acquired unique knowledge about the properties and uses of wild plants, most of which are unknown to the outside world. The surrounding plants for these people form an integral part of their culture, and the information about the plants gets passed on from generation to generation only through oral folklore, although many times it is kept secret.

3.3 Observations

3.3.1 Healthcare

The art of herbal healing is deeply rooted in Indian culture and folklore. Even today in most rural areas, people depend on the local traditional healing system for their primary healthcare. For tribals, the ambient vegetation is the main source of drugs for their ailments. Their expertise is vast and they have cure for all ailments, including deadly diseases. But it is very difficult to extract information from them because some tribes strongly believe that the efficacy of the drug would be lost if details were divulged.

The traditional healers of Rajasthan possess an impressive knowledge of the medicinal virtues of the plants that grow in their surroundings. A few important medicinal plants of the tribals, along with their family, plant parts used, and mode of administration for the effective treatment of various ailments, are given as follows.

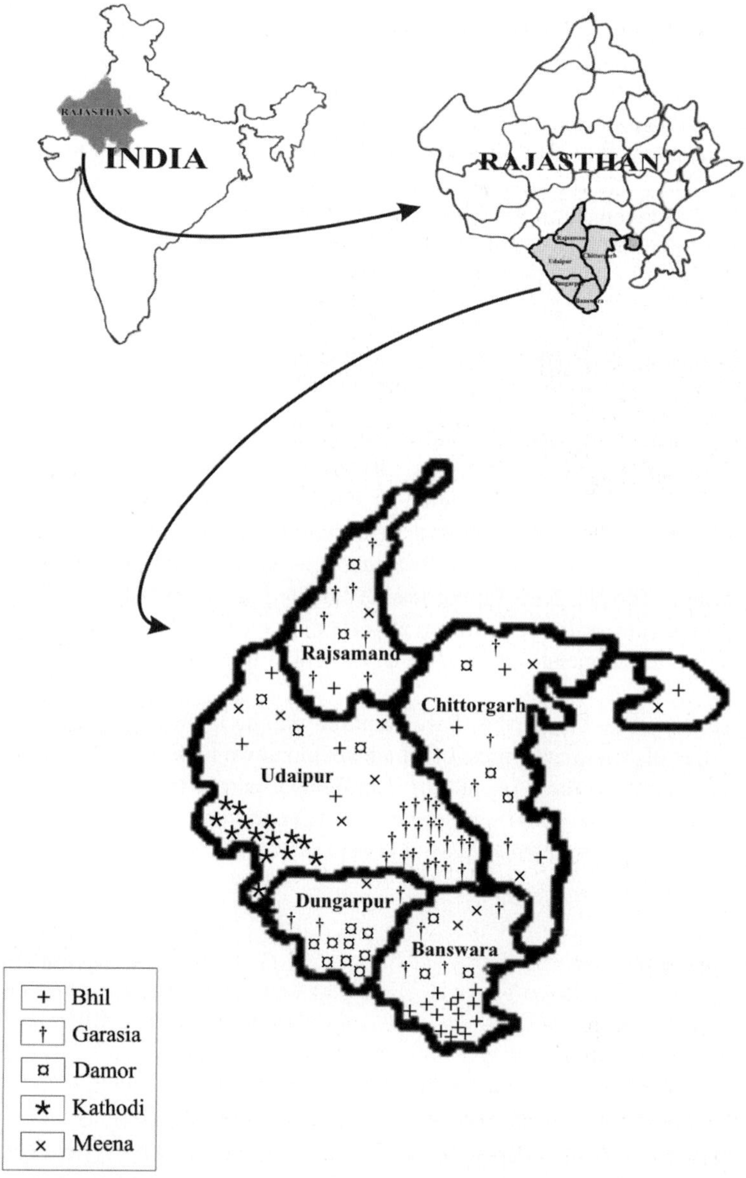

Fig. 3.1 Relative concentration of different tribes in Southern Rajasthan

Abrus precatorius Linn. (Fabaceae) (Fig. 3.2)

Fig. 3.2 *Abrus precatorius* Linn

Ethnomedicinal uses: fresh leaves are chewed to cure mouth blisters. Seed powder is used as an antifertility drug by both males and females. A high dose of seed powder is considered fatal.

Ethnoveterinary medicinal uses: the powder of two to three seeds is given to animals with bread two to four times a day for the disposal of the placenta following the birth of a baby.

Achyranthes aspera Linn. (Amaranthaceae)

Ethnomedicinal uses: decoction of two to three teaspoon of leaf powder is taken for stomachache and to ease constipation. The powder of roasted seeds is taken to cure whooping cough. Seeds are boiled in milk and taken for 3 d as a tonic to cure sexual debility and hydrophobia. The ash of the dry plant is mixed with honey and taken orally for cough, asthma, urinary complaints and stomachache. Root powder is boiled with mustard oil and used for massage by the tribals in rheumatic pain. Infusion of the whole plant is taken in liver and renal complaints. Root extract is taken orally as an antidote to scorpion bite.

Ethnoveterinary medicinal uses: a half cup of root extract is given to animals to increase milk production. It acts as a lactogogue. Root extract is also given to sick animals as a tonic. Seed powder is given to animals to treat dysentery.

Ampelocissus latifolia (Roxb.) Planch (Vitaceae)

Ethnomedicinal uses: a half-cup decoction of the tuber is taken orally for the treatment of fractured bone. The paste of the tuber is applied on abscesses for early cure.

A fresh tuber is crushed, boiled in the seed oil of *Ricinus communis* and applied externally for the treatment of gout. The extract of the tuber is considered by tribals to be a blood purifier and diuretic. One gram of dried-tuber powder is mixed with 50 g curd and taken orally twice a day for 10 d to cure chronic dysentery. The dried tuber of *Ampelocissus latifolia*, leaves of *Abutilon indicum*, flowers of *Butea monosperma*, stem bark of *Moringa oleifera* and *Acacia catechu* are mixed in equal amounts and made into a fine powder and 5 to 10 g of this powder is taken orally with water twice a day for 3 d to cure leucorrhoea.

Ethnoveterinary medicinal uses: an extract of tuber is given to animals for the treatment of fractured bone. This extract is also given to animals to alleviate flatulence.

Argemone mexicana **Linn. (Papaveraceae)**

Ethnomedicinal uses: the yellowish latex of this plant is applied externally to treat various skin diseases, especially ringworm. Latex is also rubbed on body parts affected by rheumatism and poured into the eyes to cure conjunctivitis. One fourth teaspoon of powdered seeds and roots with two drops of mustard oil is used as toothpaste by those suffering from pyorrhoea. The root paste of *Argemone mexicana* is applied locally twice a day for 10 d to cure piles.

Arisaema tortuosum **(Wall.) Schott (Araceae) (Fig. 3.3)**

Fig. 3.3 *Arisaema tortuosum* (Wall.) Schott

Ethnomedicinal uses: an extract of the tuber is used by tribals as an antidote to poisonous snake bite and also as a blood purifier. The powder of the tuber is divided into three equal doses, and each dose is taken orally once a day for 3 d to cure liver complaints and stomachache.

Ethnoveterinary medicinal uses: the fresh tuber is crushed and given to domestic animals along with green fodder for the treatment of fractured bone. A decoction of the tuber is given to animals orally in gastric disorders.

Aristolochia bracteolata Lam. (Aristolochiaceae) (Fig. 3.4)

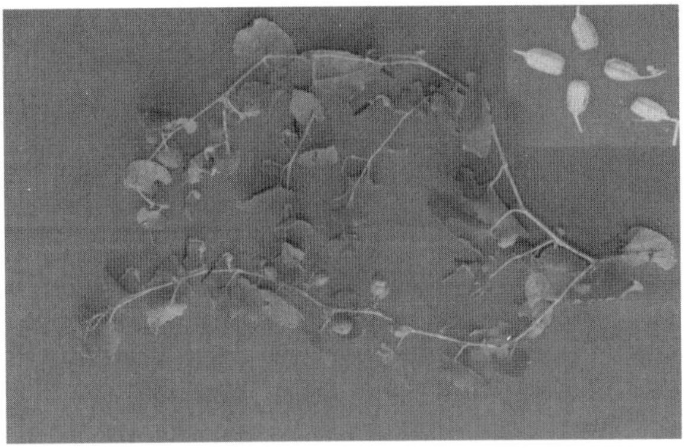

Fig. 3.4 *Aristolochia bracteolata* Lam

Ethnomedicinal uses: an infusion of the leaves is taken orally by tribals as antidote to snake bite. Rural women take the decoction of plant orally for menstrual problems. A paste of the seeds is applied to soften stiff hair.

Asparagus racemosus Willd. (Liliaceae) (Fig. 3.5)

Fig. 3.5 *Asparagus racemosus* Willd

Ethnomedicinal uses: the tuber powder is frequently taken orally with milk as an appetizer. An extract of the tuber is mixed with honey and taken orally to cure dyspepsia. Rural people of the Shekhawati region (north-east Rajasthan) take the tuber powder orally to cure dysentery, acidity, tuberculosis, seminal weakness, leucorrhoea, burning micturition, anorexia and peptic ulcer. In winter tubers are boiled in mustard oil till they are completely charred. This oil is consumed by patients suffering from rheumatoid arthritis, either by preparing sweet pudding or as a cooked vegetable. It is very effective in reducing joint pain.

Asphodelus tenuifolius Cav. (Liliaceae)

Ethnomedicinal uses: a decoction of the leaves is taken orally by the rural people of the Shekhawati region to treat toxeamia and to remove kidney stone. A paste of the leaves is applied externally to treat swelling on any part of the body.

Balanites aegyptiaca (Linn.) Delile (Balanitaceae)

Ethnomedicinal uses: an herbal bath of root and leaf extracts is taken by tribals as an antiseptic. A decoction of leaf powder is used for washing hair to get rid of lice. The powdered seed kernel is mixed with jaggery and taken orally to get relief from sciatica pain.

Ethnoveterinary medicinal uses: a paste of the bark is given orally to animals as an antidote to snake bite.

Bombax ceiba Linn. (Bombacaceae) (Fig. 3.6)

Ethnomedicinal uses: young root tips are dried in shade and cooked as a vegetable for patients suffering from impotency. This vegetable is considered to be as good as the leaves of *Adansonia digitata* to increase the amount of sperm in semen. A half-cup extract of bark and flowers is taken for 3 d to treat sexual diseases such as hydrocele, leucorrhoea and gonorrhoea and to treat an irregular menstrual cycle.

Buchanania lanzan Spreng. (Anacardiaceae)

Ethnomedicinal uses: a decoction of the roots of *Buchanania lanzan* and *Gardenia turgida* is used to treat phthisis. An infusion of the whole plant is taken orally in dysuria. A bark paste is applied locally to treat snake bite.

Caesalpinia bonduc (Linn.) Roxb. (Caesalpiniaceae) (Fig. 3.7)

Ethnomedicinal uses: a powder of the roasted leaves of *Caesalpinia bonduc* and *Azadirachta indica* is given three times a day for 4 d to treat malarial fever. A half

Fig. 3.6 *Bombax ceiba* Linn

Fig. 3.7 *Caesalpinia bonduc* (Linn.) Roxb

teaspoon of seed powder is mixed in boiled egg and eaten for 1 week to cure piles and ulcer. The crushed seeds are given orally to children to remove worms from the intestine.

Cassia tora Linn. (Caesalpiniaceae)

Ethnomedicinal uses: An extract of fresh leaves is taken orally to reduce obesity. Tea prepared from the seeds is taken to cure asthma and bronchitis, whereas a seed paste is applied locally as an anthelmintic. A seed powder is mixed with curd and applied externally on abscesses, cuts, wounds, boils and leprosy. Seed powder, curd and cow urine are mixed together and a paste is prepared which is massaged over the entire body to get relax from itching. A one-fourth-cup extract of the leaves is given three times a day for 3 d to treat jaundice.

Ethnoveterinary medicinal uses: boiled seeds are given to domestic animals to treat hypogalactia.

Calligonum polygonoides Linn. (Polygonaceae) (Fig. 3.8)

Ethnomedicinal uses: an extract of the plant is taken in cases of typhoid, whereas a decoction of the plant is gargled to cure sore gums.

Ethnoveterinary medicinal uses: an extract of the plant is given to animals to treat colic, whereas a decoction of the whole plant is given to treat dysuria.

Citrullus colocynthis (Linn.) Schard. (Cucurbitaceae) (Fig. 3.9)

Ethnomedicinal uses: the fruit of this plant is stuffed along with seeds of *Trachyspermum ammi* and common salt and the whole thing is put in sunlight till it is dried and then powdered. One teaspoon of this powder is taken orally to treat constipation and stomachache. After removing the pulp of the fruit, goat's milk is poured into the hollow fruit and kept for 12 h. This milk is then given to patients suffering from kidney stones. Two to three drops of a root decoction is poured into the ear to cure earache.

Ethnoveterinary medicinal uses: roasted fruits and a decoction of roots is given to domestic animals to cure constipation, digestive disorders and gastritis.

Clitoria ternatea Linn. (Fabaceae) (Fig. 3.10)

Ethnomedicinal uses: tribals of southern Rajasthan crush the seeds with water, warm them slightly and apply them on the penis to treat syphilis. One fourth of the seed, i.e. a very small quantity of seed, is crushed in water and given orally to children to treat colic. An extract of the root is also taken orally by tribal women to treat uterine diseases.

Fig. 3.8 *Calligonum polygonoides* Linn

Fig. 3.9 *Citrullus colocynthis* (Linn.) Schard

Fig. 3.10 *Clitoria ternatea* Linn

Costus speciosus (Koen.) Sm. (Costaceae) (Fig. 3.11)

Fig. 3.11 *Costus speciosus* (Koen.) Sm

Ethnomedicinal uses: one teaspoon of root powder is taken twice a day for 3 to 4 d to treat rheumatism, asthma and sexual dysfunction; two teaspoon of root powder are taken early in the morning as an antinematodal. A decoction of root powder is given to children twice a day for 2 d to treat diarrhoea, dysentery and stomachache. Two to three drops of root extract is poured into the ear to treat earache. A rhizome is used to treat dropsy and oedema.

Ethnoveterinary medicinal uses: a rhizome is given orally to domestic animals to treat rheumatism.

***Curculigo orchioides* Gaertn. (Hypoxidaceae) (Fig. 3.12)**

Fig. 3.12 *Curculigo orchioides* Gaertn

Ethnomedicinal uses: 100 g of powder of dried tuber is mixed in "Khowa" (concentrated milk) prepared from 5 L of buffalo milk and eaten early in the morning by tribals for 7 d as an eye tonic and also to treat fatigue. One teaspoon of root powder is taken orally by tribal women for 7 to 10 d to treat leucorrhoea and menorrhagia. A powder of the tuber is given orally to children to treat rickets. A tuber extract is applied locally to treat gonorrhoea and syphilis. A decoction of the tuber is given to children as a tonic and also to treat filariasis.

***Curcuma amada* Roxb. (Zingiberaceae) (Fig. 3.13)**

Fig. 3.13 *Curcuma amada* Roxb

Ethnomedicinal uses: a poultice of fresh crushed tuber or paste of dried tuber is tied locally for treatment of fractured bone, wounds, sores and abscesses, whereas an extract of the tuber is taken orally to treat abdominal pain and constipation. A half teaspoon of rhizome powder is taken with milk to treat internal injuries, to purify the blood and also to treat cough and cold. The tuber powder of this plant is mixed with an equal quantity of seed powder of *Trachypermum ammi* and 1 teaspoon of this powder is taken orally along with water once a day for 3 d to treat rickets. The tribals put the leaves in new leather shoes to protect the feet from shoe bite.

Ethnoveterinary medicinal uses: an extract of the tubers is given to animals orally to treat flatulence.

Eulophia ochreata Lindl. (Orchidaceae) (Fig. 3.14)

Fig. 3.14 *Eulophia ochreata* Lindl

Ethnomedicinal uses: one teaspoon of powder of the bulbs is taken orally twice a day for 1 month to cure leukaemia. An equal amount of powder of *Eulophia ochreata* bulbs and the tuber of *Chlorophytum borivilianum* are mixed together, and 1 teaspoon of this mixture is taken orally with milk for 1 month to boost the body's immune system and to treat rheumatism. The bulbs of *E. ochreata* (100 g), the stem bark of *Sterculia urens* (100 g) and the tubers of *C. borivilianum* (50 g) are mixed and powdered. A half teaspoon of this powder is taken orally with milk twice a day for 15 to 30 d to treat anaemia and general fatigue.

Hemidesmus indicus (Linn.) R.Br. (Periplocaceae)

Ethnomedicinal uses: a root infusion is used to treat menstrual complaints and to provide relief from hypertension. A one-fourth-cup extract of the root is taken once a day orally to treat dysentery; a decoction of the whole plant is taken orally once a day for 3 d to treat rheumatism.

Leea macrophylla Roxb. ex. Hornem. (Leeaceae) (Fig. 3.15)

Fig. 3.15 *Leea macrophylla* Roxb. ex. Hornem

Ethnomedicinal uses: a root powder is taken once a day for 7 d to treat sexual dysfunction by men, and a decoction of the root is taken for the treatment of fractured bone. The leaves are cooked and eaten as a medicated vegetable. A poultice of the leaves is tied over a wounded body part as a coagulant and an antiseptic. Bark powder is taken orally to treat cancer. A paste of the flowers is thoroughly mixed with the warm seed oil of *Ricinus communis* and applied over body parts to treat muscular pain. One teaspoon of leaf powder is taken along with honey twice a day for 15 to 30 d to treat blood cancer.

Ethnoveterinary medicinal uses: a decoction of the tuber (one cubic inch) is given to animals to treat food poisoning.

Leptadenia reticulata (Retz.) Wt. & Arn. (Asclepiadaceae)

Ethnomedicinal uses: a paste of the leaves and roots is taken orally with water by the tribals of southern Rajasthan to cure gangrene.

Pedalium murex Linn. (Pedaliaceae) (Fig. 3.16)

Ethnomedicinal uses: about eight to ten fresh leaves are rotated in half a litre of water, and half a cup of this water is taken once a day for 7 d to treat gonorrhoea. A fruit powder strained in cloth is taken orally with milk to treat sexual dysfunction, while a powder of the seeds is mixed in sweets and eaten to cure rheumatism and

Fig. 3.16 *Pedalium murex* Linn

lumbago. A decoction of the fruits is taken to treat dysuria and other urinary complaints and also to stop the discharge of semen with urine. A powder of the whole plant of *Withania somnifera* is mixed with the powder of dry fruits of *Pedalium murex* and taken orally as a health tonic.

Ethnoveterinary medicinal uses: the plant is dipped in water for a certain amount of time, and this water is given to animals suffering from dysentery and diarrhoea.

Plumbago zeylanica Linn. (Plumbaginaceae) (Fig. 3.17)

Fig. 3.17 *Plumbago zeylanica* Linn

Ethnomedicinal uses: a half teaspoon of leaf and fruit extract is given to children to treat skin allergies. The root paste mixed with milk and vinegar is applied locally to treat leucoderma and other skin diseases like eczema, warts, carbuncles, etc. The roots are used as a toothbrush to treat toothache. One-fourth teaspoon of root powder is taken orally three times a day for 3 d to treat stomachache and abdominal pain. A paste is prepared by mixing the dried root powder of *Plumbago zeylanica* (150 g) with curd (100 g). This paste is stored in a copper container and applied locally to treat eczema.

Pueraria tuberosa (Roxb. ex. Willd.) DC. (Fabaceae) (Fig. 3.18)

Fig. 3.18 *Pueraria tuberosa* (Roxb. ex. Willd.) DC

Ethnomedicinal uses: the powder of the roots of *Plumbago zeylanica* and *Pueraria tuberosa* and the whole plant of *Centella asiatica* is mixed in equal amounts, and a half teaspoon of this mixture is taken orally with milk daily up to 1 month to increase memory. A half teaspoon of tuber powder is taken by tribal women as a daily contraceptive, and 2 teaspoon are taken early in the morning for 3 d as an abortifacient. Pills made from the powder of the tuber are taken orally by tribal men as a tonic to treat general fatigue. The pills are also taken to treat obesity.

Rhus mysurensis G. Don (Anacardiaceae)

Ethnomedicinal uses: the ripe fruits are edible and tasty due to their acrid taste. The fruits are eaten by tribal women to increase lactation. The fruits are also consumed to improve digestion.

Sauromatum venosum (Ait.) Kunth (Araceae) (Fig. 3.19)

Fig. 3.19 *Sauromatum venosum* (Ait.) Kunth

Ethnomedicinal uses: the stem and leaves are cooked as vegetable and considered highly nutritious by the tribals. A paste prepared from the tuber and stem is applied locally on burns and inflammation, and an extract of the tuber is taken orally to treat throat swelling and breathing difficulties. A paste of the tuber is applied locally to treat snake bite and scorpion bite, while an extract of the tuber is given orally to treat dog bite. The root powder of *Sauromatum venosum* (10 g) and *Corallocarpus epigaeus* (10 g) and the alum powder (20 g) are mixed in 50 g of curd and kept in a copper container for 1 h. This paste is applied locally once a day for a week to treat ringworm.

Tecomella undulata (Sm.) Seem. (Bignoniaceae)

Ethnomedicinal uses: oil extracted from the bark is used to treat syphilis. Oil is also applied externally to treat eczema and skin eruptions. The root powder is taken with milk to treat leucorrhoea.

Tinospora cordifolia (Willd.) Miers (Menispermaceae) (Fig. 3.20)

Fig. 3.20 *Tinospora cordifolia* (Willd.) Miers

Ethnomedicinal uses: to treat piles, the swollen portion of the rectum is first washed with a leaf extract of Neem (*Azadirachta indica*) and then smeared with a bark paste of *Tinospora cordifolia*. Stem pieces are eaten raw to treat rheumatism and diabetes. A decoction of the whole plant is taken orally three times a day for 3 d to treat hepatic diseases, pneumonia, diarrhoea and periodic fever. A decoction is also taken to prevent dehydration and to increase appetite. A decoction of fresh leaves is taken orally and considered very effective in treating leucorrhoea. A powder of the whole plant of *Enicostemma axillare* (50 g) and stem powder of *Tinospora cordifolia* (50 g) are mixed with jaggery (100 g), and a bolus of 1 g each is made. One bolus is taken three times a day for 3 d to treat malarial fever.

Ethnoveterinary medicinal uses: an extract of the whole plant is given to bulls to increase sexual power.

Trichosanthes bracteata (Lam.) Voigt (Cucurbitaceae)

Ethnomedicinal uses: one teaspoon of mature seed powder is taken orally by tribal women once a day to develop sterility. The juice of fresh plants along with mustard oil is poured into the ear to treat earache. A paste of the root of *Trichosanthes bracteata* is applied locally for 4 to 5 d to treat piles. During this remedy a gum extract of *Sterculia urens* is also taken orally once a day for 4 to 5 d. A paste prepared by mixing the dried root powder of *Trichosanthes bracteata* (5 g), seeds of *Azadirachta indica* (5 g), a kernel of *Corallocarpus epigaeus* (1 kernel) and 5 g alum powder with curd (150 g) is applied locally once a day for 7 d to treat ringworm.

Ethnoveterinary medicinal uses: a root extract is given to animals to treat flatulence.

Trichosanthes cucumerina Linn. (Cucurbitaceae) (Fig. 3.21)

Fig. 3.21 *Trichosanthes cucumerina* Linn

Ethnomedicinal uses: crushed seeds are mixed with milk and taken on an empty stomach to treat diabetes. A decoction of the roots and inflorescence is taken to treat bronchitis and heart diseases. One teaspoon of the tuber powder is taken orally by tribals once a day for 3 d to treat colic. An extract of 1 g leaf powder is taken orally as an antidote to snake bite.

Ethnoveterinary medicinal uses: a fruit extract is given to cattle to treat urinary ailments.

***Xanthium strumarium* Linn. (Asteraceae)**

Ethnomedicinal uses: the leaf juice is poured in the ear poured in earache and to stop pus formation. The oil obtained from the fruits is applied locally to treat eczema and scabies, while a leaf paste is applied to treat leucoderma. The seed powder is mixed with lemon juice and water and given orally to treat dysuria. The seed oil is used for massage in the treatment of joint pains and is also applied on the forehead to relieve headache pain. An infusion of the root is applied to ulcers and boils for fast healing. The tender roots of *Xanthium strumarium* (10 g) are chewed twice a day to treat toothache.

3.4 Wild Food Plants

Wild foods are a part of rural people's diet, not only during periods of food shortages, but on a daily basis as well [31–32]. It is this daily consumption of wild products which contributes to the overall nutritional well-being of tribals. In Rajasthan, famine and food scarcity are not rare events. Under such circumstances, the role of wild edible plants becomes significant. Tribals use various parts of these wild edible plants for food, in either raw or cooked form (food grains, edible oils, edible gums), or for making pickle and marmalade, in paste as flavouring/souring agents, and for curd preparation.

A total of 162 plant species are reportedly consumed by tribals as food, of which 30 are used as grains/seeds, 34 as leafy vegetables, 23 as root/tubers, 31 as other vegetables, 3 as oilseeds, 6 as miscellaneous food, 10 as flavouring/souring agents, 52 as fruits, 5 as gum and 4 for instant curdling of milk [28] (Fig. 3.22).

Apart from the traditional use of wild plants for food and medicine, tribals also use the plants for a number of other needs, such as in religious and superstitious beliefs, tools and as cosmetics and shampoo.

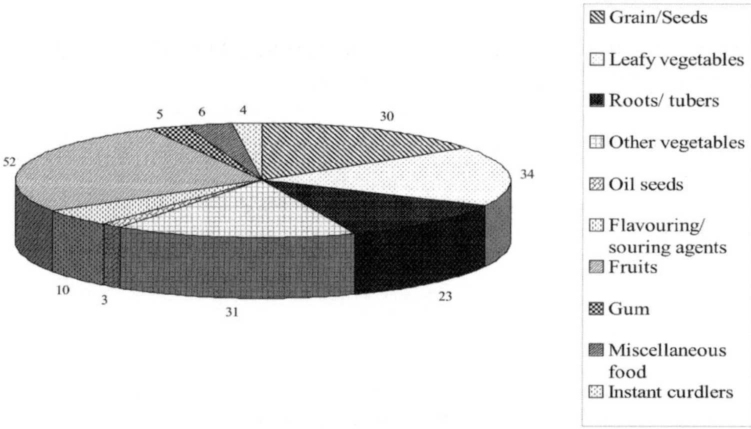

Number of ethno-food plant species reported under different food groups

Fig. 3.22

3.5 Conclusions

The present study shows that Rajasthan is home to a vast diversity of medicinal plants. Despite a gradual sociocultural transformation, local communities still possess substantial knowledge of plants and their uses. The reliance on folk medicines for healthcare is associated with the lack of modern medicines and medication, poverty and traditional belief in their effectiveness. Since there is a complete lack of phytotherapeutic evidence for many of the species, it is recommend that phytochemical and pharmacological studies be carried out in order to confirm the validity of properties attributed to these species. This is particularly relevant for species with market potential beyond the state of Rajasthan. If plans for their extraction were drawn up, these medicinal resources could provide for both subsistence needs and income. This, however, requires a detailed assessment of resource quantities, productivity potential, sustainable harvesting methods, domestication possibilities and the market value of potentially promising species and, importantly, equitable benefit-sharing regimes. The knowledge of the use of species reported here belongs to the tribal and rural people of the state. Any benefits derived from the use of this knowledge must be shared with the inhabitants of the state.

Ethnoveterinary medicines are of specific value in developing countries, where allopathic veterinary medicines are often beyond the reach of livestock producers. It can play an important role in rural farm development. The present paper is also intended to contribute to this growing body of knowledge by supplying information on the plant-based ethnoveterinary curative techniques found in tribal areas of Rajasthan.

Only a small percentage of about 400,000 plant species on the earth have been phytochemically investigated, and the fraction submitted to biological or pharmacological screening is even smaller [33]. The plant kingdom thus represents an enormous reservoir of pharmacologically valuable material to be discovered [34] which will likely be of great benefit for human welfare.

Acknowledgements The author acknowledges the tremendous amount of cooperation and assistance that he received from his students Dr. (Mrs.) Anita Jain, Dr. Praveen Galav and tribals for readily sharing their knowledge. The author is also thankful to the CSIR, DST and Ministry of Environment and Forests, Government of India, New Delhi for financial assistance.

References

1. Jain SK (1991) Dictionary of Indian Folk Medicine and Ethnobotany. Deep Publication, New Delhi
2. Fransworth NR (1990) The role of ethnopharmacology in drug development. In: Bioactive Compounds from Plants. Wiley, Chichester (Ciba Foundation Symposium), p 2
3. Ramawat KG, Sonie KC, Sharma MC (2004) Therapeutic potential of medicinal plants. In: Ramawat KG (ed) Biotechnology: Medicinal Plants. Science Publishers, Enfield, CT, p 1
4. Ramawat KG, Goyal S (2008) Indian herbal drugs scenario in global perspectives. In: Ramawat KG, Merillon JM (eds) Biotechnology: Bioactive Molecules. Springer, Berlin Heidelberg New York, p 323
5. Trivedi PC, Nehra S (2004) Traditional herbal medicines for health care and management of human diseases. In: Trivedi PC (ed) Herbal Drugs and Biotechnology. Pointer Publishers, Jaipur, India, p 1
6. Pushpangadan P (1994) Ethnobiology in India – a status report, All India Coordinated Research project on Ethnobiology (AICRPE), Ministry of Environment and Forests, Government of India, New Delhi
7. Krishna Kumar PR (1996) Ancient Sci Life 15:161
8. Bhandari MM (1974) Eco Bot 28:73
9. Singh V, Pandey RP (1998) Ethnobotany of Rajasthan. Scientific Publishers, Jodhpur, India
10. Sebastian MK, Bhandari MM (1984) Indian J Ethnopharmacol 12:223
11. Khan SM, Chaghtai SA, Khan SS (1988) Indian J Appl Pure Biol 3:115
12. Joshi P, Joshi SC (1992) Ancient Sci Life 11:163
13. Joshi P (1993) Ethnomedicine of kathodias – a monkey eating tribe in Rajasthan. In: Singh VK, Govil JN (eds) Glimpses of Plant Research, vol X, Medicinal Plants — New Vistas in Research. Today and Tomorrow's Printers and Publishers, New Delhi, p 23
14. Singh VK, Ali ZA (1996) Folk medicinal plants used for family planning in India. In: Jain SK (ed) Ethnobiology in Human Welfare. Deep Publication, New Delhi, p 23
15. Katewa SS, Arora A (1997) Ethnobotany 9:48
16. Katewa SS, Sharma R (1998) Ethnobotany 10:46
17. Katewa SS (2003) The Indian Forester 129:1117
18. Arya HC, Ramawat KG (1988) Phenylalkylamines (*Ephedra* alkaloids). In: Constabel F, Vasil IK (eds) Phytochemicals in Plant Cell Cultures, vol 5. Academic, New York, p 237
19. Bhardwaj L (1991) Ph.D. Thesis, University of Jodhpur, Jodhpur, India
20. Guar A, Bhardwaj L, Merillon JM, Ramawat KG (1993) Indian J Exp Biol 31:987
21. Valls J, Rchard T, Larronde F, Leblais V, Muller B, Delauney JP, Ramawat KG, Merillon JM (2006) Fitoterapia 77:416
22. Mathur M, Dass S, Suthar S, Ramawat KG (2007) Proc Natl Symp CIMAP, Lucknow, p 410
23. Tanwar YS, Mathur M, Ramawat KG (2007) Plant Growth Regul 51:93
24. Roat C, Kumar N, Ramawat KG (2008) J Herb Med Toxicol 2(1):11
25. Vaishava K, Goyal S, Ramawat KG (2006) Indian J Exp Biol 44:1012
26. Tanwar YS, Goyal S, Ramawat KG (2008) J Herb Med Toxicol 2:21

27. Katewa SS, Guria BD, Jain A (2001) J Ethnopharmacol 76:293
28. Katewa SS, Jain A (2006) Traditional Folk Herbal Medicines. Apex Publishing House, Udaipur, India
29. Chaudhary BL, Katewa SS, Galav PK (2008) Indian J Tradit Knowl 7:11
30. Jain A, Katewa SS, Choudhary BL, Galav P (2004) J Ethnopharmacol 90:171
31. FAO (1989) Forestry and nutrition. Swedish International Development Authority and FAO, Rome
32. Kingamokono R, Lindestrom J (eds) (1990) Forests and Fields. Ministry of Land, Natural Resources and Tourism, Tanzania
33. Hostettmann K, Potterat O, Wolfender JL (1998) Chimia 52:10
34. Potterat O, Hostettmann K (1995) Plant source of natural drugs and compounds. In: Encyclopedia of Environmental Biology, vol 3. Academic, San Diego, p 139

Chapter 4
Ginseng and Male Sexual Behavior

Laura L. Murphy and James S. Ferraro

Abstract Ginseng is widely used in Asian countries as a tonic to promote and maintain good health and as a constituent in herbal medicines used to treat various diseases, including liver dysfunction, hypertension, cerebrovascular disease, atherosclerosis, cancer, postmenopausal symptoms, and impotence. Very few controlled clinical studies have been performed to validate the medicinal use of ginseng or its constituents in humans. However, laboratory studies, primarily using rodents, have elucidated potential medical uses for ginseng and ginsenosides in the treatment of a number of human disorders, including impotence and loss of libido. *Panax ginseng* and *Panax quinquefolius* have both been shown to enhance male copulatory behavior in laboratory studies. Human studies have suggested that ginseng ingestion may be a safe and effective alternative method for treatment of erectile dysfunction. Ginsenosides have been shown to interact with steroid receptors and, through nongenomic pathways, activate NO production. How ginseng and its ginsenosides may act centrally to modulate sexual desire and copulatory performance is not known, but could potentially involve activation of NO pathways, perhaps in brain areas involved in sexual behavior.

Keywords Ginseng · Impotence · Erectile dysfunction · Sex behavior · Nitric oxide

Abbreviations

cGMP	Cyclic guanosine-3′,5′-monophosphate
GTP	Guanosine triphosphate
IIEF	International Index of Erectile Function
MPOA	Medial preoptic area of the hypothalamus
NANC	Nonadrenergic-noncholinergic neurons

L.L. Murphy and J.S. Ferraro (✉)
Department of Physiology, Southern Illinois University School of Medicine, Carbondale, IL 62901-6512 USA,
e-mail: lmurphy@siumed.edu

K.G. Ramawat (ed.), *Herbal Drugs: Ethnomedicine to Modern Medicine*,
DOI 10.1007/978-3-540-79116-4_4,

NO Nitric oxide
NOS Nitric oxide synthase, endothelial (e), neuronal (n), inducible (i)
PDE Phosphodiesterase
PVN Paraventricular nuclei
VIP Vasoactive intestinal peptide

4.1 Introduction to Ginseng

Ginseng most commonly refers to the plant *Panax ginseng* C.A. Meyer (Asian ginseng) or *Panax quinquefolius* L. (North American ginseng). Ginseng is widely used in Asian countries as a tonic to promote and maintain good health, and as a constituent in herbal medicines used to treat various diseases, including liver dysfunction, hypertension, cerebrovascular disease, atherosclerosis, cancer, postmenopausal symptoms, and impotence [1–8]. Elsewhere, ginseng supplements are marketed as a natural stimulant that will increase endurance and vitality and improve overall health. It is the root of the ginseng plant that is harvested for its reported medicinal effects. The root is typically dried (white ginseng) or steamed (red ginseng) and can also be extracted to concentrate its bioactive components, the ginsenosides. Ginsenosides are unique to ginseng and are saponin glycosides containing one or more sugar moieties that form side chains off the aglycone ginsenoside structure [9]. There are over 20 ginsenosides found in ginseng, and these are mainly categorized as being a protopanaxadiol (glycoside side chains off C_3 and C_{20}) or protopanaxatriol (side chains off C_6 and C_{20}). In Fig. 4.1, ginsenoside Rg1 represents a classic protopanaxatriol ginsenoside with properties that will be discussed later in this chapter. Very few controlled clinical studies have been performed to validate the medicinal use of ginseng or its constituents in humans. However, laboratory studies, primarily using rodents, have elucidated potential medical uses for ginseng and ginsenosides in the treatment of a number of human disorders, including impotence and loss of libido [10].

4.2 Physiology of an Erection

Impotence, or erectile dysfunction, can be defined as the persistent inability to develop and/or maintain an erection adequate for sexual intercourse. Vascular diseases, including atherosclerosis, hypertension, diabetes, and high cholesterol, can cause restricted blood flow to the penis and may account for approximately 70% of physically related causes of erectile dysfunction [11]. The arterial blood supply to the penis is mainly provided by the internal pudendal artery, which branches into several smaller arteries that contribute to vascularization of the erectile tissues. The erectile tissue of the human penis consists of two dorsal corpora cavernosa and a ventral corpus spongiosum. Reduced sympathetic input combined with stimulation

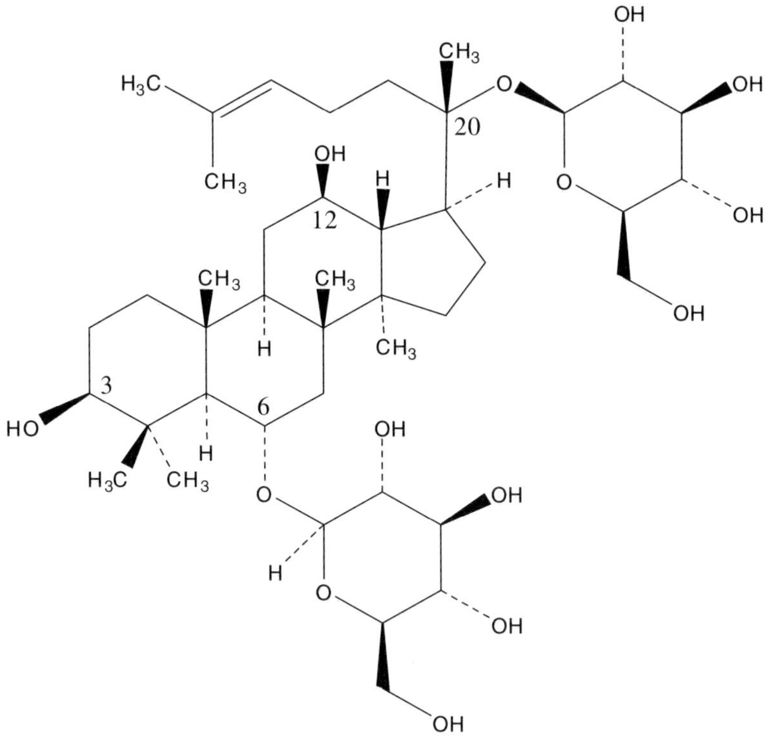

Fig. 4.1 Chemical structure of ginsenoside Rg1, a classical protopanaxatriol found in both *Panax ginseng* and *Panax quinquefolius*

of parasympathetic pathways induces vascular smooth muscle relaxation and vessel dilation, causing increased arterial blood flow to the penis, engorgement of the erectile tissues, reduced venous outflow, and erection.

The physiological regulation of erectile function has been well reviewed [12–15]. To understand how ginseng may affect the erection process, a brief overview of ginseng-relevant processes will be described. The dilation of penile vascular smooth muscle is dependent on several neuronal pathways and vasoactive substances. The flaccid state of the penis is maintained by sympathetic nerve stimulation and the release of the adrenergic neurotransmitter norepinephrine, as well as by production of the vasoconstrictor endothelin-1 by vascular endothelial cells. Attenuated sympathetic input to the penis will decrease vasoconstrictive tone and allow vessel dilation. With increased parasympathetic nerve stimulation and the release of the cholinergic neurotransmitter acetylcholine and the peptide vasoactive intestinal peptide or VIP, there is a further increase in vasodilation. Another vasoactive substance, nitric oxide, is produced by nonadrenergic-noncholingeric (NANC) neurons that innervate penile vasculature. Acetylcholine-induced vasodilation is also mediated by nitric oxide. Nitric oxide (NO) is produced from L-arginine and oxygen by the enzyme nitric oxide synthase (NOS). Three isoforms of NOS have been elucidated: neuronal NOS (nNOS or NOS-1), inducible NOS (iNOS or NOS-2), and endothelial NOS (eNOS or NOS-3). Neurons that innervate the corpus cavernosa contain nNOS, while eNOS is localized in the endothelium and smooth muscle of the cavernosa. Inducible NOS does not appear to be involved in the physiology of erections, but may be a factor in the deleterious effects of aging on the delicate erectile tissues of the corpora cavernosa [16, 17]. Nitric oxide directly increases the enzyme guanylate cyclase, which catalyzes the conversion of guanosine triphosphate (GTP) to cyclic guanosine-3′,5′-monophosphate (cGMP). The cyclic nucleotide cGMP is a second messenger that activates cGMP-dependent protein kinase, leading to decreased intracellular calcium, smooth muscle relaxation of the penile arterioles, and consequent increased blood flow into the erectile tissues. The shear forces induced by this increased blood flow against the endothelial cell surface and the increased pressure within the erectile tissues trigger eNOS and the production of more nitric oxide, thus reinforcing the turgidity of the penis. The cyclic nucleotide phosphodiesterases (PDE) located within the penile vasculature degrade cGMP by hydrolyzing cGMP into the inactive 5′-GMP. Oral phophodiesterase inhibitors can be prescribed to treat erectile dysfunction. These drugs include sildenafil, vardenafil, and tadalafil, and when taken orally they selectively inhibit cGMP-specific phosphodiesterase type 5, the predominant phosphodiesterase isoform in the penis. By preventing cGMP breakdown and, consequently, increasing cGMP levels, these drugs help to promote penile vasodilation and can prolong and maintain erection.

4.3 Ginseng and Copulatory Behavior – Animal Studies

Panax ginseng and *Panax quinquefolius* have both been shown to enhance male copulatory behavior in laboratory studies (Table 4.1) [18–20]. Copulatory behavior can be examined in male rodents by placing them with sexually receptive female conspecifics and measuring the following parameters: mount latency: time from introduction of the female until the first mount with pelvic thrusting; intromission latency: time from introduction of the female until first mount with pelvic thrusting and vaginal penetration; ejaculation latency: time from the first intromission until ejaculation; and postejaculatory interval or the refractory period:time from ejaculation until next intromission. A reduced latency to mount the female would indicate increased male sexual arousal, whereas a decreased latency to intromit or ejaculate would be indicative of increased copulatory performance. In male mice whose normal copulatory behavior was compromised as a result of prolonged individual housing, the daily intraperitoneal administration of *Panax ginseng* extract (25 to $100\,\mathrm{mg\,kg^{-1}}$ body weight) or ginsenoside Rg1 (2.5 to $10\,\mathrm{mg\,kg^{-1}}$) produced a dose-related increase in the number of mice displaying copulatory behavior [18]. In normal male rats treated subcutaneously with *Panax ginseng* extract ($20\,\mathrm{mg\,kg^{-1}}$ body weight) for 5 d, there was a significant reduction in ejaculation latency and a decreased refractory period [19]. Similarly, male rats treated orally with an emul-

Table 4.1 Ginseng preparations and their effects on copulatory behaviors in animals and on erectile dysfunction (ED) in humans

	Treatment paradigm	Results		Ref.
Animal studies				
Panax ginseng Root extract	20 mg kg^{-1} sc, daily for 5 d	Male rats: ↓ ejaculation latency, ↓ refractory period		[19]
Panax quinquefolius root powder	10 to 100 mg kg^{-1} po, daily for 1 to 28 d	Male rats: ↓ mount, intromission, ejaculation latencies		[20]
Panax ginseng red root extract	25 to 100 mg kg^{-1} ip, daily for 35 d	Male mice: ↑ mount, intromission frequencies		[18]
Ginsenoside Rg1	2.5 to 10 mg kg^{-1} ip, daily for 35 d	Male mice: ↑ mount, intromission frequencies		[18]
Ginsenosides Rb1, Rb2, Ro	2.5 to 10 mg kg^{-1} ip, daily for 35 d	Male mice: no effect on copulatory behaviors		[18]
Human studies				
Panax ginseng Red root extract	900 mg po, tid for 8 weeks	Double-blind and placebo-controlled $n = 45$ men w/ED	↑ erectile function, sexual desire; ↑ mean IIEF score	[24]
	1000 mg po, tid for 12 weeks	Double-blind and placebo-controlled $n = 60$ men w/ED	↑ erectile function ↑ mean IIEF score	[25]

sion of powdered *Panax quinquefolius* root (10, 50, or 100 mg kg^{-1} body weight) for 1, 14, or 28 d demonstrated reduced mount, intromission, and ejaculation latencies (Fig. 4.2) that were significantly influenced by all the ginseng doses as early as 1 d posttreatment [20].

4.4 Ginseng and Erectile Function

Erectile dysfunction is commonly associated with diabetes, a condition that has well-characterized effects on peripheral tissue innervation and vascularization [21]. Penile intracavernous pressure after nerve stimulation was significantly lower in long-term diabetic rats relative to age-matched controls [22]. In diabetic rats treated with Korean red ginseng (30 mg kg^{-1}) for one month, however, erectile function was not different from nondiabetic age-matched controls [22]. Ginseng use has also been shown to produce beneficial effects in clinical studies (Table 4.1) that evalu-

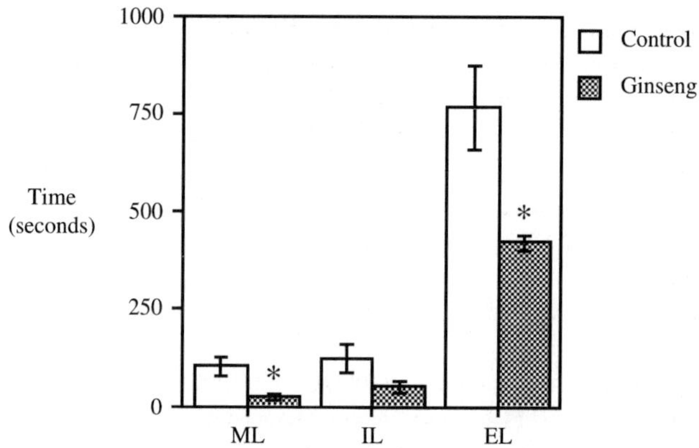

Fig. 4.2 Ginseng treatment enhances copulatory behaviors in male rats. Animals were treated daily with either a sesame oil vehicle (control) or 100 mg kg^{-1} *Panax quinquefolius* for 28 d. When placed with sexually receptive female rats, latency (in seconds) to mount (ML), intromit (IL), and ejaculate (EL) were determined. $^{*} = p < 0.05$ relative to control. (Figure adapted from [20])

ated the efficacy of ginseng use in men with erectile dysfunction [6, 10, 23–25]. In double-blind, placebo-controlled studies, men with diagnosed erectile dysfunction took 900 to 1000 mg of Korean red ginseng or placebo by mouth three times daily for either 8 or 12 weeks [23–25]. Using an International Index of Erectile Function (IIEF) self-assessment, ginseng treatment produced significant improvement in IIEF scores when compared to placebo. The mean scores for erectile function, sexual desire, and intercourse satisfaction were significantly higher following ginseng treatment when compared to the placebo group after 8 weeks of treatment [25]. Individually, scores for penis rigidity, penetration, and maintenance of erection were significantly higher in the ginseng group. There was no effect of ginseng treatment on serum testosterone levels [25, 26], which was consistent with findings in animal studies [20].

4.5 Nitric Oxide

Laboratory studies indicate a role for NO in mediating many of the actions of ginseng and ginsenosides on copulatory performance. In rats treated for 3 months with oral ginseng extract versus no treatment, there was significantly increased intracavernosal pressure following stimulation of pelvic nerves innervating the corpus cavernosum [27]. Ginsenoside-containing extracts of *P. ginseng* produced a dose-related increase in cGMP and endothelium-dependent dilation and relaxation of penile corpus cavernosum in rabbit [28, 29]; treatment with the NOS inhibitor N-nitro-L-arginine blocked the ability of ginseng to induce this relaxation. Ginseng extract also enhanced acetylcholine-induced and transmural nerve-stimulation-elicited relaxations in rabbit corpus cavernosum tissue and reversed phenylephrine-induced corpus cavernosum smooth muscle vasoconstriction [30]. Both the neurogenic and endothelium-mediated relaxations were enhanced by superoxide dismutase (SOD), a scavenger of the superoxide anion that inactivates NO, and were blocked by N-nitro-L-arginine and oxyHb, compounds that inhibit NOS and trap NO, respectively [30]. Ginseng/ginsenosides modulate vasoactivity in a number of other systems as well [2]. Ginseng extract and ginsenosides Rg3, Rg1, Re, and Rb1 have been shown to increase NOS activity and NO production and increase cGMP levels in cardiomyocytes [31] and in aortic, pulmonary, cerebral and umbilical endothelial cell cultures [32] and have been postulated to offer both neuro- and cardiovascular protection [33, 34].

The exact mechanisms by which ginseng/ginsenosides stimulate NOS and NO release remain unclear. As shown in Fig. 4.1, the ginsenoside structure contains polar and nonpolar regions that may allow ginsenosides to intercalate into cell plasma membranes and alter membrane fluidity and cell signaling [1, 35]. There is also evidence that ginsenosides may interact with steroid hormone receptors and stimulate NO production through nongenomic, transcription-independent mechanisms of action [34]. Steroidal estrogens, androgens, and corticosteroids are potent vasoactive modulators that can influence NO production through binding to their respective receptors [36–38]. Several ginsenosides have been reported to exert steroidlike activity, notably Rb1, Re, and Rg1 [34]. Through binding to estrogen receptors, ginsenoside Re increased PI3-kinase/Akt and eNOS activity in arterial smooth muscle A10 cells, and the estrogen receptor antagonist ICI 182,780 blocked this ginsenoside Re action [39]. While ginsenoside Re induced eNOS activity and NO production in cardiac myocytes, Re did not activate gene transcription and its effects on NO were blocked by concomitant treatment with estrogen, progesterone, and androgen receptor antagonists indicating a nongenomic action of Re on sex-steroid receptors [40]. Similarly, ginsenosides Re and Rg1 have been shown to be functional ligands for the glucocorticoid receptor in human umbilical vein endothelial cells [34], producing a rapid Ca^{2+} influx, activation of the PI3-kinase/Akt pathways, and NO production that was abolished by treatment with a glucocorticoid receptor antagonist or siRNA targeting of the receptor [41]. Through a nongenomic action on androgen receptors, ginsenoside Rb1 produce a rapid increase in eNOS expression and NO production in human aortic endothelial cells that was prevented by treatment with an androgen-receptor antagonist or inhibitors for the PI3-kinase/Akt and MEK/ERK pathways [42].

4.6 Central Nervous System Actions of Ginseng

Besides the potential direct effects of ginseng/ginsenosides on penile nerve and vascular tissues, ginsenosides may also exert actions on the brain to modulate neurotransmission. Ginsenosides, notably Rb1, Rc, Rg1, and Rg3, facilitate acetylcholine activity [43, 44], are neuroprotective [45–48], and alter neurotransmitter availability [49–52]. The neurotransmitter dopamine plays a central role in sexual behavior [53, 54], and treatment with dopaminergic drugs has long been shown to facilitate masculine sexual behavior [55, 56]. In the paraventricular nuclei (PVN) and medial preoptic area of the hypothalamus (MPOA), brain areas critical for male sexual behavior [57], dopamine levels are increased before and during copulation [58, 59]. The microinjection of dopamine receptor antagonists directly into the MPOA decreases libido and impairs copulatory performance [60, 61]. NO may be an important mediator of hypothalamic dopamine release [62]. Treatment with NOS inhibitors decreased hypothalamic dopamine levels [63] and reduced male mounting activity [64, 65], libido [64], and penile erection [66, 67]. Although there is no direct evidence to show that ginseng/ginsenosides modulate hypothalamic NO levels, the injection of ginsenoside Rc or Rg3 into the cerebral ventricles inhibits stress-induced hypothalamic activation in mice through a NO-mediated pathway [68], suggesting that ginsenosides directly induce NO production in the brain. Moreover, oral ginseng treatment was shown to increase striatal dopamine activity [69] and decrease plasma prolactin levels [70] in young adult male rats, consistent with a putative role for dopamine in ginseng-induced enhancement of copulatory behaviors.

4.7 Conclusions

Human studies have suggested that ginseng ingestion may be a safe and effective alternative method for treatment of erectile dysfunction [10]. Animal studies indicate that ginseng/ginsenosides mediate the release and/or the modification of release of NO from endothelial cells and perivascular nerves, and the consequent vasodilatation of the penile corpus cavernosum may contribute to the copulatory performance-enhancing action of ginseng. Ginsenosides have been shown to interact with steroid receptors and, through nongenomic pathways, activate NO production. How ginseng and its ginsenosides may act centrally to modulate sexual desire and copulatory performance is not known, but it could potentially involve activation of NO pathways, perhaps in brain areas involved in sexual behavior.

Acknowledgements We thank Dr. Rich Clough for his careful review and critique of this chapter; Dr. Yuqing Ho for the ginsenoside structure he prepared for us; and, to the authors of those relevant papers that were not included in this chapter, we thank you for understanding our space limitations and that omission does not mean lack of appreciation.

References

1. Attele AS, Wu JA, Yuan CS (1999) Biochem Pharmacol 58:1685
2. Achike FI, Kwan C-Y (2003) Clin Exp Pharmacol Physiol 30:605
3. Kiefer D, Pantuso T (2003) Am Fam Physician 68:1539
4. McKay D (2004) Altern Med Rev 9:4
5. Carroll DG (2006) Am Fam Physician 73:457
6. Tamler R, Mechanick JI (2007) Endocrinol Metab Clin N Am 36:533

7. Wang CZ, Mehendale SR, Yuan CS (2007) Am J Chin Med 35:543
8. Helms S (2004) Altern Med Rev 9:259
9. Soldati F, Sticher O (1980) Planta Med 39:348
10. Price A, Gazewood J (2003) J Fam Pract 52:20
11. Guay AT (2007) Endocrinol Metab Clin North Am 36:453
12. Lugg JA, Gonzalez-Cadavid NF, Rajfer J (1995) J Androl 16:2
13. Andersson KE (2001) Pharmacol Rev 53:417
14. Prieto D (2007) Int J Impot Res 20:17
15. Saenz de Tejada I (2002) Int J Impot Res 14:S6
16. Ghalayini IF (2004) Int J Impot Res 16:459
17. Gonzalez-Cadavid NF, Rajfer J (2005) Curr Pharm Des 11:40–41
18. Yoshimura H, Kimura N, Sugiura K (1998) Methods Find Exp Clin Pharmacol 20:59
19. Kim C, Choi H, Kim CC, Kim JK, Kim MS (1976) Am J Chin Med 4:163
20. Murphy LL, Cadena RS, Chavez D, Ferraro JS (1998) Physiol Behav 64:445
21. Musicki B, Burnett AL (2007) Int J Impot Res 19:129
22. Ryu JK, Lee T, Kim D-J, Park I-S, Yoon S-M, Lee H-S, Song SU, Suh J-K (2005) Urology 65:611
23. Choi HK, Seong DH, Rha KH (1995) Int J Impot Res 7:181
24. Hong B, Ji YH, Hong JH, Nam KY, Ahn TY (2002) J Urol 168:2070
25. de Andrade E, de Mesquita AA, de Almeida Claro J, de Andrade PM, Ortiz V, Paranhos M, Srougi M (2007) As J Androl 9:241
26. Kang HY, Kim SH, Lee WJ, Byrne HK (2002) J Strength Cond Res 16:179
27. Choi YD, Rha KH, Choi HK (1999) J Urol 162:1508
28. Choi YD, Xin ZC, Choi HK (1998) Int J Impot Res 10:37
29. Kim HJ, Woo DS, Lee G, Kim JJ (1998) Br J Urol 82:744
30. Chen X, Lee TJ (1995) Br J Pharmacol 115:15
31. Chen X (1996) Clin Exp Pharmacol Physiol 23:728
32. Toda N, Ayajiki K, Fujioka H, Okamura T (2001) J Ethnopharmacol 76:109
33. Gillis CN (1997) Biochem Pharmacol 54:1
34. Yue PYK, Mak NK, Cheng YK, Leung KW, Ng TB, Fan DTP, Yeung HW, Wong RNS (2007) Chin Med 2:6
35. Nah SY, Kim DH, Rhim H (2007) CNS Drug Rev 13:381
36. White RE (2002) Vascul Pharmacol 38:73
37. Liu PY Death AK, Handelsman DJ (2003) Endoc Rev 24:313
38. Yang S, Zhang L (2004) Curr Vasc Pharmacol 2:1
39. Nakaya Y, Mawatari K, Takahashi A, Harada N, Hata A, Yasui S (2007) J Med Invest 54:381
40. Furukawa T, Bai C-X, Kaihara A, Ozaki E, Kawano T, Nakaya Y, Awais M, Sato M, Umezawa Y, Kurokawa J (2006) Mol Pharmacol 70:1916
41. Leung KW, Leung FP, Huang Y, Mak NK, Wong RN (2007) FEBS Lett 582:2423
42. Yu J, Eto M, Akishita M, Kaneko A, Ouchi Y, Okabe T (2007) Biochem Biophys Res Commun 353:764
43. Benishin CG, Lee R, Wang L, Liu HJ (1991) Pharmacol 42:223
44. Benishin CG (1992) Neurochem Int 21:1
45. Yamaguchi Y, Haruta K, Kobayashi H (1995) Psychoneuroendo 20:645
46. Yamaguchi Y, Higashi M, Kobayashi H (1996) Eur J Pharmacol 26:149
47. Lim JH, Wen TC, Matsuda S, Tanaka J, Maeda N, Peng H, Aburaya J, Ishihara K, Sakanaka M (1997) Neurosci Res 28:191
48. Radada K, Gilleb G, Moldzioa R, Saitoc H, Rausch W (2004) Brain Res 1021:41
49. Tsang D, Yeung HW, Tso WW, Peck H (1985) Planta Med 3:221
50. Kimura T, Saunders PA, Kim HS, Rheu HM, Oh KW, Ho IK (1994) Gen Pharmacol 25:193
51. Yuan CS, Attele AS, Wu JA, Liu D (1998) J Ethnopharmacol 63:215
52. Xue J-F, Liu Z-J, Hu J-F, Chen H, Zhang J-T, Chen N-H (2006) Brain Res 1106:91
53. Pleim ET, Matochik JA, Barfield RJ, Auerbach SB (1990) Brain Res 524:160
54. Hull EM, Lorrain DS, Du J, Matuszewich L, Lumley LA, Putnam SK, Moses J (1999) Behav Brain Res 105:105

55. Bitran D, Hull EM (1987) Neurosci Biobehav Rev 11:365
56. Melis MR, Argiolas A (1995) Neurosci Biobehav Rev 19:19
57. Hull, EM, Wood RI, McKenna KE (2006) Neurobiology of male sexual behavior. In: Neill JD (ed) Physiology of Reproduction, vol 2. Raven, New York, p 1729
58. Hull EM, Du J, Lorrain DS, Matuszewich L (1995) J Neurosci 15:7465
59. Melis MR, Succu S, Mascia MS, Cortis L, Argiolas A (2003) Eur J Neurosci 17:1266
60. Pehek EA, Warner RK, Bazzett TJ, Bitran D, Band LC, Eaton RC, Hull EM (1988) Brain Res 443:70
61. Warner RK, Thompson JT, Markowski VP, Loucks JA, Bazzett TJ, Eaton RC, Hull EM (1991) Brain Res 540:177
62. Lorrain DS, Hull EM (1993) Neuro Rep 5:87
63. Lorrain DS, Matuszewich L, Howard RV, Du J, Hull EM (1996) Neuro Rep 8:31
64. Sato Y, Horita H, Kurohata T, Adachi H, Tsukamoto T (1998) Am J Physiol 274:R243
65. Ratnasooriya WD, Dharmasiri MG, Wadsworth RM (2000) Int J Androl 23:187
66. Melis MR, Stancampiano R, Argiolas A (1995) Naun Sch Arch Pharmacol 351:439
67. Melis MR, Succu S, Mauri A, Argiolas A (1998) Eur J Neurosci 10:1968
68. Kim DH, Jung JS, Suh HW, Huh SO, Min SK, Son BK, Park JH, Kim ND, Kim YH, Song DK (1998) Neuro Report 9:2261
69. Watanabe H, Ohta H, Imamura L, Asakura W, Matoba Y, Matsumoto K (1991) Jpn J Pharmacol 55:51
70. Murphy LL, Lee TJ-F (2002) Ann NY Acad Sci 962:372

Chapter 5
Herbal Treatments for Erectile Dysfunction

Jyoti Shah

Abstract Evidence for herbal treatments for ED is contradictory. This could be due to differing definitions of ED, non-standardized outcomes and different procedures used in the preparation of plant extracts. Additionally, animal studies are not accurate because they do not allow cerebral aspects of sex to be evaluated and instead rely solely on basic mechanical or instinctive sexual function.

Preliminary research on a few drugs such as ginseng, PLC and DHEA is encouraging, but well-designed randomized controlled clinical trials are lacking. It is likely that the use of medicinal plants will increase in popularity as men seek subtle methods of treating themselves. However, whilst it is easy to purchase many 'natural Viagra'-like substances over the Internet, the safety and reliability of many of these drugs is poor, and patients should be cautious when acquiring these products. Information about how these herbal products interact with prescribed drugs is also limited.

Keywords Erectile dysfunction · Herbal therapies · Impotence

5.1 Introduction

Erectile dysfunction (ED) is defined as 'the inability to achieve an erection sufficient for intercourse to the mutual satisfaction of both partners' [1]. The exact prevalence of ED is unknown. This is largely due to variations in the definition of ED and because many publications focus on select population groups. However, it is estimated that ED affects 15 to 30 million men worldwide [2]. According to the National Ambulatory Medical Care Survey (NAMCS), for every 1,000 men in thc United States, 7.7 physician office visits were made for ED. Almost 15 years later, that figure had increased three-fold to 22.3 in 1999 [3]. In countries like the USA where education and awareness are high and statistics are easy to compile, such estimates can be safely arrived at. However, in many parts of the world, men with ED suffer silently

J. Shah (✉)
Department of Urology, St. George's Hospital, Blackshaw Road, London SW17 OQT
e-mail: jyoti.shah@imperial.ac.uk

K.G. Ramawat (ed.), *Herbal Drugs: Ethnomedicine to Modern Medicine*,
DOI 10.1007/978-3-540-79116-4_5, © Springer-Verlag Berlin Heidelberg 2009

because of fear, embarrassment, lack of access to medical facilities, and perhaps because ED is a low medical priority in many parts of the world. Therefore the real prevalence may never be determined.

Although ED does not affect life expectancy, it can have a significant negative impact on an individual's well-being and quality of life [3]. There are many causes of ED but most important is the fact that this is an age-related phenomenon. The most reliable evidence for this comes from the Male Massachusetts Aging Study. This study reported that the probability of ED increased from 5.1% at 40 years of age to 15% at 70 years [4].

Other conditions such as diabetes, renal disease, chronic alcoholism, multiple sclerosis, atherosclerosis, vascular disease and neurological diseases account for approximately 70% of ED cases [2]. NIH also reported that between 35 and 50% of men with diabetes suffer from ED [2]. Furthermore, men taking many commonly used medications such as antihypertensive drugs, antihistamines, antidepressants, tranquillizers, appetite suppressants and cimetidine will experience ED as a side effect [2]. Psychological factors can also cause ED, either in isolation or concurrently with an organic cause.

With an ageing male population, an increase in comorbid conditions, greater treatment-seeking behaviour, and rising costs of treatment, the burden of ED will undoubtedly increase worldwide.

Conventional treatment methods for ED included psychosexual counselling, androgen replacement therapy, intraurethral agents, intracavernous injections, mechanical devices and surgery. The last decade has seen a huge paradigm shift in ED treatments.

The introduction of Viagra as an effective and reliable oral treatment for ED in 1998 has allowed access to treatment for many couples across the world. However, despite the success of modern treatments for ED, many men across the globe still rely on herbal treatments.

Reasons for this could be reluctance in seeking treatment for a private and personal condition, financial concerns about the side effect profile of drugs and their lack of efficacy. In one study, the long-term satisfaction of current ED therapies ranges from 40 to 70% [5].

Herbal treatments are health supplements that are chemically rich in plant or plant extracts and contain single or multiple ingredients in the form of tablets, capsules, creams or tinctures [6]. In the USA, from 1990 to 1997, there was a 380% increase in the use of herbal treatments for any condition, and in 2001, over 50% of the American population were regularly taking herbal supplements [7]. This trend is also reflected in the 47.3% increase in total visits to alternative practitioners from 427 million in 1990 to 629 million in 1997 [8]. This figure exceeds the total visits to all US primary care physicians [8].

5.2 Herbal Treatments for ED

There are many herbal drugs that have been used by men with ED with varying degrees of success. Now that the Internet is one of the most readily accessible and comprehensive methods of obtaining medical information and treatment options, the general public may try many such herbal therapies. What follows is an alphabetical overview of herbal treatments for ED with evidence, if any, of their efficacy and safety.

5.2.1 *Cola acuminata*

Cola acuminata belongs to the family Sterculiaceae and its fruits are harvested from the forests of the Democratic Republic of Congo. The fruits are rough, mottled and up to 8 inches long and contain large, flat and bright red coloured seeds. Commonly known as cola nut, this fruit is used widely for the treatment of ED.

The fruits contain about 2% catechine-caffeine (colanine), which is believed to have a greater level of alkaloids (caffeine), thereby increasing the stimulatory effect. They are roasted, pounded or chewed and can also be added to drinks such as tea or milk or cereal such as porridge [9]. In West Africa this product is sold as a treatment for ED, although evidence for its effectiveness is limited.

5.2.2 *Damiana*

Damiana (*Turnera diffusa*) has been used as an aphrodisiac in Mexico for centuries. From 1888 to 1947, Damiana leaf and its elixirs were even listed in the United States National Formulary.

It is a small shrub that is native to Central and South America and belongs to the family Turneraceae. When this shrub blooms, small yellow flowers emerge with a smell similar to camomile. Its dried leaves and stems are used as a sexual stimulant in the form of capsules, liquid extract and tea.

Damiana contains flavonoids, resins, tannins and a glycoside called arbutin, which may reflect its many apparent therapeutic actions. In the renal tract arbutin is converted to hydroquinone, which in large quantities can cause the side effects of Damiana (nausea, vomiting, tinnitus, sudden death) [10].

In scientific studies, Damiana has been found to relax the small muscles within the arterial walls of the penis, thereby allowing greater blood flow in the erect state. This relaxation has been found to be greater than that caused by Viagra (90% compared to 46%) and is thought to be one mechanism of how Damiana may work in ED [11].

In rat studies, Damiana was found to increase the sexual performance of impotent rats and increase the number of rats achieving ejaculation. The authors concluded that their study provided evidence for the folk reputation of Damiana [12].

In animal studies, Damiana has demonstrated hypoglycaemia and therefore caution is advised in men with diabetes.

5.2.3 DHEA

Dehydroepiandrosterone (DHEA) is a hormone, and men with low levels of it have been reported to suffer from ED. In one double-blind trial, 40 men with low levels of DHEA were administered 50 mg of DHEA per day compared to placebo. Men taking DHEA showed a significant improvement in both libido and erectile function [13].

5.2.4 Fo-Ti

Fo-Ti (*Polygonum multiflorum*) is also called He shou wu, which translated from Chinese means 'black-haired Mr. He'. According to Chinese folk legend, Mr. He was a village elder who took Fo-Ti, thereby restoring his black hair, youthful looks and vitality.

To date there have been no controlled human studies investigating the effects of Fo-Ti in men. In animal studies Fo-Ti has been found to enhance learning and memory and decrease the degeneration of nigrostriatal dopaminergic neurons in the brain [14]. How Fo-Ti actually works in ED is unknown.

5.2.5 Gamma-Butyrolactone (GBL)

GBL is also known by the chemical names 2,(3H)-furanone di-hydro, butyrolactone, 4-butyrolactone, dihydro-2(3H)-furanone, 4-butanolide, 2(3H)-furanone, dihydro, tetrahydro-2-furanone, and butyrolactone γ [15]. GBL is widely available on the Internet as Verve or Jolt. These are marketed as products that enhance sexual performance. Other GBL product names include Longevity, Revivarant, G.H. Revitalizer, Gamma G, Blue Nitro, Insom-X, Remforce, Firewater, and Invigorate [16].

GBL is a precursor to the compound gamma-hydroxybutyrate (GHB), to which it is rapidly metabolized once ingested. GHB is thought to be a partial agonist at the gamma-aminobutyric acid receptor and can lead to rapid respiratory arrest, hypotension, bradycardia and death. It has been linked to 58 deaths and more than 5700 recorded overdoses [17].

There is no evidence for GBL as a safe and efficacious treatment for ED.

5.2.6 Ginkgo biloba

Ginkgo is among the oldest living species of trees and is therefore often referred to as a 'living fossil'. The name biloba means two lobes and refers to the unique two-lobed leaves; its culinary and medicinal uses can be traced back for centuries. It is one of the best-selling herbal drugs in Europe and the tree is known to survive only in China. It is believed that China produces approx. 63,000 kg of dried ginkgo seeds each year [18, 19].

Ginkgo is believed to increase blood flow to the penis, although the evidence is contradictory. In an early double-blind study published in 1991, 50 men with arterial ED were administered 240 mg of ginkgo leaf extract daily for a period of 9 months. The men were divided into two groups based on their earlier response to conventional drug therapy for ED. The first group consisted of 20 men who had previously experienced some success with earlier treatments. The second group of 30 men had not experienced an erection with these treatments [20].

After 6 months of treatment with gingko leaf extract, all men in the first group regained spontaneous erections sufficient for penetration. This benefit continued throughout the 9-month study period. In the second group of 30 men, 19 responded positively. No side effects were reported [20].

The same group investigated the effect of 240 mg daily of ginkgo in 32 men diagnosed with vasculogenic ED in a 24-week placebo-controlled, double-blind, randomized trial [21]. The authors had a high dropout rate for unknown reasons, leaving 23 men for follow-up. They found no significant benefit from taking ginkgo, compared to their earlier results [21].

Cohen and Bartlik conducted a trial of ginkgo in men and women suffering from selective serotonin reuptake inhibitor (SSRI)-induced ED after observations of improvement in sexual performance in an elderly patient. This patient was taking SSRI for major depression and reported improved libido and erections after taking ginkgo. When he stopped taking the ginkgo, his ED returned [22].

In their study, all patients (men and women) were taking antidepressants, with the majority of patients taking regular SSRIs. Seventy-six percent of patients reported decreased libido, 54% reported inhibited or delayed orgasm and 19% had difficulty with erections. The authors prescribed an average dose of 207 mg a day to 63 patients while maintaining their regular antidepressant therapy. They found that 84% of patients experienced a positive effect on their sexual function, with more women reporting relief (30/33) than men (23/30). Again, there were no reported side effects [22].

Since that study, Ashton et al. have investigated the effect of 300 mg of ginkgo three times a day for a month in 22 patients (9 men, 13 women) with SSRI-induced ED [23]. They also reported no side effects but showed little positive effect on their ED – only 3 of 13 women reported improvement, with no improvement in the 9 men [23].

5.2.7 Ginseng

Ginseng (*Panax ginseng*) is a perennial plant that is native to the damp woodlands of northern China and Korea. After the plant is harvested, the root is prepared in one of two ways, resulting in either white or red ginseng. When Panax ginseng is peeled and dried, the result is hard and yellowish-brown. This is white ginseng. When the root is steamed, peeled and then dried, the result is a deep red coloured red ginseng [24]. In the Orient, red ginseng is preferred, whereas in the West, white ginseng is preferentially used in ginseng products [25].

Ginseng has been long alluded to in Chinese folk medicine as a sexual stimulant, and there appears to be some scientific truth to this. In animal studies, a concentration of 1 mg/mL of ginseng extract relaxed corpus cavernosal tissue in rabbits. This action was mediated by an increase in the sequestration of intracellular calcium and in the corpus cavernosal sinusoids by an increase in the release of nitric oxide (NO) [26]. The increase in NO is a positive finding because NO is the main mediator of penile erections and cyclic GMP, which mediates the relaxing effect of NO on penile vascular smooth muscle [27].

The same group investigated the effect of 50 mg/kg dose of Korean red ginseng in normal male rats and rabbits [28]. They found that the intracavernosal pressure was greater in the ginseng-treated rats and rabbits compared to the placebo-treated group, resulting in significantly increased rabbit and rat cavernosal smooth muscle relaxation after 3 months of treatment [28].

Clinical studies have also supported this effect. A double-blind, placebo-controlled study investigated the effect of 1,800 mg/d dose of red ginseng in 90 men with ED [29]. Of this group, 81 men had psychogenic ED and 9 men had a mild vasculogenic cause for their ED. After 3 months of treatment, they found no change in patients with premature ejaculation, frequency of intercourse, and duration and frequency of early morning erections. However, men treated with ginseng showed an increase in libido in 50% of cases, greater sexual satisfaction in 43.3%, increased penile tumescence during erection in 53.5%, greater penile rigidity in 50% and decreased early detumescence in 23.3% (all $p < 0.05$) (29). The authors gave the ginseng-treated group a therapeutic efficacy rate of 60% compared to 30% for the placebo group.

From studies to date, ginseng may have the ability to improve ED. It is well tolerated with few side effects, although excessive intake is reported to cause sleeplessness and hypertension. For this reason it is contraindicated in patients with hypertension [30].

5.2.8 Horny Goat Weed

Horny goat weed (*Epimedium brevicornum*) is also known as Yin Yan Huo and has long been used in traditional Chinese medicine to treat disorders such as dementia, fatigue, ED and arthralgia [31]. It is derived from a leafy plant that is native to Asia and the Mediterranean.

According to Chinese folk legend, horny goat weed's use for ED started after a Chinese goat herder noticed increased sexual activity in his flock after they ingested horny goat weed.

In one study testing the oestrogenic activity of 32 herbs, horny goat weed was one of the herbs to have greatest oestrogenic activity, suggesting a potential use in menopausal patients [32].

Animal studies have suggested that horny goat weed may elicit a penile erection in the rat. In that study, researchers administered different doses of horny goat weed extract (30, 100, 300, 1000, 3000, 6500, and 10,000 μg/0.1 mL of extract with 0.1 mL saline) intracavernosally to rats and measured their intracavernosal pressure (ICP) [33]. They found no effect on ICP with 30 and 100 μg/0.1 mL, but found a significant rise in ICP, with the greatest peak at $99.7 + / - 0.3$ mmHg after application of 6500 μg/0.1 mL (resting $7.8 + / - 1.0$ mmHg) [33]. The authors suggested that horny goat weed might work by increasing NO levels and thereby relaxing smooth muscle.

As yet, there is insufficient evidence for its safe use in men with ED. Reports have also shown that horny goat weed can cause tachyarrhythmia and hypomania [34].

5.2.9 L-Arginine

L-arginine is an essential amino acid found naturally in meat, dairy, poultry and fish, It is also available as an herbal supplement and is often marketed as a ‘natural Viagra’.

Nitric oxide (NO) is produced by all tissues in the body and plays a crucial role in smooth muscle relaxation. NO is formed from L-arginine using the enzyme nitric oxide synthetase (NOS).

It is well known that ED is commonly found in men with ischaemic heart disease. Vasculogenic ED may be a manifestation of atherosclerosis, and thus the L-arginine-NO pathway is hypothesized as an explanation for the association. There are only a few human studies evaluating the role of L-arginine in ED, and these have yielded mixed results.

One study investigated the effects of L-arginine in 32 patients (mean age 51.6 years) with mixed-type ED in a randomized, placebo-controlled, crossover trial [35]. Patients were administered 3×500 mg L-arginine/d or placebo for two 17-d courses. After a 7-d washout period, patients who initially received the placebo for 17 d were switched to L-arginine and vice versa.

Five patients (17%) reported a significant improvement in their sexual performance as measured by a validated questionnaire after the L-arginine phase, compared to 6 patients (20%) who reported a benefit after the placebo period. There was no statistically significant difference between improvements with L-arginine compared to placebo (56% vs. 43%). They concluded that L-arginine at 3×500 mg per day dose was no better for ED than placebo [35].

Other studies have shown positive results but have tended to use larger doses of L-arginine. One group conducted a double-blind trial in 50 men with ED [36]. Patients were given L-arginine (5 g per day) or a matching placebo for 6 weeks.

At the end of the study, 9 of the 29 arginine group reported a significant subjective improvement compared to only 2 of the 17 placebo group [36].

Another study administered Pycnogenol orally, which is known to increase the production of NO by stimulating activity of NOS, together with L-arginine as a substrate for NOS [37].

The group consisted of 40 men with ED who were given three ampoules of Sargenor a day for the first month. This is a drinkable solution of the dipeptide arginyl aspartate (equivalent to 1.7 g of L-arginine per day). During the second month, patients were additionally supplemented with 40 mg of Pycnogenol twice a day, and during the third month the daily dosage was increased to three 40-mg Pycnogenol tablets.

At the end of the first month with L-arginine alone, a statistically non-significant number of 2 patients (5%) experienced a normal erection. At the end of the second month of receiving a combination of L-arginine and Pycnogenol, the number of men with restored sexual ability increased to 80%. After the 3-month trial period, 92.5% of the men experienced a normal erection. These men also experienced a decrease in time for an erection to develop after stimulation and an extended duration of erection. The authors concluded that oral administration of L-arginine in combination with Pycnogenol causes a significant improvement in sexual function in men with ED without any side effects [37].

Although none of the studies reported side effects of receiving regular L-arginine, high doses of the drug can stimulate the production of gastrin, resulting in increased gastric acid production. Hence, it may cause gastro-intestinal problems and has been reported to induce necrotizing pancreatitis in rats [38].

5.2.10 Maca

Maca (*Lepidium meyenii*) has a radish-like root and is often called 'Peruvian ginseng'. Although unrelated to the ginseng family, its use is similar to ginseng in that it can increase energy, libido and sexual performance. Maca has been used as a food in Peru for many centuries but has only recently gained popularity as an herbal supplement.

In one study involving 9 adult men, 4 months of treatment with maca tablets resulted in increased seminal volume, sperm count and sperm motility [39].

Human studies of maca are limited. Another 12-week double-blind placebo-controlled, randomized, parallel study investigated the effect of different doses of maca compared to placebo [40]. The participants received maca in one of two doses: 1500 mg or 3000 mg or placebo. Self-perception on sexual desire was measured at 4, 8 and 12 weeks of treatment. After 8 weeks of treatment, there was a reported improvement in sexual performance, although there were no differences in serum testosterone and oestradiol levels between the treatment and placebo groups. The authors concluded that maca improves sexual desire at 8 and 12 weeks of treatment [40].

The evidence for a beneficial effect of maca in humans has not been supported by animal studies. A recent study looked at the effect of maca on sexual behaviour in rats [41]. The authors administered two doses of maca orally (25 and 100 mg/kg) for 30 d and monitored male sexual behaviour after acute, 7 and 21 d of treatment.

Maca did not produce large changes in male sexual behaviour, although there was a small effect on rat male sexual behaviour after acute and short-term administration of maca. After 21 d of treatment maca showed no effect on the rats. The researchers did, however, report an increase in ejaculation latency and postejaculatory interval [41]. None of the studies reported any side effects.

5.2.11 Muira Puama

Muira puama (*Ptychopetalum olacoides*) is a small tree that is native to the Amazon river basin. It has long been used for sexual problems in Brazil at a dose of 2 to 4 mg of liquid extract per day. To date there is no evidence to support the use of muira puama for ED [42].

5.2.12 Propionly-L-Carnitine

One study examined the effect of administering testosterone undecanoate versus propionyl-L-carnitine plus acetyl-L-carnitine and placebo in the treatment of male ageing symptoms for a period of 6 months [43]. The researchers enrolled 120 patients (mean age 66 years) and randomized them into three groups. Group 1 received testosterone undecanoate 160 mg/d, group 2 received propionyl-L-carnitine 2 g/d plus acetyl-L-carnitine 2 g/d, and the third group were given a placebo (starch).

The testosterone and carnitine groups (1 and 2) showed significantly improved peak systolic velocity, end-diastolic velocity, resistive index, nocturnal penile tumescence and International Index of Erectile Function (IIEF) score. This effect was noted for as long as the drugs were administered, with a suspension resulting in a reversal to baseline parameters. Placebo was ineffective in improving any of the above parameters. The researchers concluded that testosterone and, especially, carnitines proved to be active drugs for the therapy of symptoms associated with male ageing [43].

Another double-blind, fixed-dose study examined the efficacy and tolerability of oral propionyl-L-carnitine (PLC) plus sildenafil in men with ED and diabetes unresponsive to sildenafil monotherapy [44].

Patients with documented ED of organic or mixed aetiology and diabetes unresponsive to at least eight doses of Viagra alone were randomized to receive oral PLC (2 g/d) plus sildenafil (50 mg twice weekly) (20 patients, group 1) or sildenafil alone (20 patients, group 2).

After 24 weeks, group 1 (PLC and Viagra) showed a significantly greater percentage of improved erections (68% vs. 23%) and successful intercourse attempts (76% vs. 34%) compared to Viagra alone [44]. No patient discontinued treatment, suggesting that PLC is well tolerated.

Cavallini and colleagues investigated the use of two forms of carnitine – PLC and acetyl-L-carnitine (ALC) – in 96 patients who had undergone bilateral nerve-sparing radical retropubic prostatectomy [45]. One group were given a placebo (33 men), another group were given PLC 2 g/d plus ALC 2 g/d plus 100 mg of sildenafil when needed (32 patients), and the third group received sildenafil alone (35 men).

Placebo was found to be ineffective. The IIEF score of patients receiving PLC plus ALC and sildenafil were significantly greater than that of patients receiving sildenafil alone. The authors concluded that PLC and ALC were safe and reliable in improving the efficacy of sildenafil after surgery without increasing the side effect profile [45].

5.2.13 Reishi

Reishi (*Ganoderma lucidum*) is a fungus that grows from the tops of stumps or submerged logs and is found in most parts of the world [46]. Historically in Japan, reishi used to hang in the hallways of all homes to ward off evil spirits and was carried by brides into their new homes to guard against 'grave' matters in the new family [47].

Reishi has many uses in traditional Chinese medicine, including ED. In a study in China of 60 men with ED, there was a significant benefit after drinking a decoction made using fruit bodies of a member of the Ganoderma family for 1 month [48]. Only 4 patients reported no improvement; 17 patients reported slight improvement; 25 men claimed marked improvement in their sexual function, and 14 men reported that they were cured. No other drugs were taken during the period of the study [48].

5.2.14 Tongkat Ali

Tongkat ali (*Eurycoma longifolia*) is a tree found in Malaysia, Thailand and Indonesia. The root of this tree is reported to enhance male sexual performance as well as having medicinal properties for use in the treatment of dysentery, fever and other ailments.

Years ago the roots had to be brewed for long hours to obtain the active extract. However, it is now widely available in the form of a pill or a tea bag and is marketed as 'Malaysia's Viagra' [49].

Although human studies are sparse, its effect on rats has been investigated. In one study, middle-aged rats were administered 0.5 g/kg of various fractions of *E. longifolia*, whilst the control group received 3 ml/kg of normal saline daily for 12 weeks [50]. Results showed that *E. longifolia* enhanced the sexual qualities of the middle-aged male rats by decreasing their hesitation time compared to controls throughout the study time [50]. The authors used the outcome of this study to support the folk use of *E. longifolia* as an aphrodisiac.

Another study evaluated the effect of tongkat ali in sexually sluggish old male rats, 24-month-old rats and retired breeders [51]. The rats received 200, 400, or 800 mg/kg of various fractions of *E. longifolia* Jack, twice daily, for 10 d. The control rats received 3 ml/kg of normal saline. Ang et al. monitored the effect of treatment by the act of yawning and stretching. This is because yawning, either alone or associated with stretching, is considered an ancestral vestige surviving throughout evolution that promotes sexual arousal. They showed that 800 mg/kg of *E. longifolia* Jack increased yawning by 50% and stretching by 16.7% in sexually sluggish old male rats, by 676 to 719% and 31to 336%, respectively, in sexually active male rats, and by 22 to 44% and 75 to 100%, respectively, in middle-aged rats and retired breeders [51].

There are claims that tongkat ali increases testosterone by 48% and should therefore not be used by pregnant women or children [49]. Additionally, it should not be used for the same reason in men with prostate cancer.

5.2.15 *Tribulus*

Tribulus (*Tribulus terrestris*) is also called puncturevine and has traditionally been used in China and India for infertility and ED. Recently its use has increased as a way of improving sports performance. Prior claims that Tribulus increases serum testosterone has not been substantiated by recent studies. One group found that after 4 weeks of Tribulus supplements at 10 to 20 mg/kg body weight daily, testosterone was unaffected [52].

In rat studies, the administration of Tribulus extract heightened sexual behaviour and resulted in increased ICP. The authors claimed that this effect was due to increases in testosterone [53]. For this reason, Tribulus is contraindicated in pregnant women and men with prostate cancer. It has also been found to be toxic in sheep, eventually leading to death [54], and there are reports of gynaecomastia in men taking Tribulus [55].

5.2.16 *Yohimbine*

Yohimbine is derived from the bark of the West African yohimbe tree. It is an indole alkaloid that has been reputed to have aphrodisiac properties for over 70 years. Pharmacologically it is an alpha-2-adrenergic antagonist, and its activity is mediated by blocking these receptors in the brain.

In one study, yohimbine was shown to improve sexual function in men with psychogenic ED but had little effect in men with organic ED [56].

Men with psychogenic ED showed a 46% response to yohimbine compared with a placebo response of only 16%. However, only 50% of the men actually achieved an erection sufficient for penetration [56].

However, another study assessed the clinical response at baseline and after two different doses of yohimbine in 18 non-smoking men with organic erectile dysfunction [57]. Of the 18 men, 9 (50%) were successful in completing intercourse in more than 75% of attempts, although the authors note that the yohimbine responders were men with less severe ED [57]. They claimed that yohimbine is an effective therapy for organic ED, contradicting earlier studies.

Ernst and Pittler reviewed and meta-analysed all randomized, placebo-controlled trials of yohimbine monotherapy for erectile dysfunction to determine its therapeutic efficacy [58]. They identified seven trials that demonstrated that yohimbine is superior to placebo (odds ratio 3.85, 95% confidence interval 6.67 to 2.22) and suggested that yohimbine could be considered a reasonable therapeutic option for ED and should be considered as an initial pharmacological intervention [58].

Another meta-analysis of four independent trials from (i) controlled clinical trials of yohimbine (when used alone), (ii) uncontrolled trials examining yohimbine (alone), (iii) controlled trials of yohimbine when used in combination with other drugs and (iv) uncontrolled trials of yohimbine plus other drugs reported a consistent tendency for yohimbine, and for other medications containing yohimbine, to enhance erectile functioning relative to placebo [59].

Lebret and colleagues reported a double-blind, placebo-controlled, three-way crossover, randomized clinical trial to compare the efficacy and safety of the combination of 6 g of L-arginine glutamate and 6 mg of yohimbine hydrochloride with that of 6 mg of yohimbine hydrochloride alone and that of placebo alone [60]. They included 45 patients in their study and demonstrated the on-demand oral administration of the 6 g L-arginine glutamate and 6 mg yohimbine combination is effective in improving erectile function in patients with mild to moderate ED [60].

Whilst many studies have shown improvement in sexual function after taking yohimbine, there are at least an equal number of publications claiming that it has no effect.

Kunelius et al. conducted a randomized, double-blind, placebo-controlled, crossover comparison of a placebo with high-dose yohimbine hydrochloride (36 mg a day orally) in 29 men with mixed-type ED [61]. They found no benefit to yohimbine over placebo as a first-line treatment for mixed-type impotence [61].

Similarly, Mann and colleagues conducted a double-blind, placebo-controlled study in 31 men with ED [62]. They found a therapeutic effect of yohimbine in men with non-organic ED, with a significantly greater improvement in the yohimbine group compared to the placebo group. However, they could not demonstrate any superiority of yohimbine over placebo in the organic group [62].

The usual dose of yohimbine is 15 to 30 mg a day in divided doses. However, larger doses (20 to 30 mg) can be used 'on demand'. A tincture of yohimbine bark has been used as 5 to 10 drops three times a day. There are many side effects even in small doses, and its use is not really recommended. It is also contraindicated in men with cardiovascular and neurological diseases.

Despite encouraging scientific evidence for its effectiveness in ED, yohimbine lacks medical and public acceptance. There are many reasons for this: firstly, dose – response studies are lacking; secondly, other forms of administration, such as sublingual, have not been explored; and continuous versus 'on-demand' yohimbine has not been studied. One expert has suggested that this may be because yohimbine is an old drug that has no patent protection or commercial viability [63].

References

1. NIH Consensus Conference: impotence (1993) NIH Consensus Development Panel on Impotence. JAMA 270:83
2. National Institutes of Health (NIH) Erectile Dysfunction. The National Institute of Diabetes and Digestive and Kidney Diseases (NIDDK); (eds: Melman A, Hirsch M. NIH Publication, 2003; No. 04-3923. http://kidney.niddk.nih.gov/kudiseases/pubs/impotence/index.htm)
3. Wagner G, Fugl-Meyer KS, Fugl-Meyer AR (2000) Int J Impot Res 12 Suppl 4:S144
4. Feldman HA, Goldstein I, Hatzichristou D, Krane RJ, McKInlay JB (1994) J Urol 151:54
5. Jarrow JP, Nana-Sinkam P, Sabbagh M, Eskew A (1996) J Urol 155:1609
6. Barnes J (2003) Br J Clin Pharmacol 55:226
7. Morris CA (2003) JAMA 290:1505
8. Eisenberg DM, Davis RB, Ettner SL, Appel S, Wilkey S, Van Rompay M, Kessler RC (1998) JAMA 280:1569
9. Hirt HM, M'Pia B (2001) Natural Medicine in the Tropics. Anamed. 2nd edn. Druckerei Bauer, Winnenden
10. Piacente S, Camargo EE, Zampelli, Gracioso JS, Broti AR, Pizza C, Vilegas W (2002) Z Naturforsch [C] 57(11–12):983
11. Hnatyszyn O, Moscatelli V, Garcia J, Rondina R, Costa M, Arranz C, Balaszczuk A, Ferraro G, Coussio JD (2003) Phytomedicine 10(8):669
12. Arletti R, Benelli A, Cavazzuti E, Scarpetta G, Bertolini A (1999) Psychopharmacology 143(1):15
13. Reiter WJ, Pycha A, Schatzl G, Pokorny A, Gruber DM, Huber JC, Marberger M (1999) Urology 53(3):590
14. Li RW, David Lin G, Myers SP, Leach DN (2003) J Ethnopharmacol 85(1):61
15. FDA warns about products containing gamma butyrolactone or GBL and asks companies to issue a recall. Food and Drug Administration talk paper. April 13, 1999
16. Report serious adverse events associated with dietary supplements containing GBL, GHB, or BD. UD FDA Medwatch. August 25, 1999
17. "Date-rape" drug curbs become law. Statistics from Drug Enforcement Agency as reported in the Washington Post/Associated Press. February 19, 2000:A02
18. Hobbs C (1991) Gingko: Elixir of Youth. Botanical Press, Capitola, CA
19. He SA, Yin G, Pang ZJ (1997) Resources and prospects of Gingko Biloba in China. In: Hori T, Ridge RW, Tulecke W, Del Redicini P, Tremouillaux-Guiller J, Tobe H (eds) Gingko Biloba – A Global Treasure. Springer, Tokyo, p 373
20. Sohn M, Sikora R (1991) J Sec Educ Ther 17:53
21. Sikora R, Sohn MH, Engelke B, Jaske G (1998) J Urol 159(S5):240
22. Cohen AJ, Bartlik B (1998) J Sex Marital Ther 24:139
23. Ashton AK, Ahrens K, Gupta S, Masono PS (2000) Am J Psych 157:836
24. Foster S, Yue C (1992) Herbal Emissaries. Rochester, VT, Healing Arts, p 102
25. Stitcher O (1998) Chemtech 28:26

26. Choi YD, Xin ZC, Choi HK (1998) Int J Impot Res 10:37
27. Gillis CN (1997) Biochem Pharmacol 54:1
28. Choi YD, Rha KH, Choi HK (1999) J Urol 162:1508
29. Choi HK, Seong DH, Rha KH (1995) Int J Impot Res 7:181
30. Bradley PR (1992) British Herbal Compendium, 1. Bournemouth, Dorset, England p 115
31. Natural Medicines Comprehensive Database 2000, 3rd edn, 403
32. Zhang CZ, Wang SX, Zhang Y, Chen JP, Liang XM (2005) J Ethnopharmacol 98(3):295
33. Chen KK, Chiu JH (2006) Urology 67(3):631
34. Partin JF, Pushkin YR (2004) Psychosomatics 45(6):536
35. Klotz T, Mathers MJ, Braun M, Bloch W, Engelmann U (1999) Urol Int 63(4):220
36. Chen J, Wollman Y, Chernichovsky T, Chernichovsky T, Iaina A, Sofer M, Matzkin H (1999) BJU Int 83(3):269
37. Stanislavov R, Nikolova V (2003) J Sex Marital Ther 29(3):207
38. Saka M, Tuzun A, Ates Y, Bagci S, Karaeren N, Dagalp K (2004) Turk J Gastroenterol 15(1):56
39. Gonzales GF, Cordova A, Gonzales C, Chung A, Vega K, Villena A (2001) Asian J Androl 3(4):301
40. Gonzales GF, Cordova A, Vega K, Chung A, Villena A, Gonez C, Castillo S (2002) Andrologia 34(6):367
41. Lentz A, Gravitt K, Carson CC, Marson L (2007) J Sex Med 4(2):332
42. da Silva AL, Bardini S Nunes DS, Elisabetsky E (2002) Phytother Res 16(3):223
43. Cavallini G, Caracciolo S, Vitali G, Modenini F, Biagiotti G (2004) Urology 63(4):541
44. Gentile V, Vicini P, Prigiotti G, Koverech A, Di Silverio F (2004) Curr Med Res Opin 20(9):1377
45. Cavallini G, Modenini F, Vitali G, Koverech A (2005) Urology 66(5):1080
46. Hobbs C (1995) Medicinal Mushrooms: An Exploration of Tradition, Healing, and Culture, 2nd edn. Botanical, Santa Cruz, CA
47. Matsumoto K (1979) The Mysterious Reishi Mushroom. Woodbridge, Santa Barbara, CA
48. Liu GT (1999) Int J Medicinal Mushrooms 1:63
49. Tongkat Ali – Malaysia's very own 'Viagra'. Asian Economic News, Jan 3, 2000
50. Ang HH, Ngai TH, Tan TH (2003) Phytomedicine 10:590
51. Ang HH, Lee KL, Kiyoshi M (2004) J Basic Clin Physiol Pharmacol 15:303
52. Neuchev VK, Mitev VI (2005) J Ethnopharmacol 101:319
53. Gauthaman K, Ganesan AP, Prasad RN (2003) J Alt Complement Med 9(2):257
54. Bourke CA (1987) Research in Veterinary Science 43(3):347
55. Jameel JK, Kneeshaw PJ, Rao VS, Drew PJ (2004) Breast 13(5):428
56. Reid K, Surridge DH, Morales A, Condra M, Harris C, Owen J, Fenemore J (1987) Lancet 2(8556):421
57. Guay AT, Spark RF, Jacobson J, Murray FT, Geisser ME (2002) Int J Impot Res 14(1):25
58. Ernst E, Pittler MH (1998) J Urol 159(2):433
59. Carey MP, Johnson BT (1996) Arch Sex Behav 25(4):341
60. Lebret T, Herve JM, Gorny P, Worcel M, Botto H (2002) Eur Urol 41(6):608
61. Kunelius P, Hakkinen J, Lukkarinen O (1997) Urology 49(3):441
62. Mann K, Klingler T, Noe S, Roschke J, Muller S, Benkert O (1996) Arch Sex Behav 25(1):1
63. Morales A (2001) World J Urol 19(4):251

Chapter 6
Harpagophytum procumbens – Traditional Anti-inflammatory Herbal Drug with Broad Therapeutic Potential

G.P. McGregor*,**

Abstract *Harpagophytum procumbens* (Devil's Claw) is a traditional African herbal drug used by the natives of the Kalahari and Savannah desert regions to treat a variety of ailments. In Europe, the anti-inflammatory properties of *Harpagophytum procumbens* extracts (Hp) have been the basis for its popular use in the treatment of inflammatory disorders of the musculoskeletal system and of low back pain. This is the subject of a draft monograph recently issued by the European drug authorities (EMEA). There is good clinical evidence for Hp's use in treating painful osteoarthritis, but more evidence (such as from phase-III trials) is needed before Hp can be considered a standard treatment of osteoarthritis and other chronic inflammatory diseases. Here, the published evidence is reviewed that reveal the anti-inflammatory effects of Hp including inhibition of key mediators and promoters of inflammation. The relevance of these effects for the therapeutic use of Hp in osteoarthritis is discussed. The main constituents of Hp, glycosides of harpagoside and of acteoside, have been shown to act on molecular pathways that have become recognised as targets in the development of new anti-inflammatory drugs. The pharmacological profile of Hp is that of a modern anti-inflammatory agent. Today there is great interest in new anti-inflammatory drugs as toxicological doubts have arisen concerning standard drugs but also as it has become apparent that the therapeutic potential of novel anti-inflammatory drugs goes well beyond the treatment of recognised chronic inflammatory diseases. Inflammation is now seen as a primary process in osteoarthritis but also as a key pathophysiological factor in many common modern diseases (cardiovascular disease, diabetes, osteoporosis, neurodegenerative diseases, dementia and Alzheimer's). The therapeutic potential of Hp, therefore, demands attention.

G.P. McGregor (✉)
Philipps-University of Marburg, Faculty of Medicine, Deutschhaus Strasse 2, 35037, Marburg, Germany, e-mail: mcgregor@staff.uni-marburg.de; and PASCOE pharmazeutische Präparate GmbH, Medical Science Department, Schiffenberger Weg 55, D35394 Giessen, Germany

*Current address: HPM (Geneva), 8, Aves de Grandes-Communes, Geneva, 1213, Switzerland, e-mail: gerard.mcgregor@hpmgeneva.com

**Gerard P McGregor worked for PASCOE pharmazeutische Präparate GmbH, which markets a preparation (PASCOE®-Agil 240mg film tablets) that contains, as sole active ingredient, an extract of *Harpagophytum procumbens*.

K.G. Ramawat (ed.), *Herbal Drugs: Ethnomedicine to Modern Medicine*,
DOI 10.1007/978-3-540-79116-4_6,

Keywords *Harpagophytum procumbens* · Devil's Claw · Inflammation · Osteoarthritis · Harpagoside · Acteoside

Abbreviations

Hp	*Harpagophytum procumbens*
NSAID	Non-steroidal anti-inflammatory drug
iNOS	Inducible nitric oxide synthetase
COX	Cyclooxygenase
PG	Prostaglandin
LOX	Lipoxygenase
LT	Leukotriene
TNF	Tumour necrosis factor
LPS	Lipopolysaccharide
MMP	Matrix metaloprotease

6.1 Introduction

Extracts of the secondary roots of the southern African plant *Harpagophytum procumbens* (Devil's Claw) provide an herbal drug with a variety of traditional indications. It is a member of the sesame family (Pedaliaceae), and the genus *Harpagophytum* contains the species (Fig. 6.1) *H. zeyheri*, extracts of which were also mixed with Hp extracts in earlier studies. Monographs state Hp as the sole source of the medicinal drug.

The natives of the Kalahari and Savannah desert regions of southeast and southern Africa used the dried secondary roots of the plant to treat a variety of ailments including fever, malaria and indigestion. Since its introduction into Europe during

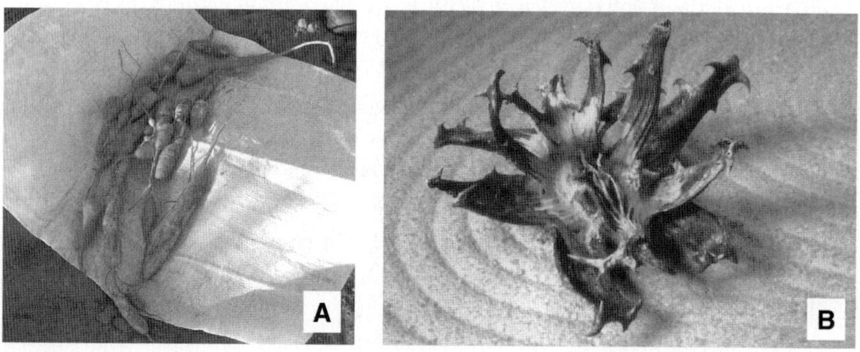

Fig. 6.1 **a** Harvested secondary roots of *Harpagophytum procumbens* from which the medicinal extracts are made. **b** The fruit of *H. procumbens* with its hook-like appendages and which give the plant the name "Devil's Claw"

the middle of the last century, the anti-inflammatory properties of extracts of Devil's Claw [1] have attracted the most attention. The European Scientific Cooperative on Phytotherapy (ESCOP) issued a monograph recommending its use in the symptomatic treatment of painful osteoarthritis, relieving lower back pain, loss of appetite and dyspepsia [2], and the EMEA Committee on Herbal Medicinal Products (HMPC) have recently issued a draft monograph for discussion in which an indication for treatment of mild articular pain and mild digestive disorders is proposed on the basis of traditional use [3]. Its use in the treatment of inflammatory disorders of the musculoskeletal system and of lower back pain has become particularly popular in recent years. The clinical evidence of Devil's Claw use in treating painful osteoarthritis is generally good and has been the topic of several recent reviews. However, more studies are required in order to establish this drug as a definitive therapeutic option. Issues such as the optimal dose and duration of treatment need to be resolved [4].

This chapter will review the published evidence that Hp extracts exert anti-inflammatory effects that are of relevance to the therapeutic use of Hp extracts in osteoarthritis. Although osteoarthritis is generally considered a degenerative disease, recent new insight into the pathophysiology of osteoarthritis reveals that inflammatory processes are involved in the early stages of the disease and are not only secondary processes following tissue degeneration. The breakdown of cartilage is probably the initial problem in osteoarthritis, but there is good evidence from human studies that this process involves the release of proinflammatory cytokines such as TNF and IL1 from chondrocytes within the cartilage [5].

There is good in vitro and in vivo pharmacological evidence of the anti-inflammatory and analgesic properties of Hp, although some negative findings have also been reported [1]. Generally, the pharmacological profile of Hp extracts is supportive of its therapeutic potential as an anti-inflammatory agent. More evidence is, however, needed in more clinically relevant models. Also, more information on the mode of action of extracts of Hp and of its constituents, and especially the identification of the cellular and molecular targets, is required before the full potential of this highly promising therapeutic plant can be appreciated. Its anti-inflammatory properties certainly warrant further research.

Treatment of painful inflammatory diseases has received a major blow with the "fall from grace" of the cyclooxygenase (COX)2 inhibitor drugs due to problems of cardiovascular risk. Thus, there is a desperate need for new anti-inflammatory drugs. The therapeutic potential of novel anti-inflammatory drugs appears to extend beyond chronic inflammatory diseases such as osteoarthritis, rheumatoid arthritis and chronic inflammatory bowel disease. Inflammatory processes are seen today as playing a significant pathophysiological role in many common modern diseases such as cardiovascular disease, diabetes, osteoporosis, neurodegenerative diseases, dementia and Alzheimer's disease, as well as in carcinogenesis.

6.2 The Status and Use of Hp-Containing Products

In Germany and some other European countries, Hp-containing products are fully-licensed drugs on the basis of their well-established use. Such products have, in general, been authorised for the supportive treatment of painful degenerative joint and muscle conditions. The UK drug authorities issued in January 2007 a traditional-use registration for a proprietary drug containing, as sole active ingredient, an aqueous extract of Hp. This over-the-counter product was issued the therapeutic indication of "A traditional herbal medicinal product used for the relief of backache, rheumatic or muscular pain, and general aches and pains in the muscles and joints, based on evidence of traditional use only". The licensing and registration of Hp products requires that the extracts of the herb be "standardised". This a basic principle in herbal medicine and aims to ensure the quality of herbal products, and to ensure that a single medicinal plant exists in the market in the form of one or more extracts that are essentially chemically similar and based on traditional processes. Hp extracts are standardised to a content of 1 to 3% of the iridoid glycoside, harpagoside, which is an active pharmaceutical marker but is considered not to be the sole active ingredient.

The recommendations for the standardisation of herbal extracts such as those set out by the German Commission E are unfortunately not universally followed by manufacturers of herbal medicinal preparations. This has led to much confusion and has undermined the image of herbal medicine. In the USA, Hp-containing products have been less stringently categorised, as are all other plant-derived medicinal products, as foods and are considered food supplements or so-called nutraceuticals [6]. Pharmacological treatments of osteoarthritis are exclusively aimed at relieving symptoms and not modifying the disease process, and alternative treatments of this chronic condition are frequently sought. There is a rate of self-medication in osteoarthritis is high, and Hp products belong to a variety of herbal preparations and food supplements that are widely used by such patients [7].

The traditional medicinal use of herbal extracts cannot be patented, but Hp extracts are found in patent applications describing Hp extracts in combination with other extracts or vitamins [8]. A novel medicinal use of a defined aqueous ethanol extract of Hp is the topic of an interesting patent application. The patent application [9], made by a German small-to-medium enterprise, describes the use of the Hp extract in the treatment of endometriosis, which is a painful inflammatory disorder.

6.3 Chemical Constituents of Hp Extracts

The major chemical constituents of Hp have been characterised and described in a number of published studies from different groups [10–12]. In their recent investigations, Clarkson and colleagues used an HPLC-SPE (solid-phase extraction)-NMR tandem methodology to confirm that iridoid glycosides and glycosides of the phenylethonoid acteoside are the major constituents of ethanol extracts of *Harpagophytum* [10]. They also showed, using the same technology, that two commercial preparations of Hp contained a similar spectrum of constituents as their laboratory extract. So it can be assumed that some, or all, of these main chemical constituents of ethanol extracts of Hp are responsible for the medicinal and pharmacological effects of Hp. In their analysis of a water extract of Hp, Boje and colleagues [12] found glycosides of harpagoside, and of acteoside (and acteoside analogues), in similar amounts and to be the major ingredients.

Extracts of Hp are characterised by the presence of iridoid glycosides, and especially harpagoside, which was first identified in Hp extracts back in the 1960s. Acteoside and other related polyphenolic glycosides were first isolated from Hp by Berger et al. [13] and represent a major constituent of Hp extracts. Clarkson and colleagues [10] were able to determine that the ethanol extract contained a high proportion (about 13%) of the original dried plant material and that the glycosides of acteoside represented more than 30% of the extracted material. Therefore, this group of chemicals appears to represent a major fraction of Hp extracts in addition to the characteristic iridoid glycosides. Acteoside, shown to exert anti-inflammatory effects, is also known as verbascoside and is present as aglycoside or glycoside in other herbal drug extracts (see further discussion below).

The HPLC-SPE-NMR methodology was described by Clarkson and colleagues [10] to be a paradigm shift in the analysis of biologically derived extracts in the sense that it allows novel natural products to be identified with improved rapidity and sensitivity. For example, using this methodology, the researchers identified novel chemical entities in the aqueous ethanol Hp extract. These novel natural products were chicane-type tricyclic diterpenes and are of interest in drug discovery as novel chemical entities. However, for specialists in herbal medicine who are interested in the pharmacological mode of action of the herbal drugs such as Hp, they are of no interest given their presence in the extract in such small quantities.

6.4 Pharmacological Properties

Anti-inflammatory and analgesic effects have been revealed in several pharmacological studies of extracts of Hp and of compounds isolated from Hp [1, 14]. These studies were carried out over the years using in vivo and in vitro models for anti-inflammatory or analgesic drug screening. More recently, studies of Hp extracts and constituent compounds have been performed with the aim of establishing the molecular targets within the network of cytokines, intracellular signal pathways and effector molecules that promote or mediate inflammation. The various pharmacological investigations of Hp are reviewed here. They generally support the notion that Hp extracts exert anti-inflammatory effects and provide some plausible insight into the possible mode of action of Devil's Claw as an anti-inflammatory drug.

There are a number of important caveats due to the design of these studies that hinder interpretation of the data with respect to the medicinal use of Hp. Firstly, the test extract or harpagoside is generally applied before induction of the inflammatory challenge, which is a standard pharmacological paradigm. But this does not reflect the therapeutic paradigm of application to the pre-existing challenged or pathophysiological state. Exceptions were two early negative studies [15, 16] in which Hp was administered orally at very high doses to animals with established inflammation and no effect was observed. Frequently, the effect of parenteral administration of the test drug is investigated whereas Hp preparations are usually for oral use. Most of the models involved acute and severe inflammation and short-term treatment, whereas Hp has therapeutic potential in chronic, and frequently mild inflammatory, conditions. The majority of the studies demonstrated anti-inflammatory effects, but used doses of Hp are much higher than doses traditionally used in humans. Despite these caveats, the pharmacological studies described are worth consideration since they provide some important information. The efficacy of very high doses in acute severe inflammatory models can have the corollary of efficacy at lower doses in chronic mild inflammation.

6.4.1 *In Vivo Anti-Inflammatory Effects of Hp Extracts and Harpagoside*

Most of the standard anti-inflammatory in vivo models depend on chemically induced tissue damage and measurement of the inflammatory response in oedema. In the rat model of formalin-induced arthritis, parenteral administration of an aqueous extract of Hp (20 mg/kg) was as effective as the non-steroidal anti-inflammatory drug (NSAID) phenylbutazone, but harpagoside was ineffective [17]. In Freund's adjuvant model of arthritis, again in rat, injections of an ethanolic extract of Hp, at 25, 50 and 100 mg/kg, was demonstrated to exert a significant analgesic and anti-inflammatory (anti-oedematous) effect following both acute and chronic treatment (20 days) [18]. A number of studies reported significant anti-inflammatory effects in the carageenan-induced rodent oedema model following parenteral administration of an aqueous ethanolic (60%) extract [19], of an aqueous extract [20, 21] and of a non-characterised extract [22]. However, in the same model, oral administration of various extract forms of Hp produced contradictory results with regard to anti-oedematous effects. Positive dose-dependent anti-inflammatory effects were reported [23–25], while other studies failed to demonstrate effects even at doses as high as 2 g/kg [15, 16, 19, 21, 22, 26]. An anti-inflammatory effect of oral Hp (extract characteristics not given) reduced oedema after provocation with adriamycin [27]. No effects were measured for harpagoside on rat paw oedema evoked by injection of ovalbumin [17].

The granuloma pouch test, a popular model of inflammation in rats, is performed by injecting air under the dorsal skin to form the pouch into which an chemical irritant (usually croton oil) is applied. In this model, a methanolic extract of Devil's Claw (3.7% harpagoside), which was administered orally for 11 days, reduced the exudate volume significantly and was equally effective as phenylbutazone [24]. In the same model, parenterally applied harpagoside (20 mg/kg) significantly inhibited exudate formation, granuloma weight and tissue granulation and was as effective as phenylbutazone.

In two studies in which arthritis was induced by injection of either *Mycobacterium butyrium* [15] or *M. tuberculosis* [16], no anti-inflammatory effects were observed with Devil's Claw applied orally at concentrations up to 2 g/kg. In the study by Whitehouse and colleagues [15], the carageenen-induced mouse inflammatory model was also used, and Hp administered orally was ineffective even at such a high dose as 2 g/kg.

6.4.2 In Vivo Analgesic Effects of Hp Extracts and Harpagoside

The analgesic properties of Hp have been demonstrated in several studies using various models. Two studies using the writhing test reported significant analgesic effects after intraperitoneal administration of aqueous extracts [20, 23], whereas only slight analgesic effects were reported with the same model for both aqueous and methanolic extracts by Erdös and colleagues [24]. In the Randall–Soletto test, a 60% aqueous ethanol extract administered intraperitoneally exerted a dose-dependent analgesic effect. The extract (200 mg/kg) was significantly more effective than the comparator drug, diclofenac (80 mg/kg) [19]. In a heat-induced pain model, an aqueous methanolic extract of Hp exerted analgesic effects [24]. Intraperitoneal harpagoside exhibited analgesic effects in the rabbit ear model [17].

6.4.3 Effects of Hp Extracts and Harpagoside on Pro-Inflammatory Molecular Targets – Eicanosoids, Cytokines, Second-Messenger Pathways and Effector Molecules

Several in vitro studies of Hp have sought to identify the molecular and cellular targets of Hp's action. Whereas inhibition of the biosynthesis of the eicosanoids (prostaglandins, leukotrienes, thromboxane) has been, for many years, the most prominent target in anti-inflammatory drug development, pro-inflammatory cytokines now attract much attention, along with the many intracellular mediators of inflammation that have now been identified [28, 29]. There is pharmacological evidence that Hp extracts and their constituents influence the biosynthesis of certain eicosanoids and cytokines, and the data are consistent with anti-inflammatory effects. However, it remains to be determined which of these effects are relevant for the medicinal use of Hp extracts as well as for identifying constituent substances as potential leads in the development of novel anti-inflammatory drugs.

Hp extracts with about 20% and 30% content of harpagoside (prepared using CO_2 extraction) were compared with a conventional aqueous ethanol (60%) extract [30] for effects on isolated 5-lipoxygenase activity (responsible for generating the leukotrieners) and on cyclooxygenase-(COX)2 [responsible for generating the pro-inflammatory prostaglandin-(PG)E2]. The activity of these enzymes was measured in isolated human polymorphonuclear leukocytes [30]. The novel extracts were clearly more active (and in a harpagoside-content-dependent way) than the conventional extract in inhibiting 5-lipoxygenase activity and leukotriene biosynthesis as well in inhibiting COX2 activity. An oddity in this study was that the effect on COX2 activity was tested on unstimulated human polymorphonuclear leukocytes. Also the most effective extract exerted complete inhibition of 5-lipoxygenase activity at a concentration of 52 mg/L and indicates only moderate potency. In 1992, Moussard and colleagues reported the absence of any effect of a non-specified extract of Hp on PGE2 and leukotriene blood levels following 21 days' oral intake in humans [31]. This reveals no direct effect on COX1 and LOX activity in healthy and unchallenged human subjects.

Other studies demonstrated that Hp extracts have fairly potent inhibitory effects on the biosynthesis of leukotrienes and thromboxanes, which mediate certain pro-inflammatory processes. Various Hp extracts and harpagoside were shown to inhibit the basal and stimulated cysteinyl-leukotriene (LTC) and thromboxane synthesis in human whole blood samples [32]. An extract, with a specified harpagoside content of 7.3%, had a stronger selective inhibitory effect on cysteinyl-leukotriene than on thromboxane B_2 synthesis and was more potent than an extract with 2.1% harpagoside content than harpagoside alone. Similar results were reported by Tippler et al. [33]. In a similar study, several iridoid glycosides were compared for their inhibitory activities in model cell systems generating the metabolites of the enzymes COX and LOX [34]. Harpagoside inhibited release of leukotrienes LTC_4, but their effects did not reach statistical significance as detected for other iridoid glycosides. In the test for effects on thromboxane synthesis, harpagoside inhibited TXB_2 release with an IC_{50} value of 10 μM (ibuprofen IC_{50}: 7 μM).

Since COX2 is not constitutively expressed in most tissues and is only induced during inflammation, it is only relevant to test anti-inflammatory drugs for effects on COX2 following appropriate stimulus. Lipopolysaccharides (LPSs), otherwise referred to as endotoxins, are a constituent of the outer wall of gram-negative bacteria that are able, in their isolated form, to provoke the innate immune system and cause inflammation. In vitro it is used to activate inflammatory cells in order to investigate the molecular processes of inflammation and for testing mechanisms of anti-inflammatory drug activity, e.g. LPS induces COX2 in various cell types. In a mouse fibroblast cell line (L929), an aqueous extract of Hp inhibited LPS-induced COX2 and the inducible form of nitric oxide synthetase, which is another pro-inflammatory enzyme induced by LPS [35]. Concentrations of 100 and 1000 μg/ml were effective in decreasing the level of the mRNAs for COX2 and iNOS as determined using reverse transcriptase polymerase chain reaction (RT-PCR). A significant decrease in PGE2 and nitrate formation was also measured. Interesting findings were reported by Kundu and colleagues who tested Hp in mouse skin for effects on COX2 gene expression induced by the phorbol ester (TPA) [36]. The mode of applying TPA to the skin is a carcinogenic model. Pro-inflammatory mediators, such as COX2, are implicated in skin carcinogenesis, and pharmacological inhibition of COX2 has a potential chemopreventive effect. Methanolic extracts of Hp were applied topically, which reduced the level of expression of COX2 enzymes (as measured by western blot analysis). Additionally, the Hp extract reduced the levels of two gene transcriptional proteins (AP-1 and CREB) that mediate the TPA-induced up-regulation of the *COX-2* gene. The study also provided evidence that Hp inhibited the activation of the protein kinase, ERK, and did not activate NFκB, which indicates a specific effect on one of the major intracellular pathways that regulate pro-inflammatory gene expression. These two recent studies indicate that Hp extracts have the ability to suppress the expression of inducible pro-inflammatory genes by actions on intracellular pathways. Thus inhibition of COX2 is by means of suppression of gene expression and not via direct enzyme inhibition.

Another important pro-inflammatory gene that is up-regulated during inflammation is the inducible form of nitric oxide synthetase (iNOS). The inhibitory effect of Hp extracts and harpagoside on iNOS expression has been reported. As discussed above, an Hp water extract decreased the level of iNOS mRNA in LPS-stimulated fibroblasts [35]. Kaszkin et al. [37] used isolated rat kidney mesengial cells activated by application of the pro-inflammatory cytokine, interleukin (IL)1b, as a model for kidney inflammation and investigated the effect of two "special" extracts of Hp on IL1b-induced iNOS expression. Both extracts were water extracts containing high levels of harpagoside (8.8% and 27% respectively) and inhibited IL1b-induced iNOS expression as evidenced by a dose-dependent reduction in the levels of iNOS mRNA (measured by Northern blot), iNOS protein (measured by western blot) and nitrite (product of iNOS) synthesis. This study also demonstrated that the inhibitory effects of Hp extracts were via inhibition of the transcription of the iNOS gene (as measured in a reporter gene assay). Also, the study provided evidence that the two extracts inhibit IL1b activation of the NFκB pathway, which is a putative intracellular mediator of IL1b-induced iNOS expression. However, further experiments revealed that this effect was not due to harpagoside, which itself also inhibited IL1b-induced iNOS expression, but with insufficient potency to explain the effect of the extracts. The authors concluded that constituents of Hp extracts, in addition to harpagoside, contribute to the effects observed and presumed that the effect on the pro-inflammatory NFκB pathway might be due to an antioxidant property of the extracts, not to the presence of harpagoside. More recently, Huang and colleagues [38] convincingly demonstrated that harpagoside inhibited LPS-induced iNOS expression and NFκB pathway activation, as well as COX2 expression in two human macrophage cell lines. This led these authors to speculate that harpagoside and related iridoid glycosides should be considered as lead substances for the development of novel NFκB inhibitors.

Induction of the pro-inflammatory cytokine, tumour necrosis factor (TNFα), by proinflammatory cells such as monocytes and macrophages is a very early event in activation of the innate immune system. TNFα activates and initiates many of the pro-inflammatory processes, including the induction of other pro-inflammatory cytokines. It is considered a central target in treating chronic inflammation [29]. Therefore, the discovery that Hp extracts possess a TNFα-blocking ability is an interesting finding. An aqueous ethanolic (60%) extract of Hp attenuated LPS-induced synthesis of TNFα by isolated human monocytes in a dose-dependent manner (IC_{50} of about 100 μg/ml) while inhibiting the other cytokines (IL-6, IL-1β) and PGE2 less potently. In the same study, harpagoside had less pronounced or no effects [38].

In osteoarthritis, pro-inflammatory cytokines induce the biosynthesis by chondrocytes of matrix metaloproteases that are responsible for connective-tissue breakdown during inflammation. Two standardised aqueous ethanolic (35% and 65%) Hp extracts attenuated Il1ß-induced MMP activity in isolated primary (non-rheumatic) human chondrocytes in vitro [39]. Single doses (1 mg/ml) of each extract were applied for 48 hours. Immunohistochemical and western blot analysis revealed a reduction in the IL1ß-induced level of each of the three MMP enzymes (MMP1, 3 and 6) that were investigated.

Neutrophil elastase is another protease that plays an important role in inflammation. Novel iridoid and phenylethanoid constituents of Hp revealed moderate potency in vitro in inhibiting this enzyme [12].

6.4.4 Possible Mode of Action of Hp

The preceding discussion indicates that Hp extracts influence a variety of pro-inflammatory processes. More work is required to determine if these effects are due to actions on multiple targets by one or more of the constituents of Hp. Alternatively, acting via a single target may have multiple effects. There is some evidence outlined above that Hp extracts can inhibit key intracellular pathways that promote inflammation. Many of these pathways are common to different inflammatory cells and, when inhibited, may attenuate different inflammatory processes simultaneously.

The notion that Hp inhibits pro-inflammatory second-messenger pathways suggests that this traditional herbal drug can compete with modern drugs. Pro-inflammatory second-messenger pathways are currently the target of much effort and investment in the development of novel anti-inflammatory drugs. Such drugs are potentially more effective anti-inflammatory agents as they act up-stream in the inflammatory cascade and will exert pleiotropic effects. For example, NSAIDs target the production of just one pro-inflammatory effector, while inhibition of a key intracellular pathway will influence an array of effector molecules.

One key intracellular promoter of inflammation is NFκB, which activates several pro-inflammatory genes. As discussed above, there is some contradictory evidence concerning Hp and inhibition of this key promoter of inflammation. In mouse fibroblasts an Hp extract, but not harpagoside, inhibited the activated NFκB pathway [35], whereas in human macrophage cell lines, harpagoside was inhibitory [37]. It is possible differences between species or cell type determine these differences.

Further studies should, perhaps, focus on this pathway as a potential target for the anti-inflammatory effects of Hp.

Also relevant in this regard is the inhibitory effects of Hp on the NFκB pathway and the possible anti-oxidant effects of Hp extracts, for which there is some evidence. Antioxidants are potential inhibitors of NFκB, which is activated by reactive oxygen species generated during inflammation, and are recognised as key promoters of inflammation and inducers of pro-inflammatory genes. In vitro assessment of a water extract of Devil's Claw showed moderate antioxidant capacity (Trolox assay), which was not due to the content of iridoids, which exert minimal antioxidant activity [40]. However, the constituent acteoside has antioxidant properties, as has been shown in a number of studies.

Acteoside is rarely discussed with regard to the pharmacology of Hp. It is present in Hp extracts in amounts similar to that of harpagoside [10], which is the more characteristic and most often discussed constituent of Hp. Acteoside and acetylated derivatives are present in many plants and have been identified as an active ingredient of, for example, *Plantago lanceolata*. In its isolated form, acteoside exhibits an anti-inflammatory property in vitro due to its antioxidant effects [41–44], which also include inhibition of eicanosoids production [45, 46]. Also in in vivo models of inflammation, it down-regulates expression of ICAM-1 [47, 48], which is involved in leucocyte recruitment within inflamed tissue and of iNOS expression [49]. More recently, it has been shown that parenteral acteoside exerts significant dose-dependent anti-inflammatory effects in a rodent model of inflammatory bowel disease (dextran sulphate sodium-induced colitis in mice) when administered after induction of chronic inflammation [50]. The study demonstrated that acteoside reduced the degree of inflammation, local lymph node cytokine production, and intestinal myeloperoxidase activity. The authors speculated that the anti-inflammatory effects were due to an antioxidant effect causing a reduced oxidative burst of local inflammatory cells.

6.5 Secondary Pharmacology of Hp

As mentioned above, Hp is used traditionally to enhance digestion, and this is acknowledged in the literature. The Hp constituent, acteoside, has been described in a short report of animal studies as a new hypertensive agent [51] and has been shown to have cardioprotective effects in vitro [52]. However, these studies were not well designed and lacked adequate control experiments. But interestingly, extracts of Hp have also been shown to have antiarythmic properties [53–56]. In light of the suggestion that Hp may be an alternative to NSAIDs, these positive effects on gastrointestinal and cardiovascular function are encouraging. However, it should be mentioned that because Hp activates the upper gastrointestinal tract, there may be cause for concern for administration to individuals prone to peptic ulcer. But such problems have never been reported despite the widespread use of Hp extracts.

6.6 Clinical Studies of Hp

The efficacy of Hp-containing products in the treatment of lower back pain or osteoarthritis has been tested in several clinical trials, and the results have been the topic of a number of recent reviews [4,8,57,58]. These reviews unanimously concluded that good quality evidence of efficacy exists, but that there are a number of issues that need to be addressed in further studies. The general conclusion of these reviews was that evidence of efficacy and good tolerability exists for the short-term treatment of painful osteoarthritis, and large phase-III type studies are justified. In particular, good comparative studies are required. The existing data do indicate equivalence with NSAIDs and superior tolerability. The authors of these reviews also emphasise the need to establish Hp extracts as an effective and safe alternative to NSAIDs, which are associated with significant gastric side effects and which remain the standard treatment option following the "demise" of the COX2 inhibitors due to fears of cardiovascular risks.

Gagnier et al. [59] in a Cochrane Review included ten trials that tested the efficacy of Hp in lower back pain. Two of these provided high-quality evidence of short-term improvements in pain and rescue medication, while one trial provided high-quality evidence of relative equivalence with 12.5 mg rofecoxib (Vioxx). These studies involved very small groups of patients and were too short to assess long-term efficacy or safety in chronic conditions. However, one report of a follow-up study indicated an equivalence between a water extract of Hp and a NSAID in 54 patients during 1 year of treatment, and recorded just 3 patients with mild adverse effects [60].

The Cochrane reviewers highlighted the absence of evidence of efficacy and safety during long-term use. An additional concern was the issue of authorship of the clinical trials and the absence of conflict-of-interest statements and the fact that the majority of reports were from the same group. Also, the poor standard of reporting of the Hp trials was seen as a problem, albeit one shared with reports of herbal drug trials in general. Poor reporting may not reflect poor study quality but simply a failure to meet the modern standards of transparency that are currently expected. New standards [Consolidated Standards of Reporting Trials (CONSORT)] have been set for reporting clinical trials. These have been laid out in the form of a statement by a group of senior editors of major clinical journals [61]. The CONSORT checklist aims to ensure a standard of transparency and consistency in reporting clinical trials but also serve to emphasize the required procedures for conducting high-quality trials. This checklist has been adapted for herbal drugs [62] by a group of international herbal drug experts, who additionally consulted a wider group of experts. Important issues that are specific to clinical investigation of herbals drugs included a full description of pharmaceutical characteristics and quality assurance of the extract under investigation, a description of any placebo used, details and rationale of dosage, how dosage of the herbal extract was standardised, name and status of the product, full description of the training and experience of the practitioners involved in the intervention.

6.7 Toxicological Considerations

No systematic toxicological testing of Hp has been reported. However, it is to be expected that some preparation-specific toxicological data, such as genotoxicity and potential for drug-drug interaction data, have been submitted to the authorities by manufacturers. Unger and Frank [63] reported significant inhibition by Hp of the liver drug metabolising cytochrome P450 (CYP) isoenzymes 2C9, 2D6 and 3A4 and measured IC_{50} values, respectively, of 121, 1044 and 335 μg/ml. The implication of these data is that the tested Hp extract may exert clinically important interactions with co-medications. There are a number of shortcomings in the report from Unger and Frank [63]. They aimed to validate a novel high-throughput procedure, and the drug interaction potential of the Devil's Claw extract was not the primary aim. The study contains no details of the nature of the Devil's Claw extract. The assays were not performed under GLP conditions and employed commercial preparations of the CYP isoenzymes. The report also contains no details on the source and method of enzyme preparations – for example it is not clear whether the enzymes were of human origin, although this is implied in the abstract. Finally, no recognised CYP isoenzyme-specific inhibitors or inducers were included, as is required according to accepted guidelines in order to have a comparative control. It needs to be pointed out that there are no reports of drug-drug interactions relating to Hp, despite the widespread use of Hp extracts and the particular concern in recent years to highlight this potential problem with the use of herbal drugs.

6.8 Concluding Remarks and Outlook

Hp extracts probably contain more than one active anti-inflammatory ingredient. Current pharmacological studies indicate that this traditional herbal drug may target pro-inflammatory intracellular pathways and so represents a "modern" anti-inflammatory drug. Inflammatory processes are now acknowledged as being ubiquitous in human disease, which broadens the therapeutic potential of anti-inflammatory drugs. Further pharmacological studies of Hp are warranted in order to assess whether Hp has potential beyond the treatment of osteoarthritis. Recent clinical interest in Hp has focused on osteoarthritis, and there is positive evidence of good quality. This is the unanimous conclusion of recently published systematic reviews, which are a key part of establishing evidence-based medicine. More good-quality evidence with improved reporting would enhance the case for Hp becoming a rational option in the treatment of osteoarthritis. Unfortunately, the resources for definitive phase-III trials have been lacking. One can only hope that such resources will soon be made available. Osteoarthritis is a major health problem that urgently requires new therapeutic options [64]. Current therapy targets symptoms and not the disease. Inflammation is now recognised as a key pathophysiological factor in the pathogenesis of osteoarthritis, and anti-inflammatory drugs are considered more than a symptomatic therapy.

The UN named the decade 2001–2010 the "Bone & Joint" Decade. This international initiative (which includes the UN, WHO and several other concerned international and national organisations) aims at improving the treatment and management of diseases of the musculoskeletal system. This initiative indicates that the world has a significant problem with diseases such as osteoarthritis, which is one of the most common musculoskeletal diseases. The incidence of osteoarthritis is increasing as the number of elderly persons in the population increases. Across Europe, North America and Japan, more than 30 million people suffer from this disease, which means that it is more than ten times more common than rheumatoid arthritis, which attracts much more scientific and public attention.

NSAIDS are revered by many as the most successful drugs ever. They have their origin in herbal medicine. As there is an almost drastic need for new anti-inflammatory drugs, these roots should not be forgotten, and the potential of herbal drugs and Hp in particular demands attention.

References

1. McGregor G, Fiebich B, Wartenberg A, Brien S, Lewith G, Wegener T (2005) Phytochem Rev 4:47
2. ESCOP (2003) Harpagophyti radix – Devil's Claw Root. European Scientific Cooperative on Phytotherapy, Exeter, UK
3. EMEA (2008) EMEA/HMPC/476255/07
4. Brien S, Lewith GT, McGregor G (2006) J Altern Complement Med 12:981
5. Saxne T, Lindell M, Mansson B, Petersson IF, Heinegard D (2003) Rheumatology (Oxford) 42:903
6. Ameye LG, Chee WS (2006) Arthritis Res Ther 8:R127
7. Darlington LG, Ramsey NW (1994) Compr Ther 20:490
8. Brendler T, Gruenwald J, Ulbricht C, Basch E (2006) J Herb Pharmacother 6:89
9. WO (2006) 114422
10. Clarkson C, Staerk D, Hansen SH, Smith PJ, Jaroszewski JW (2006) J Nat Prod 69:1280
11. Munkombwe NM (2003) Phytochemistry 62:1231
12. Boje K, Lechtenberg M, Nahrstedt A (2003) Planta Med 69:820
13. Berger TJ (1987) J Cardiothorac Anesth 1:500
14. Wenzel P, Wegener T (1995) Deutsche Apotheker Zeitung 135:1131
15. Whitehouse LW, Znamirowska M, Paul CJ (1983) Can Med Assoc J 129:249
16. McLeod DW, Revell P, Robinson BV (1979) Br J Pharmacol 66:140P
17. Eichler O, Koch C (1970) Arzneimittelforschung 20:107
18. Andersen ML, Santos EH, Seabra Mde L, da Silva AA, Tufik S (2004) J Ethnopharmacol 91:325
19. Morgenstern E, Pollex S (2000) Phytopharmakaforschung: Poster P18; Abstractband: 92
20. Lanhers MC, Fleurentin J, Mortier F, Vinche A, Younos C (1992) Planta Med 58:117
21. Soulimani R, Younos C, Mortier F, Derrieu C (1994) Can J Physiol Pharmacol 72:1532
22. Catelan SC, Belentani RM, Marques LC, Silva ER, Silva MA, Caparroz-Assef SM, Cuman RK, Bersani-Amado CA (2006) Phytomedicine 13:446
23. Baghdikian B, Lanhers MC, Fleurentin J, Ollivier E, Maillard C, Balansard G, Mortier F (1997) Planta Med 63:171
24. Erdos A, Fontaine R, Friehe H, Durand R, Poppinghaus T (1978) Planta Med 34:97
25. Manez S, Alcaraz MJ, Paya M, Rios JL, Hancke JL (1990) Planta Med 56:656
26. Abe S, Ishibashi H, Masuo K, Tanaka S-i, Yamaguchi H (2002) Pharmacometrics 62:27

27. Jadot G, Lecomte A (1992) Lyon Méditerranée médical médecine du sud-est 28:833
28. O'Neill LA (2006) Nat Rev Drug Discov 5:549
29. Westacott CI, Sharif M (1996) Semin Arthritis Rheum 25:254
30. Gunther M, Laufer S, Schmidt PC (2006) Phytochem Anal 17:1
31. Moussard C, Alber D, Toubin MM, Thevenon N, Henry JC (1992) Prostaglandins Leukot Essent Fatty Acids 46:283
32. Loew D, Mollerfeld J, Schrodter A, Puttkammer S, Kaszkin M (2001) Clin Pharmacol Ther 69:356
33. Tippler B, Syrovets T, Loew D, Simmet T (1996) In: Loew D, Rietbrock N (eds) Phytopharmaka II – Forschung und klinische Anwendung, p 96
34. Benito PB, Lanza AM, Sen AMS, Galindez JDS, Matellano LF, Gomez AS, Martinez MJA (2000) Planta Med 66:324
35. Jang MH, Lim S, Han SM, Park HJ, Shin I, Kim JW, Kim NJ, Lee JS, Kim KA, Kim CJ (2003) J Pharmacol Sci 93:367
36. Kundu JK, Mossanda KS, Na H-K, Surh Y-J (2005) Cancer Lett 218:21
37. Kaszkin M, Beck KF, Koch E, Erdelmeier C, Kusch S, Pfeilschifter J, Loew D (2004) Phytomedicine 11:585
38. Huang TH, Tran VH, Duke RK, Tan S, Chrubasik S, Roufogalis BD, Duke CC (2006) J Ethnopharmacol 104:149
39. Fiebich BL, Heinrich M, Hiller KO, Kammerer N (2001) Phytomedicine 8:28
40. Schulze-Tanzil G, Hansen C, Shakibaei M (2004) Arzneimittelforschung 54:213
41. Betancor-Fernandez A, Perez-Galvez A, Sies H, Stahl W (2003) J Pharm Pharmacol 55:981
42. Pan N, Hori H (1996) Redox Rep 2:149
43. Wang P, Kang J, Zheng R, Yang Z, Lu J, Gao J, Jia Z (1996) Biochem Pharmacol 51:687
44. Li W, Zheng R, Su B, Jia Z, Li H, Jiang Y, Yao S, Lin N (1996) Int J Radiat Biol 69:481
45. Herbert JM, Maffrand JP, Taoubi K, Augereau JM, Fouraste I, Gleye J (1991) J Nat Prod 54:1595
46. Kimura Y, Okuda H, Nishibe S, Arichi S (1987) Planta Med 53:148
47. Ravn H, Nishibe S, Sasahara M, Xuebo L (1990) Phytochemistry 29:3627
48. Hayashi K, Nagamatsu T, Ito M, Yagita H, Suzuki Y (1996) Jpn J Pharmacol 70:157
49. Hayashi K, Nagamatsu T, Ito M, Hattori T, Suzuki Y (1994) Jpn J Pharmacol 66:47
50. Lee JY, Woo ER, Kang KW (2005) J Ethnopharmacol 97:561
51. Hausmann M, Obermeier F, Paper DH, Balan K, Dunger N, Menzel K, Falk W, Schoelmerich J, Herfarth H, Rogler G (2007) Clin Exp Immunol 148:373
52. Ahmad M, Rizwani GH, Aftab K, Ahmad VU, Gilani AH, Ahmad SP (1995) Phytother Res 9:525
53. Pennacchio M, Syah YM, Alexander E, Ghisalberti EL (1999) Phytother Res 13:254
54. Costa De Pasquale R, Busa G, Circosta C, Iauk L, Ragusa S, Ficarra P, Occhiuto F (1985) J Ethnopharmacol 13:193
55. Circosta C, Occhiuto F, Ragusa S, Trovato A, Tumino G, Briguglio F, de Pasquale A (1984) J Ethnopharmacol 11:259
56. Clarkson C, Campbell WE, Smith P (2003) Planta Med 69:720
57. Mahomed IM, Ojewole JA (2006) Brain Res Bull 69:57
58. Gagnier JJ, van Tulder MW, Berman B, Bombardier C (2007) Spine 32:82
59. Denner SS (2007) Holist Nurs Pract 21:203
60. Gagnier J, Vantulder M, Berman B, Bombardier C (2006) Cochrane Database Syst Rev:CD004504
61. Chrubasik S, Kunzel O, Thanner J, Conradt C, Black A (2005) Phytomedicine 12:1
62. Moher D, Schulz KF, Altman DG (2001) Ann Intern Med 134:657
63. Gagnier JJ, Boon H, Rochon P, Moher D, Barnes J, Bombardier C (2006) Ann Intern Med 144:364
64. Unger M, Frank A (2004) Rapid Commun Mass Spectrom 18:2273
65. Wieland HA, Michaelis M, Kirschbaum BJ, Rudolphi KA (2005) Nat Rev Drug Discov 4:331

Chapter 7
The Role of Curcumin in Modern Medicine

Gautam Sethi, Bokyung Sung and Bharat B. Aggarwal

Abstract Curcumin (diferuloylmethane) is an orange-yellow component of turmeric (*Curcuma longa)*, a spice often found in curry powder. Since the time of Ayurveda numerous therapeutic activities have been assigned to turmeric for a wide variety of diseases and conditions, including those of the skin, pulmonary, and gastrointestinal systems, aches, pains, wounds, sprains, and liver disorders. Curcumin has been shown to exhibit antioxidant, anti-inflammatory, antiviral, antibacterial, antifungal, and thus has a potential against various malignant cancers, diabetes, allergies, arthritis, Alzheimer's disease, and other chronic illnesses. These effects are mediated through the regulation of various transcription factors, growth factors, inflammatory cytokines, protein kinases, and other enzymes. Thus, curcumin, by virtue of its effect on multiple cell signaling pathways, could prove to be a panacea for modern human diseases.

Keywords *Curcuma longa* · Curcumin · Antioxidant · Anticancer

7.1 Introduction

The turmeric (*Curcuma longa*) plant, a perennial herb belonging to the ginger family, is cultivated extensively in south and southeast tropical Asia. The rhizome of this plant is also referred to as the "root" and is the most useful part of the plant for culinary and medicinal purposes. The most active component of turmeric is curcumin, which makes up 2 to 5% of the spice. Curcumin was first isolated in 1815, obtained in crystalline form in 1870 [1], and identified as 1,6-heptadiene-3,5-dione-1,7-bis(4-hydroxy-3-methoxyphenyl)-(1E,6E) or diferuloylmethane. The feruloylmethane skeleton of curcumin was subsequently confirmed in 1910 by the initial work and synthesis by Lampe [2, 3]. Curcumin is a yellow-orange powder

B.B. Aggarwal (✉)
Cytokine Research Laboratory, Departments of Experimental Therapeutics, The University of Texas, M.D. Anderson Cancer Center, 1515 Holcombe Blvd., Houston, TX 77030, USA,
e-mail: aggarwal@mdanderson.org

K.G. Ramawat (ed.), *Herbal Drugs: Ethnomedicine to Modern Medicine*,
DOI 10.1007/978-3-540-79116-4_7,

that is insoluble in water and ether but soluble in ethanol, dimethylsulfoxide, and acetone. Curcumin has a melting point of 183°C, molecular formula of $C_{21}H_{20}O_6$, and molecular weight of 368.37 g/mol. Turmeric contains curcumin along with other chemical constituents known as the "curcuminoids" [4].

7.2 Isolation and Chemical Properties of Curcumin

The major curcuminoids present in turmeric are demethoxycurcumin (curcumin II), bisdemethoxycurcumin (curcumin III), and the recently identified cyclocurcumin [5]. Commercial curcumin contains curcumin I (∼77%), curcumin II (∼17%), and curcumin III (∼3%) as its major components. Spectrophotometrically, curcumin has a maximum absorption (λmax) in methanol at 430 nm, with a Beer's law range from 0.5 to 5 μg/mL [6]. It absorbs maximally at 415 to 420 nm in acetone, and a 1% solution of curcumin has 1650 absorbance units. Curcumin has a brilliant yellow hue at pH 2.5 to 7 and takes on a red hue at pH > 7. The spectral and photochemical properties of curcumin have been studied in different solvents by Chignell and coworkers [7]. In toluene, the absorption spectrum of curcumin contains some structure, which disappears in more polar solvents such as ethanol and acetonitrile. Curcumin is also reported to be able to photogenerate superoxide in toluene and ethanol [7].

7.3 Antioxidant Properties of Curcumin

Unnikrishnan and Rao [8, 9] studied the antioxidative properties of curcumin and its three derivatives (demethoxycurcumin, bisdemethoxycurcumin, and diacetyl curcumin). The authors demonstrated that these substances provide a protection of hemoglobin from oxidation except the diacetyl curcumin, which has little effect in the inhibition of nitrite-induced oxidation of hemoglobin. The effect of curcumin on lipid peroxidation has also been studied in various models by several authors. Curcumin is a good antioxidant and inhibits lipid peroxidation in rat liver microsomes, erythrocyte membranes, and brain homogenates. The antioxidant activity of curcumin could be mediated through antioxidant enzymes such as superoxide dismutase, catalase, and glutathione peroxidase. Curcumin has been shown to serve as a Michael acceptor, reacting with glutathione and thioredoxin [10]. Reaction of curcumin with these agents reduces intracellular GSH in the cells. In fact, curcumin has been found to be at least ten times more active as an antioxidant than even vitamin E [11]. In curcumin, the phenolic and the methoxy group on the phenyl ring and the 1,3-diketone system seem to be important structural features that can contribute to these effects. Another fact presented in the literature is that the antioxidant activity increases when the phenolic group with a methoxy group is at the ortho position [12].

7.4 Molecular Targets of Curcumin

Various studies have shown that curcumin modulates multiple targets (Table 7.1). These include the growth factors, growth factor receptors, transcription factors, cytokines, enzymes, and genes regulating apoptosis.

Table 7.1

Molecular Targets of Curcumin
Transcription factors
Activator protein-1 ↓
β-Catenin ↓
CREB-binding protein ↓
Early growth response 1 ↓
Electrophile response element ↑
Hypoxia inducible factor-1 ↓
Notch-1 ↓
Nuclear factor-kappa B ↓
Nuclear factor erythyroid 2-related factor (Nrf2)
Peroxisome preoliferator-activated receptor-γ ↑
Signal transducers and activators of transcription-1 ↓
Signal transducers and activators of transcription-3 ↓
Signal transducers and activators of transcription-4 ↓
Signal transducers and activators of transcription-5 ↓
Wilms' tumour gene 1 ↓
Inflammatory cytokines
Interleukin-1 ↓
Interleukin-2 ↓
Interleukin-5 ↓
Interleukin-6 ↓
Interleukin-8 ↓
Interleukin-12 ↓
Interleukin-18 ↓
Monocyte chemoattractant protein ↓
Migration inhibition protein ↓
Macrophage inflammatory protein ↓
Tumor necrosis factor α ↓
Enzymes
Arylamine N-acetyltransferases-1 ↓
ATFase ↓
ATPase ↓
Cyclooxygenase-2 ↓
Desaturase ↓
DNA polymerase ↓
Farnesyl protein transferase ↓
Gluthathione S-transferase ↑

Table 7.1 (continued)

Molecular Targets of Curcumin
Glutamyl cysteine ligase ↑
Hemeoxygenase-1 ↑
Inducible nitric oxide synthase ↓
Lipoxygenase ↓
Matrix metalloproteinase ↓
NAD(P)H:quinone oxidoreductase ↓
Ornithine decarboxylase ↓
Phospholipase D ↓
Telomerase ↓
Tissue inhibitor of metalloproteinase-3 ↓
Kinases
Autophosphorylation-activated protein kinase ↓
Ca^{2+} -dependent protein kinase ↓
EGF receptor-kinase ↓
Extracellular receptor kinase ↓
Focal adhesion kinase ↓
IL-1 receptor-associated kinase ↓
Janus kinase ↓
c-jun N-terminal kinase ↑
Mitogen-activated protein kinase ↓
Phosphorylase kinase ↓
Protamine kinase ↓
Protein kinase A ↓
Protein kinase B ↓
Prorein kinase C ↓
Protein tyrosine kinase ↓
Growth factors
Connective tissue growth factor ↓
Epidermal growth factor ↓
Fibroblast growth factor ↓
Hepatocyte growth factor ↓
Nerve growth factor ↓
Platelet derived growth factor ↓
Transforming growth factor-β1 ↓
Vascular endothelial growth factor ↓
Receptors
Androgen receptor ↓
Aryl hydrocarbon receptor ↓
Chemokine (C-X-C motif) receptor 4 ↓
Death receptor-5 ↑
EGF-receptor ↓
Endothelial protein C-receptor ↑
Estrogen receptor-alpha ↓
Fas receptor ↑
Histamine (2)- receptor ↓
Human epidermal growth factor receptor-2 ↓
Interleukin 8-receptor ↓
Inositol 1,4,5-triphosphate receptor ↓

Table 7.1 (continued)

Molecular Targets of Curcumin
Integrin receptor ↓
Low-density lipoprotein-receptor ↑
Adhesion molecules
Endothelial leukocyte adhesion molecule-1 ↓
Intracellular adhesion molecule-1 ↓
Vascular cell adhesion molecule-1 ↓
Antiapoptotic proteins
B-cell lymphoma protein 2 ↓
Bcl-xL ↓
Inhibitory apoptosis protein-1 ↓
Others
Cyclin D1 ↓
DNA fragmentation factor 40-kd subunit ↑
Heat-shock protein 70 ↑
Multi-drug resistance protein ↓
Urokinase-type plasminogen activator ↓
p53 ↑

7.4.1 Cytokines and Growth Factors

Numerous growth factors have been implicated in the growth and promotion of tumors. Curcumin has been shown to down-regulate expression of several cytokines, including tumor necrosis factor (TNF), interleukin (IL)-6, IL-8, IL-12, and fibroblast growth factor 2 [13, 14].

7.4.2 Receptors

Down-regulation of epidermal growth factor receptor (EGFR)-1 and EGFR-2 (also called HER2/neu) activity represents one possible mechanism by which curcumin suppresses the growth of breast cancer cells. Almost 30% of breast cancer cases overexpress the *HER2/neu* proto-oncogene [15], and both HER2 and EGF receptors stimulate proliferation of breast cancer cells. Overexpression of these two proteins correlates with progression of human breast cancer and poor patient prognosis [15]. Curcumin has been shown to down-regulate the activity of EGFR [16, 17] and HER2/neu [16, 17] and to deplete the cells of the HER2/neu protein [18].

Prostate cancer cell lines LNCaP and PC-3 were treated with curcumin, and its effects on signal transduction and expression of androgen receptor (AR) and AR-related cofactors were analyzed. The results showed that curcumin down-regulated transactivation and expression of AR and CREB (cAMP response element-binding protein)-binding protein. It also inhibited the transforming activities of both cell lines as evinced by reduced colony-forming ability in soft agar. These findings suggest that curcumin has a potential therapeutic effect on prostate cancer cells through down-regulation of AR and AR-related cofactors [18].

7.4.3 Transcription Factors

Curcumin may operate through suppression of various transcription factors, including nuclear factor-kappa B (NF-κB), signal transducer and activator of transcription 3 (STAT3), early growth response protein 1, activator protein 1 (AP-1), peroxisome proliferators-activated receptor gamma (PPAR-γ), and β-catenin [13, 14, 19]. These transcription factors play essential roles in various diseases. The constitutively active form of NF-κB has been reported in a wide variety of cancers. NF-κB is required for the expression of genes involved in cell proliferation, cell invasion, metastasis, angiogenesis, and resistance to chemotherapy. Bharti et al. demonstrated that curcumin inhibited IL-6-induced STAT3 phosphorylation and consequent STAT3 nuclear translocation [20]. Activation of PPAR-γ inhibits the proliferation of nonadipocytes. Xu et al. demonstrated that curcumin dramatically induced expression of the *PPAR-γ* gene and activated PPAR-γ [21]. AP-1, another transcription factor that has been closely linked with proliferation and transformation of tumor cells, has been shown to be suppressed by curcumin. Studies also suggest that curcumin has a potential therapeutic effect on prostate cancer cells through down-regulation of AR and AR-related cofactors [13].

7.4.4 Proinflammatory Enzymes

Cyclooxygenase-2 (COX-2) and lipoxygenase (LOX) are important enzymes that mediate inflammation through production of prostaglandins and leukotrienes, respectively. Curcumin has been shown to suppress the expression of both COX-2 and LOX proteins as well as their enzymatic activities, most likely through the down-regulation of NF-κB, which is needed for COX-2 expression. Several groups have shown that curcumin down-regulates the expression of COX-2 protein in various tumor cells [22, 23]. Chun et al. reported that curcumin inhibited phorbol-ester-induced expression of COX-2 in mouse skin through suppression of extracellular signal-regulated kinase activity and NF-κB activation [24]. Plummer et al. measured COX-2 protein induction and prostaglandin E2 production in human blood after incubation with lipopolysaccharide. When 1 μM curcumin was added *in vitro* to blood from healthy volunteers, lipopolysaccharide-induced COX-2 protein levels and concomitant prostaglandin E2 production were reduced by 24% and 41%, respectively [25].

7.4.5 Protein Kinases

Curcumin suppresses a number of protein kinases, including mitogen activated protein kinase (MAPK), Jun N-terminal kinase, protein kinase A (PKA), protein kinase C (PKC), src tyrosine kinase, phosphorylase kinase, IκBα kinase, janus kinase (JAK), and the growth factor receptor protein tyrosine kinases. Treatment with curcumin inhibited highly purified PKA, PKC, protamine kinase, phosphorylase kinase, autophosphorylation-activated protein kinase, and pp60c-src tyrosine kinase [26]. Phorbol-myristate-acetate-induced activation of cellular PKC is suppressed by curcumin [27]. Treatment of cells with curcumin inhibited tetradecanoylphorbol-acetate-induced PKC activity without affecting the level of PKC. Curcumin inhibited the PKC activity in vitro as well as in the cells [28].

7.4.6 Cell Cycle

Curcumin modulates cell-cycle-related gene expression. Specifically, curcumin induced G_0/G_1 and/or G_2/M phase cell cycle arrest, up-regulated cyclin-dependent kinase inhibitors p21WAF1/CIP1, p27KIP1, and p53, and slightly down-regulated cyclin B1 and cdc2 [13]. We found that curcumin can down-regulate cyclin D1 expression at the transcriptional and posttranscriptional levels [29].

7.4.7 Adhesion Molecules

The expression of various cell surface adhesion molecules such as intercellular cell adhesion molecule 1, vascular cell adhesion molecule 1, and endothelial leukocyte adhesion molecule 1 on endothelial cells is absolutely critical for tumor metastasis [30]. Curcumin inhibits inflammation by blocking the adhesion of monocytes to endothelial cells by inhibiting activation of these cell adhesion molecules. The expression of these molecules is in part regulated by NF-κB [31]. We have shown that treatment of endothelial cells with curcumin blocks the cell surface expression of adhesion molecules, and this accompanies the suppression of tumor cell adhesion to endothelial cells [32]. We also have demonstrated that down-regulation of these adhesion molecules is mediated through down-regulation of NF-κB activation [32]. Gupta and Ghosh reported that curcumin inhibits TNF-induced expression of adhesion molecules on human umbilical vein endothelial cells [33]. Jaiswal et al. showed that curcumin treatment causes p53- and p21-independent G_2/M phase arrest and apoptosis in colon cancer cell lines [34]. Their results suggest that curcumin treatment impairs both Wnt signaling and cell-cell adhesion pathways, resulting in G_2/M phase arrest and apoptosis in HCT-116 cells.

7.4.8 Antiapoptotic Proteins

Curcumin is known to down-regulate expression of apoptosis suppressor proteins such as Bcl-2 and Bcl-X_L in several cancer cell lines. We found that curcumin induces apoptosis through a mitochondrial pathway involving caspase-8, Bid cleavage, cytochrome c release, and caspase-3 activation. Our results also suggest that Bcl-2 and Bcl-X_L are critical negative regulators of curcumin-induced apoptosis [35].

Curcumin suppresses the constitutive expression of Bcl-2 and Bcl-X_L in mantle cell lymphoma [36] and multiple myeloma [37] cell lines. It also activates caspase-7 and caspase-9 and induces polyadenosine-5′-diphosphate-ribose polymerase cleavage in both cell lines. Thus, curcumin induces apoptosis by targeting several apoptotic pathways, inducing cytochrome c release, Bid cleavage, and caspase-9 and -3 activation, and by down-regulating the antiapoptotic proteins Bcl-2 and Bcl-X_L.

7.4.9 Multidrug Resistance

Multidrug resistance is a phenomenon that often is associated with enhanced drug efflux and thus decreased intracellular drug accumulation in tumor cells. It often is related to overexpression of P-glycoprotein on the surface of tumor cells, which reduces drug cytotoxicity. Curcumin has been shown to suppress the overexpression of P-glycoprotein in the multidrug-resistant human cervical carcinoma cell lines [38]. Curcumin also down-regulates drug resistance by inhibiting expression of the *mdr* gene, which is responsible for this phenomenon [39].

7.5 Disease Targets of Curcumin

The various traditional uses of curcumin are shown in (Fig 1.) and its disease targets are discussed in detail below:

7.5.1 Anticancer Effects

The anticancer potential of curcumin has been demonstrated in various *in vitro* and *in vivo* models [40]. Curcumin has been shown to block transformation, tumor initiation, tumor promotion, invasion, angiogenesis, and metastasis. In *in vivo* studies, curcumin suppressed carcinogenesis of the skin, forestomach, colon, and liver in mice. Curcumin has been shown to inhibit proliferation of a wide variety of tumor cells, including B-cell and T-cell leukemias [41–44], colon carcinoma [45], epidermoid carcinoma [17], head and neck squamous cell carcinoma [46], multiple

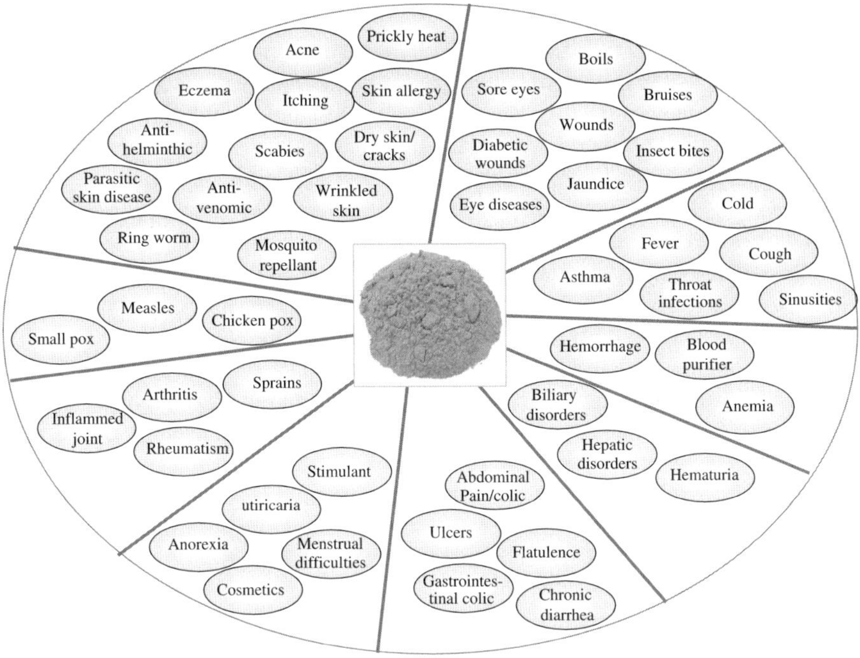

Fig. 7.1 Traditional Uses of Curcumin

myeloma [20], and mantle cell lymphoma [36]. It has also been shown to suppress proliferation of various breast carcinoma cell lines in culture [47–49].

Mehta et al. examined the antiproliferative effects of curcumin against several breast tumor cell lines, including hormone-dependent and -independent and multidrug-resistant lines [47]. All the cell lines tested, including the multidrug-resistant ones, were highly sensitive to curcumin. The growth-inhibitory effect of curcumin was time and dose dependent, and correlated with its inhibition of ornithine decarboxylase activity. Curcumin preferentially arrested cells in the G_2/S phase of the cell cycle.

Fang et al. reported that rat thioredoxin reductase activity in thioredoxin-dependent disulfide reduction was inhibited by curcumin [50]. By using mass spectrometry and blotting analysis, they showed that this irreversible inhibition by curcumin was caused by alkylation of both residues in the catalytically active site (Cys (496)/Sec (497)) of the enzyme. Kang et al. reported that exposure of human hepatoma cells to curcumin led to a significant decrease of histone acetylation [51]. Curcumin can selectively downregulate transcription of human papillomavirus type 18, which is etiologically associated with development of cancer of the uterine cervix in women, as well as activator protein 1 (AP-1) binding activity in HeLa cells. Most interestingly, curcumin can reverse the expression dynamics of c-fos and fra-1 in this tumorigenic cell line [52].

Curcumin had synergic activity with chemotherapeutic agent vinorelbine in suppressing the growth of human squamous cell lung carcinoma H520 cells [53]. It significantly inhibited the growth of human gastric carcinoma AGS cells in a dose- and time-dependent manner [54]. Using time-lapse video and immunofluorescence labeling methods, Holy et al. demonstrated that curcumin significantly altered microfilament organization and cell motility in PC-3 and LNCaP human prostate cancer cells *in vitro* [55]. Chemoresistance is a major problem in the treatment of patients with multiple myeloma due to constitutive expression of NF-κB and STAT3. Bharti et al. showed that suppression of NF-κB and STAT3 activation in multiple myeloma cells by *ex vivo* treatment with curcumin resulted in decreases in adhesion to bone marrow stromal cells, secretion of cytokines, and viability of cells [56].

Helicobacter pylori is a Group 1 carcinogen that is associated with the development of gastric and colon cancers. Curcumin inhibited the growth of all strains of *H. pylori in vitro* with a minimum inhibitory concentration range of 6.25 to 50 μg/ml [57]. Chen et al. used microarray analysis of gene expression profiles to characterize the anti-invasive mechanisms of curcumin in highly invasive lung adenocarcinoma cells (CL1-5) [58]. In these studies, curcumin significantly reduced the invasive capacity of CL1-5 cells in a concentration range far below its levels of cytotoxicity (20 μM), and this anti-invasive effect was concentration dependent. Kim et al. evaluated the antiangiogenic activity of demethoxycurcumin, a structural analog of curcumin, and found that nine angiogenesis-related genes were down-regulated by at least fivefold in response to this agent [59].

Numerous studies have evaluated the cancer-chemopreventive properties of curcumin. The anticancer potential of curcumin was examined *in vivo* in mice using Dalton's lymphoma cells grown as ascites [60]. When curcumin was administered in liposomal formulations at a concentration of 1 mg/animal, all animals survived 30 days and only two of the animals developed tumors and died before 60 days. Similarly, Busquets et al. showed that systemic administration of curcumin for 6 consecutive days to rats bearing the highly cachectic ascites hepatoma resulted in a significant inhibition of tumor growth [61]. Interestingly, curcumin was able to reduce *in vitro* tumor cell content by 24% at concentrations as low as 0.5 μM without promoting any apoptotic events.

Menon et al. reported curcumin-induced inhibition of B16F10 melanoma lung metastasis in mice [62]. Oral administration of curcumin at concentrations of 200 nmol/kg body weight reduced the number of lung tumor nodules by 80%. The life span of the animals treated with curcumin was increased by 143.85% [62]. Curcumin treatment (10 μg/ml) significantly inhibited the invasion of B16F10 melanoma cells by inhibition of matrix metalloproteinases (MMP), thereby inhibiting lung metastasis.

Curcumin decreases the proliferative potential and increases the apoptotic potential of both androgen-dependent and androgen-independent prostate cancer cells in vitro, largely by modulating the apoptosis suppressor proteins and by interfering with the growth factor receptor signaling pathways as exemplified by the EGFR [63]. The chemopreventive activity of curcumin was observed when it was administered prior to, during, and after carcinogen treatment as well as when it was given only during the promotion/progression phase of colon carcinogenesis [64]. The chemopreventive effect of curcumin was also examined on the development of adenomas in the intestinal tract of a mouse model of human familial adenomatous polyposis coli [65]. Curcumin at 0.2% and 0.5% of diet reduced adenoma multiplicity by 39% and 40%, respectively.

Odot et al. showed that curcumin was cytotoxic to B16-R melanoma cells resistant to doxorubicin [66]. Treatment with a prophylactic immune preparation of soluble proteins from B16-R cells, in combination with curcumin, resulted in substantial inhibition of growth of B16-R melanoma and a significant increase in the median survival time of the animals.

7.5.2 Skin Diseases

Curcumin has been shown to be effective against various skin conditions [67], including skin carcinogenesis, psoriasis [68], scleroderma [69], and dermatitis. Numerous reports suggest that curcumin accelerates wound healing. Sidhu et al. examined the wound-healing capacity of curcumin in rats and guinea pigs [70]. Punch wounds closed faster in curcumin-treated animals than in untreated animals. Biopsies of the wounds showed re-epithelialization of the epidermis and increased migration of various cells, including myofibroblasts, fibroblasts, and macrophages, in the wound beds of curcumin-treated wounds. Curcumin-treated animals showed extensive neovascularization in multiple areas within the dermis and greater collagen deposition in the wounds. Curcumin seemed to have potent effects in inhibiting proliferation and contraction of excessive scar-derived fibroblasts [71].

7.5.3 Diabetes

Administration of curcumin reduced the blood sugar, and glycosylated hemoglobin levels significantly in an alloxan-induced rat model of type 2 diabetes. Diabetic rats maintained on a curcumin diet for 8 weeks excreted less albumin, urea, creatinine, and inorganic phosphorus than rats not fed curcumin. Dietary curcumin also partially reversed abnormalities in plasma albumin, urea, creatinine, and inorganic phosphorus in diabetic animals [13].

7.5.4 Rheumatoid Arthritis

Curcumin has been shown to possess antirheumatic and antiarthritic effects, most likely through down-regulation of COX-2, TNF, and other inflammatory cytokines. Deodhar et al. were the first to report on the antirheumatic activity of curcumin in human subjects [72]. They performed a short-term double-blind cross-over study in 18 patients with rheumatoid arthritis to compare the antirheumatic activity of curcumin (1200 mg/day) with that of phenylbutazone (300 mg/day). Subjective and objective assessment in patients who were taking corticosteroids just prior to the study showed significant improvements in morning stiffness, walking time, and joint swelling following 2 weeks of curcumin therapy.

Liacini et al. examined the effects of curcumin in articular chondrocytes. Interleukin (IL)-1, the main cytokine instigator of cartilage degeneration in arthritis, induces MMP-3 and MMP-13 RNA and protein in chondrocytes through activation of MAPK, AP-1, and NF-κB transcription factors [73]. Curcumin achieved 48–99% suppression of MMP-3 and 45–97% suppression of MMP-13 in human chondrocytes and 8–100% (MMP-3) and 32–100% (MMP-13) suppression in bovine chondrocytes. Inhibition of IL-1 signal transduction by these agents could be useful for reducing cartilage resorption by MMPs in arthritis.

7.5.5 Multiple Sclerosis

Multiple sclerosis is characterized by the destruction of oligodendrocytes and myelin sheath in the central nervous system (CNS). Curcumin inhibits experimental allergic encephalomyelitis (EAE), a model for multiple sclerosis, by blocking IL-12 signaling in T cells, suggesting that it would be effective in the treatment of multiple sclerosis. Natarajan and Bright investigated the effect of curcumin on the pathogenesis of CNS demyelination in EAE [74]. *In vivo* treatment of SJL/J mice with curcumin significantly reduced the duration and clinical severity of active immunization and adoptive transfer in EAE [74]. Curcumin inhibited EAE in association with a decrease in IL-12 production from macrophage/microglial cells and differentiation of neural antigen-specific Th1 cells. *In vitro* treatment of activated T cells with curcumin inhibited IL-12–induced tyrosine phosphorylation of Janus kinase 2, tyrosine kinase 2, and STAT3 and STAT4 transcription factors. Inhibition of the Janus kinase-STAT pathway by curcumin resulted in a decrease in IL-12–induced T-cell proliferation and Th1 differentiation. These findings show that curcumin inhibits EAE by blocking IL-12 signaling in T cells and suggest a rationale for its use in the treatment of multiple sclerosis and other Th1 cell–mediated inflammatory diseases.

Verbeek and coworkers examined the effects of curcumin on autoimmune T-cell reactivity in mice and on the course of EAE. Continuous oral administration of curcumin significantly affected antigen-specific proliferation and interferon-gamma production by lymph node–derived T cells following immunization with an EAE-inducing peptide [75]. The overall effects of oral curcumin were mild but beneficial.

7.5.6 Alzheimer's Disease

Brain inflammation in Alzheimer's disease is characterized by increased cytokines and activated microglia. Curcumin can suppress oxidative damage, inflammation, cognitive deficits, and amyloid accumulation in Alzheimer's disease [76]. Lim et al. found that curcumin reduces oxidative damage and amyloid pathology in an Alzheimer transgenic mouse model [77].

7.5.7 Inflammatory Bowel Disease

Inflammatory bowel disease is characterized by oxidative and nitrosative stress, leukocyte infiltration, and up-regulation of proinflammatory cytokines. Ukil et al. recently investigated the protective effects of curcumin on inflammatory bowel disease induced in a mouse model. Pretreatment of mice with curcumin for 10 days significantly ameliorated the appearance of diarrhea and the disruption of colonic architecture [78]. Curcumin is able to attenuate colitis in the dinitrobenzene/sulfonic acid-induced murine model of colitis [79]. When given before the induction of colitis, it reduced macroscopic damage scores and NF-κB activation, reduced myeloperoxidase activity, and attenuated the dinitrobenzene-induced message for IL-1β. Thus, curcumin attenuates experimental colitis through a mechanism that also inhibits activation of NF-κB and effects a reduction in the activity of p38 MAPK. The authors proposed that this agent might have therapeutic implications for human inflammatory bowel disease.

7.5.8 Cystic Fibrosis

Cystic fibrosis, the most common lethal hereditary disease in the white population, is caused by mutations in the cystic fibrosis transmembrane conductance regulator (*CFTR*) gene. *CFTR* mutation disrupts the surface localization and/or gating of the CFTR chloride channel. The most common cystic fibrosis mutant is ΔF508-CFTR, which inefficiently traffics to the surfaces of most cells. In a recent report, Egan et al. demonstrated that curcumin corrected the cystic fibrosis defects in ΔF508 cystic fibrosis mice [80]. Most likely, curcumin exerts these effects by directly stimulating the CFTR chloride channels [81].

7.5.9 Others

Curcumin has been shown to have other activities that suggest potential clinical applications, as follows:

- Curcumin was found to be a potent and selective inhibitor of HIV-1 long terminal repeat–directed gene expression, which governs the transcription of HIV-1 provirus.
- Curcumin has been shown to prevent cataractogenesis in an *in vitro* rat model.
- Treatment with curcumin prevented experimental alcoholic liver disease.
- Curcumin has a protective effect on cyclophosphamide-induced early lung injury.
- Chemotherapy-induced nephrotoxicity can be prevented with curcumin [13].

7.6 Structure Activity Relationship of Curcumin

To elucidate which portion of the molecule is critical for the activity, a large number of structural analogs of curcumin have been synthesized. Some analogues are more active than native curcumin, while others are less active [82]. It was found that the phenolic analogues were more active than the nonphenolic analogues [83]. The highest antioxidant activity was obtained when the phenolic group was sterically hindered by the introduction of two methyl groups at the ortho position. The phenolic group is essential for the free-radical scavenging activity, and the presence of the methoxy group further increases the activity [84]. Curcumin shows both antioxidant and pro-oxidant effects. Ahsan et al. have shown that both these effects are determined by the same structural moieties of the curcuminoids [85]. Dinkova–Kostova showed that the presence of hydroxyl groups at the ortho position on the aromatic rings and that beta-diketone functionality was required for high potency in inducing Phase 2 detoxification enzymes [86]. Curcumin is a noncompetitive inhibitor of rat liver microsomal delta 5 desaturase and delta 6 desaturase. Kawashima et al. [87] have shown that only half the structure is essential for the desaturase inhibition. A 3-hydroxy group of the aromatic ring is essential for the inhibition and a free carboxyl group at the end opposite to the aromatic ring interferes with the inhibitory effect. Simon et al. found that the presence of the diketone moiety in the curcumin molecule seems to be essential for its ability to inhibit the proliferation of MCF-7 human breast tumor cells [49]. The aromatic enone and dienone analogs of curcumin have been demonstrated to have potent antiangiogenic property in an in vitro SVR assay [88].

7.7 Conclusions

The medicinal properties of curcumin have been known for centuries, although the scientific basis of its actions has been investigated only over the last couple of decades. Curcumin has been shown to target numerous molecules inside the cell that are known to modulate several signaling pathways. Curcumin is a potent antioxidant, anti-inflammatory, and anticancer agent and has therapeutic efficacy in numerous diseases. Despite its proven safety over the ages, its efficacy has been limited by its poor solubility, and poor bioavailability. Several analogs of curcumin have been designed to overcome these limitations. Some of these analogs have yielded promising results in in vitro and animal studies, but whether these more bioavailable forms are equally safe is still unknown. Therefore, it is important to evaluate the curcumin analogs for bioavailability and clinical efficacy.

Acknowledgements This work was supported by the Clayton Foundation for Research and a PO1 grant (CA91844) from the National Institutes of Health on lung cancer chemoprevention to BBA.

References

1. Daube FV (1870) Uber Curcumin, den Farbstoff der Curcumawurzel Ber 3:609
2. Lampe V (1910) Ver Dtsch Chem Ger 43:2163
3. Lampe V Milobedzka J (1913) Ver DtschChem Ges 46:2235
4. Srinivasan KR (1952) Curr Sci 21:311
5. Kiuchi F, Goto Y, Sugimoto N, Akao N, Kondo K, Tsuda Y (1993) Chem Pharm Bull (Tokyo) 41:1640
6. Prasad NS (1997) Indian Drugs 34:227
7. Chignell CF, Bilski P, Reszka KJ, Motten AG, Sik RH, Dahl TA (1994) Photochem Photobiol 59:295
8. Sharma OP (1976) Biochem Pharmacol 25:1811
9. Ruby AJ, Kuttan G, Babu KD, Rajasekharan KN, Kuttan R (1995) Cancer Lett 94:79
10. Adams BK, Cai J, Armstrong J, Herold M, Lu YJ, Sun A, Snyder JP, Liotta DC, Jones DP, Shoji M (2005) Anticancer Drugs 16:263
11. Khopde SM, Priyadarsini KI, Venkatesan N, Rao MNA (1999) Biophys Chem 80:83
12. Motterlini R, Foresti R, Bassi R, Green CJ (2000) Free Radic Biol Med 28:1303
13. Shishodia S, Chaturvedi MM, Aggarwal BB (2007) Curr Prob Cancer 31:243–305.
14. Jagetia GC, Aggarwal BB (2007) J Clin Immunol 27:19
15. Slamon DJ, Godolphin W, Jones LA, Holt JA, Wong SG, Keith DE, LevinWJ, S Stuart G, Udove J, Ullrich A, et al. (1989) Science 244:707
16. Korutla L, Cheung JY, Mendelsohn J, Kumar R (1995) Carcinogenesis 16:1741
17. Korutla L, Kumar R (1994) Biochim Biophys Acta 1224:597
18. Hong RL, Spohn WH, Hung MC (1999) Clin Cancer Res 5:1884
19. Aggarwal BB, Sundaram C, Malani N, Ichikawa H (2007) Adv Exp Med Biol 595:1
20. Bharti AC, Donato N, BB Aggarwal (2003) J Immunol 171:3863
21. Xu J, Fu Y, Chen A (2003) Am J Physiol Gastrointest Liver Physiol 285:G20
22. Plummer SM, Holloway KA, Manson MM, Munks RJ, Kaptein A, Farrow S, Howells L (1999) Oncogene 18:6013
23. Chen H, Zhang ZS, Zhang YL, Zhou DY (1999) Anticancer Res 19:3675
24. Chun KS, Keum YS, Han SS, Song YS, Kim SH, Surh YJ (2003) Carcinogenesis 24:1515
25. Plummer SM, Hill KA, Festing MF, Steward WP, Gescher AJ, Sharma RA (2001) Cancer Epidemiol Biomarkers Prev 10:1295
26. Reddy S, Aggarwal BB (1994) FEBS Lett 341:19
27. Liu JY, Lin SJ, Lin JK 919930 Carcinogenesis 14:857
28. Hasmeda M, Polya GM (1996) Phytochemistry 42:599
29. Mukhopadhyay A, Banerjee S, Stafford LJ, Xia C, Liu M, Aggarwal BB (2002) C Oncogene 21:8852
30. Ohene-Abuakwa Y, Noda M, Perenyi M, Kobayashi N, Kashima K, Hattori T, Pignatelli M (2000) J Pathol 192:433
31. Iademarco MF, Barks JL, Dean DC (1995) J Clin Invest 95:264

32. Kumar A, Dhawan S, Hardegen NJ, Aggarwal BB (1998) Biochem Pharmacol 55:775
33. Gupta B, Ghosh B (1999) Int J Immunopharmacol 21:745
34. Jaiswal AS, Marlow BP, Gupta N, Narayan S (2002) Oncogene 21:8414
35. Anto RJ, Mukhopadhyay A, Denning K, Aggarwal BB (2002) Carcinogenesis 23:143
36. Shishodia S, Amin HM, Lai R, Aggarwal BB (2005) Biochem Pharmacol 70:700
37. Bharti AC, Donato N, Singh S, Aggarwal BB (2003) Blood 101:1053
38. Anuchapreeda S, Leechanachai P, Smith MM, Ambudkar SV, PN Limtrakul (2002) Biochem Pharmacol 64:573
39. Limtrakul P, Anuchapreeda S, Buddhasukh D (2004) BMC Cancer 4:13
40. Aggarwal BB, Kumar A, Bharti AC (2003) Anticancer Res 23:363
41. Kuo ML, Huang TS, Lin JK (1996) Biochim Biophys Acta 1317:95
42. Ranjan D, Johnston TD, Reddy KS, Wu G, Bondada S, Chen C (1999) J Surg Res 87:1
43. Piwocka K, Zablocki K, Wieckowski MR, Skierski J, Feiga I, Szopa J, Drela N, Wojtczak L, Sikora E (1999) Exp Cell Res 249:299
44. Han SS, Chung ST, Robertson DA, Ranjan D, Bondada S (1999) Clin Immunol 93:152
45. Chen A, Davis BH, Bissonnette M, Scaglione-Sewell B, Brasitus TA (1999) J Biol Chem 274:35505
46. Aggarwal S, Takada Y, Singh S, Myers JN, Aggarwal BB (2004) Int J Cancer 111:679
47. Mehta K, Pantazis P, McQueen T, Aggarwal BB (1997) Anticancer Drugs 8:470
48. Ramachandran C, You W (1999) Breast Cancer Res Treat 54:269
49. Simon A, Allais DP, Duroux JL, Basly JP, Durand-Fontanier S, Delage C (1998) Cancer Lett 129:111
50. Fang J, Lu J, Holmgren A (2005) J Biol Chem 280:25284
51. Kang J, Chen J, Shi Y, Jia J, Zhang Y (2005) Biochem Pharmacol 69:1205
52. Prusty BK, Das BC (2005) Int J Cancer 113:951
53. Sen S, Sharma H, Singh N (2005) Biochem Biophys Res Commun 331:1245
54. Koo JY, Kim HJ, Jung KO, Park KY (2004) J Med Food 7:117
55. Holy J (2004) Cell Motil Cytoskeleton 58:253
56. Bharti AC, Shishodia S, Reuben JM, Weber D, Alexanian R, Raj-Vadhan S, Estrov Z, Talpaz M, Aggarwal BB (2004) Blood 103:3175
57. Mahady GB, Pendland SL, Yun G, Lu ZZ (2002) Anticancer Res 22:4179
58. Chen HW, Yu SL, Chen JJ, Li HN, Lin YC, Yao PL, Chou HY, Chien CT, Chen WJ, Lee YT, Yang PC (2004) Mol Pharmacol 65:99
59. Kim JH, Shim JS, Lee SK, Kim KW, Rha SY, Chung HC, Kwon HJ (2002) Jpn J Cancer Res 93:1378
60. Kuttan R, Bhanumathy P, Nirmala K, George MC (1985) Cancer Lett 29:197
61. Busquets S, Carbo N, Almendro V, Quiles MT, Lopez-Soriano FJ, Argiles JM (2001) Cancer Lett 167:33
62. Menon LG, Kuttan R, Kuttan G (1995) Cancer Lett 95:221
63. Dorai T, Cao YC, Dorai B, Buttyan R, Katz AE (2001) Prostate 47:293
64. Kawamori T, Lubet R, Steele VE, Kelloff GJ, Kaskey RB, Rao CV, Reddy BS (1999) Cancer Res 59:597
65. Perkins S, Verschoyle RD, Hill K, Parveen I, Threadgill MD, Sharma RA, Williams ML, Steward WP, Gescher AJ (2002) Cancer Epidemiol Biomarkers Prev 11:535–540.
66. Odot J, Albert P, Carlier A, Tarpin M, Devy J, Madoulet C (2004) Int J Cancer 111:381
67. Thangapazham RL, Sharma A, Maheshwari RK (2007) Adv Exp Med Biol 595:343
68. Heng MC, Song MK, Harker J, Heng MK (2000) Br J Dermatol 143:937
69. Tourkina E, Gooz P, Oates JC, Ludwicka-Bradley A, Silver RM, Hoffman S (2004) Am J Respir Cell Mol Biol 31:28
70. Sidhu GS, Singh AK, Thaloor D, Banaudha KK, Patnaik GK, Srimal RC, Maheshwari RK (1998) Wound Repair Regen 6:167
71. Phan TT, Sun L, Bay BH, Chan SY, Lee ST (2003) J Trauma 54:1212
72. Deodhar SD, Sethi R, Srimal RC (1980) Indian J Med Res 71:632
73. Liacini A, Sylvester J, Li WQ, Zafarullah M (2002) Matrix Biol 21:251
74. Natarajan C, Bright JJ (2002) J Immunol 168:6506

75. Verbeek R, van Tol EA, van Noort JM (2005) Biochem Pharmacol 70:220
76. Yang F, Lim GP, Begum AN, Ubeda OJ, Simmons MR, Ambegaokar SS, Chen PP, Kayed R, Glabe CG, Frautschy SA, Cole GM (2005) J Biol Chem 280:5892
77. Lim GP, Chu T, Yang F, Beech W, Frautschy SA, Cole GM (2001) J Neurosci 21:8370
78. Ukil A, Maity S, Karmakar S, Datta N, Vedasiromoni JR, Das PK (2003) Br J Pharmacol 139:209
79. Salh B, Assi K, Templeman V, Parhar K, Owen D, Gomez-Munoz A, Jacobson K (2003) Am J Physiol Gastrointest Liver Physiol 285:G235
80. Egan ME, Pearson M, Weiner SA, Rajendran V, Rubin D, Glockner-Pagel J, Canny S, Du K, Lukacs GL, Caplan MJ (2004) Science 304:600
81. Berger AL, Randak CO, Ostedgaard LS, Karp PH, Vermeer DW, Welsh MJ (2005) J Biol Chem 280:5221
82. Shishodia S, Sethi G, Aggarwal BB (2005) Ann N Y Acad Sci 1056:206
83. Venkatesan P, Rao MN (2000) J Pharm Pharmacol 52:1123
84. Sreejayan N, Rao MN (1996) Arzneimittelforschung 46:169
85. Ahsan H, Parveen N, Khan NU, Hadi SM (1999) Chem Biol Interact 121:161
86. Dinkova-Kostova AT, Talalay P (1999) Carcinogenesis 20:911
87. Kawashima H, Akimoto K, Jareonkitmongkol S, Shirasaka N, Shimizu S (1996) Biosci Biotechnol Biochem 60:108
88. Robinson TP, Ehlers T, Hubbard IR, Bai X, Arbiser JL, Goldsmith DJ, Bowen JP (2003) Bioorg Med Chem Lett 13:115

Chapter 8
Proprietary Herbal Medicines in Circulatory Disorders: Hawthorn, Ginkgo, Padma 28

Jörg Melzer and Reinhard Saller

Abstract A look at the available clinical evidence of herbal preparations from hawthorn (leaves, flowers, fruits), Padma 28 (Swiss-Tibetan herbal preparation with 20 herbal drugs) and ginkgo (leaves) in terms of circulatory disorders shows the following: in chronic heart failure New York Heart Association (NYHA) II a meta-analysis showed that hydroethanolic extracts from hawthorn leaves and flowers, given at a daily dosage of 300 to 900 mg, can increase the maximum workload to up to 7 W when given concomitantly with standard therapy. The same was seen in one RCT with an extract from hawthorn fruits and flowers. The herbal preparations seem to be well tolerated and no interaction is known so far. The data on a possible decrease on blood pressure are inconclusive.
A meta-analysis on Padma 28 showed that two tablets given twice or three times a day over 16 weeks can increase the maximum walking distance by more than 100 m in patients with claudicatio intermittens. The preparation is well tolerated and safe. Research evidence from a meta-analysis on an extract from ginkgo shows that its use in the treatment of intermittent claudication can result in a significant but clinically modest improvement in pain-free walking for distances up to 34 m given at a dosage of 160 mg per day. Despite the general evidence that ginkgo preparations are relatively safe, physicians and therapists should be cautious when anticoagulants (i.e. warfarin) are given as well.

Keywords Circulatory disorder · Chronic heart failure · Peripheral arterial occlusive disease · Claudicatio intermittens · Hawthorn · Ginkgo · Padma 28

8.1 Introduction

To aid understanding of this chapter, some introductory remarks might be beneficial to help the reader appreciate the authors' point of view on herbal medicine.

J. Melzer (✉)
Institute of Complementary Medicine, Department of Internal Medicine, University Hospital Zurich, Raemistr. 100, 8091 Zurich,
e-mail: joerg.melzer@usz.ch

K.G. Ramawat (ed.), *Herbal Drugs: Ethnomedicine to Modern Medicine*,
DOI 10.1007/978-3-540-79116-4_8,

In our understanding, ethno-medicine is linked to cultures, traditions and empirical evidence. Modern medicine is guided by scientific results from experimental and clinical studies as well as – especially in recent years – by the method of evidence-based medicine (EBM) and regulatory policies. All these aspects might give an idea of the complex diversity that influences herbal medicine today and the medical system as such [1]. Our selection of the medicinal plants described in this chapter on circulatory disorders is somehow a subjective choice and might only highlight some of the above-mentioned aspects to a limited degree. Nevertheless, our choice of plants/drugs is clearly guided by the clinical point of view from the context of a policlinic setting at a university hospital in Switzerland and research evidence.

With respect to terminology we stick to current definitions: *Herbal drug*: "whole, fragmented or cut plants, parts of plants[1], algae, fungi, lichen in an unprocessed state, usually in dried form but sometimes fresh" [2–4]. *Herbal drug preparation*: "preparations obtained by subjecting [herbal drugs] to treatments such as extraction, distillation, expression, fractionation, purification, concentration or fermentation" [2–4].

The herbal preparations given in this chapter come from different countries and cultures: in the Eastern tradition, as in Tibetan medicine, herbal drugs are often only cut and ground (powdered) and pressed into tablets (e.g. Padma 28). In the Western tradition, such as in European herbal medicine, herbal drugs are often used to prepare hydroalcoholic extracts, which themselves are processed into tablets or drops, for instance hydroethanolic hawthorn extracts (e.g. Faros, Crataegisan). Each of the herbal preparations mentioned in this chapter can be called an *herbal medicinal product* (HMP) according to the European Agency for the Evaluation of Medicinal Products (EMEA) [4]. As HMPs are approved by their respective health authorities, their efficacy and quality records cannot be compared to that of supplements (synonym: nutraceuticals, food supplements), for which different standards of approval are necessary, but have a tradition in other regions of the world (e.g. North America, Asia), too.

8.2 Hawthorn

8.2.1 Plant

Hawthorn (Crataegus) is a member of the family Rosaceae (subfamily Maloideae). The plant is a thorny shrub or small tree and grows in temperate zones especially of the Northern but also Southern Hemisphere [5].

Of over 100 species regarded as genuine, two are regularly used for officinal herbal preparations in Europe: *Crataegus monogyna* – hawthorn with one style or *C. laevigata* – with two styles (synonym: *C. oxyacantha* L.). Other species are *C.*

[1] I.e.: flower (flos – in botanical terms), leaf (folium), herb (herba), fruit (fructus) root (radix, rhizome).

pentagyna – with five styles – or those well known in China or Japan, *C. cuneata* and *C. pinnatifida* [5–7]. Besides these botanical terms there exist various names in the different countries which are often used synonymously: China: Shem zha; Germany: Weissdorn; England: White thorn; France: Aubépin; Italy: Biancospino; Norway: Hagtorn; Poland: Glóg [5], [8–11].

Note that some of the different Crataegus species may be preferred traditionally in one of the various medical systems in the world and that the parts of the plant used as an herbal drug may differ as well. For example, in modern European herbal medicine most often the leaves and flowers of the two species mentioned above are used today, but traditionally their fruits are used as well [5, 12].

8.2.2 Tradition

Knowledge of the use of hawthorn in ancient times is limited. According to some authors, the plant was first mentioned in Chinese medicine in the pharmacopoeia *Tang Ben Cao* in 659 A.D [13]. Although ancient Greek writers such as Dioscorides mention Oxyakantha in the first century A.D., it remains questionable whether he was really referring to hawthorn [14, 15]. In medieval times the fruits were said to be eaten to treat digestive disorders and the roots to treat small injuries [16]. Then, in 1753 Linnaeus named hawthorn *Crataegus oxyacantha L.* and Jacquin divided the species *C. oxyacantha* and *C. monogyna* [5]. But it was only in the late 1800s that European doctors set up the first experiments on the clinical use of hawthorn for heart disease [13, 16]. Beringer is said to have suggested the use of the extract of berries as a tonic for the heart [8, 17]. In traditional medicine Crataegus is still used as a kind of heart tonic as well as heart and vascular remedy or to regulate blood pressure. Finally, it has some standing as a 'calmative agent for heart and soul' (reduction of nervous heart complaints). One can still find a variety of combinations of hawthorn with other herbal plants which derive from this tradition (e.g. calming effects with hop, passion flower or valerian). For these combination preparations an additive effect of hawthorn has also been discussed in the literature [18–20].

In Traditional Chinese Medicine (TCM) berries from Chinese hawthorn are used to treat disturbed digestion, diarrhoea, feeling of fullness, inappetence of children, arteriosclerosis or hypercholesterolaemia for example [5, 13], [21–23].

For many of these traditional uses no clinical trials are available.

8.2.3 Chemistry and Pharmacology

8.2.3.1 Compounds

The main constituents of officinal hawthorn preparations are thought to be flavonoids (flavons: e.g. hyperosid, vitexinrhamnosid; flavonols: e.g. rutin) and flavanols (e.g.

catechin, epicatechin or procyanidins which, depending on degree of polymerisation, are divided into oligomeric: n = 2 to 8 and polymeric: n > 8). Flavonoids have been found in all parts of the plant of those species examined so far. But as the pattern of flavonoids can be very different, the species can be distinguished by their main flavonoid pattern (thin-layer chromatography) [5, 9, 24]. The spectrum of flavanols, i.e. oligomeric procyanidins (OPC), is similar between the different species, yet considerable quantitative differences can be seen between species as well as parts of the plants [5].

Further constituents of hawthorn species are triterpens (e.g. ursolic acid, oleanolic acid, crataegolic acid) and phenolic acids (e.g. chlorogenic acid, caffeic acid) or amines (e.g. cholin, acetylcholin, phenylethylamin) [5, 25].

The herbal drug from flowers and leaves can contain 1 to 3% OPCs and 1 to 2 % flavonoids [5, 26]. The various herbal drugs from hawthorn can be processed into various herbal preparations, and their chemical compositions can differ greatly depending on factors such as vegetation period of the plant, extraction process and extractant (e.g. with water the OPCs can be extracted easily, whereas with alcohol the polymeric procyanidins and triterpens are more easily extracted).

The hydro-ethanolic hawthorn extracts, that is to say the herbal preparations made from them, can be standardized to a certain amount of constituents. Those preparations most widely studied in clinical trials are standardized to contain 18.75% OPCs or 2.2% flavonoids [25].

8.2.3.2 Pharmacology

To date, the mode of action of hawthorn preparations has not been fully explained, although quite a number of experimental studies [9, 13, 22, 25], [27–29] and reviews do exist [5, 13, 25, 30]. Different in vitro and in vivo models showed the following pharmacological properties of hawthorn (extracts from leaves, flowers, fruits or fractions of flavonoids or procyanidins): positive inotropic and negative bathmotropic action, increased coronary blood flow and myocardial circulation, decrease of peripheral vascular resistance and hypotensive action as well as triglyceride- and cholesterol-lowering or antioxidative activities [22, 25], [27–29], [31]. According to these studies, procyanidins and flavonoids seem to be involved in the effects of extracts from crataegus [25, 27, 32].

The mechanisms of action discussed on a molecular level are the inhibition of phosohodiesterase activity with increase of intracellular cyclic adenosine monophosphate (cAMP), inhibition of membranic Na^+/K^+-ATPase activity of the heart muscle and the angiotensin-converting enzyme, and extension of the effective refractory time [22, 25, 27, 28, 30, 33, 34].

Whether and to what extent central nervous system effects contribute to the efficacy of herbal preparations of hawthorn has not been examined. Some clinical observations indicate that the effects of herbal preparations from hawthorn might depend on the neuro-vegetative state of the patient, i.e. they might vary according to the parasymathico- or sympathicotonus [22, 28, 34].

Concerning the daily dosage, a monograph of the European Scientific Cooperative on Phytotherapy (ESCOP) mentions, among other things, 1 to 1.5 g of the herbal drug as tea infusion 3 to 4 times or 160 to 900 mg of hydro-alcoholic extracts with a drug-to-extract ratio (DER) of 4 to 7:1 with a defined content of OPCs and flavonoids [25].

8.2.3.3 Pharmacokinetic Properties

Human studies on resorption, distribution, metabolism and elimination of herbal preparations of hawthorn or their constituents like OPCs or flavonoids are not available [13, 25, 32, 35].

Experimental studies on animals indicate that some constituents of hawthorn widely distribute into the organism including the brain [36].

Information on the influence of age and diseases (e.g. kidney or liver disorders) do not exist. Yet therapeutic evidence does not show a limitation for the use of hawthorn preparations for the elderly or multimorbid patient [16, 18, 37].

One pharmacokinetic trial examined the possible interactions between hawthorn and a synthetic drug. In that study, the concomitant administration 0.25 mg digoxin and 900 mg hawthorn extract in healthy volunteers did not show a significant difference in any measured pharmacokinetic parameter when compared to the administration of digoxin alone [38].

8.2.4 Clinical Evidence

In the 20th century hawthorn preparations were used as a kind of heart tonic in European herbal medicine – especially in German-speaking countries. Over the last two decades quite a number of clinical trials on hawthorn extracts have become available in the treatment of cardiovascular diseases. Due to scientific and economic interests the focus in clinical research has shifted mainly towards the examination of extracts from the flowers and leaves of hawthorn (Table 8.1) for the indication of chronic or so-called congestive heart failure (CHF).

Due to the differing quality of clinical trials according to changing scientific and regulatory standards in recent decades, it is worth pointing out that when we do not mention if an analysis has been performed based on the intention to treat (ITT) or per protocol (PP), then this information was lacking in the publication.

8.2.5 Hawthorn Leaves and Flowers in CHF

An overview of the various clinical trials using standardized hydro-alcoholic hawthorn extracts to treat patients with chronic heart failure reveals several changes

Table 8.1 Herbal preparations of hawthorn used in the studies summarised in the meta-analysis and the systematic review

Medication	Drug-extract ratio	Extract	Daily dosage	Compound thought to be responsible for efficacy
Cardiplant 450	4–6.6:1	Ethanol 45% -dry-	1–2 tablets (450 to 900 mg dry extract)	18.75% procyanidins (78–90.6 mg OPC) -standardised-
Faros 300	4–7:1	Methanol 70% -dry-	3 × 1 coated tablets (900 mg dry extract)	2,2 % flavonoids -standardised-
Crataegisan	1:3.2	Ethanol 49% -liquid-	3 × 30 drops (approx. 2.25 ml)	6.4 mg procyanidins
				12.7 mg total of phenols

in the past two decades: a gradual rise in the stage of heart failure (from I to III according NYHA) and the intervention itself, that is to say in the daily dosage (from 160 to 1800 mg) as well as the treatment period (from 3 to 24 weeks and a randomized controlled trial – RCT over 2 years – not yet published).

Some of these clinical trials have been analysed in a meta-analysis which facilitates a close look at the evidence from randomized controlled trials (RCTs) [39, 40]. In that publication, from the quite high number of 26 potentially relevant studies, 13 could be included because they met the inclusion criteria (e.g. RCT, chronic heart failure, monopreparations). From the remaining 13 trials the data of 5 were not sufficient for pooling for meta-analytic evaluation. However, 8 trials could be analysed regarding efficacy but had to be divided into two groups according to their primary endpoint. In 4 trials [41–44] this was a change in maximal workload measured in watts (W) using bicycle ergometry. Here the pooled data of 310 patients with CHF NYHA I-III showed a significant increase of 7 W ([95 % CI 3–11]; $p < 0.01$) under verum (dosage: 180 to 1800 mg daily, duration: 3 to 16 weeks) compared to placebo. In four further trials [45–48] the primary outcome was a different target: the pressure-heart rate product which was also measured in one [41] of the previous four trials, bringing the total number of trials to five. Here the pooled data of 264 patients with CHF NYHA I-II showed a significant reduction of minus 20 mm Hg/min ([95 % CI −32 to −8]; p-value not given) under verum (dosage: 160 to 240 mg daily, duration: 6 to 12 weeks). The third primary outcome evaluated in the meta-analysis was the exercise tolerance in 98 patients in two of the trials already mentioned above [41, 43], but the result was not significant.

One must take into account that the study results are mostly based on a concomitant treatment of patients with hawthorn preparations in addition to other medications to treat CHF (e.g. diuretics, calcium antagonists, ACE inhibitors). Reflecting this, the results show the additional value of hawthorn along with standard treatment.

In what follows, we discuss trials which were not evaluated in the meta-analysis. Yet some of the older studies will not be discussed here due to an unclear definition of patient groups [49–53] or herbal preparation [54].

Iwamoto et al. 1981 conducted a controlled double-blind study in 102 patients with heart disease of ischaemic and/or hypertensive origin (mostly NYHA Stages II and partly III) to evaluate the efficacy of Crataegutt vs. placebo (3×2 dragées of an extract from leaves, flowers and fruits or placebo over 6 weeks). The analysis of 80 patients (14 excluded for non-adherence to the test plan, 8 dropped out) showed a general improvement and improvement in cardiac function under verum vs. placebo ($p < 0.01$) as well as improvement in subjective symptoms ($p < 0.001$; for individual subjective symptoms verum vs. placebo: dyspnoea, palpitation $p < 0.01$, cardiac oedemas $p < 0.05$). No group differences were noted for ECG. Side effects were observed in one patient under verum [55].

Eichstädt showed in patients with NYHA II, after 4 weeks of treatment with an extract from leaves and flowers (3×2 dragée), a significant (one-sided test) increase of the left ventricular ejection fraction from 40.2% to 43.5% at rest and 41.5% to 46.6% under exercise (radionuclide-ventriculography), a slight reduction in blood pressure at rest and reduction in the systolic pressure under exercise (188 mm Hg to 177 mm Hg); the frequency remained unchanged [56].

Schmidt et al. analysed 78 patients with NYHA II in a RCT (daily dosage: 600 mg hawthorn extract, duration: 8 weeks) [57]. After the 1-week wash-out phase cardial medication (e.g. heart glycosides, ß-blocker, ACE inhibitor) was not allowed except for diuretics if the dosage was stable for the 4 weeks prior to the study. The analysis showed a significant increase in the primary parameter, maximal workload on bicycle ergometry, of 12.5 W ($p < 0.001$) vs. placebo at the end of the trial. The improvements in the secondary parameters like RR, heart rate, and subjective symptom reduction were significant, too.

Tauchert et al. performed the only available direct comparison between hawthorn and an ACE inhibitor. In a RCT with 132 patients suffering from CHF NYHA II, they showed that the treatment over 8 weeks with a daily dosage of either 900 mg hawthorn extract (leaves and flowers) or 37.5 mg Captopril led to a significant improvement in the maximum workload in each group (hawthorn: 14; Captopril: 16 W). Yet, the group difference was not significant, suggesting a comparable efficacy of both interventions [58].

Until recently no long-term data have been available for the treatment of patients with CHF with hawthorn extracts. Yet in 1998 the SPICE (Survival and Prognosis Investigation of Crataegus Extract) trial started. Aside from the study protocol [59], only some preliminary results, presented at the meeting of the American College of Cardiology, are available. After 2 years the treatment of 2681 patients with NYHA II to III with daily 900 mg hawthorn extract concomitantly with standard medication showed no benefit over placebo in the primary outcome (composite of cardiovascular mortality such as sudden cardiac death, heart failure death, fatal myocardial infarction) [37]. Yet there seems to be evidence for efficacy in a subgroup of patients. However, the concomitant use of hawthorn with standard therapy was regarded as safe.

Hawthorn preparations are mentioned in connection with elevated blood pressure (RR). To date, the clinical evidence on that topic remains inconclusive: Tauchert et al. examined RR as one of the secondary parameters in the above-mentioned comparison trial [58]. This target did not show a significant reduction in the hawthorn group but under Captopril.

Schmidt et al. found a significant reduction in RR, which was a secondary parameter in their above-mentioned RCT [57].

The pilot study of Walker et al. ($n = 36$; daily dosage: either 500 mg hawthorn extract or 600 mg Mg or a combination of both or placebo; duration: 10 weeks) showed a reduction in RR which was insignificant [60]. Nevertheless the authors regarded the results as promising and suggested to perform a randomised trial.

Yet, in that RCT Walker et al. [61] examined a different hawthorn extract than the one they have analysed in the pilot study. 79 hypertensive patients with type 2 diabetes (70% under hypotensive treatment) were treated (daily dosis: 1200 mg extract; duration: 16 weeks). The study showed a small but significant decrease in diastolic blood pressure in the hawthorn group vs. placebo (−2.6 mmHg; $p = 0.035$).

8.2.6 Hawthorn Berries in CHF

Although clinical research has focused on hawthorn preparations from leaves and flowers, some trials on berries are available and have been summarized within a systematic review [14]. This review showed that three clinical trials examined pharmacological aspects referring to orthostatic dysregulation and two were on chronic heart failure NYHA II, but only one of them with a monopreparation. This RCT evaluated the efficacy and safety of a hydroethanolic extract of hawthorn berries ($n = 143$; 30 drops 3 times daily for 8 weeks). The efficacy analysis in the ITT population showed a significant improvement in the exercise tolerance of 8.3 W in the verum group vs. placebo ([95% CI −16.3 to −0.3]; $p = 0.045$), which was confirmed in the PP population [62]. It is noteworthy that in this trial standard medication to treat chronic heart failure was not allowed. The safety analysis revealed that only mild to moderate adverse events occurred and were statistically insignificant between verum ($n = 9$) and placebo ($n = 11$). They ranged from gastrointestinal to musculoskeletal, respiratory and psychiatric events and were unlikely to be related to the medication.

Even in the herbal preparations from hawthorn which have been examined in RCTs (not to mention the variety of other preparations freely available on the market), the question of phyto-equivalence seems to be an unsolved one.

8.2.7 Safety

Concerning safety, the authors of the meta-analysis conclude that dizziness and vertigo ($n = 8$) were the most common ones, but that five trials did not report adverse events [40]. However, the adverse event profile of Crataegus species has been evaluated separately in a systematic review taking into account 24 clinical trials (e.g. RCT, observational, cohort study). This analysis of the data of 5577 patients indicates that hawthorn preparations are generally well tolerated: 8 serious adverse events not related to hawthorn and 166 adverse events which cover the above-mentioned spectrum (e.g. gastrointestinal complaints, dizziness/vertigo) [63].

Some other safety aspects might be worth mentioning: up-to-date effects on responsiveness (ability to drive, use of machines) are not known for hawthorn preparations [25]. No serious toxic effects are known for Crataegus and its preparations neither in the short nor in the long term use [16, 23, 25, 64, 65]. Nevertheless, the hypersensibility to Crataegus or one of its compounds should be regarded as a contraindication. Finally, no human data are available on whether or how hawthorn crosses over into the placenta or breast milk or on its use during pregnancy and lactation [25]. In animal models a reduction of tonus and motility of the uterus was seen with extracts from hawthorn [64, 65].

8.2.8 Summary

Although a variety of RCTs, a meta-analysis and a systematic review contribute to a broad basis of clinical evidence concerning the use of hawthorn in CHF, it is difficult to translate all this knowledge into practice due to the wide range of daily dosage and treatment duration. However, based on the above-mentioned research evidence and our clinical experience, some conclusions can be drawn. In the case of CHF up to NYHA II a daily dosage of 300 to 900 mg seems to be appropriate. Although most evidence is available on hydroethanolic hawthorn extracts from leaves and flowers, one RCT shows a comparable efficacy for an extract from berries. As hawthorn extracts have been used together with standard medication in the treatment of CHF, in most of the studies its additive efficacy remains the one with quite robust data, although one trial indicates that monotherapy might be possible as well and, one would like to add, at least in the early stages of CHF treatment. The clinical trials seem to show that it might be necessary to treat CHF over a period of 3 to 8 weeks to be able to draw a conclusion on its efficacy. However, the herbal preparations tested so far can be given safely together with standard medication in the treatment of CHF even in the long run – up to 2 years. Yet, even in the herbal preparations from hawthorn which have been examined in RCTs (not to mention the variety of other preparations freely available on the market), the question of phyto-equivalence seems to be an unsolved one (Table 8.1).

When discussing herbal medicine, one should take into account that a treatment decision is largely based on the same procedure as with conventional medicine such as anamnesis and diagnosis. But before talking about treatment options the patient's view on the disease and his or her subjective expectations need to be considered in order to respect them as far as possible in an individual treatment schedule. This includes discussing the pros and cons of a monotherapy with an herbal preparation or a possible combination with conventional drugs in terms of the monotherapy's possible benefits and limitations.

8.3 Padma 28

8.3.1 Plants

In the European medical tradition, an herbal preparation often consists of one or two herbal drugs or preparations made out of them (i.e. mostly extracts). Nevertheless, fixed combinations are known which can contain preparations of up to five to nine herbal drugs and have shown their clinical efficacy [66, 67]. In other medical traditions like Tibetan Medicine or TCM, combination preparations with up to 15 or 20 herbal drugs are used. Padma 28 (synonym: Padmed Circosan) is such a fixed combination of Tibetan origin and contains 20 herbal drugs, a mineral and an essential oil (Table 8.2). On the one hand it would exceed the space limitations of this chapter to give an overview of all the herbal drugs of this preparation. On the other hand the daily dosages of the single herbal drugs combined in Padma 28 are far below the recommended effective daily dosage if each herbal drug would be used alone (Table 8.2). Thus one comes to realize that combination preparations like Padma 28 are somewhat like an own pharmacological entity, that is to say as a formula acting as a single agent.

8.3.2 Tradition

In the 8th century Yuthog Yongten Gonpo compiled the four main standard texts, known as *gso rig rgyud bzhi* (medical tantras), which still form the basis of the Tibetan medicinal system called Sowa Rigpa. In later centuries Tibetan Medicine spread to China, Mongolia and Russia, where the Siberian physician Sultim Badma founded what was probably the first Tibetan pharmacy and practice in Europe [68]. In the 1960s one of his descendants, Wlodzimierz Badmajeff, ran a pharmacy and practice in Poland from where a collection of Tibetan formulas came to Switzerland. This included Padma 28, which has been used in circulatory problems and as an anti-inflammatory agent [69].

The preparation was registered in Switzerland for the first time in 1977 for symptomatic treatment of circulatory disorders. The product information approved by the Swiss Agency for Therapeutic Products confirmed the efficacy of Padma 28

Table 8.2 Herbal drugs and compounds of Padma 28

Herbal drug (HD)	Medicinal plant	HD per table in Padma 28	Daily dose (DD)	DD of HD recommended (e.g. in monographs if used single)	% of HDs in Padma 28 compared to recommended DD
Aconiti tuber	Aconitum napellus L.	1 mg	6 mg	10–15 mg	40–60%
Aegle sepiar fructus	Aegle marmelos L.	20 mg	120 mg	3000–5000 mg	2.4–4%
Amomi fructus	Pimenta officinalis Lindl.	25 mg	150 mg	1000–3000 mg	
Aquilegiae vulgaris herba	Aquilegia vulgaris L.	15 mg	90 mg	Not known	
Calcii sulfas ad usum chirurgicum	Calcii sulfas pulvis	20 mg	120 mg	Not known	
Calendulae flos	Calendula officinalis L.	5 mg	30 mg	1000–3000 mg (tea)	1–3%
D-camphora	from Cinnamomum camphora L.	4 mg	24 mg	100–200 mg	12–24%
Cardamoni fructus	Elattaria cardamomum L.	30 mg	180 mg	1000–3000 mg	6–8%
Caryophylli flos	Syzygium aromaticum L.	12 mg	72 mg	1000–3000 mg	2.4–7.2%
Costi amari radix	Saussurea lappa Decne.	40 mg	240 mg	250–1250 mg	19.2–96%
Hedychii rhizoma	Hedychium spicatum Buch.	10 mg	60 mg	1000–3000 mg	2–6%
Lactucae sativae folium	Lactuca sativa var. capitata L.	6 mg	36 mg	Not known	
Lichen islandicus	Cetraria islandica L.	40 mg	240 mg	3000–6000 mg	4–8%
Liquiritiae radix	Glycyrrhiza glabra L.	15 mg	90 mg	3000–5000 mg	1.8–3%
Meliae Tousend fruct.	Azadirachta indica Juss.	35 mg	210 mg	2000–4000 mg	5.25–10.5%
Myrobalani fructus	Terminalia chebula Retz.	30 mg	180 mg	3000–6000 mg	3–6%
Plantaginis herba	Plantago lanceolata L.	15 mg	90 mg	5000–7000 mg	1.3–1.8%
Polygoni avicul. herba	Polygonum aviculare L.	15 mg	90 mg	4500–7500 mg (tea)	1.2–2%
Potentillae aureae herba	Potentilla aurea L.	15 mg	90 mg	Not known	
Santali rubri lignum	Pterocarpus santalinus L.	30 mg	180 mg	3000–6000 mg	3–6%
Sidae cordifol. herba	Sida cordifolia L.	10 mg	60 mg	1000–8000 mg	0.75–6%
Valerianae radix	Valeriana officinalis L.	10 mg	60 mg	2000–3000 mg (tea)	2–3%

in symptoms of peripheral arterial occlusive disease (PAOD) (Fontaine stage II) in 2002 [70], and the preparation is produced today in compliance with the guidelines of Good Manufacturing Practice.

8.3.3 Chemistry and Pharmacology

8.3.3.1 Compounds

The active constituents in Padma 28 can be divided into several therapeutically important groups: essential oils with the main terpene components, e.g. D-camphor (in higher doses a circulatory stimulant) and eugenol (antibacterial and local analgesic) as well as flavonoids (anti-inflammatory, antioxidative, metal chelating), polysaccharides (e.g. immune-modulatory), saponines (e.g. anti-inflammatory, antihistaminergic) and tannins (locally anti-inflammatory, antioxidative) [71, 72].

8.3.3.2 Pharmacology

An overview of the effects of Padma 28 in pharmacological studies is given in Table 8.3. Note the following effects: in vitro data consistently show an inhibition of respiratory bursts or production of reactive oxygen species (ROS), protection against oxidants and against pro-inflammatory agonists. Ex vivo data demonstrate several effects on leukocytes compatible with an inhibition of the generation of free radicals and a facilitation of thrombolysis.

A review summarises the in vitro and ex vivo models on those direct and indirect anti-inflammatory effects of Padma 28 which might play a role in the complex process of atherosclerosis [72].

The latest study regarding anti-inflammatory mechanisms on human aortic endothelial cells shows that an aequeous solution of Padma 28 completely prevented the expression of the cell adhesion molecule E-selectin, which was induced by C-reactive protein. Additionally, the vascular protective enzyme H0-1 was upregulated [90].

Brunner–La Rocca et al. examined the effect of Padma 28 on lipids in 60 otherwise healthy participants with mild hypercholesterolaemia in a mixed clinical and in vitro trial [91]. Whereas after a 4-week intervention (two tablets three times daily) vs. placebo the blood lipids (total, HDL or LDL-cholesterol, triglycerides, apo-lipprotein A1 and B) did not change, the *in vitro* evaluation showed a dose-dependent reduction in blood lipid oxidisability in the verum group which persisted 1 week after cessation of the trial.

To date, no further data are available on the clinical pharmacokinetic or pharmacodynamic properties of Padma 28. Interactions are not yet known [70].

Table 8.3 Summary of pharmacological data on Tibetan herbal preparation Padma 28

In vitro and ex vivo (human and animal data)
- Decrease in plasminogen activator inhibitor PAI-1, decrease in the activation (oxidative burst) ex vivo of monocytes with opsonised zymosan [73]
- Ex vivo: granulocytes show increased migratory activity and phagocytic activity; lymphocytes show decrease in B-cell and an increase in activated T-cells [74]
- Inhibition of NEM-stimulated platelet MDA production [75, 76]
- Inhibition of ADP and arachidonate-induced platelet aggregation [77]
- Inhibition of respiratory bursts, inhibition of intralipid peroxidation, protection against oxidants and pro-inflammatory agonists [78]
- Radical scavenging, iron-chelating, and anti- and prooxidative properties in biochemical systems [79]
- Inhibition of generation of free radicals in human neutrophils [78]
- Inhibition of inducible nitric oxide synthesis [80]
- Moderate inhibition of O_2^- production and lysozyme release by neutrophils [81]
- Antibacterial against Gram-positive bacteria but not against most Gram-negative bacteria and yeast tested [82]
- Ex vivo: minor inhibition of platelet aggregation (patients) [83]
- Ex vivo: minor inhibition of platelet aggregation (rabbit) [84]
In vivo (acute and short-term trials)
- Antinociceptive action (intraperitoneal, mice) [85]
In vivo (long-term trials)
- Inhibition of high-fat diet induced changes in MDA, plasma, platelets, total cholesterol, triglycerides, beta-lipoproteins (rats) [86, 87]
- Inhibition of high-fat diet induced plaque formation and anatomic and lipidemic changes (rabbit) [88, 89]

8.3.4 Clinical Evidence

A systematic review and meta-analysis summarise the evidence from clinical trials performed with Padma 28. They revealed that 19 clinical studies are available with regard to this herbal preparation [92]. While most studies examined the efficacy in patients with PAOD, that is to say claudicatio intermittens, individual studies were performed in other diseases (e.g. angina pectoris, hepatitis B, respiratory infections) but are not discussed here.

From the studies on PAOD (stage II according to Fontain) seven controlled trials were included in the systematic review. Six of them examined the same primary outcome (i.e. maximum walking distance), and one examined ankle pressure. The authors of the meta-analysis received the raw data of the single trials and therefore were able to perform a thorough re-analysis on an ITT basis. The re-analysis revealed non-comparability in the baseline data in one trial. Finally, the raw data of five trials could be pooled for meta-analysis thanks to the homogeneity of the data (272 patients: 155 verum, 117 placebo). The general analysis showed that the treatment with two tablets twice or three times a day for 16 weeks resulted in a significant prolongation of the maximal walking distance compared to placebo. To understand the clinical relevance of this finding, the authors chose the criterion for prolonging the maximum walking distance of more than 100 m. This criterion was significant as well and fulfilled by 18.2% of the patients under verum and by only 2.1% under placebo ($P < 0.001$; odds ratio: 10 [95% CI 3.03, 33.33]; number needed to treat: 6.2; RR: 0.12) [92]. In addition, as a secondary result a small but significant reduction in the blood levels of triglycerides and blood sugar as well as in systolic and diastolic blood pressure was observed in the patients of the verum group compared to placebo.

As regards safety, not only were the data of the 5 trials on PAOD analysed but all the available safety data from all 19 trials reporting on adverse events were as well (555 patients: 371 verum, 184 placebo). Serious adverse events (verum 2.4%, placebo 3.3%) were not associated with the treatment. Non-serious adverse events associated with the treatment were comparable in both treatment groups (verum 3.2%, placebo 2.7%; the most frequent were gastrointestinal side effects).

Data on other safety aspects are limited [70]: no studies have been done on the effects on responsiveness (ability to drive, use of machines). No risk potential was seen in experimental studies on animals concerning acute or chronic toxicity or in in vitro studies on gentoxicity. No human data are available on the use of Padma 28 during pregnancy. However, the hypersensibility to one of the compounds of Padma 28 should be regarded as a contraindication [70].

8.3.5 Summary

The current evidence shows that treatment with Padma 28 (two tablets twice or three times daily for 16 weeks) results in a statistically significant and clinically relevant increase in the maximum walking distance by more than 100 m. Following a clear improvement in symptoms, a reduction in the daily dosage is possible to two tablets a day as stated in the package insert.

According to a meta-analysis, the efficacy of Padma 28 in the symptomatic treatment of PAOD is greater than that of pentoxifylline [93] or Ginkgo [94–97] and at least comparable with naftidrofuryl [93]. The product is well tolerated and safe.

In view of the experimental effects and clinical data, the use of the complex herbal preparation can be regarded as an example of a multitarget [98] approach in herbal medicine. With this in mind the mostly small dosages of its herbal drugs could be seen as additional or synergetic in terms of biological effects and efficacy, while possible side effects are limited because of the small amounts of the individual herbal drugs (maybe because certain enzyme systems or receptors are only partly affected) [99]. If one considers PAOD as a kind of marker for systemic atherosclerosis [100] and if one keeps in mind that patients with PAOD often suffer from other chronic diseases (e.g. metabolic syndrome, diabetes, high blood pressure) [101], such an approach seems rewarding. Moreover it would partly be in concordance with theories in Tibetan Medicine [68] as well as latest research in pharmacology [98].

If the anti-oxidative properties of the preparation play a prominent role needs to be evaluated in robust studies.

8.4 Ginkgo

8.4.1 Plant

The species Ginkgo (*Ginkgo biloba* L.) belongs to the family Ginkgoaceae (the only one in the class of Ginkgopsida). The plant grows as a tree, reaches 30 to 40 m in height and up to 200 years of age. The species ginkgo was spread around the world in the mesozoikum (100 to 250 million years ago). *Ginkgo biloba* L. derives from the mountain forests of eastern and western China, entered Europe in the first half of the 18th century and America at the end of the century [102–105].Today the tree is planted in plantations in Japan, South Korea, South Carolina (USA), around Bordeaux (France) and in the temperate zones of New Zealand and Argentina [102].

The Ginkgo plant is diocious, which means there are female and male trees. The male pollen is transmitted by wind. Ginkgo trees first blossom after 20 to 30 years. In Europe and America they are in bloom around May. The herbal drugs used from the Ginkgo are the leaves (Ginkgo bilobae folium), which are harvested and dried between August and September. Yet, in autumn the semen (Semen Ginkgo) drop 2 to 3 weeks after the leaves fall. Traditionally they are collected and separated from the outer shell peel, which smells unpleasant due to its butter-acid or valerian-acid content, and prepared to be eaten [102], [104–107].

Although the botanical term Ginkgo is most widespread, it is worth mentioning its synonyms (e.g. *Salisburia adiantifolia* SMITH) or common names in certain countries: China: Bai guo; Germany: Fächerblattbaum; England: Maidenhair tree; France: Arbre aux quarante ecus; Japan: Gin kyo [104].

8.4.2 Tradition

Clues on the medical use of ginkgo have been traced back to 2800 B.C. [102, 108]. In medical books its use is mentioned in connection with lung, stomach, circulation or skin ailments or in states of nervousness and as a kind of elixir in Hindu medicine [102, 109]. In TCM, similar conditions are mentioned for the use of Semen Ginkgo (synonym: Gingko nuts) such as cough, asthmatic states, bladder problems or in nervousness [108].

The cooked nuts are used for digestive or nutritional reasons in Asia, and the roasted nuts are known as a delicatessen under names like "pa-kewo" or "bai-guo" [102, 108, 110].

The ginkgo tree or its leaves are used in the arts worldwide and sometimes are linked to cultural activities and rites (e.g. coloured nuts are eaten during certain feasts or used during weddings [108].

8.4.3 Chemistry and Pharmacology

8.4.3.1 Compounds

Characteristic constituents of Ginkgo leaves are flavonoids such as flavonols (e.g. glycosides of quercetin, kaempferol, isorhamnetin), flavons (e.g. luteolin), biflavons (e.g. bilobetin, ginkgetin), flavanols (e.g. catechin, epicatechin, procyanidins) and terpens (e.g. the diterpene trilactones ginkgolides A, B, C, J).

Most research has been done with two Ginkgo extracts (i.e. EGb 761 and Li1370) from Germany. They are in accordance with the standards mentioned in the German Pharmacopeia: DER 35–67:1, flavonoids 22 to 27% (i.e. flavonglycosides), terpene lactons 5 to 7 % (i.e. 2.8 to 3.4% ginkgolide A, B and C and 2.6 to 3.2 % bilobalids, ginkgol acid $\leq$ 5 ppm [104, 111].

The extract is made from dried ginkgo leaves with an acetone and water mixture, and undesirable compounds are eliminated (e.g. ginkgol acid).

8.4.3.2 Pharmacology

In numerous experimental models, ginkgo extracts have shown free radical-scavenging activities, relaxation of vascular smooth muscle via a nitric oxide pathway, protection of ischaemia-reperfusion damage (shown in animal models as well as in a human study), a reduction in platelet aggregation by inhibiting platelet activating factor (PAF) and the formation of platelet thromboxane A2 [104, 111, 112]. The latest research has shown a reduction in atherosclerotic nanoplaque formation after a 2-month treatment regimen with a ginkgo extract in a pilot study on eight patients following aortocoronary bypass surgery [113].

8.4.3.3 Clinical Evidence

Herbal preparations of *Ginkgo bilobae folium* are among the more extensively researched in herbal medicine. To date, results from numerous trials in dementia or PAOD, that is to say intermittent claudication, have been evaluated in several reviews and meta-analyses. In what follows the focus will be on intermittent claudication:

a. Dementia: A review of studies of ginkgo extracts vs. cholinesterase inhibitors showed that the two had similar response rates [114]. A review of older meta-analyses showed a significant benefit on cognition for cholinesterase inhibitors but for ginkgo preparations only when all dosages (greater and less than 200 mg/d) were taken into account [115]. The latest meta-analysis confirmed significant results for ginkgo vs. placebo in clinical global improvement at a dosage of > 200 mg/d after 24 weeks, for cognition after 12 but not 24 weeks and for activities of daily life after 12 and 24 weeks. These results were interpreted as inconsistent and unconvincing [116].

b. Vertigo: A systematic review of RCTs stated the efficacy of a standardised ginkgo preparation (120 to 240 mg over 1 to 4 months) in vestibular and non-vestibular vertigo [117].
c. Tinnitus: A meta-analysis of RCTs failed to show a significant difference when treating patients suffering from tinnitus with ginkgo preparations (30–200 mg over 2 to 56 weeks) compared to placebo treatment [118].
d. Intermittent claudication: Several reviews [94, 95, 119, 120] and meta-analyses [96, 97] demonstrate the efficacy and tolerability of preparations of ginkgo compared to placebo. However, the latest meta-analysis provides a more thorough evaluation:

For this meta-analysis 12 RCTs met the inclusion criteria (i.e. double-blind, placebo-controlled) but 3 had to be excluded because they did not assess walking distance and 1 was a double publication [96]. The published data of the remaining 8 trials were pooled for statistical evaluation (patients: $n = 415$ entered, $n = 385$ analysed). Pain-free walking distance was the criterion to be assessed as it was the primary outcome in four trials but was also measured in the other four trials. The pooled data showed a significant increase in pain-free walking distance of 34 m [95% CI 26, 43] after treatment with ginkgo extracts (daily dosage: 120–160 mg extract; duration: 6–24 weeks) compared to placebo This result was confirmed in a subanalysis of six trials with high internal quality (i.e. 4–5 point in the Jadad score): 37 m [95% CI 26, 47]. Next to pain-free walking distance the maximum walking distance was reported in seven trials and achieved significant improvement in six of them (ranging from 36 to 189 m), whereas this criterion deteriorated in one trial. Yet no further data are available as the authors did not pool these results. This result was confirmed in a recent review [97].

Concerning safety, the authors summarise that five trials reported adverse effects under Ginkgo with abdominal complaints, nausea and dyspepsia being most frequent complaints. They conclude that *Ginkgo biloba* extracts seem relatively safe.

To gain further information on safety, a look at the data from the meta-analysis on the effect of ginkgo preparations on cognitive impairment and dementia might be useful, although the indications are not comparable [116]. Here the authors conclude that ginkgo appears to be safe with no excess side effects compared to placebo. A closer look at the data shows no significant difference between the number of serious (not mentioned) or non-serious adverse events (e.g. nausea/vomiting, diarrhoea, hypertensive crisis) between verum and placebo.

8.4.3.4 Summary

The research evidence from clinical trials with standardised herbal preparations from leaves of *Ginkgo biloba* shows that their use in the treatment of intermittent claudication can result in a significant but clinically modest improvement in the pain-free walking distance. A direct comparison of ginkgo preparations with Padma 28 cannot be made as data on the maximum walking distance are not yet available. However, ginkgo extracts can be regarded as a treatment option especially if Padma cannot be given.

The antioxidant properties of herbal preparations from ginkgo, as well as their positive effects on blood lipids, make them a sensible component in a complex treatment setting in atherosclerosis and can – according to recent studies – be given concomitantly with statines. Our clinical evidence suggests that a daily dosage of at least 160 mg standardised extract seems appropriate.

Despite the general evidence that ginkgo preparations are relatively safe, physicians and therapists should be cautious when anticoagulants (i.e. warfarin) are given as well. Although this might be possible in individual patients following rigorous counselling and careful consideration, it cannot be advised generally. This might be different when acetylsalicylic acid (ASS) is used instead.

References

1. Saller R, Melzer J (2004) Forsch Komplementarmed Klass Naturheilkd 11:264
2. Europarat: Europäische Pharmakopöe (2005) Stuttgart, Deutscher Apotheker Verlag, vol 2
3. EMEA (2004) Recording of core-data for Herbal substances / Herbal preparations, HMPWP, vol 2007
4. EMEA, HMPC, CHMP, CVMP (2007) Guideline on declaration of herbal substances and herbal preparations in Herbal Medicinal Products/Traditional Herbal Medicinal Products in the SPC, vol 2007
5. Bauer I, Hölscher U, Juretzek W, Spiess E, Schulz V (2006) Crataegus. In: Hänsel R, Keller K, Rimpler H, Schneider G (eds) Hagers Handbuch der Pharmazeutischen Praxis: CD-ROM. Springer, Berlin Heidelberg New York
6. DAB10
7. Weiss R (1985) Lehrbuch der Phytotherapie 206:214
8. Madaus G (1979) Crataegus; Lehrbuch der biologischen Heilmittel. Olms, Hildesheim, New York, vol 2, p 1113
9. Schüssler M, Hölzl J (1992) Dtsch Apoth Ztg 132:1327
10. Wichtl M (2002) Weißdornblätter mit Blüten. In: Wichtl M (ed) Teedrogen und Phytophone. Wissenschaftliche Verlagsgesellschaft, Stuttgart, p 152
11. WHO (1999) Folium cum Flore Crataegi. In: WHO (ed) WHO Monographs on Selected Medicinal Plants. WHO, Geneva, vol 2, p 66
12. Paulus E, Yu-he D (1987) Handbuch der traditionellen chinesischen Heilpflanzen. K-F Haug, Heidelberg
13. Chang Q, Zuo Z, Harrison F, Chow MS (2002) J Clin Pharmacol 42:605
14. Melzer J, Iten F, Saller R (2003) Perfusion 16:358
15. Bruneton J (2002) Coeur: "palpitations". In: Phytothérapie. Les données de l'evaluation. Edition medicales internationales, Paris, p 71
16. Hobbs C, Foster S (1990) Herbal Gram 22:20

17. Lenz W, Arends G (eds) (1919) Hagers Handbuch der Pharmazeutischen Praxis. Springer, Berlin, New York, Supplement, p 240
18. Meier B (1989) Z Phytother 10:182
19. Hanus M, Lafon J, Mathieu M (2004) Curr Med Res Opin 20:63
20. Bourin M, Bougerol T, Guitton B, Broutin E (1997) Fundament Clin Pharmacol 11:127
21. Leung A, Foster S (1996) Encyclopedia of Common Natural Ingredients Used in Food, Drugs and Cosmetics, 2 edn. Wiley-Interscience, New York
22. Siegel G, Casper U, Walter A, Hetzer R (1994) Münch med Wschr 136:47
23. Duke J (1986) Crataegus Oxyacantha L, CRC Handbook of Medicinal Herbs. CRC, Boca Raton, FL, p146
24. Petereit F, Nahrstedt A (2005) Pharm Unserer Zeit 34:22
25. European Scientific Cooperative on Phytotherapy (ESCOP) (2003) Crataegi folium cum flore. In: ESCOP (ed) ESCOP Monographs: The Scientific Foundation for Herbal Medicinal Products. Thieme, Stuttgart, New York, p 98
26. van Wyk B-E, Wink C, Wink M (2003) Handbuch der Arzneipflanzen: ein illustrierter Leitfaden, wissenschaftl. Verlagsgesellschaft, Stuttgart
27. Ammon H, Kaul R (1994) Dtsch Apoth Ztg 134:2631
28. Siegel G, Casper U (1995) Crataegi folium cum flore. In: Loew D, Rietbrock N (eds) Phytopharmaka in Forschung und klinischer Anwendung. Steinkopf, Darmstadt, p 1
29. Tauchert M, Loew D (1995) Crataegi folium cum flore bei Herzinsuffizienz. In: Loew D, Rietbrock N (eds) Phytopharmaka in Forschung und klinischer Anwendung. Steinkopf, Darmstadt, p 137
30. Brixius K, Willms S, Schwinger RH (2005) Pharm Unserer Zeit 34:48
31. Ernst F, Reuther G, Walper A (1994) Münch med Wschr 136:57
32. Ammon H, Händel M (1981) Planta Med 43:313
33. Höltje H (1994) Münch med Wschr 136:61
34. Schüssler A, Hölzl J, Fricke U (1995) Arzneim – Forsch/Drug Res 45:842
35. Rigelsky JM, Sweet BV (2002) Am J Health Syst Pharm 59:417
36. Hecker-Niediek A (1983) Dissertation, University of Marburg
37. Cleland JG, Abdellah AT, Khaleva O, Coletta AP, Clark AL (2007) Eur J Heart Fail 9:1070
38. Tankanow R, Tamer HR, Streetman DS, Smith SG, Welton JL, Annesley T, Aaronson KD, Bleske BE (2003) J Clin Pharmacol 43:637
39. Pittler MH, Schmidt K, Ernst E (2003) Am J Med 114:665
40. Pittler MH, Schmidt K, Ernst E (2005) Pharm Unserer Zeit 34:66
41. Zapfe G (2001) Phytomedicine 8:262
42. Bödigheimer K, Chase D (1994) Münch med Wschr 136:S7
43. Hanak T, Brückel MH (1983) Therapiewoche 33:4331
44. Tauchert M (2002) Am Heart J 143:910
45. O'Connolly M, Bernhöft G, Bartsch G (1987) Therapiewoche 37:3587
46. O'ConnollyM (1986) Fortschr Med 104:805
47. Weikl A, Assmus K, Neukum-Schmidt A, Schmitz J, Zapfe Gj, Noh H, Siegrist J (1996) Fortschritte der Medizin 114:291
48. Leuchtgens H (1993) Fortschr Med 111:352
49. Böhlau V, Böhlau E (1968) Beiheft Fortschritte der Therapie 15 (Fortschr. Med.)15(86):1
50. Pozenel H (1986) Z Allgemeinmed 62:526
51. Kümmell H, Schreiber K, Koenen J (1982) Herzmedizin 5:157
52. Koppermann E (1956) Arztl Forsch 10:I/585
53. Förster A, Förster K, Bühring M, Wolfstädter H (1994) Münch med Wschr 136:21
54. Bercker R (1984) Medwelt 35:1567
55. Iwamoto M, Sato T, Ishizaki T (1981) Planta Med 42:1
56. Eichstädt H (1989) Therapiewoche 39:3288
57. Schmidt U, Kuhn U, Ploch M, Hübner W (1994) Münch Med Wschr 136:13
58. Tauchert M, Ploch P, Hübner W-D (1994) Munch Med Wochenschr 136 (Suppl 1):S27
59. Holubarsch CJ, Colucci WS, Meinertz T, Gaus W, Tendera M (2000) Eur J Heart Fail 2:431
60. Walker AF, Marakis G, Morris AP, Robinson PA (2002) Phytother Res 16:48

61. Walker AF, Marakis G, Simpson E, Hope JL, Robinson PA, Hassanein M, Simpson HC (2006) Br J Gen Pract 56:437
62. Degenring FH, Suter A, Weber M, Saller R (2003) Phytomedicine10:363
63. Daniele C, Mazzanti G, Pittler MH, Ernst E (2006) Drug Saf 29:523
64. Ammon H, Händel M (1981) Planta Med 43:105
65. Ammon H, Händel M (1981) Planta Med 43:209
66. Melzer J, Rosch W, Reichling J, Brignoli R, Saller R (2004) Aliment Pharmacol Ther 20:1279
67. Melzer J, Saller R, Schapowal A, Brignoli R (2006) Forsch Komplement Med 13:78
68. Schwabl H, Geistlich S, McHugh E (2006) Forsch Komplement Med 13 (Suppl 1):1
69. Rgya mtsho S (1994) Tibetan Medical & Astro Institute, Dharamsala, India
70. Arzneimittelkompendium der Schweiz: Padma 28®, 2008
71. Saller R, Kristof O, Reichling J (1999) Schweizerische Zeitschrift für Ganzheitsmedizin11:28
72. Ueberall F, Fuchs D, Vennos C (2006) Forsch Komplement Med 13 (Suppl 1):7
73. Winther K, Kharazmi A, Himmelstrupp H, Drabaek H, Mehlsen J (1994) Fibrinolysis 8:47
74. Schleicher P (1990) Therapy of PAOD with PADMA28. Study II – Influence of PADMA28 on the immune system – lymphocytic subpopulations, PMNs' adhesion, motility and phagocytic activity, Zytognost, Munich
75. Altermatt R, von Felten A (1992) Schweizerische Zeitschrift für Ganzheitsmedizin 4(Suppl 1):7
76. Altermatt R (1991) Dissertation, University of Zurich, Switzerland
77. von Felten A (1994) Dissertation, Coagulation laboratory, University Hospital, Zurich
78. Ginsburg I, Sadovnik M, Sallon S, Milo-Goldzweig I, Mechoulam R, Breuer A, Gibbs D, Varani J, Roberts S, Cleator E, Singh N (1999) Inflammopharmacology 7:47
79. Suter M, Richter C (2000) Redox Rep 5:17
80. Moeslinger T, Friedl R, Volf I, Brunner M, Koller E, Spieckermann PG (2000) Can J Physiol Pharmacol 78:861
81. Matzner Y, Sallon S (1995) J Clin Lab Immunol 46:13
82. Weseler A, Saller R, Reichling J (2002) Forsch Komplementarmed Klass Naturheilkd 9:346
83. Smulski HS (1991) Ann Acad Med Stetin 37:191
84. Perren S (1975) Schweizerisches Forschungsinstitut
85. Jagiello-Wojtowicz E, Kleinrok Z, Szponar J, Chodkowska A, Brzosko W (1998) J Pharmacology 50:
86. Wojcicki J, Gonet B, Samochowiec L, Gnacinska K, Juzwiak S (1989) Phytother Res 2/3:54
87. Samochowiec L, Wojcicki J (1987) Herba Polonica 33:219
88. Wojcicki J, Samochowiec L, Ceglecka M, Z J, Tustanowski S, Kadlubowska D, Juzwiak S (1988) Phytother Res 2:199
89. Gieldanowski J, Dutkiewicz T, Samochowiec L, Wojcicki J (1992) Arch Immunol Ther Exp (Warsz) 40:291
90. Exner M, Raith M, Holzer G, Gmeiner B, Wagner O, Kapiotis S (2006) Forsch Komplement Med 13 (Suppl 1):13
91. Brunner-La Rocca H, Schindler R, Schlumpf M, Saller R, Suter M (2005) Vasa 34:11
92. Melzer J, Brignoli R, Diehm C, Reichling J, Do D-D, Saller R (2006) Atherosclerosis 189:39
93. Girolami B, Bernardi E, Prins MH, Ten Cate JW, Hettiarachchi R, Prandoni P, Girolami A, Buller HR (1999) Arch Intern Med 159:337
94. Linde K, ter Riet G, Hondras M, Vickers A, Saller R, Melchart D (2003) Forsch Komplementarmed Klass Naturheilkd 10 (Suppl 1):17
95. Pittler MH, Ernst E (2005) Atherosclerosis 181:1
96. Pittler MH, Ernst E (2000) Am J Med 108:276
97. Horsch S, Walther C (2004) Int J Clin Pharmacol Ther 42:63
98. Csermely P, Agoston V, Pongor S (2005) Trends Pharmacol Sci 4:178
99. Saller R (2006) Forsch Komplement Med 13:VII
100. Faxon DP, Creager MA, Smith SC, Jr, Pasternak RC, Olin JW, Bettmann MA, Criqui MH, Milani RV, Loscalzo J, Kaufman JA, Jones DW, Pearce WH (2004) Circulation 109:2595

101. Diehm C, Schuster A, Allenberg JR, Darius H, Haberl R, Lange S, Pittrow D, von Stritzky B, Tepohl G, Trampisch HJ (2004) Atherosclerosis 172:95
102. Corrigan D (1995) Ginkgo biloba – Ancient medicine, 2 edn. Amberwood, East Horsley
103. Gäbler H (1988) Ginkgo biloba – Rückblick nach vorn, Dr. Willmar Schwabe (Selbstverlag), Karlsruhe
104. Spieß E, Juretzek W, Schulz V (2006) Ginkgo. In: Hänsel R, Keller K, Rimpler H, Schneider G (eds) Hagers Handbuch der Drogen und Arzneistoffe. Hager ROM, Springer, Berlin Heidelberg New York.
105. WHO (1999) Folium Ginkgo. In WHO (ed) WHO Monographs on Selected Medicinal Plants. WHO, Geneva, vol 1, p 154
106. DeFeudis F (1991) Ginkgo-biloba-Extrakt. Pharmacological Activities and Clinical Applications. Elsevier, Amsterdam
107. Kemper F, Schmid-Schönbein H (1991) Rökan Ginkgo biloba EGb 761, Pharmakologie. Springer, Berlin, New York, vol 1
108. Schmid M, Schmoll-Eisenwerth H (1994) Ginkgo. Urbaum und Arzneipflanze – Mythos, Dichtung und Kunst. Wissenschaftliche Verlagsgesellschaft, Stuttgart
109. Paulus E, Yu-he D (1987) Handbuch der traditionellen chinesischen Heilpflanzen, KF Haug, Heidelberg
110. Saller R, Hellenbrecht D (1993) Ginkgobaum. In: Saller R, Feiereis H (eds) Beiträge zur Phytotherapie. Marseille Verlag, München, vol 1, p 155
111. European Scientific Cooperativeon Phytotherapy (ESCOP) (2003) Ginkgo folium, In: ESCOP (ed): ESCOP Monographs: The scientific foundation for Herbal Medicinal Products, Thieme, Stuttgart, New York, p 178
112. Dugoua JJ, Mills E, Perri D, Koren G (2006) Can J Clin Pharmacol 13:e277
113. Siegel G, Schafer P, Winkler K, Malmsten M (2007) Wien Med Wochenschr 157:288
114. Wettstein A (2000) Phytomedicine 6:393
115. Kurz A, Van Baelen B (2004) Dement Geriatr Cogn Disord 18:217
116. Birks J, Grimley Evans J (2007) Cochrane Database Syst Rev CD003120.
117. Hamann KF (2007) HNO 55:258
118. Rejali D, Sivakumar A, Balaji N (2004) Clin Otolaryngol Allied Sci 29:226
119. Ernst E (1996) Fortschr Med 114:85
120. Jacoby D, Mohler ER (2004) Drugs 64:1657

Chapter 9
The Effects of the Green Tea Polyphenol Epigallocatechin Gallate on the Central Nervous, Endocrine, and Innate Immune Systems

Lisa A. Beltz

Abstract The central nervous, endocrine, and immune systems interact at several levels, including cell surface and intracellular molecules as well as compounds facilitating intercellular communication. Molecules produced by components of the immune system, such as reactive oxygen and nitrogen species and some cytokines secreted by brain microglia, may induce disorders of the nervous and endocrine systems if present in excess. Pathologic conditions of these three systems, including neurodegenerative diseases, anxiety, memory loss, ischemia, multiple sclerosis, alterations in weight and metabolism, diseases of insulin dysregulation, and several autoimmune disorders, have several common themes such as changes in iron/Ca^{+2} levels, altered proteosomal activity, mitochondrial dysfunction, and activation of the caspase cascade, leading to apoptosis. The green tea polyphenol epigallocatechin gallate (EGCG) has beneficial activity in a number of human diseases. It mitigates some of the above-listed damage in part by altering intracellular signal transduction pathways, by scavenging reactive oxygen and nitrogen species and iron, and by affecting cytokine production and expression of neurotransmitters or their receptors. This chapter presents these activities of EGCG and how they affect diseases of these related physiological systems.

Keywords Epigallocatechin gallate · Neurodegenerative diseases · Autoimmune diseases · Endocrine diseases · Reactive oxygen species · Cytokines

9.1 Introduction

Polyphenolic compounds from a variety of plants commonly utilized as food or beverage sources have been the focus of great interest in both the general populace and the scientific community. They bestow many benefits to human health and

L.A. Beltz (✉)
Department of Biology, University of Northern Iowa, 901 Campbell Cr., NE Massillon, OH 44646, USA
e-mail: lisa.beltz@uni.edu

K.G. Ramawat (ed.), *Herbal Drugs: Ethnomedicine to Modern Medicine*,
DOI 10.1007/978-3-540-79116-4_9,

thus have great potential as alternative medicine sources. Green tea extract, derived from leaves of *Camelia senensis*, contains several such polyphenols, the most abundant of which is epigallocatechin gallate (EGCG). This compound has favorable activity in a wide variety of human disease states, including cancer, cardiovascular disorders, obesity, bacterial and viral infections, neurological disorders, and other more localized conditions. This chapter focuses on the effects of EGCG upon the hypothalamus-pituitary-adrenal axis, which encompasses the central nervous system (CNS), the endocrine system, and the immune system, with particular emphasis upon pathological conditions.

9.2 The Effects of EGCG on the Central Nervous System

The pathology of neurodegenerative diseases has several common, interrelated features, including excessive cytoplasmic Ca^{+2} levels, increased production of reactive oxygen species (ROS), dysfunction of the ubiquitin-proteosomal degradative system, mitochondrial dysfunction, cytochrome *c* release/initiation of the caspase cascade, and neuronal injury or death by apoptosis. Neurotransmitters or their receptors are also commonly involved. One major contributory factor is abnormal iron regulation; protective factors include protein kinase C (PKC), mitogen-activated protein kinase (MAPK), and phosphatidylinositol 3′-OH kinase (PI3K)/AKT activation [1]. The effect of EGCG on these processes will be discussed throughout the chapter (also see Table 9.1).

A number of factors are directly involved in initiating apoptosis in neurons, including cytochrome *c* release from mitochondria, caspase activation, and proapoptotic (Bad, Bax, Bid, Bim)/antiapoptotic (Bcl-2, Bcl-$_X$L) proteins. Proapoptotic proteins increase mitochondrial permeability, resulting in decreased membrane potential and cytochrome *c* release. Levels of these proteins are regulated by proteosomal degradation. Exogenous EGCG rapidly decreases intracellular Bad levels in human neuroblastoma cells via a pathway utilizing proteosomes and PKC [2]. Oral administration of EGCG increases murine hippocampal levels of PKC and induces its membrane translocation in neuroblastoma cells [1, 3]. EGCG additionally decreases Bax expression and increases the Bcl-2:Bax ratio and Bad phosphorylation by signals involving MAPK and Akt in other cells [1, 4]. MAPK may function by activating stress-related genes, including the phase II drug-metabolizing enzyme glutathione-s-transferase, known to be stimulated by EGCG [5]. Decreased neuronal apoptosis by EGCG may involve decreased expression of proapoptotic genes, including caspase-1, mdm2, p21, and TNF-related apoptosis-inducing ligand, rather than increased antiapoptotic gene activity [3].

Table 9.1 Processes involving diseases of the CNS affected by EGCG

Processes decreased by EGCG*	Processes increased by EGCG*
Lipid peroxidation (1, 2)**	Scavenging ROS (1, 2, 4, 5, 7, 8)
Xanthine oxidase activity (2)	Catalase activity (1, 4, 7, 8)
LDL oxidation (2)	Superoxide dismutase activity (1, 4, 7, 8)
Oxidative DNA damage (3)	Iron chelation (1, 4, 5, 7, 8)
nNOS and iNOS activation (4)	Iron regulatory proteins (1, 4)
Aggregation of hyperphosphorylated tau (1)	Mitochondrial complex IV activity (2)
Holo-APP activity (1)	5-Amyloid precursor protein formation (1)
β-amyloid formation (1)	α-Secretase activity (1)
Catechol-O-methyltransferase activity (4)	HIF-1 production (1)
α-synuclein (4)	Dopamine levels (4)
Glutamate receptor activity (1, 4, 5, 6, 7)	Tyrosine hydrolase activity (4)
Excessive cytoplasmic Ca^{+2} (1, 4, 5, 7, 8)	Citrate synthase activity (2)
NF-κB activation (4, 5)	$GABA_A$ receptor activity (9)
JAK/STAT activation (2, 6)	PKC activity (1)
TNF-α production (5, 6)	IκB-α levels (5)
Proteosomal activity (5)	PI3K activity (7)
Plasma cortisone levels (9)	pAKT activity (7)
	pGSK-3β activity (7)

* Processes involved in disease causation in bold; those involved in protective activity in normal font.
** Diseases affected by these processes. 1 = Alzheimer's disease, 2 = ischemia, 3 = age-associated memory loss, 4 = Parkinson's disease, 5 = multiple sclerosis, 6 = HAD, 7 = ALS, 8 = HD, 9 = anxiety

9.2.1 Alzheimer's Disease

Alzheimer's disease is an age-related cognitive disorder associated with oxidative brain damage. Those with Alzheimer's or other cognitive disorders have lower levels of antioxidants in their blood [6]. Individuals with Alzheimer's and Parkinson's diseases have decreased levels of reduced glutathione and increased levels of lipid peroxidation [3].

Accumulation of β-amyloid is associated with Alzheimer's disease development; its toxicity in cultures of hippocampal neurons is mediated via ROS, lipid peroxidation, activation of the caspase cascade, and apoptosis. When such cells were co-exposed to EGCG, incidence of the latter three events decreased in a manner independent of p53, Bax, Bcl-xL, and cyclooxygenase (COX) [7]. The cytotoxicity of amyloid proteins appears to rely largely upon the formation of well-ordered fibrillar assemblies. Polyphenols inhibit this formation independently of their antioxidant activity [8]. Iron chelation by EGCG also reduces the aggregation of a major component of neurofibrillary tangles, hyperphosphorylated tau, in the brains of these patients [9].

β-amyloid is formed after processing of amyloid precursor protein (APP) via a pathway requiring iron, which EGCG effectively chelates (see below). APP may alternatively be processed to form soluble APP, which blocks β-amyloid production;

EGCG promotes this pathway in a PKC-dependent fashion, increasing α-secretase activity [3]. EGCG also blocks the generation of β-amyloid-derived diffusible neurotoxin ligands and the associated apoptosis [10].

Malfunctioning of the glutamate neurotransmitter system is also involved in this disease [11]. β-amyloid-mediated neuronal death may involve the *N*-methyl-D-aspartate (NMDA) receptor for glutamate; receptor antagonists are clinically effective for Alzheimer's disease. EGCG effects both glutamate production and the increased cytoplasmic Ca^{+2} levels resulting from its binding to this receptor (see below).

ROS generation is partially dependent upon processes requiring intracellular iron, including the Fenton reaction. Dysregulation of cellular iron homeostasis, including uptake, distribution, transport, and storage, is a causal factor in the pathogenesis of neurodegenerative diseases, including Alzheimer's, Parkinson's, and Huntington's diseases and multiple sclerosis [12]. In the former, increased levels of iron and the transferrin receptor are found in the hippocampus and cerebral cortex, which may induce β-amyloid deposition as well as regulate APP mRNA posttranscriptionally, via iron regulatory proteins (IRP), or its translation, via an iron-responsive element. Several iron-chelating compounds are neuroprotective; however, some are toxic and others do not easily penetrate the brain. EGCG is a highly effective iron chelator – but is nontoxic and readily enters the brain [12]. It reduces levels of holo-APP in mouse hippocampus and β-amyloid in neuroblastoma cells [12]. Iron chelation by EGCG increases levels of hypoxia inducible factor-1 (HIF-1), key to regulating induction of genes that protect against the deleterious effects of hypoxia, by interfering with its degradation by an iron-dependent proteosomal pathway [12]. Degradation of both HIF-1 and IRP is triggered by prolyl hydroxylase, an enzyme responsive to high levels of O_2 and iron.

9.2.2 *Parkinson's Disease*

Parkinson's disease is a progressive disorder involving the degeneration of dopamine-producing neurons in the substantia nigra, resulting in symptoms such as resting tremor, bradykinesia, rigidity, gait disturbance, and postural instability [13]. Tyrosine hydrolase (TH) is critical to the formation of L-dopa, and thus dopamine, and decreases in its activity occur in animal models of Parkinson's disease. In a murine model of Parkinson's disease, oral administration of EGCG inhibited damage to dopaminergic neurons, preserving numbers of TH-positive cells and TH activity in the striatum and preventing loss of dopamine and its metabolites [13]. Dopamine is a substrate of catechol-O-methyltransferase. Since EGCG inhibits this enzyme, it may additionally increase dopamine levels in synapses by this mechanism [14].

Reactive nitrogen species (RNSs) generated by neuronal nitric oxide synthase (nNOS) are involved in Parkinson's disease pathogenesis since enzyme inhibition or absence of its gene promotes disease resistance [14]. EGCG inhibits activity of nNOS in the CNS and inducible NOS (iNOS) in macrophages (discussed below), perhaps by negatively regulating nuclear factor-κB (NF-κB), a transcription factor to which these genes' promoters respond [5, 13].

A number of polyphenols, including EGCG, function as antioxidants at low concentrations [5]. Some of their actions involve scavenging ROSs (discussed below) and induction of endogenous antioxidants via PKC [16]. ROS generation via iron dysregulation plays a role in Parkinson's disease. Toxic α-synuclein aggregates form after exposure to redox-active iron; both occur within Lewy bodies, characteristic of Parkinson's disease [12]. As in Alzheimer's disease, iron also regulates degradation of IRP in Parkinson's disease. Decreased amounts of IRP lead to decreased transcription of the transferrin receptor, and thus increases in the iron transport protein, ferritin. Mice lacking the IRP gene accumulate iron in their substriatia and develop symptoms of Parkinson's disease, including tremor and bradykinesia. EGCG prevents accumulation of α-synuclein and decreases removal of IRP in murine models of Parkinson's disease [1]. Iron chelation may thus contribute to EGCG's neuroprotective effects in Parkinson's disease.

Epidemiological studies support the importance of green tea and EGCG in reducing incidence of Parkinson's disease. Studies in Hong Kong found that regular green tea consumption (particularly at least 2 cups per day) correlated with decreased risk for disease development [13, 17]. Rates of Parkinson's disease are five to ten times lower in China and Japan (high consumption of green tea) than the Western world (black tea favored). Green tea contains 10- to 20-fold more EGCG than black tea, derived from the same plant.

9.2.3 Huntington's Disease

Huntington's disease is a dominantly inherited CNS disorder characterized by severe cognitive, motor, and psychiatric manifestations. A prime feature of this disease is the expansion of CAG repeats in the amino terminus of the huntingtin protein (htt), resulting in glutamine-expanded htt. Wild-type htt is involved in many cellular processes, including neurotransmission. The altered protein forms aggregates in neuronal nuclei in addition to contributing to altering synaptic function by inducing abnormal responses to stimulation via the NMDA glutamate receptor [18]. Animal models containing "knockin" glutamine-expanded htt experience loss of neurons in the striatum and cortex. In such a *Drosophila* model, increased synaptic neurotransmitter release efficiency occurred in association with elevated cytoplasmic Ca^{+2} levels. Blocking synaptic transmission or voltage-gated Ca^{+2} channels reduces neurodegeneration [18]. EGCG alters neurotransmitter levels and activity of their receptors, as well as regulates Ca^{+2} homeostasis (see below).

As stated above for Alzheimer's and Parkinson's diseases, iron dysregulation and the resultant ROS generation are also involved in neuronal injury via oxidative stress in Huntington's disease. The iron-chelating and antioxidant properties of EGCG thus may potentially be beneficial for this disease as well [12].

9.2.4 Amyotropic Lateral Sclerosis (ALS)

Excessive Ca^{+2} levels, ROS, and abnormal iron disposition are features of ALS. Elevated Ca^{+2} influx in ALS is at least partially due to glutamate binding the α-amino-3-hydroxy-5-methyl-4-isoxazolepropionic (AMPA) receptor [11]. EGCG can alter glutamate production/receptor interactions, inhibit abnormal accumulation of cytoplasmic Ca^{+2}, chelate iron, and decrease oxidative stress. It may thus be of benefit in the treatment of this disease as well as the previously mentioned neurological disorders [12].

A murine model for human ALS uses transgenic mice expressing mutated Cu/Zn-superoxide dismutase gene. These mice are considered to be symptomatic when their limbs shake while suspended in air; disease progresses to the point at which they can no longer right themselves. Mice ingesting EGCG had delayed symptom onset and prolonged lifespans [19]. Their spinal cords contained higher levels of survival signals, such as PI3K, pAKt, and pGSK-3β.

9.2.5 Ischemic Conditions/Stroke

ROS and lipid peroxidation play causal roles in neuronal injury following ischemia and reperfusion of the brain. When transient global ischemia was induced in Mongolian gerbils or Wistar rats by occlusion of the carotid arteries, hippocampal CA1 region pyramidal neurons were severely damaged, becoming pyknotic with chromosomal condensation [20, 21]. Intraperitoneal administration of EGCG immediately following ischemic insult reduced the damage and infarct size. The protective mechanism is believed to involve the ability of EGCG to inhibit xanthine oxidase and reduce the activity and protein levels of iNOS (see below). On the other hand, nNOS and endothelial NOS activities increased; the latter may have a protective effect via increased cerebral circulation. In the rat system, EGCG also maintained activities of mitochondrial complex IV and citrate synthase. Ischemia-related memory dysfunction also decreased in mice [22].

Arteriosclerosis is a major factor in ischemic cerebrovascular disease. Oxidative alterations of cholesterol-containing low-density lipoproteins (LDL) occur; these oxidized particles are subsequently ingested by macrophages, which transform into foam cells. EGCG prevents LDL oxidation [23].

In a large study of nondrinking/nonsmoking women aged ≥ 40 years in Japan, daily green tea consumption lowered risks of stroke occurrence and mortality, but not hypertension, even in those who ingested high Na^{+2} levels daily [24]. Activation of the JAK/STAT pathway may increase neuronal pathology in stroke; this pathway is inhibited by EGCG [25].

9.2.6 Multiple Sclerosis

Multiple sclerosis (MS) is an autoimmune CNS disorder characterized by axonal demyelination. T lymphocytes reactive to components of myelin basic protein proliferate and secrete cytokines underlying the neuropathology. Following demyelination, 7-ketocholesterol, a major breakdown product of myelin, activates microglia to stimulate apoptosis of brain stem motor neurons [26]. Asian populations, consuming high levels of green tea, have a lower incidence of MS [27].

Experimental autoimmune encephalomyelitis (EAE) is a common model of MS. Oral administration of EGCG lowered production of inflammatory lesions in the brain stem and spinal cord, and decreased relapse occurrence and development of severe disease in murine EAE [26]. T cells from treated mice had less proliferation, and lower levels of the inflammatory cytokine tumor necrosis factor-α (TNF-α), but not interferon-γ or interleukin-4, were associated with decreased activity of 20S/26S proteosomal complexes, intracellular accumulation of IκB-α, and decreased NF-κB activation.

Excitoneurotoxicity from glutamate stimulation of AMPA/kainate receptors, with the resultant rise in intracellular Ca^{+2} and ROS levels, contributes to neuronal death in EAE [11]. EGCG decreases this toxicity and inhibits increased Ca^{+2} and ROS levels in neurons (see below).

9.2.7 Anxiety

Anxiety may be attenuated via actions of γ-aminobutyric acid (GABA), the major inhibitory neurotransmitter. Several flavonoids, including quercetin from red wine, bind to the benzodiazepine $GABA_A$ receptor [28]. Since EGCG is structurally related to these compounds, it may also exert effects via this receptor. At low concentrations, it enhances benzodiazepine's activity on $GABA_A$-receptor-mediated currents in *Xenopus* oocytes and partially reverses the effects of β-carboline, a negative benzodiazepine modulator, in rat hippocampal neurons without enhancing glutamate-mediated spontaneous excitatory synaptic transmission [29, 30]. Both chlorodiazepoxide and EGCG induced the anxiolytic activity of mice in an elevated plus maze test [30]. EGCG reduced anxiety and was sedative for chicks experiencing social separation stress, partially by decreasing plasma corticosterone levels, perhaps via the $GABA_A$ receptor [31].

9.2.8 Memory

Memory loss during aging and brain senescence may be at least partially due to oxidative DNA damage. Impaired learning and memory appear to be causally linked to atrophy in the hippocampus and forebrain. In a murine model of aging, cerebral weight decreased with age. The rate of decrease slowed in mice ingesting green tea catechins, and oxidative DNA damage in the cerebrum was reduced [32]. Memory loss was also inhibited, but no effect was seen with regard to learning time or lifespan. Thus, tea catechins appear to primarily improve quality of life rather than its length.

Green tea polyphenols were beneficial to cognitive memory in aged mice and rats [33, 34]. Epidemiological studies in humans also found negative correlations between green tea consumption and cognitive impairment [35].

9.2.9 HIV-Associated Dementia (HAD)

Retroviral invasion of the CNS and establishment in microglia are common occurrences in HIV-infected individuals and may result in severe neuropyschiatric disease, such as HAD, which is resistant to highly active antiretroviral therapy. Survival time after this diagnosis averages 6 months. Production of interferon-γ (IFN-γ) increases during HIV infection; this cytokine enhances neuronal damage induced by viral gp120 and Tat via the JAK/STAT pathway [36]. EGCG reduced this IFN-γ-augmented damage both *in vitro* and in vivo by inhibiting JAK/STAT signaling.

Viral proteins bind NMDA glutamate receptors and can pathologically dysregulate calcium concentrations and mitochondrial functions [36]. As described below, ECGC also affects these processes. Another factor enhancing gp120 neurotoxicity is the cytokine TNF-α, whose production is altered by EGCG (see below).

9.2.10 Neuronal Activity

The addition of EGCG to primary cultures of rat medial vestibular nuclear neurons lowered their spontaneous firing rate and hyperpolarized their membrane potential. This is believed to result from alteration of the potassium currents. No effects were noted on the amplitude of the afterhyperpolarization or the action potential's spike width [37]. EGCG also depolarized membranes of myenteric neurons in small intestines of guinea pigs [38].

9.2.11 Neurotransmitters and Their Receptors

Glutamate is a major neurotransmitter involved in cognitive functions. Its release by vesicular exocytosis is regulated in part by levels of free cytoplasmic Ca^{+2}. EGCG increases this release in response to the potassium channel blocker 4-aminopyridine via an increased Ca^{+2} influx through N- and P/Q-type voltage-dependent Ca^{+2} channels in rat cerebral cortex nerve terminals [6]. This process involves phosphorylation of PKC, which subsequently phosphorylates myristoylated alanine-rich C kinase substrate (MARCKS), inducing actin cytoskeleton disassembly prior to vesicle mobilization and exocytosis. Cytochalasin D, which inhibits actin polymerization, blocked EGCG-enhancement of glutamate release. EGCG did not function in this system by depolarizing the synaptosomal plasma membrane or activating protein kinase A.

There are two basic types of glutamate receptors, NMDA and non-NMDA types. The latter type responds primarily to AMPA and kainate, and their stimulation releases the Mg^{+2} blockade of NMDA-receptor ion channels. This, in turn, allows glutamate to gate the NMDA receptors and permits Ca^{+2} influx from extracellular sources. Triggering both types of receptors normally raises the levels of free intracellular Ca^{+2}, activating calmodulin-dependent protein kinases. Excessive stimulation by either receptor type may be deadly for cells. For example, the snake-derived neurotoxin, β-bungarotoxin, binds directly to NMDA receptors in primary cultures of cerebellar granule neurons and induces excessive Ca^{+2} influx, leading to ROS production, caspase-3 activation, and apoptosis [39]. Exceedingly high Ca^{+2} levels may trigger this cascade by overloading mitochondria, inducing overproduction of ROS, or by triggering production of xanthine oxidase and superoxide via proteolysis of xanthine dehydrogenase. Ca^{+2} may also act by stimulating calcineurin and NOS. NO may then interact with superoxide to generate peroxynitrite, which induces lipid peroxidation [40]. EGCG and several other antioxidants decrease excessive Ca^{+2} release, ROS production, and lipid peroxidation in response to β-bungarotoxin or AMPA stimulation of glutamate receptors, protecting neurons from death [39, 40].

Kainate triggering of the non-NMDA glutamate receptors is involved in damage to cortex and hippocampal neurons in Alzheimer's and Parkinson's disease. This process involves iron, since its removal is protective. EGCG's iron-chelation activity thus may exert positive benefits in this manner as well [12].

GABA, an inhibitor transmitter, binds to either $GABA_A$ or $GABA_B$ receptors. The former are primarily found postsynaptically and are responsible for the majority of the inhibitory synaptic signal transmissions in the CNS. EGCG binds these receptors in vitro [29]. The ability of EGCG to reduce anxiety in acutely stressed chicks was decreased by picrotoxin, a $GABA_A$ receptor antagonist [31].

9.3 The Effects of EGCG on the Endocrine System (Table 9.2)

Table 9.2 Processes involving diseases of the endocrine system affected by EGCG

Processes decreased by EGCG*	Processes increased by EGCG*
Testosterone levels (1)**	Plasma adiponectin levels (4)
17 β-estradiol levels (2)	Fatty acid oxidation (4)
Luteinizing hormone levels (2)	Glucose uptake (4)
GDH (3)	PI3K activity (4)
Gluconeogenesis (4)	
5′-AMP-activated protein kinase activity (4)	
Ca^{+2}/calmodulin-dependent protein kinase activity (4)	
ROS levels (4)	
Insulin levels (4)	
Glucose levels (4)	
PEPCK (4)	
FOXO transcription factors (4)	
Triacylglycerol levels (5, 6)	
Cholesterol levels (5, 6)	

* Processes involved in disease causation in bold; those involved in protective activity in normal font.

** Diseases affected by these processes. 1 = Prostate cancer, 2 = breast cancer, 3 = HI/HA syndrome, 4 = diabetes, 5 = obesity, 6 = cardiovascular diseases

9.3.1 Production of Hormones and Consequent Events

EGCG decreases plasma levels of several hormones, including the sex hormones testosterone, 17β-estradiol, and luteinizing hormone, as well as leptin, insulinlike growth factor I (IGF-1), and insulin [41]. (For other such effects of EGCG on the endocrine system, see Table 9.3.) Weights of androgen-/estrogen-sensitive organs (prostate, seminal vesicles, uterus, ovaries) decreased within 7 d of intraperitoneal EGCG injection in rats. Most effects in males were not solely dependent upon lowered testosterone levels since exogenous androgens did not reverse them. Decreased levels of sex hormones may lower risk of development of prostate and breast cancer. Several Asian countries do indeed report low incidence. Additionally, low levels of EGCG increase estrogen stimulation of the estrogen receptor α, expressed on the ovaries, uterus, vagina, and mammary glands, while higher amounts are inhibitory [42].

Serum levels of glucose, lipids, triglycerides, and cholesterol decrease after EGCG injection, as do subcutaneous and abdominal fat deposits [41]. Weight decrease in EGCG-treated rats, with or without a functional leptin receptor, is partially due to reduced food intake, suggesting effects on a leptin receptor-independent appetite-control pathway. Plasma levels of other factors related to the control of food

Table 9.3 Processes Involving diseases of the innate immune system affected by EGCG

Processes decreased by EGCG*	Processes increased by EGCG*
ROS levels (1)**	Scavenging nitric oxide (2)
Iron dysregulation (1)	Scavenging peroxynitrite (2)
Nitric oxide levels (2)	IκB-α activity (2)
Macrophage iNOS activity (2)	Macrophage GPx activity (2)
IL-6 production (1)	Phosphorylation of p38 MAPK (4)
TNF-α production (1, 3, 4)	
IFN-γ production (3)	
NF-κB activation (1, 2)	
COX-2 activity (3)	

* Processes involved in disease causation in bold; those involved in protective activity in normal font.
** Diseases affected by these processes. 1 = Inflammatory reactions, 2 = cellular aging/senescence, 3 = rheumatoid arthritis, 4 = Sjogren's syndrome

intake (ACTH, neuropeptide Y, CRF, urocortin, galanin) were unaltered by EGCG. Control rats on restrictive diets demonstrated hormonal changes similar to those injected with EGCG. Oral administration was not very effective in inducing these short-term changes.

9.3.2 Diseases Involving Insulin Alterations

Insulin production/secretion occurs in β cells of the islets of Langerhans of the pancreas. This hormone regulates the entry of glucose into cells, and its secretion is regulated in part by glucose, amino acids, free intracellular Ca^{+2}, and glutamate dehydroxygenase (GDH). GDH dysregulation is associated with hyperinsulinemia/hyperammonemia (HI/HA) syndrome, a genetic disorder involving hypoglycemia. EGCG inhibits GDH in a reversible, allosteric manner at nanomolar levels independently of its antioxidant activities [43]. This only occurs under conditions in which GDH is not inhibited by high-energy metabolites. EGCG also effectively inhibits mutant forms of GDH found in HI/HA. While EGCG blocks the enhancement of glutamine oxidation by GDH by an activating compound, it does not affect basal levels of glutaminolysis or glucose-stimulated insulin secretion.

For centuries, green tea has been used as a folk remedy for diabetes and is effective in the prevention/treatment of Type I/II diabetes in rodents [44]. Type II diabetes is characterized by insulin resistance and glucose intolerance, often in older populations, and associated with obesity. EGCG consumption tends to lead to weight loss (see above). This may result from higher plasma adiponectin levels, which improves insulin resistance by elevating fatty acid oxidation, reducing gluconeogenesis, and increasing glucose uptake in muscle. Adiponectin also decreases production of the cytokines TNF-α and interleukin-6 (IL-6) in adipose tissue. EGCG treatment increases adiponectin and lowers triacylglycerol levels in obese and nonobese rodent models, and consumption of oolong tea (containing lower levels of EGCG than green tea) increases plasma adiponectin and lowers cholesterol levels in Japanese populations [45]. It also decreases gluconeogenesis in purified hepatocytes via stimulation of 5′-AMP-activated protein kinase by Ca^{+2}/calmodulin-dependent protein kinase in an ROS-dependent fashion [44].

EGCG/green tea extract reduces blood insulin and glucose levels and raises glucose metabolism by adipocytes [46]. Like insulin, it inhibits phosphoenol-pyruvate carboxykinase (PEPCK), which increases glucose production during diabetes. PEPCK expression is regulated by FOXO transcription factors, and inhibited by insulin and IGF-1 by phosphorylation via PI3K. EGCG also induces PI3K phosphorylation of FOXO1a and PEPCK inhibition but, unlike insulin/IGF-1, requires ROS [46].

9.4 Effects of EGCG on Aspects of the Innate Immune System

The innate immune system consists of white blood cell types, such as polymorphonuclear neutrophils, monocyte/macrophages (MO/MΦ), dendritic cells, natural killer cells, eosinophils, and basophils. This chapter focuses upon the effects of EGCG on MO/MΦ and the manner in which these relate to diseases of the CNS and endocrine systems. MO/MΦ are the same phagocytic cell type at different stages of maturation. MO are the immature form found in the circulation, while MΦ are an older, mature form primarily in tissues. One specialized type of MΦ is the microglia, the major inflammatory cell in the CNS. Activated microglia produce several types of inflammatory compounds, including cytokines, ROS, and RNS, which contribute to CNS damage in neurodegenerative diseases. Compounds such as EGCG that decrease inappropriate stimulation of these cells may have significant neuroprotective effects. For a summary of EGCG-induced changes to MΦ/MΦ functions, see Table 9.3).

9.4.1 Cytokine Production

Cytokines play important roles in the immune system, serving as messenger molecules for leukocyte communication, as signals increasing or decreasing immune activity, or by directly killing microbes or malignant cells. One major group of cytokines has inflammatory activity and includes MO/MΦ products IL-6 and TNF-α. The latter is an "endogenous pyrogen," since it acts upon the hypothalamus to induce fever. It is alternatively named "cachexin" for its role in inducing wasting. While inflammation is beneficial in eliminating microbial threats, it can be destructive to normal host cells and result in severe pathology, shock, and/or death. Iron, ROS, and nuclear translation of NF-κB aid in the production of inflammatory cytokines. NF-κB activation in dopaminergic neurons of Parkinson's patients is inhibited by EGCG [3].

As described above, deletion of dopaminergic neurons in the substantia nigra is important to the development of Parkinson's disease. These neurons are particularly vulnerable to the deleterious effects of inflammatory products from activated microglia, including TNF-α. Compounds stimulating microglia activity, such as lipopolysaccharride (LPS) from cell walls of Gram-negative bacteria, selectively kill these neurons in mice, inducing pathology similar to that of humans with Parkinson's disease, while compounds inhibiting microglia activation protect dopaminergic neurons [47]. Microglia isolated from rat brains and exposed to LPS shifted their morphology from a "resting" unipolar/bipolar form to an "activated" round form with amoeboid cell bodies. EGCG partially blocked this morphological shift [47]. A conditioned medium from LPS-stimulated microglia contained high levels of TNF-α and nitric oxide (NO), and its addition to neuroblastoma cells reduced their viability, and selectively depleted TH^+ cells in rat mesencephalon. EGCG reduced production of TNF-α and NO, blocked transcription of TNF-α mRNA, and protected neuron viability [47].

Together with TNF-α, inflammatory mediators such as COX-2 and IFN-γ cause various autoimmune diseases. Ingestion of green tea polyphenols reduced arthritic symptoms (cartilage/bone erosion) in an animal model of rheumatoid arthritis, accompanied by decreased levels of COX-2, TNF-α, and IFN-γ, and lowered cellular infiltration of the joints [48]. In this system, the majority of the TNF-α-producing cells were MΦ, which can be stimulated by IFN-γ. TNF-α is also toxic to salivary acinar cells in Sjogren's syndrome, another autoimmune condition, decreasing their secretory activity. EGCG protects these cells from TNF-α-mediated destruction in a manner involving phosphorylation of p38 MAPK [49].

9.4.2 The Role of Reactive Oxygen Species

ROS include molecules such as superoxide, H_2O_2, the hydroxyl radical, singlet oxygen, and hypochloric acid, and play many important physiological roles. They aid in the destruction of microbes and tumor cells, but may also alter and damage several components of normal cells, including membrane proteins/lipids and DNA, and may result in cancer induction. Many ROS effects depend upon the levels of the free radicals. EGCG, like other "antioxidants," decreases ROS levels at low concentrations and increases them at high concentrations [5].

Superoxide functions as a key ROS. It is converted into H_2O_2 by superoxide dismutase within cells. H_2O_2 is degraded into water and O_2 by catalase. EGCG scavenges superoxide and elevates the activity of both protective superoxide dismutase and catalase in neurons [23, 50]. All phagocytic cells, including brain microglia, produce and release ROS. These cells also contain protective enzymes to inactivate the free radicals, including glutathione peroxidase (GPx), which is expressed at higher levels in microglia than neurons. During aging in mice, activity levels of

GPx, but not catalase, declined. When mice ingested green tea catechins, however, their GPx activity was similar to that of much younger animals [51]. Amounts of GPx protein itself were not significantly decreased by aging or affected by tea consumption. Levels of NO and nNOS do, however, increase in aged mice, and this RNS has been found to inhibit GPx activity. As discussed below, EGCG blocks NO generation.

Much of the pathology in the CNS disorders described earlier in this chapter involves ROS. At low concentrations, EGCG decreases ROS production [5]. It does so in neurons even in the presence of buthionine sulfoximine, an inhibitor of glutathione synthase [26].

9.4.3 The Role of Reactive Nitrogen Species

As described above, LPS-activated microglia secrete the inflammatory compound, NO. EGCG decreases NO secretion and induction of iNOS protein in MΦ [52]. Since the iNOS promoter contains a NF-κB-binding region, EGCG may act by its ability to inhibit nuclear translocation of NF-κB via inhibition of the proteosomal degradation of IκB. It also scavenges NO [5].

Peroxynitrite is produced by the interaction of NO with superoxide. It is an oxidizing/nitrating species that induces lipid peroxidation, modifies amino acids, causes DNA strand breakage/oxidation, and stimulates COX-2 activity. Peroxynitrite increases Comet scores, indicative of DNA damage. EGCG decreases this damage and effectively scavenges peroxynitrite [5].

9.5 Conclusions

Much current interest is focused on interactions between the central nervous, endocrine, and immune systems. The latter is vital in protecting against microbial invasion and tumor growth; however, overstimulation or inappropriate targeting of immune effecter molecules can be pathological. Such potentially pathogenic molecules include ROS, RNS, and several cytokines, many of which are produced by MΦ, including microglia of the brain. Various CNS diseases involve the formation of pathogenic aggregates, abnormal functioning of neurotransmitters or their receptors, altered iron or Ca^{+2} homeostasis, impaired proteosomal activity, and mitochondrial dysfunction, resulting in cytochrome *c* release, activation of the caspase cascade, and apoptosis. EGCG affects all of these processes at least partially by altering the activity or production of key enzymes, via elements of intracellular signal transduction pathways such as PI3K, MAPK, and NF-κB, or by scavenging ROS, RNS, or iron.

EGCG is also beneficial for treating endocrine and autoimmune disorders. This polyphenol alters the production of a number of hormones, including those involved in cancers of the reproductive system, those regulating metabolism and weight, and those affecting disorders related to insulin. The immune system and its cytokines are also involved in the latter, particularly diabetes mellitus. EGCG also effects production of cytokines, including those which cause inflammatory responses, in ways that are only recently being explored.

In summary, EGCG makes multiple contributions to human health using a variety of mechanisms and via multiple intracellular pathways. It is only one of a number of polyphenols present in plant-derived alternative medicinal materials. These compounds and the foods/beverages/plant extracts that contain them promise to continue to help mankind to overcome diseases and improve the quality of life well into the future.

References

1. Mandel SA, Avramovich-Tirosh Y, Reznichenko L, Zheng H, Weinreb O, Amit T, Youdim MBH (2005) Neurosignals 14:46–60
2. Kalfon L, Youdim, MBH, Mandel SA (2007) J Neurochem 100:992
3. Weinreb O, Mandel S, Amit T, Youdim MBH (2004) J Nutr Biochem 15:506
4. Chung JH, Han JH, Hwang EJ, Seo JY, Cho KH, Kim KH, Youn JI, Eun HC (2003) FASEB J 17:1913
5. Beltz LA, Bayer DK, Moss AL, Simet IM (2006) Anti-Cancer Agents Medicin Chem 6:389
6. Chou C-W, Huang W-J, Tien L-T, Wang S-J (2007) Synapse 61:889
7. Choi YT, Jung CH, Lee SR, Bae JH, Baek WK, Suh MH, Park J, Park CW, Suh SI (2001) Life Sci 70:603
8. Porat Y, Abramowitz A, Gazit E (2006) Chem Biol Drug Des 67:27
9. Yamamoto A, Shin RW, Hasegawa K, Naiki H, Sato H, Yoshimatsu F, Kitamoto T (2002) J Neurochem 82:1137
10. Bastianetto S, Yao Z-X, Papadopoulos V, Quirion R (2006) Eur J Neurosci 23:55
11. Mizuno T, Zhang G, Takeuchi H, Kawanokuchi J, Wang J, Sonobe Y, Jin S, Takada N, Komatsu Y, Suzumura A (2008) FASEB J 22:online 1/15/08
12. Mandel S, Amit T, Reznichenko L, Weinreb O, Youdim MBH (2006) Mol Nutr Food Res 50:229
13. Choi J-Y, Park C-S, Kim D-J, Cho M-H, Jin B-K, Pie J-E, Chung W-G (2002) Neuro Toxicol 23:367
14. Lu H, Meng X, Yang CS (2003) Drug Metabol Disposal 31:572
15. Hantraye P, Brouillet E, Ferrante R, Palfi S, Dolan R, Matthews RT, Beal MF (1996) Nat Med 2:1017
16. Maher P (2001) J Neurosci 21:2929
17. Chan DKY, Woo J, Ho SC, Pang CP, Law LK, Ng PW, Hung WT, Kwok T, Hui E, Orr K, Leung MF, Kay R (1998) J Neurol Neurosurg Psyc 65:781
18. Romero E, Cha G-H, Versteken P, Ly CV, Hughes RE, Bellen HJ, Botas J (2008) Neuron 57:27
19. Koh S-H, Lee SM, Kim HY, Lee K-Y, Lee YJ, Kim H-T, Kim J, Kim M-H, Hwang MS, Song C, Yang K-W, Lee KW, Kim SH, Kim O-H (2006) Neurosci Lett 395:103
20. Lee S-R, Suh S-I, Kim S-P (2000) Neurosci Lett 287:191
21. Sutherland BA, Shaw OM, Clarkson AN, Jackson DM, Sammut IA, Appleton I (2005) FASEB J 19:258
22. Matsuoka Y, Hasegawa H, Okuda S, Muraki T, Uruno T, Kubota K (1995) J Pharmacol Exp Ther 274:602

23. Kakuda T (2002) Biol Pharmacol Bull 25:1513
24. Sato Y, Nakatsuka H, Watanake T, Hisamichi S, Shimizu H, Fujisaku S, Ichinowatari Y, Ida Y, Suda S, Kato K, Ikeda M (1989) Tohoku J Exp Med 157:337
25. Townsend PA, Scarabelli TM, Pasini E, Gitti G, Menegazzi M, Suzuki H, Knight RA, Latchman DS, Stephanou A (2004) FASEB J 18:1621
26. Aktas O, Prozorovski T, Smorodchenko A, Savaskan NE, Lauster R, Kloetzel P-M, Infante-Duarte C, Brocke S, Zipp F (2004) J Immunol 173:5794
27. Kira, J-I (2003) Lancet Neurol 2:117
28. Medina JH, Viola H, Wolfman C, Marder M, Wasowski C, Calvo D, Paladini AC (1997) Neurochem Res 22:419
29. Campbell EL, Chebib M, Johnston GA (2004) Biochem Pharmacol 68:1631
30. Vignes M, Maurice T, Lanté F, Nedjar M, Thethi K, Guiramand J, Récasens M (2006) Brain Res 1110:102
31. Adachi N, Tomonaga S, Tachibana T, Denbow DM, Furuse M (2006) Eur J Pharmacol 531:171
32. Unno K, Takabayashi F, Yoshida H, Choba D, Fukutomi R, Kikunaga N, Kishido T, Oku, N, Hoshino M (2007) Biogerontology 8:89
33. Unno K, Takabayashi F, Kishido T, Oku N (2004) Exp Gerontol 39:1027
34. Haque AM, Hashimoto M, Karakura M, Tanabe Y, Hara Y, Shido O (2006) J Nutr 136:1043
35. Kuriyama S, Hozawa A, Ohmori K, Shimazu T, Matsui T, Ebihara S, Awata S, Hagatomi R, Arai H, Tsuji I (2006) Am J Clin Nutr 83:355
36. Giunta B, Obregon D, Hou H, Zeng J, Sun N, Nikolic V, Ehrhart J, Shytle D, Fernandez, F, Tan J (2006) Brain Res 1123:216
37. Jeong H-S, Kim Y-S, Park J-S (2005) Brain Res 1047:267
38. Homma T, Hirai K, Hara Y, Katayama Y (2001) Neurosci Lett 309:93
39. Tseng W-P, Lin-Shiau S-Y (2003) Neurochem Int 42:333
40. Bae JH, Mun KC, Park WK, Lee S-R, Suh S-I, Baek WK, Yim M-B, Kwon TK, Song D-K (2002) Biochem Biophy Res Comm 290:1506
41. Kao Y-H, Hiipakka R A, Liao S (2000) Endocrinol 141:980
42. Kuruto-Niwa R, Inoue S, Ogawa S, Muramatsu M, Nozawa R (2000) J Agric Food Chem 48:6355
43. Li C, Allen A, Kwagh J, Doliba NM, Qin W, Najah H, Collins HW, Matschinsky FM, Stanley CA, Smith TJ (2006) J Biol Chem 281:10214
44. Collins QF, Liu H-Y, Pi J, Liu Z, Quon MJ, Cao W (2007) J Biol Chem 282:30143.
45. Shimada M, Mochizuki K, Sakurai N, Goda T (2007) Biosci Biotechnol Biochem 71:2079
46. Anton S, Melville L, Rena G (2007) Cellular Signal 19:378
47. Li R, Huang Y-G, Fang D, Le W-D (2004) J Neurosci Res 78:723
48. Haqqi TM, Anthony DD, Gupta S, Ahmad N, Lee M-S, Kumar GK, Mukhtar H (1999) Proc Natl Acad Sci USA 96:4524
49. Hsu SD, Dickinson DP, Qin H, Borke J, Ogbureke KUE, Winger JN, Camba AM, Bollag WB, Stöppler HJ, Sharawy MM, Schuster GS (2007) Autoimmunity 40:138
50. Levites Y, Weinreb O, Maor G, Youdim MB, Mandel S (2001) J Neurochem 78:1073
51. Kishido T, Unno K, Yoshida H, Choba D, Fukutomi R, Asahina S, Iguchi K, Oku N, Hoshino M (2007) Biogerontology 8:423
52. Lin YL, Yin JK (1997) Mol Pharmacol 52:465

Chapter 10
Natural Products in Cancer Chemoprevention and Chemotherapy

K.G. Ramawat and Shaily Goyal

Abstract Medicinal plants are an important source of diverse chemical compounds that have been used for the past several centuries in the treatment of cancer. About 25% of drugs in the modern pharmacopoeia are derived from plants, including several anticancer drugs currently in clinical use such as vincristine, vinblastine, paclitaxel, podophyllotoxin, camptothecin and combretastatin. These natural products, their derivatives and analogues based on these drugs constitute an arsenal against various types of neoplasms. The traditional use of plants provides a lead for cancer chemopreventive molecules. The development of new derivatives from bioactive compounds of food origin has been a viable way to reduce toxicity and increase their effectiveness against cancer. The combined efforts of botanists, pharmacologists, chemists and biologists are required to discover new effective drugs to fight cancer. An evaluation of the mode of action of these bioactive molecules will be helpful in designing novel drugs targeting mitosis. This article discusses natural products currently in clinical use, and under clinical trials, for cancer chemotherapy and chemoprevention.

Keywords Betulinic acid · Chemoprevention · Genistein · Podophyllotoxin · Resveratrol · Taxol · Vincristine

10.1 Introduction

About 12.5% of the 422,000 plant species of higher plants are known as medicinal plants and constitute a principal source of bioactive molecules. As compared to this, a very low proportion (0.1 to 5%) of microorganisms has been explored for the production of secondary metabolites. These figures may increase as many plants, and most microorganisms, have not yet been screened for their biological properties [1]. The proportion of medicinal plants to the total documented species

K.G. Ramawat (✉)
Laboratory of Biomolecular Technology, Department of Botany,
M. L. Sukhadia University, Udaipur-313001, India
e-mail: kg_ramawat@yahoo.com

K.G. Ramawat (ed.), *Herbal Drugs: Ethnomedicine to Modern Medicine*,
DOI 10.1007/978-3-540-79116-4_10,

in different countries varies from 4.4 to 20% [2]. About 25% of drugs in the modern pharmacopoeia are derived from plants, and many others are synthetic analogues built on prototype compounds isolated from plants. Up to 60% of prescribed drugs in the Western world contain plant products or their derivatives [3]. In India, 1100 species are recognised as a source of raw material for Ayurvedic and Unani formulations [4]. Medicinal plants and their extracts have significant export value in some developing countries like India and China. Commerce of medicinal plants involves trading herbs, their extracts, and value-added products [5–7]. The purified active principle of several medicinal plants provides some of the most potent medicines used against several types of neoplasms, e.g. vincristine, paclitaxel [8–10]. The majority of these are complex natural products, and the chemical synthesis of chiral molecules is not economically feasible [11].

Drug discovery from medicinal plants has evolved with our civilisation. Herbs have been used to treat various types of tumours and other disorders for thousands of years [7]. The scientific basis for such use is not known to herbalists, but scientific validation has been done for many such drugs in recent years [12]. In the past few decades, significant advances in experimental methodology and molecular biology have enabled researchers to investigate the potential use of natural secondary products to treat or manage a plethora of chronic diseases, including various types of cancers [13]. Cancer represents one of the most severe health problems worldwide, and the search for more effective anticancer agents continues. Cancer alone caused over six million deaths in 2000, affecting over ten million people worldwide [14]. Cancer is the second leading cause of death in the United States, after cardiovascular diseases [15]. Carcinogenesis involves a complex interplay between genes and the environment, and multiple cumulative genetic changes are required for the transformation of normal cells into fully malignant cells; changes in several fundamental cell physiological characteristics take place during malignancy [16]. Essentially, self-sufficiency in growth signals, insensitivity to growth inhibitory (antigrowth) signals, evasion of programmed cell death (apoptosis), limitless replicative potential, sustained angiogenesis, and tissue invasion and metastasis are the main characteristics of tumourigenesis [17]. Disturbances in complex molecular events within cells involves growth-factor-signalling pathways (e.g. platelet-derived growth factor, epidermal growth factor and SOS-Ras-Raf-MAPK-cascade), transcription factors (e.g. NF-κB and AP-1), apoptotic signalling proteins (e.g. caspases, polyadenosine-5′-diphosphate ribose polymerase, Bcl-2 and Bcl-X1), protein kinases (e.g. mitogen-activated protein kinases), cell cycle proteins (e.g. cyclins, cyclin-dependent kinases and retinoblastoma proteins) and cell adhesion molecules (e.g. integrins, selections, cadherins, intercellular adhesion molecule and vascular adhesion molecule). However, despite the large amount of information on signals involved in tumourigenesis, effective drugs to control or prevent it are not available [17].

Secondary anticancer metabolites and their derivatives finding their way to clinical trials are mostly natural products of plant origin [18–20]. Generally natural products are screened against cancer to find a new weapon to fight this scourge. During the first phase (1960–1982) of evaluation of plant extracts against cancer, mainly leukaemia, the anticancerous property of taxol was established by the National Cancer Institute (NCI) in the USA. In the second phase, NCI began a new programme in 1985 to evaluate extracts prepared from microorganisms, plants and marine products against an array of 60 different types of cancer cell lines, including those from solid tumours and leukaemia [21].

10.2 Cancer Chemoprevention

Chemoprevention and functional foods are new emerging areas in the prevention of diseases. Chemoprevention is defined as the use of synthetic or natural agents, alone or in combination, to block the development of cancer in human beings. Plants, herbs, vegetables and spices, used in folk and traditional medicine, have been accepted as one of the main sources of cancer chemopreventive drug discovery and development (Fig. 10.1) [22]. Carcinogenesis is a multistage process by which a normal cell is transformed into a cancerous cell. The process of transformation involves initiation by DNA damaging agents, promotion of cell proliferation, and progression involving additional genetic alterations. Chemopreventive molecules target each of these steps including anti-initiation strategies (e.g. DNA repair, detoxification, free-radical scavenging and carcinogen metabolism) and antipromotional/antiprogression strategies (e.g. free-radical scavenging, proliferation suppression, differentiation induction, immunity enhancement, inflammation reduction, increase in apoptosis, altered gene expression and decrease in angiogenesis) [20, 23]. No toxicity is expected from a chemopreventive agent because of its long history of human consumption as herbal medicines, botanical dietary supplements or edible plants [20, 24, 25]. Bioactive components of dietary phytochemicals with chemopreventive properties include curcumin, genistein, resveratrol, diallyl sulphide, S-allyl cysteine, allicin, lycopene, capsaicin, diosgenin, (6)-gingerol, ellagic acid, ursolic acid, silymarin, anethol, catechins, eugenol, isoeugenol, isothiocyanates, indol-3-carbinol, isoflavones, phytoestrols, folic acids, b-carotene and flavonoids [26].

Several plant-derived compounds are in clinical trials as potential chemopreventive agents for various types of cancers, including curcumin (phase I colon), genistein (phase I breast and endometrial), soy isoflavonoids (phase II prostrate), indole-3-carbinol (phase I breast recurrence), perillyl alcohol (phase I breast), various forms of retinoic acid, phenethyl isothiocyanate (phase I lung), green tea/epigallocatechin gallate (phase II bladder and breast), and resveratrol (phase I and II) [20, 27, 28]. Details about these bioactive molecules are given in this book and elsewhere [29].

Dietary bioactive food components that interact with the immune response have considerable potential to reduce the risk of cancer. The reduction of chronic inflammation or its downstream consequences may represent a key mechanism whose effects can be reduced by targeting signal transduction or through antioxidant effects. Some of the most important immunomodulators are phytochemicals such as the polyphenols, epigallocatechin gallate (EGCG) and curcumin [30, 31]. Oxidative

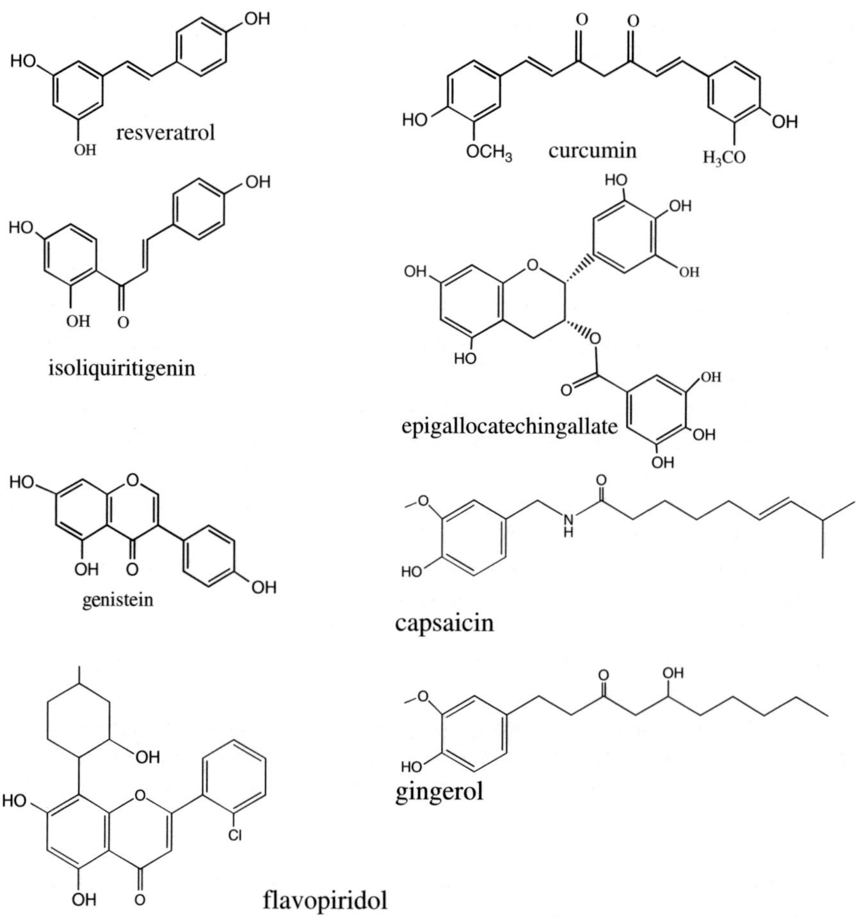

Fig. 10.1 Cancer chemopreventive compounds

stress and associated mechanisms involving inflammation, aberrant signalling pathways, and gap junction intercellular communications is increasingly associated with the pathogenesis of various chronic degenerative disorders such as atherosclerosis, neurodegeneration and cancer [32]. The consumption of fresh fruits, vegetables and teas has the capacity to reduce the risk of cancer.

Polyphenols are the most abundant antioxidants in the diet. Despite their wide distribution in plants, the beneficial health effects of dietary polyphenols have been recognised rather recently. Until the mid-1990s, the most widely recognised antioxidants were vitamins, carotenoids and minerals. Research on flavonoids and other polyphenols for their biological activity, including antioxidant properties and their chemopreventive effects, appeared after 1995 [33, 34]. One of the major difficulties in establishing a correlation between polyphenols and their health benefits is the presence of large numbers of polyphenolics in food [35] having different biological activities [36]. Major differences in the bioavailability of phenolics have, by now been well established and the influence of structural factors is better understood [37].

Resveratrol (3,5,4-trihydroxy trans-stilbene), a polyphenol, is present in black grapes and its preparations like red wine [38, 39]. The ethyl-acetate-soluble extract was found to inhibit cyclooxygenase-1 (COX-1) enzyme (88% inhibition at 69 μg/ml). Resveratrol has been isolated from several species of the families Vitaceae, Dipterocarpaceae, Gnetaceae and Leguminosae. Apart from its cardioprotective effects, resveratrol exhibits anticancer properties as demonstrated by its ability to suppress proliferation of a wide variety of neoplasms, including lymphoid and myeloid cancers, multiple myeloma, and cancers of the breast, prostate, colon, stomach and pancreas [40]. The growth inhibiting effects of resveratrol are mediated through cell cycle arrest [41]. Resveratrol is in phase I clinical trials for colon cancer (www.clinicaltrials.gov/ct/show/nct00256334) and against AIDS (www.ihv.org/clinical_trials/thera_vacc2.html). Details about the chemistry and production of resveratrol in *Vitis* cell cultures [39], epidemiological studies correlating consumption of red wine containing polyphenols like resveratrol and cardiovascular diseases [42, 43], and biological activities including anticancer properties of resveratrol are discussed elsewhere [44–46] and in this book.

Tea polyphenols are widely consumed as non-alcoholic beverages. Some of the nutrients identified as chemopreventive agents in prostate cancer are green tea polyphenols. EGCG, derived from *Camellia sinensis*, has been shown to inhibit a variety of processes involved in cancer cell growth, survival and metastasis [47]. Of the tea polyphenols, EGCG has been most extensively investigated because of its relative abundance and strong cancer-preventive properties, particularly in the chemoprevention of breast cancer [48]. Details about tea polyphenols are given elsewhere [49] and in this volume.

Genistein, an isoflavonoid found with a large number of other isoflavonoids in several members of the family Fabaceae, particularly in *Glycine max*, has been established as a prostate cancer chemopreventive agent [50]. Genistein in the diet results in increased apoptosis in the prostate [51, 52]. Genistein-induced apoptosis involves activation of calpain, caspase 7, and poly (ADP ribose) polymerase [52].

Curcumin (diferuloylmethane) is a polyphenolic compound present in rhizomes of *Curcuma longa*, commonly used in Indian cooking. Although the broad curative properties of curcumin have been known for centuries, its potential anticancer and chemopreventive properties have been established only recently [53, 54]. Details are given elsewhere in this book.

Gingerol or (6)-gingerol, the phenolic substance responsible for the spicy taste of fresh ginger (*Zingiber officinale*), has diverse pharmacological effects, such as being antioxidant, anti-apoptotic and anti-inflammatory. Though its use dates back 2500 years, its anticancer and chemopreventive properties came to light only recently [55]. Inhibition of COX2 expression has been co-related as a molecular basis for its antitumour effects [56].

Capsaicin (8-methyl-*N*-vanillyl-6-nonenamide) is responsible for the piquancy of chilli pepper (*Capsicum annuum*). In experimental animals, it is suspected to act as a carcinogen due to its irritant properties. However, other studies indicated chemopreventive and chemoprotective effects as capsaicin inhibits transcription factors (NF-κB, STAT3) [57] and angiogenesis [58].

Flavopiridol is a semisynthetic flavonoid and has been shown to be a potent inhibitor of cyclin-dependent kinases [59]. Flavopiridol also induces apoptosis, suppresses inflammation and modulates the immune response [17]. Thus, it may be concluded that the daily diet has an important role in cancer chemoprevention, and the cancer death rate could be reduced to approx. 35% by proper diet alone [60]. Bioactive molecules present in the diet influence all levels of metabolism, including epigenetic and genetic (transcription and translation levels). However, this requires scientific validation by modern tools and has become evident in plants such as ginger [61] and saffron [22].

10.3 Anticancer Drugs from Microorganisms

The search for effective compounds from microorganisms is more recent and not based on ethnobotanical records. Following the discovery of Penicillin from *Penicillium flavum*, microorganisms were explored for therapeutically significant molecules. A number of small-molecule drugs have been approved by the FDA, including 17% of microorganism origin out of 126 approved since World War II; the first of them was actinomycin D [62].

Mostly products from Actinomycetales were isolated, including daunorubicin and its derivative doxorubicin, which are still used against breast cancer. New derivatives produced were epirubicin, pirirubicin, idarubicin and lastly valrubicin. Cancer cell growth is affected by the inhibition of topoisomerase II. Bristol–Myers developed another important anticancer agent of Actinomycetales origin – bleomycins. Mitomycin C, mithramycin, streptozotocin, pentostatin and calicheamicin are the other potent compounds in use, except the last one (Fig. 10.2).

Antibiotics are defined as low-molecular-weight organic natural products (secondary metabolites or idiolites) made by microorganisms which are active at low concentrations against other microorganisms. In 1995, about 12,000 antibiotics were known; 55% were produced by Streptromyces (Actinomycetes), 11% from other Actinomycetes, 12% from non-filamentous bacteria, and 22% from filamentous fungi. Many more new products are still being discovered from microorganisms. About 350 antimicrobials are available on the market for various human diseases. The antibiotic category includes cephalosporin (45%), penicillins (15%), tetracyclines (6%), macrolides (5%), aminoglycosides, ansamycins, glycopeptides and polyenes [63]. Of the 25 top-selling drugs in 1997, 42% were natural products or derived from natural products [64], of which antibiotics contributed 67% of sales (US$55 billion world market). Besides antimicrobial activity, other properties of antibiotics have been explored. An extremely important concept for the further drug development of natural products is that a compound possesses more than one biological activity [65].

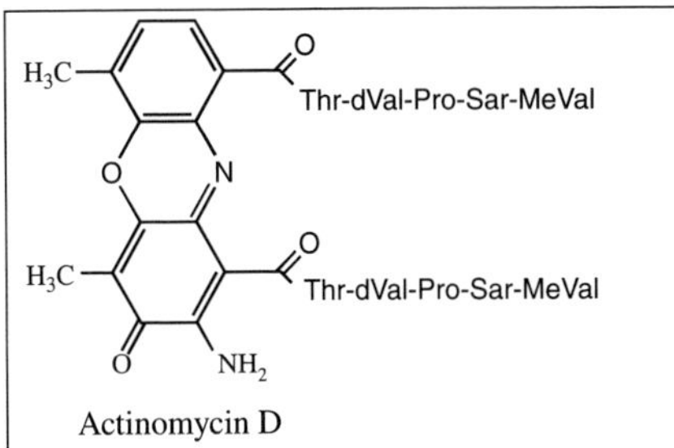

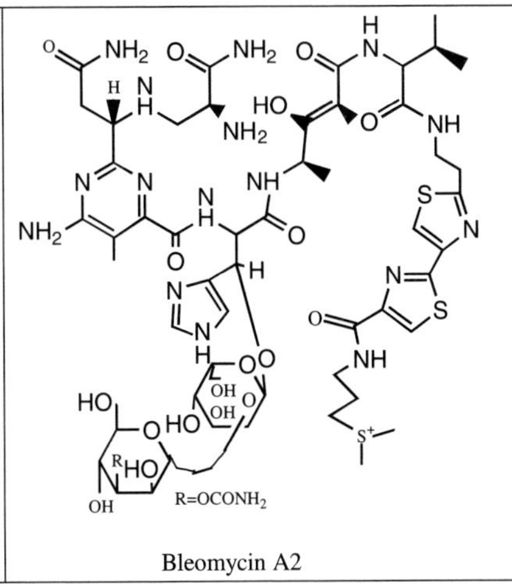

Fig. 10.2 Antibiotics used in cancer therapy

10.4 Anticancer Drugs from Plants

More than 90 drugs are commercially available for cancer therapy, of which 62% are derived from natural products, excluding compounds like interferon [19]. Cancer or neoplastic conditions are poorly defined in ancient traditional or folklore medicine. Hence, leads are not available from such literature. Anticancer agents from plants which are currently in clinical use are *Catharanthus roseus* alkaloids (vinblastine and vincristine), epipodophyllotoxins from *Podophyllum* species, taxanes from *Taxus* spp., camptothecins from *Camptotheca*, and others (Fig. 10.3) [9, 21, 62, 66].

10.4.1 Podophyllotoxins

The rhizome of *Podophyllum* species is known to contain several lignans. Lignans possess antitumour activity and are highly toxic for the treatment of neoplastic diseases in humans. These are dimerisation products of phenylpropanoid pathway intermediates linked by the central carbons of their side chain, podophyllotoxin being the most active cytotoxic compound. Therefore, two semisynthetic glycosides of the isomeric natural product epipodophyllotoxin, called etoposide and teniposide, have been developed by Sandoz (now part of Novartis). They show less cytotoxicity and act as inhibitors of microtubule assemblies, and are commercially used as therapeutic agents against several human neoplasms, including small cell carcinomas of the lung, testis neuroblastoma, Hodgkin's disease, diffuse histocytic lymphoma, and others [9, 62, 67]. Much recent synthetic work on podophyllotoxins has concentrated on the design of more water-soluble analogues. Although podophyllotoxin

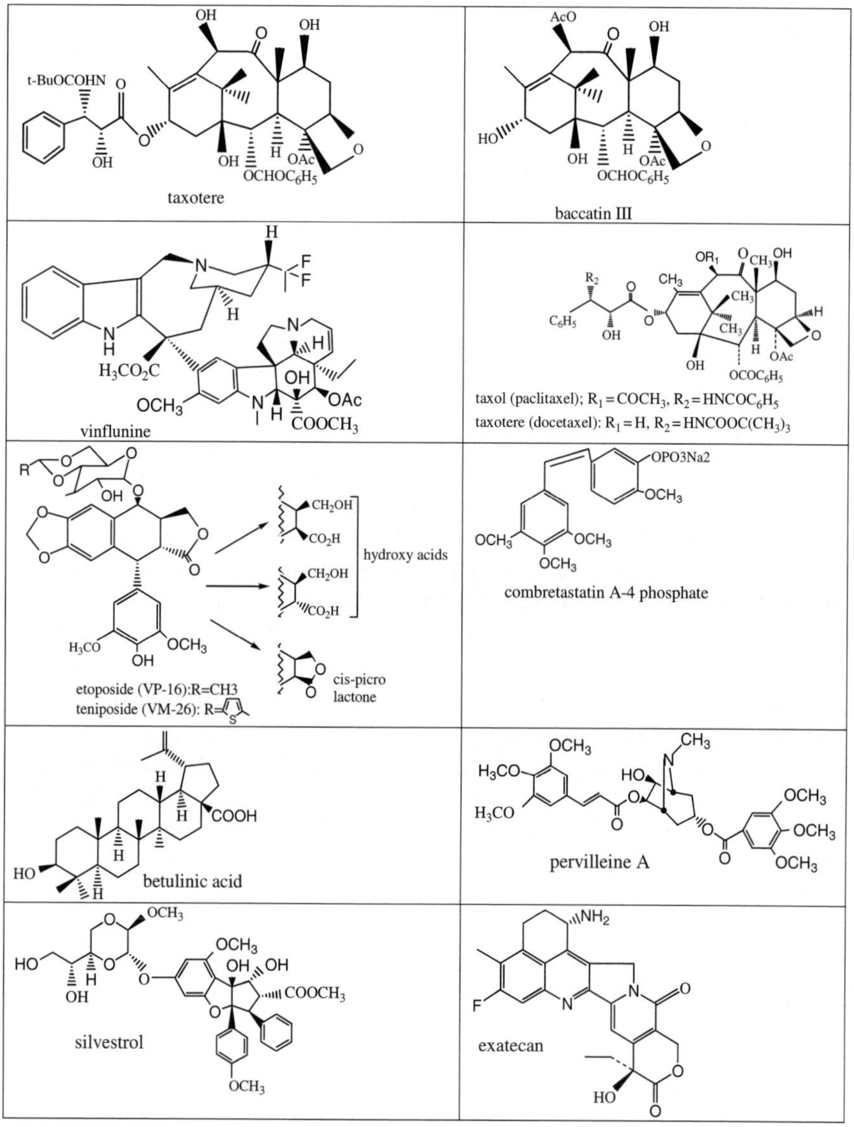

Fig. 10.3 Antineoplastic compounds and their derivatives

binds to tubulin at a site different from that for *Catharanthus* alkaloids, the drug has no effect on microtubular structure or function but inhibits topoisomerase II [67].

American podophyllum (*P. peltatum*) contains 4 to 5% podophyllum resin, whereas the Indian species (*P. hexandrum*) contains 7 to 16%. The variation in percentage of resin is attributed to seasonal differences, age of the plant and altitude [68–70]. The resin of *P. hexandrum* contains approx. 40% podophyllotoxin and low amounts of peltatin [71]. Unfortunately, the Indian species, *P. hexandrum* Royle (Berberidaceae), has become a critically endangered species. A recent study aimed at the domestication of the plant in the Himalayan region [72] reported that the plants grown at high altitude contain high podophyllotoxin content [69], and content increases with the age of the plant [70]. One of the major problems associated with the utilisation of *P. hexandrum* is its long juvenile phase and poor trait setting ability of the plants resulting in selection problems. Moreover, its seeds take a long time to germinate [72].

10.4.2 Vinblastine

The medicinal properties of *Catharanthus roseus* have been described in the traditional and folk medicine of several countries. The beneficial effects of its extract in diabetes mellitus were known, but later on active principles suppressing neoplasm were also isolated. The extracts yielded four active dimeric monoterpenoid indole alkaloids – vinblastine, vincristine, vinleurosine and vinrosidine. The catharanthus alkaloids are cell-cycle-specific agents, similar to colchicine and podophyllotoxin, block mitosis and cause metaphase arrest. Though vincristine and vinblastine have antiproliferative properties, both have different patterns of cytotoxic effect and have been used in combination for the last 40 years (FDA approval received in 1963 and 1965, respectively) [73].

Navelbine (Vinorelbine), vinblastine (Velban) and vincristine (Oncovin) are currently used clinically [74]. Vinblastine introduces a wedge at the interface of two tubulin molecules, thus interfering with tubulin assembly and inducing self-association of tubulin into spiral aggregates at the expense of microtubule growth, which may be an attractive target for drug designing [75]. Several analogues of vinblastine and vincristine are also in clinical use (Table 10.1), such as vinorelbine and vindesine [76].

10.4.3 Taxol

Taxol, a diterpene, is the latest antineoplastic drug of natural origin. Between 1960 and 1982, NCI evaluated 3500 plant samples primarily against mouse leukaemia cell lines L1210 and P388. Taxol was the outcome of that programme and got approval from the US Food and Drug Administration for the treatment of refractory ovarian cancer in 1992 and metastatic breast cancer in 1994 [77]. Taxol was first obtained from *T. brevifolia* (bark contains 0.01 to 0.03% taxol) during the 1960s and subsequently from bark of *T. baccata* and other *Taxus* spp. This long development time was due to the limited supply associated with killing of the plant upon removal of the bark. Paclitaxel and its derivatives act by binding tubulin without allowing depolymerisation or interfering with tubulin assembly [78, 79].

Bristol-Mayers Squibb, USA, is producing taxol (drug) at the commercial level. It is estimated that about 50 kg of taxol will be required every year for the treat-

Table 10.1 Antineoplastic natural compounds and their stages of clinical development

Action site	Drug derivatives	Disease type	Stage of clinical evaluation	Reference
A. Tubulin				
Vinca domain	Vinblastine (Velban)	Hodgkin's disease, testicular germ cell cancer	In clinical use, 22 combination trials	[88]
	Vincristine (Oncovin)	Leukaemia, lymphomas	In clinical use, 108 combination trials	[88]
	Vinorelbine (Navelbine)	Solid tumours, lymphomas, lung cancer	In clinical use, 29 phase I–III trials (single or combination)	[88, 89]
	Vinflunine	Bladder, breast cancer	phase III	[88, 90, 91]
Taxane site	Paclitaxel and analogues	Ovarian, breast and lung tumours, Kaposi's sarcoma, several other tumours	In clinical use, 207 phase I– II trials	[92]
	Docetaxel (Taxotere)	Prostate, brain and lung tumours	Phase I–III trials	[93, 94]
Colchicine domain	Combretastatin AVE-8062	Potential vascular targeting compounds	Phase I-II	[95, 98]
B. Other (apoptosis, protein kinase C)	Homoharrintonine	Hematologic malignancies	Phase I	[99]
	Ingenol-3-O-angelate	Skin cancer	Phase I	[100]
	Phenoxodiol	Ovarian, prostate, renal, vaginal cancer	Phase I clinical trial	[101]
	Protopanaxidiol	Multidrug-resistant tumours	Phase I	[102]

ment of approx. 12,500 women in the USA alone. Current world demand may touch 250 kg per year. The needles contain equivalent or even higher amounts of 10-deacetyl baccatin III (DAB) in four species of *Taxus, T. brevifolia, T. baccata, T. canadensis, T. cuspidata*, which is a renewable source. DAB is converted into paclitaxel and into the more potent analogue taxotere [80].

Taxol, which entered the generic drug market in the early 1990s [81], is now largely produced by Taxus cell cultures [82] or by semisynthetic means from advanced precursors (e.g. baccatin III) that are easily available from the needles of *Taxus* plants as a renewable source [77, 83]. The continuous supply of taxol and its precursors for further synthesis will continue to depend on the plant (needles) or cell-culture system [77, 84] which has become a commercially viable alternative. A paclitaxel production level of 140 to 295 mg/l has been achieved in cell cultures of *T. baccata*, making it a commercially viable production system [82, 85].

10.4.4 Camptothecin

Camptothecin (CPT) was isolated by Wall and coworkers [86] from the Chinese tree *Camptotheca acuminate* and is used to treat gastric, rectal, colon and bladder cancers. Govindachari and Viswanathan [87] isolated camptothecin and 9-methoxy camptothecin from an unrelated plant, *Mappia foetida* (later called *Nothapodytes nimmoniana*). Subsequently it was also isolated from *Merriliodendron megonpum, Nothapodytes nimmoniana* (both from the family Icacinaceae), *Ophrrohiza mugos, O. pumila* (Rubiaceae), *Eravatamia heyneana* (Apocynaceae) and *Mustuea brunonis* (Loganiaceae) [103].

Camptothecins and its derivatives (9-amino CPT, 10-hydroxy CPT, camptothecin, topotecan, and irinotecan) are potent antitumour and DNA topoisomerase I inhibiting agents. The basic molecule had poor solubility in aqueous phase and was found to be too cytotoxic; therefore, attempts were made to develop new effective soluble derivatives [74]. Topotecan was approved for use in the USA in 1996, and other derivatives are in clinical trials [18].

10.4.5 Others

Several diverse compounds were isolated under a project entitled "Novel strategies for plant derived anticancer agents" with a National Co-operative Drug Discovery Group (NCDDG) programme of the National Cancer Institute (NCI), USA. Several of these compounds are currently under investigation for their anticancer activity including betulinic acid, pervilleine A and silvestrol [20].

Cephalotaxus harringtonia (Cephalotaxaceae, Gymnosperm) contains several alkaloids – deoxyharringtonine, harringtonine, homoharringtonine, and isoharringtonine – that have shown anticancer activity against leukaemia in mice.

Homoharringtonine affects a number of cellular pathways leading to apoptosis and angiogenesis [104]. A synthetic derivative of homoharringtonine is in phase II clinical trials for the treatment of patients with chronic myeloid leukaemia that is resistant to the first line therapy, Gleevec [105].

The combretastatins obtained from *Combretum caffrum* are from a class of polyphenolics known as stilbenes, which act as anti-angiogenic agents, causing vascular shutdown in tumours and resulting in tumour necrosis [106]. Its water-soluble derivative combretastatin A4 phosphate, a disodium phosphate prodrug of combretastatin A4, is in phase II clinical trial [62]. The drug is effective against anaplastic thyroid cancer and myopic muscular degeneration and is currently in phase II clinical trials [107]. Combretastatin is a vascular targeting agent; it destroys tumour vasculature by inducing morphological changes within the endothelial cells [95, 108].

Phenoxodiol, a synthetic analogue of diadzein, an isoflavone present in members of the family Fabaceae (*Pueraria tuberosa, Glycine max*), is being developed for the treatment of cervical, ovarian, prostate, renal and vaginal cancers [109]. Phenoxodiol is a broad-spectrum drug that induces cancer cell death through inhibition of anti-apoptopic proteins, including XIAP and FLIP [110]. The drug is in phase II and phase III clinical trials in the USA and Australia, respectively [111].

Betulinic acid, a lupane type triterpene widely distributed in the plant kingdom, and its derivatives are potential anticancer and anti-HIV agents, presently in clinical trials [112]. Recent evidence indicates that betulinic acid possesses broad-spectrum anticancer activity against several cancer cell types [8].

10.5 Mechanism of Action

Most of the antineoplastic compounds act upon DNA by modifying its chemical and physical nature. Broadly speaking, all the drugs can be categorised as alkylating agents such as cisplastin, antimetabolites, e.g. 5-flurouracil and methotrexate, mitotic inhibitors which include vincristine, taxol and colchicines, and lastly DNA intercalating drugs like actinomycin-D [9].

Cancer is a complex disease. Mostly cancer is diagnosed late during the final stages of carcinogenesis, i.e. angiogenesis and metastasis. Chemically diverse compounds have different properties that act and react with cell metabolism. A better understanding of the process of cell division and how different compounds affect tubulin formation have a direct bearing on cancer treatment. The chemopreventive role of dietary molecules has been presented above. Molecules affecting all three phases of cancer, particularly metastasis, are in demand. Increasingly evidence suggests such a multidimensional role of resveratrol. Resveratrol is also able to activate apoptosis, to arrest the cell cycle, or to inhibit kinase pathways [45]. Drug designs, based on the structure of specific enzymes playing a role in carcinogenesis (tyrosine kinase) or DNA replication (topoisomerases II), have been successful at identifying novel effective anticancer drugs. In addition, many natural products are effective inhibitors of NF-κB, a cancer, indicating that the source of these compounds

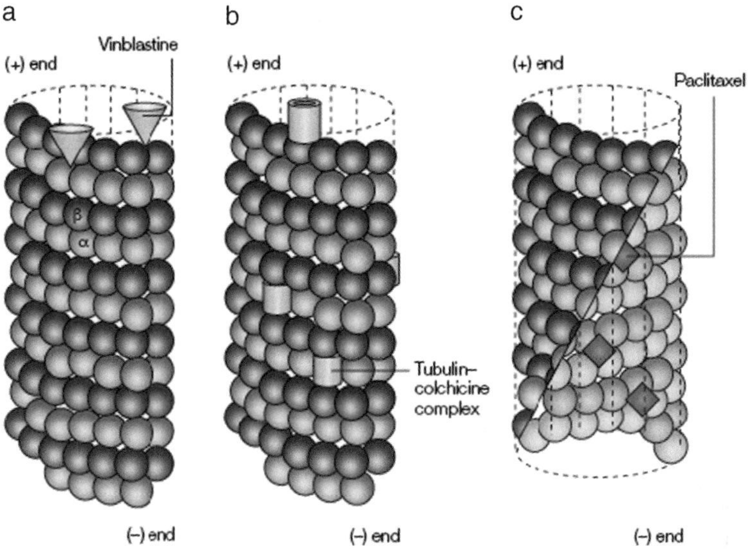

Fig. 10.4 Binding sites of antimitotic drugs to microtubules. Paclitaxel binds along interior surface of microtubule. Binding of these compounds results in suppression of microtubule dynamics (adapted from [115] with permission)

might possess antitumour properties. [113]. This has resulted in the discovery of Gleevec, an inhibitor of the bcr-abl protein tyrosine kinase, for the treatment of chronic myeloid leukaemia [114].

Microtubules, the principal components of the cytoskeleton, are long, filamentous, tubular protein polymers that are essential in all eukaryotic cells. They play a crucial role in the maintenance of cell shape, in cell signalling and in cell division. Microtubules are composed of α- and β-tubulin heterodimers (100,000 dalton in mass) arranged in the form of cylindrical tubes several microns long [115]. Microtubules are involved in the separation of duplicated chromosomes of a cell during mitosis. This makes them an important target for anticancer drugs. Chemical compounds that interfere with microtubules, such as vinca alkaloids and taxanes, are important chemotherapeutic agents for the treatment of cancer. Vincristine, taxol and other mitotic inhibitors bind to specific sites on the microtubules (Fig. 10.4). The anticancer activity of microtubule-targeting drugs lies in their inhibitory effects on spindle microtubule dynamics, rather than their effects on microtubule polymer mass [116]. There is increasing evidence showing that even minor alterations in microtubule dynamics can engage the spindle checkpoint, arresting cell cycle progression at mitosis and eventually leading to apoptotic cell death (Fig. 10.5). Microtubules as a target for anticancer drugs are discussed elsewhere in an excellent review [115].

Several new microtubule-targeting agents have shown potent activity against the proliferation of various cancer cells, including cells that show resistance to the existing microtubule-targeting drugs. Microtubule-interacting agents can be grouped

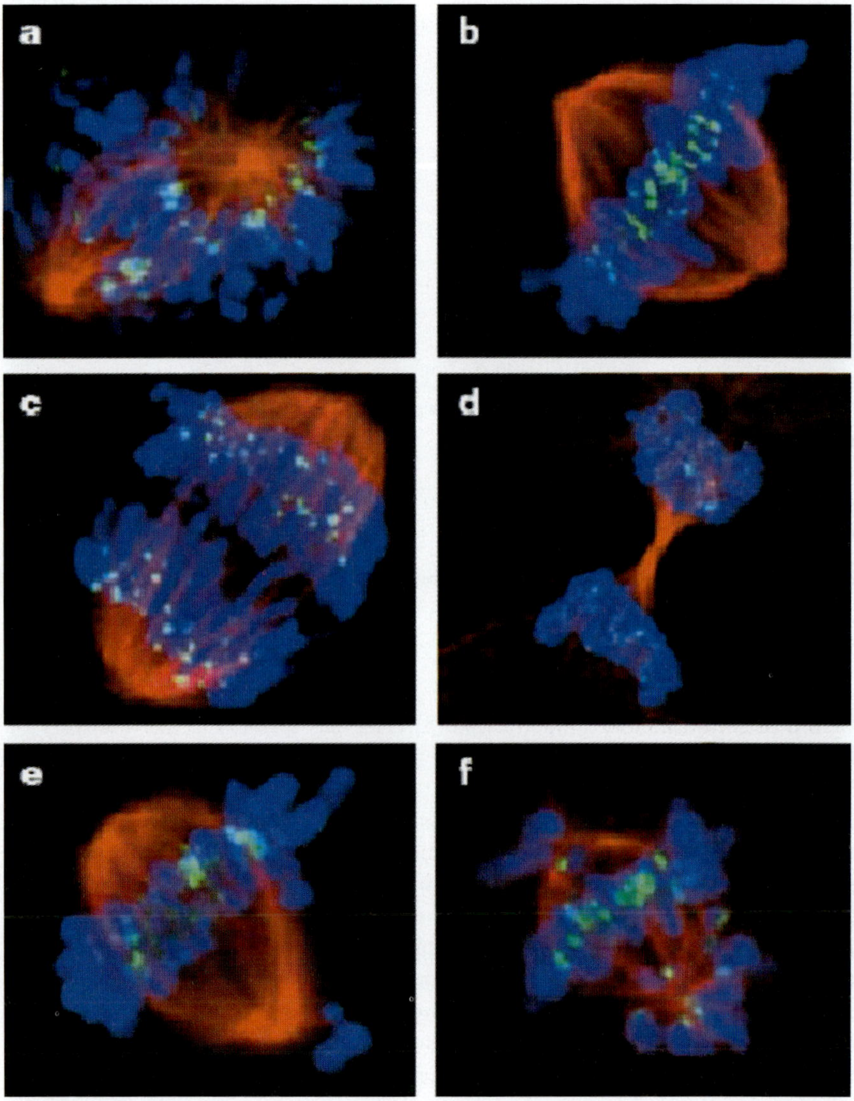

Fig. 10.5 Human osteosarcoma cells in different stages of cell cycle with and without addition of antimitotic drugs. Microtubules are shown in red, chromosomes in blue and kinetochores in green. A-D showing prophase, metaphase, anaphase and telophase stages of cell division. In presence of paclitaxel E and vinflunine F cell division is disturbed resulting in blocking of mitosis (adapted from [91] and [127])

into two distinct functional classes: (i) compounds which inhibit the assembly of tubulin heterodimers into microtubule polymers (tubulin polymerisation inhibitors, e.g. vincristine, vinblastine) and (ii) compounds which stabilise microtubules under normally destabilising conditions (microtubule stabilisers, e.g. paclitaxel). A variety of diverse natural compounds have been shown to possess a taxol-like ability to inhibit depolymerisation of microtubules, such as epothilones-A and B, discodermolide, eleutherobin, sarcodictyins-A and B, laulimalide, cyclostreptin, peloruside A and dictyostatin [73]. The effectiveness of microtubule-targeting drugs for cancer therapy has been impaired by various side effects, particularly neurological and hematological toxicities [116].

Another approach utilises the ability of several compounds, especially microtubule-targeting agents, to rapidly shut down existing tumour vasculature [95]. Since the late 1990s, the combretastatins and *N*-acetylcolchicine-*O*-phosphate, compounds that resemble colchicines and bind to the colchicine domain on tubulins, have been developed as antivascular agents and are in clinical trials, e.g. combretastatin-A-43-O phosphate, combretastatin A-1 phosphate, ZD6126, AVE 8062A, and TZT-1027 (Table 10.1) [115].

10.6 Herb-Drug Interactions

The use of complementary and alternative medicines (CAM) by cancer patients in the western world has grown rapidly in recent years [117]. Generally, cancer patients are using CAM with conventional therapy, but more than 72% of them do not inform their treating physician about CAM [118]. CAM-anticancer drug interactions can occur at the pharmaceutical, pharmacodynamic, or pharmacokinetic level [119]. Interactions at pharmacokinetic level involve changes in absorption, distribution, metabolism, or excretion of the chemotherapeutic drug. One of the best known examples of a clinically significant effect of CAM on the pharmacokinetics of chemotherapeutic drugs is the herbal product, St. John's Wort (SJW) [120].

Pharmacokinetic interactions between CAM and oncolytic drugs occur when CAM inhibit or induce the metabolising enzymes (e.g. cytochrome P450 enzymes, phase II enzymes, dihydropyrimidine dehydrogenase) or drug transporters (e.g. p-glycoprotein, breast cancer resistant protein, multidrug resistance associated proteins) involved in the pharmacokinetic disposition of chemotherapeutic drug [117].

In vitro studies have been performed to investigate the potential of CAM to activate nuclear receptors and to induce metabolising enzymes. Such activities have been shown by apigenin, curcumin, garlic, ginseng, kava-kava, quercetin, resveratrol, silymarin [121], guguggulsterone [122, 123] and several others [117]. Based on current available information Meijerman et al., 2006 suggested that CAM like SJW, grape fruit juice, vitamin E, quercetin, ginseng, garlic, β-carotene and Echinacea should be taken with care along with anticancer therapy as these CAM can cause changes in drug metabolising enzymes.

10.7 Conclusions

Drug development is a long and expensive process and due to this reason, several pharmaceutical companies reduced their expenses on research and development for drug discovery from natural sources. From the above discussion, it is very much clear that natural resources, particularly plants, are an excellent source of life saving drugs. New compounds are also being discovered from lichens, microorganisms and marine organisms. Therefore, efforts are required to collaborate internationally in an attempt to isolate, identify and establish biological/ pharmacological properties of molecules [124]. New tools like high-resolution NMR, Mass spectrophotometers, and 2-D HPLC are helpful in identifying a plethora of molecules from plants and microorganisms. HTS systems are capable of evaluating a large number of bioactive molecules on various bioassays. At the same time, it is necessary to conserve traditional and folk knowledge about use of medicinal plants, and develop appropriate technology to conserve the plants. It is urgently required to investigate plants used in traditional medicine for the complete spectra of primary and secondary metabolites, and where a cocktail of plants is used, the effect of this cocktail on molecular reorientation may be explored. What we know is still a tip of the iceberg, we have yet to explore the entire biological world for novel leads.

A combination of two and more anticancer drugs such as taxol with vinblastine has been reported to effectively control the cancer growth and improved quality of life in patients suffering from hormone refractory prostrate cancer [125]. Additive and synergistic laboratory interactions with other cytotoxic drugs have been exploited to allow development of etoposide based multidrug regimens, which are showing promising activity in several malignancies [126]. Herbs and their preparations can play important role in increasing the efficacy of established drug, reducing suffering of the patients and to certain extend, increasing the longevity of life.

Acknowledgements Research on medicinal plants in the laboratory is supported by funds from UGC under DRS programme and DST under FIST programme. SG thanks UGC for fellowship.

References

1. Demain AL, Zhang L (2005) Natural products and drug discovery. In: Zhang L, Demain AL (eds) Natural Products: Drug Discovery and Therapeutic Medicine. Humana, Totowa, NJ, p 3
2. Schippmann U, Leaman DJ, Cunnningham AB (2002) Biodiversity and the Ecosystem: Approaches in Agriculture, Forestry and Fisheries. 9th Regular Session of the Commission on Genetic Resources for Food and Agriculture. FAO, Rome, Italy 1
3. Anonymous (1994) Lancet 343:1513
4. Gupta R (1986) Integration of medicinal plants cultivation in forest and forest plantations of north-western Himalaya. In: Agroforestry Systems: A New Challenge. Indian Society of Tree Scientists, Solan, India, p 59
5. Van Wyk B-E, Wink M (2004) Medicinal Plants of the World. Briza, Pretoria, Briza

6. Merillon JM, Ramawat KG (2007) Biotechnology for medicinal plants: research need. In: Ramawat KG, Merillon JM (eds) Biotechnology: Secondary Metabolites, Plants and Microbes. Science Publishers, Enfield, NH, p 1
7. Ramawat KG, Goyal S (2008) Indian herbal drugs scenario in global perspectives. In: Ramawat KG, Merillon JM (eds) Bioactive Compounds and Medicinal Plants. Springer, Berlin Heidelberg New York, p 323
8. Salim AA, Chin Y-W, Kinghorn AD (2008) Drug discovery from plants. In: Ramawat KG, Merillon JM (eds) Bioactive Compounds and Medicinal Plants. Springer, Berlin Heidelberg New York, p 1
9. Ramawat KG (2007) Production of alkaloids. In: Ramawat KG, Merillon JM (eds) Biotechnology: Secondary Metabolites, Plants and Microbes. Science Publishers, Enfield, NH, p 179
10. Chin Y-W, Balunas MJ, Chai HB, Kinghorn AD (2006) AAPS J 8:E239
11. Wink M, Alfermann W, Franke R, Wetterauer B et al (2005) Plant Gen Resourc 3:90
12. Taylor JLS, Rabe T, McGaw LJ, Jager AK, Staden JV (2001) Plant Growth Regul 34:23
13. Issa AY, Volate SR, Wargovich MJ (2006) J Food Composit Analysis 19:405
14. Parkin DM (2001) Lancet Oncol 2:533
15. Jemal A, Muarray T, Ward E, Samuels A, Tiwari RC, Ghafoor A, Feuer EJ, Thun MJ (2005) CA-A Cancer J Clin 55:10
16. Hanahan D, Weinberg RA (2000) Cell 100:57
17. Deorukhkar A, Krishnan S, Sethi G, Aggarwal BB (2007) Expert Opin Investigat Drugs 16:1753
18. Newman DJ, GM Cragg, KM Snader (2000) Nat Prod Rep 17:215
19. Newman DJ, Cragg GM, Snader KM (2003) J Nat Prod 66:1022
20. Balunas MJ, Kinghorn AD (2005) Life Sci 78:431
21. Mann J (2002) Nat Rev 2:143
22. Abdullaev FI (2002) Exp Biol Med 227:20
23. Tsao AS, Kim ES, Hong WK (2004) CA-A Cancer J Clin 54:150
24. Park EJ, Pezzuto JM (2002) Cancer Metastasis Rev 21:231
25. Reddy L, Odhav B, Bhoola KD (2003) Pharmacol Therapeut 99:1
26. Aggarwal BB, Shishodia S (2006) Biochem Pharmacol 71:1397
27. Kelloff GJ, Crowell JA, Steele VE, Lubet RA, Malone WA, Boone CW, Kopelovich L, Hawk ET, Lieberman R, Lawrence JA, Ali L, Viner JL, Sigman CC (2000) J Nutr 130:467s
28. Greenwald P (2002) Br Med J 324:714
29. Ramawat KG, Merillon JM (eds) (2008) Bioactive Compounds and Medicinal Plants. Springer, Berlin Heidelberg New York
30. Ferguson LR, Philpott M (2007) Curr Cancer Drug Targets 7:459
31. Shukla S, Gupta S (2005) Nutr Cancer 53:18
32. Soobrattee MA, Bahorun T, Aruoma OI (2006) Free Radicals Biol Med 27:19
33. Scalbert A, Johnson IT Saltmarsh M (2006) Am J Clin Nutr 81:215s
34. Gutteridge JMC, Halliwell B (1994) Oxford University Press, Oxford, UK
35. Cheynier V (2005) Am J Clin Nutr 81:223s
36. Kuntz S, Wenzel U, Daniel H (1999) Eur J Nutr 38:133
37. Manach C, Williamson G, Morand C, Scalbert A, Remesy C (2005) Am J Clin Nutr 81:230s
38. Waffo–Teguo P, Stephanie K, Tristan R, Merillon JM (2008) Grapevine stilbenes and biological effects. In: Ramawat KG, Merillon JM (eds) Bioactive Compounds and Medicinal Plants. Springer, Berlin Heidelberg New York, p 25
39. Vitrac X, Krissa S, Decendit A, Deffieux G, Merillon JM (2004) In: Ramawat KG (ed) Biotechnology of Medicinal Plants. Science Publishers, Enfield, NH, p 33
40. Shishodia S, Aggarwal BB (2006) Resveratrol: a polyphenol for all seasons. CRC Press/Taylor and Francis, Boca Raton, FL
41. Aggarwal BB, Bhardwaj A, Aggarwal RS et al (2004) Anticancer Res 24:2783
42. DeLorgeril M, Salen P (1998) Lancet 351:1440
43. Renaud S, De Lorgeril M (1992) Lancet 339:1523

44. Delmas D, Jannin B, Latruffe N (2005) Mol Nutr Food Res 49:377
45. Delmas D, Lancon A, Colin D, Jannin B, Latruffe N (2006) Curr Drug Targets 7:423
46. Henry-Vitrac C, Desmouliere A, Girard D, Merillon J-M, Krisa S (2006) Eur J Nutr 45:376
47. Khan N, Afaq F, Saleem M et al (2006) Cancer Res 66:2500
48. Pan MH, Lin CC, Lin JK, Chen WJ (2007) J Agric Food Chem 55:5030
49. Kumar NB (2005) Evidence Based Integrative Med 2:13
50. Dixon RA, Sumner LW (2003) Plant Physiol 131:878
51. Lamartiniere CA (2006) http:/handle.dtic.mil/100.2/ada463968
52. Shim HY, Park JH, Paik HD, Nah SY, Kim DS, Han YS (2007) Anticancer Drugs 18:649
53. Aggarwal BB, Shishodia S, Young-Joon S (2007) The Molecular Targets and Therapeutic Uses of Curcumin in Health and Disease. Springer, Berlin Heidelberg New York
54. Shishodia S, Sethi G, Aggarwal BB (2005) Ann NY Acad Sci 1056:206
55. Shukla Y, Singh M (2007) Food Chem Toxicol 45:683
56. Kim SO, Chun KS, Kundu JK, Surh YJ (2004) Biofactors 21:27
57. Bhutani M, Pathak AK, Nair AS et al (2007) Clin Cancer Res 13:3024
58. Min JK, Han KY, Kim EC et al (2004) Cancer Res 64:644
59. Senderowicz AM (2005) Prog Drug Res 63:183
60. Milner JA (2006) Nutr Cancer 56:216
61. Kim EC, Min JK, Kim TY, Lee SJ, Yang HO, Han S, Kim YM, Kwon YG (2005) Biochem Biophys Res Comm 335:300
62. Newman DJ, Cragg GM (2005) The discovery of anticancer drugs from natural sources. In: Zhang L, Demain AL (eds) Natural Products: Drug Discovery and Therapeutic Medicine. Humana, Totowa, NJ, p 129
63. Strohl W (1997) Industrial antibiotics: today and the future. In: Strohl W (ed) Biotechnology of Antibiotics, 2nd edn. Dekker, New York, p 1
64. Bull AT, AC Ward, M Goodfellow (2000) Microbiol Mol Biol Rev 64:573
65. Cragg GM, D Newman (2000) Exp Opin Invest Drugs 9:2783
66. Butler MS (2005) Nat Prod Rep 22:162
67. Lee KH (2004) J Nat Prod 67:273
68. Purohit MC, Bahuguna R, Maithani UC, Purohit AN, Rawat MSM (1999) Curr Sci 77:1078
69. Nadeem M, Palni LMS, Kumar A, Nandi SK (2007) Planta Med 73:388
70. Pandey H, Nandi SK, Kumar A, Palni UT, Palni LMS (2007) Acta Physiol Plant 29:121
71. Giri A, Narasu ML (2000) Cytotechnol 34:17
72. Handa SS, Chawla AS, Maninder A (1989) Fitoterapia 60:195
73. Altmann K-H, Gertsch J (2007) Nat Prod Rep 24:327
74. Lee K-H (1999) J Biomed Sci 6:236
75. Gigant B, Wang C, Ravelli RB, Roussi F, Steinmetz MO, Curmi PA, Sobel A, Knossow M (2005) Nature 435:519
76. Fahy J (2001) Curr Pharmaceut Design 7:1181
77. Croteau R, Ketchum REB, Long RM, Kaspera R (2006) Phytochem Rev 5:75
78. Schiff PB, Fant J, Horwitz S (1979) Nature 277:665
79. Horwitz SB (2004) J Nat Prod 67:136
80. Gueritte-Voegelein F, Guenard D, Lavelle F et al (1991) J Med Chem 34:992
81. Wall ME, Wani MC (1996) J Ethnopharmacol 51:239
82. Tabata H (2004) Adv Biochem Eng Biotechnol 7:1
83. Kintzios S (2008) Secondary metabolite production from plant cell cultures: the success stories of rosmarinic acid and taxol. In: Ramawat KG, Merillon JM (eds) Bioactive compounds and Medicinal Plants. Springer, Berlin Heidelberg New York, p 85
84. Takeya K (2003) Plant tissue culture of taxoids. In: Itokawa H, Lee K-H (eds) *Taxus*-The Genus *Taxus*. Taylor & Francis, London, p 134
85. Tabata H (2006) Curr Drug Targets 7:453
86. Wall ME, MC Wani, CE Cook, KH Palmer, AT McPhail, GA Sim (1966) J Am Chem Soc 88:3888
87. Govindachari TR, Viswanathan N (1972) Indian J Chem 10:453
88. Duflos A, Kruczynski A, Barnet JM (2002) Curr Med Chem Anticancer Agents 2:55

89. Plosker GL, Figgitt D (2003) Drugs 63:803
90. Seidman AD (2003) Semin Oncol 30:6
91. Okouneva T, Hill BT, Wilson L, Jordan MA (2003) Mol Cancer Ther 2:427
92. Horwitz SB (1994) Nature 367:593
93. Belani CP, Langer C (2002) Lung Cancer 38:13
94. Fossella FV, Lynch T, Shepherd FA (2002) Lung Cancer 38:5
95. Tozer GM, Kanthou C, Parkins CS, Hill SA (2002) Int J Exp Pathol 83:2
96. Cirla A, Mann J (2003) Nat Prod Rep 20:558
97. West CML, Price P (2004) Anticancer Drugs 15:179
98. Hamel E, Lin CM, Plowman J et al (1996) Biochem Pharmacol 51:53
99. Kantarjian HM, Talpaz M, Santini V, Murgo A, Cheson B, O'Brian SM (2001) Cancer 92:1591
100. Kedei N, Lundberg DJ, Toth A, Welburn P, Garfield SH, Blumberg PM (2004) Cancer Res 64:3243
101. Kamsteeg M, Rutherford T, Sapi E, et al (2003) Oncogene 22:2611
102. Jia W, Yan H, Bu X, Liu G, Zhao Y (2004) J Clin Oncol 22:9663
103. Uma Shaanker R, Ramesha BT, Ravikanth G, Gunaga R, Vasudeva R, Ganeshaiah KN (2008) Chemical profiling of *Nothapodytes nimmoniana* for camptothecin, an important anticancer alkaloid: towards development of a sustainable production system. In: Ramawat KG, Merillon JM (eds) Bioactive Compounds and Medicinal Plants.e Springer, Berlin Heidelberg New York, p 197
104. Shiraga T, Tozuka Z, Ishimura R, Kawamura A, Kagawama A (2005) Biol Pharm Bull 28:124
105. Sandor V, Bakke S, Robey RW, Kang MH, Blagosklonny MV, Bender J, Brooks R, Piekarz RL, Tucker E, Figg WD, Chan KK, Goldspiel B, Fojo AT, Balcerzak SP, Bates SE (2002) Clin Cancer Res 8:718
106. Holwell SE, Cooper PA, Grosios K (2002) Anticancer Res 22:707
107. Bernier SG, Lazarus DD, Clark E et al (2004) Proc Natl Acad Sci USA 101:10768
108. Chun E, Han CK, Yoon JH, Sim TB, Kim Y-K, Lee K-Y (2005) Int J Cancer 114:124
109. Hayakawa Y, Rovero S, Forni G, Smyth MJ (2003) Proc Natl Acad Sci USA 100:9464
110. Cragg GM, Newman DJ (2004) J Nat Prod 67:232
111. Honore S, Kamath K, Braguer D, Wilson L, Briand C, Jordan MA (2003) Mol Cancer Ther 2:1303
112. Rajendran P, Jaggi M, Singh MK, Mukherjee R, Burman AC (2008) Invest New Drugs 26:25
113. D'Incalci M, Brunelli D, Marangon E et al (2007) Curr Pharmaceut Design 13:2744
114. Capdeville R, Buchdunger E, Zimmermann J, Matter A (2002) Nat Rev Drug Discovery 1:493
115. Jordan MA, Wilson L (2004) Nat Rev Cancer 4:253
116. Zhou J, Giannakakou P (2005) Curr Med Chem-Anticancer Agents 5:65
117. Meijerman I, Beijnen JH, Schellens JHM (2006) Oncologist 11:742
118. Sparreboom A, Cox MC, Acharya MR et al (2004) J Clin Oncol 22:2489
119. Beijinen JH, Schellens JH (2004) Lancet Oncol 5:489
120. Mathijssen RH, Verweij J, de Bruijn P et al (2002) J Natl Cancer Inst 94:1247
121. Raucy JL (2003) Drug Metab Dispos 31:533
122. Brobst DE, Ding X, Creech KL et al (2004) J Pharmacol Exp Ther 310:528
123. Owsley E, Chiang JY (2003) Biochem Biophys Res Commun 304:191
124. Cordell GA (2002) Phytochem Rev 1:261
125. Ackler JR, Tester WJ, Leighton JC (2005) J Clin Oncol 23:8171
126. Ying QL, Liu Y, Xuan T (2007) Curr Bioactive Comp 3:37
127. Kelling J, Sullivan K, Wilson L, Jordan MA (2003) Cancer Res 63:2794

Chapter 11
Artemisinin: A Versatile Weapon from Traditional Chinese Medicine

Thomas Efferth

Abstract Traditional Chinese medicine (TCM) commands a unique position among all traditional medicines because of its 5000 years of tradition. Our own interest in natural products from TCM was triggered in the 1990s by sesquiterpene lactones of the artemisinin type from *Artemisia annua* L. The first description of the Chinese herb *Artemisia annua* L. (*qinghao*, Sweet wormwood) dates back to 168 B.C.E. Artemisinin (*qinghaosu*) was identified in 1972 as the active antimalarial constituent of *Artemisia annua* L. Artemisinin and its derivatives are used for the treatment of malaria. As shown in recent years, this class of compounds also shows activity against cancer cells, schistosomiasis, and certain viruses, i.e., human cytomegalovirus, hepatitis B anc C virus, and bovine viral diarrhea virus. Interestingly, the bioactivity of artemisinin seems to be even broader and also includes the inhibition of other protozaons such as *Leishmania, Trypanosoma*, and *Toxoplasma gondii*, as well as some trematodes, fungi, yeast, and bacteria. The analysis of its complete profile of pharmacological activities, as well as the elucidation of molecular modes of action and the performance of clinical trials, will further elucidate the full potential of this versatile weapon from nature against diseases.

Keywords Artemisinin · Cancer · Chemotherapy · Malaria · Pharmacognosy · Schistosomiasis · Traditional Chinese medicine · Viral infections

Abbreviation

TCM Traditional Chinese medicine

T. Efferth (✉)
German Cancer Research Center, Pharmaceutical Biology (C015),
Im Neuenheimer Fedl 280, 69120 Heidelberg, Germany,
e-mail: t.efferth@dkfz.de

K.G. Ramawat (ed.), *Herbal Drugs: Ethnomedicine to Modern Medicine*,
DOI 10.1007/978-3-540-79116-4_11,

11.1 Introduction

Traditional Chinese medicine (TCM) comprises medicinal products from plants, animals and minerals, acupuncture, moxibustion, and other practices. Herbal prescriptions consist of a varying number of different medicinal plants and are used as extracts, decoctions, concoctions, and teas. Among all traditional medicines, TCM commands a unique position because of its 5000 years of tradition. Hence, it can be assumed that many ineffective prescriptions have disappeared over time. Until recently, TCM has been frequently regarded with some skepticism by Western academic medicine. On the other hand, prominent examples of isolated therapeutics derived from Chinese plants are established in modern medicine without being treated with the same reluctance as traditional herbal products. Among them are the ion channel blocker tetrandrine (*Stephania tetrandra*), the CNS stimulator ephedrine (*Ephedra sinica*), and the well-known anticancer agents camptothecin from *Camptotheca acuminata* and paclitaxel from *Taxus chinensis*. Since natural products represent a valuable source of drug discovery and development, there has been a recently thriving interest in chemically characterized compounds derived from TCM [1–3].

Our own interest in natural products from TCM was triggered in the 1990s by sesquiterpene lactones of the artemisinin type from *Artemisia annua* L. [4]. Apart from artemisinin, which is the focus of this chapter, we analyzed phytochemical and molecular biological aspects of natural products derived from TCM. The modes of action were studied on known compounds with still unknown cellular and molecular mechanisms such as arsenic trioxide, homoharringtonine, cephalotaxine, berberine, cantharidin, curcumin, luteolin, scopoletin, isoscopoletin, ascaridin, the quinolones 1-methyl-2-undecyl-4-quinolone, 1-methyl-2-trideca-dienyl-4-quinolone and evocarpine, the indoloquinazoline alkaloids rutaecarpine and evodiamine, and four geranylated furocoumarines [5–19]. Furthermore, novel natural products were isolated and identified from plants derived from TCM, some of which showed growth inhibitory activity against cancer cells, i.e., tetracentronsine, a new indole alkaloid (3-(2-hydroxyethyl)-1H-indole-5-O-beta-D-glucopyranoside), and two new phenol derivatives, 3-{2-[(beta-glucopyranosyl) oxy]-4,5-(methylenedioxy)phenyl} propanoic acid and methyl 3-{2-[(beta-glucopyranosyl)oxy]-4,5-(methylene-dioxy) phenyl}propanoate, two new alpha-tetralone (= 3,4-dihydronaphthalen-1(2H)-one) derivatives, berchemiaside A and B, a new flavonoid, quercetin-3-O-(2-acetyl-alpha-L-arabinofuranoside), and a diprenylated indole, (E)-3-(3-hydroxy-methyl-2-butenyl)-7-(3-methyl-2-butenyl)-1H-indole [20–22].

11.2 Use of Artemisinin in Traditional Chinese Medicine

The first description of the Chinese herb *Artemisia annua* L. (*qinghao*, sweet wormwood) dates back to 168 B.CE. The plant was mentioned in the prescriptions for 52 diseases in the Mawangdui tomb of the Han dynasty. The next historical tradition is from the year 1086, written by Shen Gua. In the “Handbook of Prescriptions for

Fig. 11.1 *Artemisia annua* seedlings for the sustainable production of artemisinin (with permission of Dr. Joachim Stadler, GE Healthcare, Bio-Sciences, Freiburg, Germany)

Emergency Treatment" Ge Hong (281–340 C.E.) recommended tea-brewed leaves to treat fever and chills. The "Compendium of *Materia Medica*" published by Li Shizen in 1596 cited Ge Hong's prescription. In the course of the Vietnam War, the Chinese government started an antimalarial research program to systematically search for antimalarial TCM plants to support the Vietnamese army. This task was certainly not easy to fulfill during the Cultural Revolution in China [23]. As a result, artemisinin (*qinghaosu*) was identified in 1972 as the active antimalarial constituent of *Artemisia annua* L. [24, 25]. Today, artemisinin is widely used around the world to combat otherwise drug-resistant *Plasmodium* strains, cerebral malaria, and malaria in children [26]. Since harvesting of *Artemisia annua* plants in the wild does not meet the requirements for a sustainable production of artemisinin, the cultivars are bred in plantations and greenhouses (Fig. 11.1). While *Artemisia annua* and artemisinin were evaluated by the World Health Organization (WHO) with much reluctance fora long time, the full potential has recently been recognized.

11.3 Mode of Action of Artemisinin

In malaria parasites, artemisinin acts by a two-step mechanism. It is first activated by intraparasitic heme-iron, which catalyzes the cleavage of the endoperoxide. The *Plasmodium* trophozoites and schizonts live within red blood cells. Hemoglobin serves as an amino acid source, being taken up by the parasites into food vacuoles where enzymatic degradation takes place [27, 28]. The release of heme-iron during hemoglobin digestion facilitates the cleavage of the endoperoxide moiety by an Fe(II) Fenton reaction. The breaking of the endoperoxide bridge results in the generation of typical reactive oxygen species such as hydroxyl radicals and superoxide anions. These damage the membranes of food vacuoles and lead to autodigestion [29, 30]. In addition, the heme-iron(II)-mediated decomposition of artemisinin generates carbon-centered radical species [31–33]. The cleavage of the endoperoxide bond of artemisinin and its derivatives leads to the alkylation of heme and some *Plasmodium*-specific proteins, including the *P. falciparum* translationally controlled tumor protein (TCTP).

As the iron storage of tumor cells is generally much less than that of erythrocytes, but is greater in tumor cells compared to normal cells [34], the question arises as to whether iron may also play a role in the inhibitory action of artemisinins toward tumor cells [35]. The growth rate of a tumor was significantly retarded by daily oral administration of ferrous sulfate followed by dihydroartemisinin. No significant tumor growth retardation effect was observed in rats treated with either dihydroartemisinin or ferrous sulfate alone. The drug treatment did not significantly affect body weight compared with untreated tumor-implanted animals, and no apparent toxic effect was observed after drug treatment [36]. Iron(II) glycine sulfate (Ferrosanol) and transferrin increased the cytotoxicity of free artesunate, artesunate microencapsulated in maltosyl-β-cyclodextrin, and artemisinin toward CCRF-CEM leukemia and U373 astrocytoma cells compared with that of artemisinins applied without iron [37]. Growth inhibition by artesunate and ferrous iron correlated with induction of apoptosis. The effect of ferrous iron and transferrin was reversed by monoclonal antibody RVS10 against the transferrin receptor, which competes with transferrin in binding to the receptor. The IC_{50} values for eight different artemisinin derivatives in the NCI cell line panel were correlated with the microarray mRNA expression of 12 genes involved in iron uptake and metabolism to identify iron-responsive cellular factors enhancing the activity of artemisinins. This analysis pointed to mitochondrial aconitase and ceruloplasmin (ferroxidase). Interestingly, exposure of artemisinins produces no or only marginal cytotoxicity to normal peripheral blood mononuclear cells (PBMC). The absorption of iron increases in growing cells and tissues, and the uptake of transferrin is related to the rate of tumor cell proliferation [38]. Cellular iron uptake and internalization are mediated by the binding of transferrin–iron complexes to the transferrin receptor (CD71), expressed on the cell surface membrane, and subsequent endocytosis. CD71 expression in normal tissues is limited, e.g., to the basal epidermis, endocrine pancreas, hepatocytes, Kupfer cells, testes, and pituitary, while most other tissues are CD71-negative [39]. In contrast, CD71 is expressed in much larger amounts in proliferating and malignant cells [40–42] and is widely distributed among clinical tumors [39]. We found that CD71 expression was much higher in CCRF-CEM and U373 tumor cells (48–95%) than in peripheral mononuclear blood cells of healthy donors ($< 2\%$) [37]. This raises the attractive possibility that tumors that express more CD71 than normal cells are preferentially affected by combination treatments of transferrin or Ferrosanol plus artemisinin derivatives. The finding that iron(II) glycine sulfate increased the action of artemisinins is interesting, since Ferrosanol has been in clinical use for many years. Hence, artemisinins might be safely applied in combination with Ferrosanol in a clinical setting.

Whether oxidative stress and iron ions also play a role for artemisinins' activity against other diseases such as viral infections or schistosomiasis remains unknown.

11.4 Activity Against Malaria

Antimalarial drug resistance has spread and intensified in recent decades and represents a severe global challenge. It is estimated that 300 to 500 million human beings are infected each year and that 1.5 to 2.5 million individuals die of malaria annually [43, 44]. The development of novel drugs has not kept pace even worsening the problem. Artemisinin and its derivates are, therefore, promising new drugs on the horizon that are expected to ease the malaria burden worldwide.

Drug combinations based on artemisinins offer an effective possibility to counteract drug resistance [45]. Combination regimes prolong the useful therapeutic life of existing antimalarial drugs. The probability that a mutant strain of *Plasmodium* simultaneously would exert resistance to two drugs with different modes of action and different therapeutic targets is low. Combinations of drugs are generally accepted to improve treatment efficacy and to delay the selection of drug-resistant parasites [46].

Despite the recommendation of WHO to use artemisinin-based combination therapies, in order to avoid the emergence of artemisinin resistance, the overall deployment of such combination regimes was still unsatisfactory [47]. For this reason, WHO banned artemisinin monotherapy in 2006.

Artemisinins proved to be valuable in drug combinations since they are able to reduce the number of parasites by approximately 10^4 per asexual cycle [26, 48]. Artemisinins are active within 48 to 72 h [49, 50]. This considerably reduces the number of parasites to be killed by a partner drug in a combination regimen. Since they inhibit the production of gametocytes, artemisinins are able to reduce transmission [51].

Another favorable feature of artemisinins is that they are active in uncomplicated as well as severe forms of malaria. Severe malaria does not stop with clearing parasitemia. Even if parasites are cleared, the clinical symptoms associated with cerebral malaria may get worse. Besides the brain, other organs such as kidneys or lungs can also be injured in severe malaria. Artemisinin has been proven as an effective antimalarial drug for the treatment of cerebral malaria [52]. Artemisinin derivatives reveal a very good tolerability [53, 54]. Mild and reversible hematological and electrocardiographic abnormalities, such as neutropenia and first-degree heart block, have been observed infrequently. Neurotoxicity, e.g., ataxia, slurred speech, and hearing loss have been reported in few patients [55]. Due to their lack of severe side effects, artemisinins are also well suited for the treatment of malaria in children [56].

The efficacy of artemisinin and its derivatives in combination with standard antimalarial drugs has been shown in numerous clinical studies. Recent meta-analyses provide convincing evidence for the success of artemisinin-containing regimens for uncomplicated *Plasmodium falciparum* malaria in terms of both treatment response and beneficial profile of side effects.

A meta-analysis by Bakshi et al. [57] investigated a total of 1869 patients treated with artemether and lumefantrine (A-L). The most commonly reported adverse effects involved the gastrointestinal (abdominal pain, anorexia, nausea, vomiting, diarrhea) and central nervous (headache, dizziness) systems. Pruritus and rash were reported by more than 2% of patients. There were no serious or persistent neurological side effects and no adverse clinical cardiac events. More than 90% of the reported adverse events, many of which overlapped considerably with the clinical symptomatology or evolution of acute malaria, were rated mild to moderate in intensity.

Omari et al. [58] reported on nine trials ($n = 4547$ patients) that tested the six-dose regimen of A-L. Total failure at day 28 for A-L was lower when compared with amodiaquine plus sulfadoxine-pyrimethamine, but not with chloroquine plus sulfadoxine-pyrimethamine. In comparisons with artemisinin derivative combinations, A-L performed better than amodiaquine plus artesunate, worse than mefloquine plus artesunate, and similar to dihydroartemisinin-napthoquine-trimethoprim. The authors conclude that the six-dose regimen of A-L appears more effective than antimalarial regimens not containing artemisinin derivatives.

Another meta-analysis of 16 randomized trials ($n = 5948$ patients) studied the effects of the addition of artesunate to standard treatment [54]. Parasitological failure was lower with 3 d of artesunate at day 14 and at day 28. Parasite clearance was significantly faster. Recrudescence and gametocyte carriage was substantially reduced. The occurrence of serious adverse events did not differ significantly between artesunate and placebo.

The meta-analysis of Bukirwa and Critchley [59] included four trials with 775 participants and compared sulfadoxine-pyrimethamine plus amodiaquine (SP plus AQ) with sulfadoxine-pyrimethamine plus artesunate (SP plus AS) for treating uncomplicated *Plasmodium falciparum* malaria. SP plus AQ performed better at controlling treatment failure at day 28, but was not as good as SP plus AS at reducing gametocyte carriage at day 7.

Seven randomized trials of 4472 children were included in the meta-analysis of Obonyo et al [60]. The authors compared the efficacy of amodiaquine plus sulfadoxine/pyrimethamine (AQ + SP) versus artemisinin-based combination therapies (ACT) in the treatment of uncomplicated malaria. Treatment failure of AQ + SP was significantly reduced compared with AS + SP, but increased compared with AL. All treatment regimens were safe and well tolerated.

11.5 Activity Against Cancer

During the past dozen years, our own efforts have been focused on the activity of artemisinin and its derivatives with respect to cancer cells. Our results suggest oxidative stress as a mechanism of artesunate against cancer cells [4], [61–67]. We found that thioredoxin reductase and catalase expression correlated significantly with the IC_{50} values for artesunate. WEHI7.2 mouse thymoma cells selected for resistance to hydrogen peroxide or transfected with thioredoxin, manganese superoxide dismutase or catalase showed resistance to artesunate as compared to the parental cell line [68]. The microarray-based mRNA expression of dihydrodiol dehydrogenase, γ-glutamylcysteine synthase (γ-GCS; *GLCLR*), glutathione S-transferases *GSTM4, GSTT2, GSTZ1*, and microsomal glutathione S-transferase *MGST3* correlated significantly with resistance to artesunate in the NCI cell line panel. A tendency for correlation ($0.05 < p < 0.1$) was observed for *GSTA1, GSTA2, GSTP1*, and *MGST1*. MSC-HL13 cells transfected with cDNAs for heavy and light subunits of γ-GCS were more resistant to ART than mock transfected MSV-PC4 cells [69]. L-buthionine sulfoximine, a γ-GCS inhibitor that depletes cellular glutathione pools, completely reversed ART resistance in MSV-HL13 cells [70].

As tumor cells contain much less iron than erythrocytes, but more than other normal tissues [34], the question arises as to whether iron may be critical for artemisinin's action with respect to tumor cells. Cellular iron uptake and internalization are mediated by binding of transferrin-iron complexes to the transferrin receptor (CD71) expressed on the cell surface membrane and subse-quent endocytosis. While most normal tissues are CD71-negative, CD71 is highly expressed in clinical tumors and is widely distributed among clinical tumors [39, 42, 71]. We found that CD71 expression was much higher in CCRF-CEM and U373 tumor cells (48 to 95%) than in peripheral mononuclear blood cells of healthy donors ($< 2\%$) [37]. Iron(II) glycine sulfate (Ferrosanol) and transferrin increased the cytotoxicity of free artesunate, artesunate microencapsulated in maltosyl-β-cyclodextrin, and artemisinin towards CCRF-CEM leukemia and U373 astrocytoma cells compared with artemisinins applied without iron [37, 72]. This effect was reversed by monoclonal antibody RVS10 against the transferrin receptor, which competes with transferrin for binding to the receptor. These results are in accordance with data from other authors [36, 73, 74].

The outgrowth of new blood vessels from preexisting ones is crucial for tumors to gain access to sufficient amounts of oxygen and nutrients [75]. If tumors reach a size where diffusion alone cannot supply enough oxygen and nutrients, a process termed angiogenesis is promoted by numerous proangiogenic or antiangiogenic factors. As a consequence, inhibitors of angiogenesis has raised considerable interest for cancer treatment [76]. Artemisinin and its derivatives inhibit angiogenesis, as shown by several groups including our own [77–82].

In 1996, we showed for the first time that artesunate induces apoptosis in cancer cells [4]. This result was subsequently confirmed by others [83–87]. Recently, we analyzed the precise apoptotic pathways induced by artesunate [88]. Using leukemic

T-cells as a model system, we showed that artesunate induces malignant T-cells to undergo apoptosis through the mitochondria pathway. This was demonstrated by inducing the release of cytochrome c from the mitochondria upon ART treatment and followed by activation of caspase-9, the main caspase involved in the intrinsic pathway. In contrast, no activation of caspase-8, the main caspase for the extrinsic pathway, was seen. Furthermore, cells deficient in either the death adapter molecule FADD or caspase-8 were at least as sensitive to ART as the parental cells. Further investigation of the molecular mechanisms by which ART triggers apoptosis revealed that ART induces the intrinsic death pathway by generation of reactive oxygen species (ROSs). This was confirmed by the fact that the antioxidant NAC could completely block ROS generation and, consequently, inhibited ART-induced apoptosis.

The activity of artemisinin and its derivatives in vivo has been demonstrated by several authors. Moore et al. [36] found that the growth of fibrosarcoma in Fisher 344 rats was significantly delayed by the daily application of the active metabolite of artemisinin, dihydroartemisinin, plus ferrous sulfate compared to untreated control animals. Chen et al. [79] applied a chorioallantoic membrane (CAM) assay in chicken eggs. This represents a well-established assay to analyze the development of blood vessels in vivo. In particular, the CAM assay is suited for the screening of angiogenesis inhibitors. Dihydroartemisinin significantly suppressed neoangiogenesis by means of this test system. These results are conceivable with results of our own investigations [80]. We soaked Matrigel plugs with vascular epithelial growth factor (VEGF), tumor necrosis factor-α (TNF-α), and heparin, which act as strong stimuli for angiogenesis. The Matrigel plugs were subcutaneously injected into nude mice. In control animals without artesunate treatment, a strong vascularization blood filling of the plugs took place after 4 d. In contrast, a statistically significant reduction of Matrigel vascularization was observed in mice fed with artesunate in the drinking water. To determine the in vivo effects of artesunate on tumor growth, we subcutaneously injected KS-IMM Kaposi sarcoma cells to nude mice [80]. Whereas a strong tumor growth was found in untreated mice, it was strongly suppressed in artesunate-treated animals. These results were subsequently confirmed by other authors. Disbrow et al. [89] found that dihydroartemisinin inhibited the virus-induced tumor formation in vivo. Dogs infected with canine oral papillomavirus developed tumors in the oral mucosa. The tumor development was, however, significantly inhibited by topical application of dihydroartemisinin. Lai and Singh [90] induced breast cancer in rats by application of 7,12-dimethylbenzo[a]anthracene (DMBA). In comparison to untreated control animals, rats fed with artemisinin showed a delayed tumor development and the tumor size was smaller. Furthermore, fewer rats showed multiple breast tumors and fewer rats developed tumors at all.

The successful treatment of human xenograft tumors in nude mice with artesunate [80] inspired us to apply artesunate in a clinical setting. We have treated two patients suffering from uveal melanoma on a compassionate-use basis after standard chemotherapy alone was ineffective in stopping tumor growth [91]. Generally, this tumor entity has a median survival of 2 to 5 months. We did not observe additional side effects that exceed those seen with standard chemotherapy, indicating that artesunate was well tolerated. One patient experienced a temporary response after the addition of artesunate while the disease was progressing under standard therapy with fotemustine alone. The patient died after 24 months. The second patient first experienced a stabilization of the disease after the addition of artesunate plus iron to the standard drug dacarbazine followed by objective regressions of splenic and lung metastases. This patient is still alive 47 months after first diagnosis of stage IV uveal melanoma. This promising result indicates that artesunate might be a valuable adjuvant drug for the treatment of melanoma and other tumors in combination with standard chemotherapy. The treatment of single cases of a laryngeal squamous cell carcinoma with artesunate [92] and a pituitary macroadenoma with artemether has been reported recently [93]. Larger clinical trails are required to establish artesunate for cancer therapy in the clinical setting.

11.6 Activity Against Schistosomiasis

It is estimated that more than 200 million people are affected with schistosomiasis caused by blood flukes (*Schistosoma haematobium, S. intercalatum, S. japonicum, S. mansoni*, and *S. mekongi*) [94–96]. While in some countries Schistosomiasis seems to be under control, i.e., Brazil, China, and Egypt, other regions suffer from this disease, i.e., sub-Saharan Africa [97, 98]. Affected hosts develop immunological reactions against the parasite eggs in the host tissues ranging from allergic reactions at early stages to considerable morbidity in chronic phases later ones. Typical symptoms of schistosomiasis are exercise intolerance, anemia, delayed cognitive development, and growth retardation. Praziquantel is a standard drug in the treatment of the disease that enables one to control morbidity in affected areas [98].

Again, Chinese scientists were the first to observe that artemisinin and its derivatives were active against *Schistosoma* infections [99] – a result that was subsequently confirmed by authors outside China (as reviewed by Utzinger et al. [100]).

In contrast to praziquantel, which shows highest activity against adult worms (and very young schistosomula), detailed studies in a mouse model revealed that 2- to 3-week-old schistosomula were killed more efficiently by artemether than adult worms [100]. Artemether is most active against juvenile *Schistosoma* parasites, and prevents morbidity associated with schistosomiasis, since the egg-laying worms do not develop [101]. In combination with praziquantel, which is most active against adult worms, all developmental parasite stages can be attacked [102].

As reviewed in a recent meta-analysis by Utzinger et al. [100], 16 randomized, placebo-controlled clinical trials were conducted with oral artesunate for the prevention of *S. japonicum* infection, which demonstrated convincingly the efficacy of the drug combating the disease. Eight randomized, placebo-controlled clinical trials were conducted with artemether. The meta-analysis of Utzinger et al. (2007) [100] also showed the utility of this drug to treat schistosomiasis. Artemisinins also revealed a good safety profile when used to prevent *Schistosoma* infections [103, 104].

11.7 Activity Against Viral Infections

11.7.1 Human Cytomegalovirus

Although ART and other artemisinin derivatives have been described as antimalaria drugs [24], their antiproliferative activity is not restricted to protozoans [4, 65]. We investigated whether ART had antiviral activity and to identify possible underlying molecular mechanisms. We found that ART effected a strong inhibition of plaque formation of HCMV AD169 and HSV-1, a partial inhibition of HIV-1, but no inhibition of influenza A virus [105]. Concerning the inhibitory potential for HCMV, it was important to demonstrate that viruses with various phenotypes, i.e., natural isolates, resistant mutants, and laboratory strains and recombinant virus clones, were all highly sensitive to treatment with ART. A possible mechanism is provided by the finding that central regulatory processes of the infected cell were inhibited by ART, thus interfering with critical requirements of HCMV replication in terms of host cell type and metabolism.

Human cytomegalovirus (HCMV) infection can cause maldevelopment of the central nervous system of embryos and neonates [106]. HCMV can be fatal to immunocompromised adults, for example, organ transplant recipients and patients with acquired immunodeficiency syndrome [107]. Furthermore, the virus is indirectly involved in the etiology of certain tumor types (e.g., by the synergistic interaction with tumor-inducing viruses), indicating its role in the coregulation of cellular proliferation [108]. HCMV infections are generally treated with the nucleoside/nucleotide analogs ganciclovir (GCV) and cidofovir (CDV) or the inorganic pyrophosphate analog foscarnet [109–111], all of which cause adverse side effects and reveal low oral bioavailability [112]. In addition, the therapeutic effectiveness is frequently compromised by the emergence of drug-resistant virus isolates. A variety of amino acid changes in the UL97 protein kinase and the viral DNA polymerase have been reported to cause drug resistance [113]. For this reason, the identification of novel drugs with activity toward drug-resistant HCMV variants with a low level of toxic side effects is urgently needed. A possible mechanism is provided by the finding that central regulatory processes of the infected cell were inhibited by ART, thus interfering with critical requirements of HCMV replication in terms of host cell type and metabolism. In particular, the replication of HCMV is tightly coregulated with cellular activation pathways, mediated by direct or indirect interaction with cellular DNA-binding factors such as NF-κB or Sp1. These factors provide a major determinant of the virus-host cell interaction. Moreover, there are several examples demonstrating that chemical compounds interfering with activation pathways of cellular transcription factors (e.g., the signal transduction pathway including mitogen-activated protein kinase p38) possess a strong inhibitory effect on the replication of HCMV. As ART and other artemisinin derivatives have captured the attention of scientists and physicians concerned with malaria treatment for its activity toward multidrug-resistant *Plasmodium* strains [51], we were interested in analyzing whether ART was active toward drug-resistant HCMV as well. Indeed, we observed that the GCV-resistant strain AD169-GFP314 was inhibited with similar efficacy as the drug-sensitive parental AD169-GFP virus. It is obvious from these experiments with GCV-resistant HCMV that the putative inhibitory mechanisms of ART must differ from those of the DNA polymerase inhibitor GCV.

We were further intrigued by the question of how the anti-HCMV effect is mediated at the cellular level. Upon the infection of cultured cells with HCMV, immediate early (e.g., IE1-72, IE2-86), early (e.g., UL84), and late genes (e.g., UL94) are expressed. IE gene products are essential for the subsequent expression of viral early and late genes, all of which are involved in viral replication. A reduction in IE2-86 expression is generally considered a critical limitation in viral replication since IE2 gene products are essential for HCMV infection [114]. It is also noteworthy that the viral immediate early promoter enhancer (in addition to other viral promoters) contains binding sites for both Sp1 and NF-κB and therefore is responsive to both factors [115]. Thus the activation pathways involving Sp1 and NF-κB are important factors in the initial onset of the viral replication cycle and might be critical in the mechanism of ART's inhibition of HCMV. In particular, the replication of HCMV is tightly coregulated with cellular activation pathways, mediated by the direct or indirect interaction with cellular DNA binding factors such as NF-κB or Sp1 [116, 117]. These regulatory events are required for viral replication and provide a major determinant of the efficiency of virus multiplication. The present investigation demonstrated that ART inhibits the HCMV-induced DNA binding activities of both NF-κB and Sp1. For both NF-κB and Sp1 we observed a drastic reduction in HCMV-induced Sp1 protein synthesis under ART treatment, which explains (at least in most part) the reduced DNA binding activity. This is a clue that ART's anticytomegaloviral activity is mediated by inhibition of transactivation of NF-κB and Sp1 transcription factors. Although we have shown that NF-κB and Sp1 binding activities are molecular targets of ART, we cannot rule out the possibility that other mechanisms may also contribute to ART's antiviral action. Infection of quiescent human fibroblasts with HCMV was found to cause a rapid activation of cellular PI3-K. PI3-K regulates phosphorylation of a number of kinases in cellular signaling cascades, including Akt (also known as protein kinase B), cyclic AMP-dependent kinase (protein kinase A), some isoforms of protein kinase C, and the ribosomal S6 kinases p70 and p85 (p70S6K and p85S6 K, respectively). Akt and p70S6K are two major downstream effectors of PI3-K and are strongly activated upon HCMV infection, in a manner similar to the PI3-K-dependent activation of NF-κB and Sp1 [118]. Akt activation is needed for HCMV DNA replication. It is very likely that the ART inhibition of HCMV replication is targeted to the initial stage of viral signaling. In particular, those kinase cascades are suggestive to be involved that are already induced by the interaction of the viral ligands with the cellular receptor. Thus inhibition of PI3-K activity inhibits viral replication and virus-induced signaling [118].

Currently, a limited number of drugs is used for the treatment of a systemic or locally reactivated HCMV infection: (1) ganciclovir (GCV), (2) its prodrug valganciclovir, (3) cidovovir (CDV), (4) foscarnet (FOS), and (5) fomivirsen. Nevertheless, each of these drugs has a number of disadvantages. First, these antiviral compounds are usually administered intravenously or intravitreally, except for valganciclovir, which possesses improved oral bioavailability. In addition, prolonged treatment with each one of these drugs is frequently accompanied by serious side effects. Moreover, GCV, valganciclovir, CDV, and FOS have similar mechanisms of action by targeting, either directly or indirectly, the viral DNA polymerase. Treatment with any of these antiviral agents may therefore ultimately result in the emergence of single or multiresistant HCMV mutants [119, 120]. These considerations have promoted an intense search for novel therapeutic agents that are safe, potent, and act as alternative antiviral targets.

We demonstrated that the antiviral activity of ART against CMV is not restricted to human strains, but also includes animal CMVs, in particular RCMV [121]. An important feature seems to be the finding that increased intracellular iron concentrations in the presence of ART significantly enhance its anti-CMV activity. This enhancing effect was demonstrated by several observations: (1) treatment of CMV-infected fibroblasts with ART combined with ferrous iron (Ferrosanol) and/or soluble transferrin resulted in enhanced suppression of viral replication; (2) the expression of a cell surface marker (Thy-1), which is associated with the proinflammatory effect of CMV infection and which is not affected by established antiviral drugs, is strongly influenced by ART treatment; (3) the antiviral activity of ART against CMV could also be demonstrated in vivo using the RCMV/rat model; and, finally, (4) the antiviral activity of ART is additive when applied in combination with conventional drugs such as GCV, CDV, and FOS. The last finding might be particularly helpful in the treatment of HCMV disease inflicted by mutant viruses that are resistant to conventional antiviral drugs [122]. GCV, CDV, and FOS are all directed at an identical target of viral replication (i.e., DNA synthesis mediated by the viral DNA polymerase) and, consequently, cross-resistance conferred by polymerase-related mutations has frequently been reported [113]. The combination of drugs with different modes of action may delay the development of drug resistance in a clinical setting. Therefore, using an antiviral drug that targets an alternative pathway and does not interfere with the activities of conventional antiviral drugs seems highly promising. Moreover, Ferrosanol is a clinically approved medication and has been in practical use for many years. It is a safe drug without severe toxicity and can, hence, be safely combined with ART.

11.7.2 Human Hepatitis B Virus (HBV)

The Hepadnaviridae family is formed by a group of highly species-specific viruses that share the presence of an endogenous DNA polymerase with reverse transcriptase activity [123–125] and whose genome in the mature virions is formed by a circular partially double-stranded DNA (pdsDNA) in which both strands are held together by hydrogen bonding between the 5′ ends of the two strands [126]. One member of this family, the human hepatitis B virus (HBV), is characterized by its high hepatotropism. This virus belongs to the genus *Orthohepadnavirus* and is not cytopathic itself, although it may cause acute fulminant hepatitis [127] or chronic liver disease that may evolve into cirrhosis and, eventually, hepatocellular carcinoma [128]. In spite of the availability of an effective and safe vaccine against HBV, infection by this virus is an important worldwide health problem [129, 130]. Although several pharmacological strategies are currently being implemented to treat affected patients, no effective antiviral therapy against HBVinfection has yet been fully developed.

In a recent investigation, a group of natural products from medicinal herbs used in TCM was assayed for their anti-HBV-activity [131]. Among them, artemisinin and, in particular, its semisynthetic derivative artesunate displayed the most interesting properties. Moreover, their interest is enhanced by the existence of synergic effects with lamivudine in the absence of drug-induced toxicity in host cells, which may be an important characteristic due to the frequent problem in clinical practice of infection by lamivudine-resistant HBV strains.

The range of concentrations at which artesunate was active against HBV ($>$ 10μM) was quite similar to that previously reported for its activity vs. human cytomegaloviruses [105]. Interestingly, these levels are close to the drug concentrations reached in the plasma of patients when this drug is used in antimalaria treatments ($\approx$ 7μM) [132].

Similarly to artemisinin and artesunate, the model compound lamivudine induced a pronounced inhibition of HBsAg release and/or viral DNA at concentrations at which host cell viability was not affected. This effect was similar to that previously reported by other authors in HepG2 2.2.15 cells [133].

Although artesunate induced a parallel inhibition in HBsAg and HBV-DNA secretion, artemisinin-induced dose-dependent inhibition in HBsAg secretion was initially accompanied by an enhanced release of HBV-DNA (mainly rcDNA forms). This paradoxical effect was previously observed when HepG2 2.2.15 cells were treated with DNA-reactive drugs, such as Bamet-UD2 or cisplatin [134]. A similar behavior, observed under different experimental circumstances, has been suggested to be due to the inhibition of complete HBV production associated with the intracellular accumulation of HBV-DNA intermediates and their subsequent release into the medium [135].

11.7.3 Human Hepatitis C Virus (HCV)

Paeshuyse et al. [136] reported that the antimalarial drug artemisinin inhibits hepatitis C virus (HCV) replicon replication in a dose-dependent manner in two HCV subgenomic replicon constructs at concentrations ineffective toward Huh 5-2 host cells. Hemin, an iron donor, inhibits HCV replicon replication by inhibition of the viral polymerase [137]. The combination treatment of artemisinin and hemin caused a pronounced synergistic antiviral activity without affecting host cells.

11.7.4 Bovine Viral Diarrhea Virus (BVDV)

The Flaviviridae family includes three different genera: Pestivirus (e. g., bovine viral diarrhoea virus, BVDV); Flavivirus (e. g., Japanese encephalitis virus); and Hepacivirus (e. g., hepatitis C virus). Flaviviridae viruses constitute a major cause of disease worldwide. Thus, infection by HCV frequently causes chronic hepatitis that may progress to cirrhosis and hepatocellular carcinoma [138]. The problem is aggravated by the absence of an efficient vaccine against HCV and because currently the standard treatment, based on pegylated IFN-α and the purine nucleoside analog ribavirin (1β-*D*-ribofuranosyl-1,2,4-triazole-3-carboxamide), in addition to having noxious side effects, is not efficient in approximately half of the infected patients [138]. This means that the search for more effective therapies is crucial.

Since all members of the Flaviviridae family share similarities in virion structure, genome organization, and replication machinery, some viruses, in particular BVDV, have been used as in vitro models for infection by these viruses [139]. The reason for using BVDV is that this virus is less hazardous than other members of this family because it is not infective for humans and BVDV replicates efficiently in cell culture [140]. An additional advantage is that there are two biotypes of BVDV: cytopathic and noncytopathic according to their effect on cell cultures. In contrast to infection with noncytopathic strains, infection with cytopathic BVDV leads to lysis of the host cell and hence represents a very useful tool in the investigation of the antiviral protective effect of drugs.

The findings of a recent study suggest that artemisinin is an inhibitor of the production of Flaviviridae viruses and that its effect is additive to those of IFN-α and ribavirin. The pharmacological interest of artemisinin and its derivatives for the treatment of infections by these viruses is enhanced by the facts that a large proportion of people infected with HCV do not respond to available pharmacological regimes (ranging from 20% of patients infected with genotype 3 to 80% of those infected with genotype 1b) [141].

Owing to the fact that the mechanisms of action of IFN-α [142, 143] and ribavirin [143, 144] against Flaviviridae viruses are probably different from those described for artemisinin [145], there exists the possibility of additive effects of these drugs, which, indeed, were observed in the study of Romero et al. [146]. IFN binds to cell surface receptors and stimulates signal pathways that lead to the activation of cellular enzymes that repress viral replication [143], whereas ribavirin, in addition to its immunomodulatory properties, has direct antiviral activities that can be ascribed to several possible mechanisms. These include the inhibition of the HCV RNA-dependent RNA polymerase NS5B and the recently described activity as an RNA mutagen able to impair viral replication [147].

11.7.5 Other Viruses

Disbrow et al. [89] reported that dihydroartemisinin inhibited papillomavirus-induced tumor formation in vivo. Human papillomavirus-16 (HPV-16) is causatively linked with carcinogenesis [148]. This raises the question of whether dihydroartemisinin acts on HPV-16, thereby preventing tumor development. Although dogs topically treated with dihydroartemisinin did not develop mucosal tumors, they developed antibodies against the viral L1 capsid protein, suggesting that dihydroartemisinin had inhibited tumor growth but not early rounds of papillomavirus replication [89].

Other authors reported that artemisinins are inactive against human herpes simplex virus-6 (HHV-6), which is causatively related to the development of Kaposi's sarcoma in immunodeficiency diseases such as AIDS [149]. Artesunate was very active against human simplex virus-1 (HSV-1), weakly active against human immunodeficiency virus-1 (HIV-1), but inactive against influenza A viruses [105].

11.8 Side Effects of Artemisinin

A clinical-safety review of 108 clinical studies enrolling 9241 patients provided ample evidence that artemisinins are safe and without serious adverse events or severe significant toxicity and especially without neurotoxicity [53].

Neurotoxic effects have been repeatedly reported in experiments with mice rats, and dogs, as recently reviewed by Toovey [150]. Affected areas in the brain stem are the reticular system with autonomic control, the vestibular system, the auditory system (trapezoid nucleus), and the red nucleus, which is important for coordination [151–157]. Longer exposure times with lower peak blood concentrations are more neurotoxic than shorter durations of exposure and higher peak blood levels [158]. These animal experiments give rise to concerns about the safety of artemisinin and its derivatives in human beings.

However, few reports point to neurotoxic effects of artemether in clinical application. Van Hensbroek et al. [159] observed delayed coma recovery times (CRT) under treatment with intramuscular artemether vs. intravenous quinine in Gambian children suffering from malaria. Due to these conflicting results, Stepniewska et al. [160] performed a meta-analysis of seven studies with 1919 malaria patients. Applying a uniform CRT definition, no significant different in CRT between artemether and quinine was found. Furthermore, no statistically significant difference with regard to neurological sequalae was observed. In a recent study by Dondorp et al. [161], malaria patients treated with artesunate were compared with quinine-treated patients. The authors did not find significant differences of neutoxic symptoms (in terms of times to speak, eat, and sit) between both treatment groups. Neurological sequelae did not occur after treatment. Interestingly, malaria patients who developed late-onset hypoglycemia had a higher incidence of death than artesunate-treated patients without hypoglycaemia. This may be an issue that deserves further investigation in the future.

11.9 Conclusion and Perspectives

After being used in TCM for two millennia, one of the "gems" of TCM's treasure box has been rediscovered in recent years. Artemisinin is certainly one of the most promising natural products of the past two decades. With respect to malaria, it has great potential to contribute to a change in the desperate situation that the world faces. Fortunately, the value of this molecule is not limited to the treatment of malaria, and a wealth of papers has been published demonstrating the activity of artemisinin and its derivatives against cancer cells, schistosomiasis, and various viral diseases. Even more, the bioactivity of artemisinin and its derivatives is much broader (Fig. 11.2). An enlarged bioactivity profile of artemisinin and its derivatives may include the inhibition of

- Protozoans in addition to the *Plasmodium* or *Schistosoma* genus, i.e., *Toxoplasma gondii* [162–169], *Leishmania major* and *L. donovani* [170–173]; *Trypanosoma cruzi* and *T. brucei rhodesiense* [174], *Neospora caninum* [175], and *Eimeria tenella* [176];
- Trematodes such as *Echinostoma caproni* [177];
- Fungi as exemplified by *Pneumocystis carinii* [178], *Candida albicans* [179], and *Cryptoccocus neoformans* [180];
- Yeast (*Saccharomyces cerevisiae*) [65, 181]; and
- Bacteria, i.e., *Leptospira serovars* [182] and some anaerobic bacteria [183].

Ironically, in an age when many scientists are searching for compounds with increased specificity to their molecular and cellular targets, artemisinin is coming up

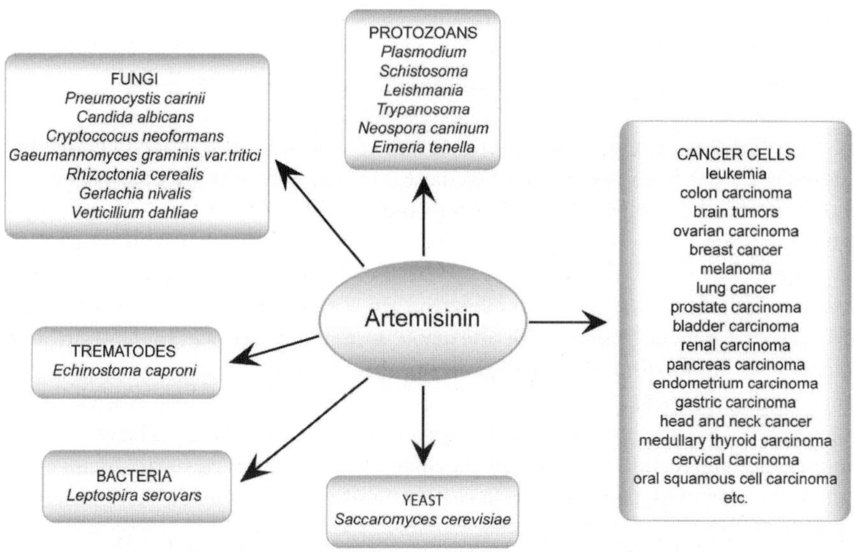

Fig. 11.2 Bioactivity of artemisinins

with its own multifunctionality. This class of compounds seems to have many different targets against different diseases. Conceptually, modern concepts in molecular pharmacology aim to increase treatment efficacy and to decrease unwanted side effects by developing compounds that attack disease-related target molecules with high affinity. It is quite obvious that the natural evolution of pharmacologically active compounds in plants took a different path. Natural products have evolved into plants as chemical weapons to protect from infections with bacteria, viruses, and other microorganisms as well as herbivores such as insects, worms, humans, etc. It comes as no surprise that multifunctional molecules might be more versatile and, hence, more successful than mono-specific ones in protecting plants from environmental harm. In the case of artemisinin, it has been shown to be active against various plant pathogenic fungi (*Gaeumannomyces graminis var. tritici, Rhizoctonia cerealis, Gerlachia nivalis*, and *Verticillium dahliae*) [179], supporting a role of artemisinin as protective agent for plants.

References

1. Efferth T, Miyachi H, Bartsch H (2007a) Cancer Genom Proteom 4:81–91
2. Efferth T, Li PC, Konkimalla VS, Kaina B (2007b) Trends Mol Med 13:353–361
3. Efferth T, Fu YJ, Zu YG, Schwarz G, Konkimalla VS, Wink M (2007c) Curr Med Chem 14:2024–2032
4. Efferth T, Rücker G, Falkenberg M, Manns D, Olbrich A, Fabry U, Osieka R (1996) Arzneimittelforschung 46:196–200
5. Efferth T, Kaina B (2004) Cancer Genom Proteom 1:363–370
6. Efferth T, Olbrich A, Sauerbrey A, Ross DD, Gebhart E, Neugebauer M (2002a) Anticancer Res 22:4221–4224
7. Efferth T, Olbrich A, Bauer R (2002b) Biochem Pharmacol 64:617–623
8. Efferth T, Sauerbrey A, Halatsch ME, Ross DD, Gebhart E (2003a) Naunyn Schmiedebergs Arch Pharmacol 367:56–67
9. Efferth T, Rauh R, Kahl S, Tomicic M, Boechzelt H, Tome ME, Briehl MM, Bauer R, Kaina B (2005a) Biochem Pharmacol 69:811–818
10. Efferth T, Chen T, Kaina B, Wang G (2005b) Cancer Genomics Proteomics 2:115–124
11. Efferth T (2005) Oncol Rep 13:459–463
12. Rinner B, Siegl V, Pürstner P, Efferth T, Brem B, Greger H, Pfragner R (2004) Anticancer Res 24:495–500
13. Adams M, Efferth T, Bauer R (2006) Planta Med 72:862–864
14. Adams M, Mahringer A, Bauer R, Fricker G, Efferth T (2007a) Planta Med 73:1475–1478
15. Adams M, Mahringer A, Fricker G, Efferth T, Bauer R (2007b) Planta Med 73:1154–1157
16. Konkimalla VB, Efferth T (2008) Anti–cancer natural product library from traditional Chinese medicine. Combinat Chem High Throughput Screening 11:7–15
17. Konkimalla VB, Suhas VL, Chandra NR, Gebhart E, Efferth T (2007) Expert Rev Anticancer Ther 7:317–329
18. Rauh R, Kahl S, Boechzelt H, Bauer R, Kaina B, Efferth T (2007) Chin Med 2:8
19. Romero MR, Serrano MA, Efferth T, Alvarez M, Marin JJ (2007) Planta Med 73:552–558
20. Wang J, Zheng Y, Efferth T, Wang R, Shen Y, Hao X (2005) Phytochemistry 66:697–701
21. Wang YF, Cao JX, Efferth T, Lai GF, Luo SD (2006a) Chem Biodivers 3:646–653
22. Wang YF, Lai GF, Efferth T, Cao JX, Luo SD (2006b) Chem Biodivers 3:1023–1030
23. Hsu E (2006) Br J Clin Pharmacol 61:666–670
24. Klayman DL (1985) Science 228:1049–1055

25. Li Y, Wu YL (1998) Med Trop 58 Suppl 3:9S–12S
26. Yeung S, Pongtavornpinyo W, Hastings IM, Mills AJ, White NJ (2004) Am J Trop Med Hyg 71(2 Suppl):179–186
27. Semenov A, Olson JE, Rosenthal PJ (1998) Antimicrob Agents Chemother 42:2554–2558
28. Shenai BR, Sijwali PS, Singh A, Rosenthal PJ (2000) J Biol Chem 275:29000–29010
29. Berman PA, Adams PA (1997) Free Radic Biol Med 22:1283–1288
30. O'Neill PM, Posner GH (2004) J Med Chem 47:2945–2964
31. Posner GH, Oh CH (1992) J Am Chem Soc 114:8328–8329
32. Meshnick SR, Yang YZ, Lima V, Kuypers F, Kamchonwongpaisan S, Yuthavong Y (1993) Antimicrob Agents Chemother 37:1108–1114
33. Butler AR, Gilbert BC, Hulme P, Irvine LR, Renton L, Whitwood AC (1998) Free Radic Res 28:471–476
34. Shterman N, Kupfer B, Moroz C (1991) Pathobiology 59:19–25
35. Payne AG (2003) Med Hypotheses 61:206–209
36. Moore JC, Lai H, Li JR, Ren RL, McDougall JA, Singh NP, Chou CK (1995) Cancer Lett 98:83–87
37. Efferth T, Benakis A, Romero MR, Tomicic M, Rauh R, Steinbach D, Hafer R, Stamminger T, Oesch F, Kaina B, Marschall M (2004a) Free Radic Biol Med 37:998–1009
38. Aulbert E, Disselhoff W, Sorje H, Schulz E, Gericke D (1980) Eur J Cancer 16:1217–1232
39. Gatter KC, Brown G, Trowbridge IS, Woolston RE, Mason DY (1983) J Clin Pathol 36:539–545
40. Judd W, Poodry CA, Strominger JL (1980) J Exp Med 152:1430–1435
41. Trowbridge IS, Omary MB (1981) Proc Natl Acad Sci USA 78:3039–3043
42. Sutherland R, Delia D, Schneider C, Newman R, Kemshead J, Greaves M (1981) Proc Natl Acad Sci USA 78:4515–4519
43. Butler D (1997) Nature 386:535–541
44. WHO (2000) Trans R Soc Trop Med Hyg 94:S1–S75
45. White NJ (2004) J Clin Invest 113:1084–1092
46. Silachamroon U, Krudsood S, Phophak N, Looareesuwan S (2002) Korean J Parasitol 40:1–7
47. Mutabingwa TK (2005) Acta Trop 95:305–315
48. White N (1999) Philos Trans R Soc London B Biol Sci 354:739–749
49. Mishra SK, Asthana OP, Mohanty S, Patnaik JK, Das BS, Srivastava JS, Satpathy SK, Dash S, Rath PK, Varghese K (1995) Trans R Soc Trop Med Hyg 89:299–302
50. Mohanty S, Mishra SK, Satpathy SK, Satpathy S, Patnaik JK (1997) Trans R Soc Trop Med Hyg 91:328–330
51. Price R, van Vugt M, Nosten F, Luxemburger C, Brockman A, Phaipun L, Chongsuphajaisiddhi T, White N (1998) Am J Trop Med Hyg 59:883–888
52. Mohanty S, Patel DK, Pati SS, Mishra SK (2006) Indian J Med Res 124:245–260
53. Ribeiro IR, Olliaro P (1998) Med Trop (Mars) 58:50–53
54. Adjuik M, Babiker A, Garner P, Olliaro P, Taylor W, White N (2004) Lancet 363:9–17
55. Davies TME, Karunajeewa HA, Ilett KF (2005) Med J Aust 182:181–185
56. Bar–Zeev N, White N (2006) J Trop Pediatr 52:78–82
57. Bakshi R, Hermeling–Fritz I, Gathmann I, Alteri E (2000) Trans R Soc Trop Med Hyg 94:419–424
58. Omari AA, Gamble C, Garner P (2005) Cochrane Database Syst Rev 19:CD005564
59. Bukirwa H, Critchley J (2006) Cochrane Database Syst Rev 25:CD004966
60. Obonyo CO, Juma EA, Ogutu BR, Vulule JM, Lau J (2007) Trans R Soc Trop Med Hyg 101:117–126
61. Efferth T (2005a) Drug Resist Updat 8:85–97
62. Efferth T (2006) Curr Drug Targets 7:407–421
63. Efferth T (2007) Planta Med 73:299–309
64. Efferth T, Oesch F (2004) Biochem Pharmacol 68:3–10
65. Efferth T, Dunstan H, Sauerbrey A, Miyachi H, Chitambar CR (2001) Int J Oncol 18:767–773

66. Efferth T, Davey M, Olbrich A, Rücker G, Gebhart E, Davey R (2002b) Blood Cells Mol Dis 28:160–168
67. Efferth T, Ramirez T, Gebhart E, Halatsch ME (2004b) Biochem Pharmacol 67:1689–1700
68. Efferth T, Briehl MM, Tome ME (2003b) Int J Oncol 23:1231–1235
69. Efferth T, Sauerbrey A, Olbrich A, Gebhart E, Rauch P, Weber HO, Hengstler JG, Halatsch ME, Volm M, Tew KD, Ross DD, Funk JO (2003c) Mol Pharmacol 64:382–394
70. Efferth T, Volm M (2005) In Vivo 19:225–232
71. Aisen P (2004) Int J Biochem Cell Biol 36:2137–2143
72. Kelter G, Steinbach D, Konkimalla VB, Tahara T, Taketani S, Fiebig HH, Efferth T (2007) PLoS ONE 2:e798
73. Singh NP, Lai H (2001) Life Sci 70:49–56
74. Lai H, Sasaki T, Singh NP, Messay A (2005) Life Sci 76:1267–1279
75. Folkman J (1992) Semin Cancer Biol 3:65–71
76. Kerbel R, Folkman J (2002) Nat Rev Cancer 2:727–739
77. Chen HH, Zhou HJ, Fang X (2003) Pharmacol Res 48:231–236
78. Chen HH, Zhou HJ, Wang WQ, Wu GD (2004a) Cancer Chemother Pharmacol 53:423–432
79. Chen HH, Zhou HJ, Wu GD, Lou XE (2004b) Pharmacology 71:1–9.
80. Dell'Eva R, Pfeffer U, Vene R, Anfosso L, Forlani A, Albini A, Efferth T (2004) Biochem Pharmacol 68:2359–2366
81. Huan-huan C, Li-Li Y, Shang-Bin L (2004) Cancer Lett 211:163–173
82. Anfosso L, Efferth T, Albini A, Pfeffer U (2006) Pharmacogenomics J 6:269–278
83. Li Y, Shan F, Wu JM, Wu GS, Ding J, Xiao D, Yang WY, Atassi G, Leonce S, Caignard DH, Renard P (2001) Bioorg Med Chem Lett 11:5–8
84. Sadava D, Phillips T, Lin C, Kane SE (2002) Cancer Lett 179:151–156
85. Wang Q, Wu LM, Zhao Y, Zhang XL, Wang NP (2002) Yao Xue Xue Bao 37:477–478
86. Singh NP, Lai HC (2004) Anticancer Res 24:2277–2280
87. Yamachika E, Habte T, Oda D (2004) Anticancer Res 24:2153–2160
88. Efferth T, Giaisi M, Merling A, Krammer PH, Li-Weber M (2007d) PLoS ONE 2:e693
89. Disbrow GL, Baege AC, Kierpiec KA, Yuan H, Centeno JA, Thibodeaux CA, Hartmann D, Schlegel R (2005) Cancer Res 65:10854–10861
90. Lai H, Singh NP (2006) Cancer Lett 231:43–48
91. Berger TG, Dieckmann D, Efferth T, Schultz ES, Funk JO, Baur A, Schuler G (2005) Oncol Rep 14:1599–1603
92. Singh N, Verma K (2002) Arch Oncol 10:279–280
93. Singh NP, Panwar VK (2006) Integr Cancer Ther 5:391–394
94. Ross AGP, Bartley PM, Sleigh AC, Olds GR, Li YS, Williams GM, McManus DP (2002) N Engl J Med 346:1212–1220
95. Gryseels B, Polman K, Clerinx J, Kestens L (2006) Lancet 368:1106–1118
96. Steinmann P, Keiser J, Bos R, Tanner M, Utzinger J (2006) Lancet Infect Dis 6:411–425
97. Engels D, Chitsulo L, Montresor A, Savioli L (2002) Acta Trop 82:139–146
98. WHO (2002) WHO Tech Rep Ser, p 912
99. Chen DJ, Fu LF, Shao PP, Wu FZ, Fan CZ, Shu H, Ren CX, Sheng XL (1980) Zhong Hui Yi Xue Zha Zhi 60:422–425
100. Utzinger J, Xiao SH, Tanner M, Keiser J (2007) Curr Opin Investig Drugs 8:105–116
101. Utzinger J, Xiao SH, Keiser J, Chen MG, Zheng J, Tanner M (2001) Curr Med Chem 8:1841–1860
102. Utzinger J, Keiser J, Xiao SH, Tanner M, Singer BH (2003) Antimicrob Agents Chemother 47:1487–1495
103. Utzinger J, Zhou XN, Chen MG, Bergquist R (2005) Acta Trop 96:69–96
104. Xiao SH (2005) Acta Trop 96:153–167
105. Efferth T, Marschall M, Wang X, Huong SM, Hauber I, Olbrich A, Kronschnabl M, Stamminger T, Huang ES (2002c) J Mol Med 80:233–242
106. Britt WJ, Alford CA (1996) Cytomegalovirus. In: Fields BN, Knipe DM, Howley PM (eds) Fields Virology, 3rd edn. Lippincott-Raven, Philadelphia, pp 2493–2523

107. Stratta RJ, Shaeffer MS, Markin RS, Wood RP, Langnas AN, Reed EC, Doonovan JP, Woods GL, Bradshaw KA, Pillen TJ, Shaw BW Jr (1992) Dig Dis Sci 37:673–688
108. Shen CY, Ho MS, Chang SF, Yen MS, Ng HAT, Huang ES, Wu CW (1993) J Infect Dis 168:449–452
109. Crumpacker CS (1996) N Engl J Med 355:721–729
110. Chrisp P, Clissold SP (1991) Drugs 41:104–129
111. Hitchcock MJM, Jaffe HS, Martin JC, Stagg RJ (1996) Antimicrob Agents Chemother 7:115–127
112. Polis MA, Spooner KM, Baird BFM, Manischewitz JF, Jaffe HS, Fisher PE, Fallon J, Davey RT Jr, Kovacs JA, Walker RE, Whitcup SM, Nussenblatt RB, Lane HC, Masur H (1995 Antimicrob Agents Chemother 39:882–886
113. Erice A (1999) Clin Microbiol Rev 12:286–297
114. Marchini A, Liu H, Zhu H (2001) J Virol 75:1870–1878
115. Ghazal P, Lubon H, Fleckenstein B, Henninghausen L (1987) Proc Natl Acad Sci USA 84:3658–3662
116. Yurochko AD, Kowalik TF, Huong SM, Huang ES (1995) J Virol 69:5391–5400
117. Yurochko AD, Mayo MW, Poma EE, Baldwin AS Jr, Huang ES (1997) J Virol 71:4638–4648
118. Johnson RA, Wang X, Ma X-L, Huong S-M, Huang E-S (2001) J Virol 75:6022–6032
119. Chou S, Marousek G, Guentzel S, Follansbee SE, Poscher ME, Lalezari JP, Miner RC, Drew WL (1997) J Infect Dis 176:786–789
120. Scott GM, Weinberg A, Rawlinson WD, Chou S (2007) Antimicrob Agents Chemother 51:89–94
121. Kaptein SJ, Efferth T, Leis M, Rechter S, Auerochs S, Kalmer M, Bruggeman CA, Vink C, Stamminger T, Marschall M (2006) Antiviral Res 69:60–69
122. Alejo E (1999) Clin Microbiol Rev 12:286–297
123. Fang CT, Nath N, Pielech M, Dodd RY (1981) J Virol Methods 2:349–356
124. Lien J, Petcu DJ, Aldrich CE, Mason WS (1987) J Virol 61:3832–3840
125. Wang GH, Seeger C (1993) J Virol 67:6507–6512
126. Molnar-Kimber KL, Summers JW, Mason WS (1984) J Virol 51:181–191
127. Bartholomeusz A, Locarnini S (2001) Hepatology 34:432–435
128. Iino S (2002) Oncology 62 (Suppl 1):18–23
129. Beasley RP, Hwang LY (1991) Overview on the epidemiology of hepatocellular carcinoma. In: Hollinger FB, Lemon SM, Margolis M (eds) Viral Hepatitis and Liver Disease. Williams and Willkins, Baltimore, pp. 532–535
130. EASL Jury (2003) J Hepatol 38:533–540
131. Romero MR, Efferth T, Serrano MA, Castano B, Macias RI, Briz O, Marin JJ (2005) Antiviral Res 68:75–83
132. Batty KT, Davis TM, Thu LT, Binh TQ, Anh TK, Ilett KF (1996) J Chromatogr B Biomed Appl 677:345–350
133. Placidi L, Faraj A, Loi AG, Pierra C, Egron D, Cretton-Scott E, Gosselin G, Perigaud C, Martin LT, Schinazi RF, Imbach JL, el Kouni MH, Bryant ML, Sommadossi JP (2001) Antivir Chem Chemother 12:99–108
134. Romero MR, Martinez-Diez MC, Larena MG, Macias RIR, Dominguez M, Garcia-Monzon C, Serrano MA, Marin JJG (2002) Antivir Chem Chemother 13:371–380
135. Chouteau P, Le Seyec J, Saulier-Le Drean B, Cannie I, Brissot P, Lescoat G, Guguen-Guillouzo C, Gripon P (2001) J Hepatol 34:108–113
136. Paeshuyse J, Coelmont L, Vliegen I, Van Hemel J, Vandenkerckhove J, Peys E, Sas B, De Clercq E, Neyts J (2006) Biochem Biophys Res Commun 348:139–144
137. Fillebeen C, Rivas-Estilla AM, Bisaillon M, Ponka P, Muckenthaler M, Hentze MW, Koromilas AE, Pantopoulos K (2005) J Biol Chem 280:9049–9057
138. Tanikawa K (2004) Hepatobiliary Pancreat Dis Int 3:17–20
139. Buckwold VE, Beer BE, Donis RO (2003) Antiviral Res 60:1–15
140. Behrens SE, Grassmann CW, Thiel HJ, Meyers G, Tautz N (1998) J Virol 72:2364–2372

141. McHutchison JG, Gordon SC, Schiff ER, Shiffman ML, Lee WM, Rustgi VK, Goodman ZD, Ling MH, Cort S, Albrecht JK (1998) N Engl J Med 339:1485–1492
142. Samuel CE (2001) Clin Microbiol Rev 14:778–809
143. Durantel D, Carrouee-Durantel S, Branza-Nichita N, Dwek RA, Zitzmann N (2004) Antimicrob Agents Chemother 48:497–504
144. Lau JY, Tam RC, Liang TJ, Hong Z (2002) Hepatology 35:1002–1009
145. Yang YZ, Little B, Meshnick SR (1994) Biochem Pharmacol 48:569–573
146. Romero MR, Serrano MA, Vallejo M, Efferth T, Alvarez M, Marin JJ (2006) Planta Med 72:1169–1174
147. Tam RC, Lau JY, Hong Z (2001) Antivir Chem Chemother 12:261–272
148. Schiffman M, Castle PE, Jeronimo J, Rodriguez AC, Wacholder S (2007) Lancet 370:890–907
149. Naesens L, Bonnafous P, Agut H, De Clercq E (2006) J Clin Virol 37(Suppl 1):S69–75
150. Toovey S (2006) Toxicol Lett 166:95–104
151. Brewer TG, Grate SJ, Peggins JO, Weina PJ, Petras JM, Levine BS, Heiffer MH, Schuster BG (1994a) Am J Trop Med Hyg 51:251–259
152. Brewer TG, Peggins JO, Grate SJ, Petras JM, Levine BS, Weina PJ, Swearengen J, Heiffer MH, Schuster BG (1994b) Trans R Soc Trop Med Hyg 88 (Suppl 1):S33–S36
153. Kamchonwongpaisan S, McKeever P, Hossler P, Ziffer H, Meshnick SR (1997) Am J Trop Med Hyg 56:7–12
154. Genovese RF, Newman DB, Li Q, Peggins JO, Brewer TG (1998a) Brain Res Bull 45:199–202
155. Jones-Brando L, D'Angelo J, Posner GH, Yolken R (2006) Antimicrob Agents Chemother 50:4206–4208
156. Panossian LA, Garga NI, Pelletier D (2005) Ann Neurol 58:812–813
157. Petras JM, Young GD, Bauman RA, Kyle DE, Gettayacamin M, Webster HK, Corcoran KD, Peggins JO, Vane MA, Brewer TG (2000) Anat Embryol 201:383–397
158. Li QC, Mog SR, Si YZ, Kyle DE, Gettayacamin M, Milhous WK (2002) Am J Trop Med Hyg 66:516–525
159. van Hensbroek MB, Onyiorah E, Jaffar S, Schneider G, Palmer A, Frenkel J, Enwere G, Forck S, Nusmeijer A, Bennett S, Greenwood B, Kwiatkowski D (1996) N Engl J Med 335:69–75
160. Stepniewska K, Day N, Babiker A, Lalloo D, Warrell D, Olliaro P, White N (2001) Trans R Soc Trop Med Hyg 95:637–650
161. Dondorp A, Nosten F, Stepniewska K, Day N, White N (2005) Lancet 366:717–725
162. Chang HR, Pechere JC (1988) Trans R Soc Trop Med Hyg 82:867
163. Ke OY, Krug EC, Marr JJ, Berens RL (1990) Antimicrob Agents Chemother 34:1961–1965
164. Genovese RF, Newman DB, Petras JM, Brewer TG (1998b) Pharmacol Biochem Behav 60:449–458
165. Berens RL, Krug EC, Nash PB, Curiel TJ (1998) J Infect Dis 177:1128–1131
166. Son ES, Song KJ, Shin JC, Nam HW (2001) Korean J Parasitol 39:133–141
167. Holfels E, McAuley J, Mack D, Milhous WK, McLeod R (1994) Antimicrob Agents Chemother 38:1392–1396
168. Nagamune K, Beatty WL, Sibley LD (2007a) Epub ahead of print
169. Nagamune K, Moreno SN, Sibley LD (2007b) Epub ahead of print
170. Yang DM, Liew FY (1993) Parasitology 106 (Pt 1):7–11
171. Avery MA, Muraleedharan KM, Desai PV, Bandyopadhyaya AK, Furtado MM, Tekwani BL (2003) J Med Chem 46:4244–4258
172. Menon RB, Kannoth MM, Tekwani BL, Gut J, Rosenthal PJ, Avery MA (2006) Comb Chem High Throughput Screen 9:729–741
173. Sen R, Bandyopadhyay S, Dutta A, Mandal G, Ganguly S, Saha P, Chatterjee M (2007) J Med Microbiol 56(Pt 9):1213–1218
174. Mishina YV, Krishna S, Haynes RK, Meade JC (2007) Antimicrob Agents Chemother 51:1852–1854

175. Kim JT, Park JY, Seo HS, Oh HG, Noh JW, Kim JH, Kim DY, Youn HJ (2002) Vet Parasitol 103:53–63
176. Allen PC, Danforth HD, Augustine PC (1998) Int J Parasitol 28:1131–1140
177. Keiser J, Brun R, Fried B, Utzinger J (2006) Antimicrob Agents Chemother 50:803–805
178. Merali S, Meshnick SR (1991) Antimicrob Agents Chemother 35:1225–1227
179. Tang HQ, Hu J, Yang L, Tan RX (2000) Planta Med 66:391–393
180. Galal AM, Ross SA, Jacob M, ElSohly MA (2005) J Nat Prod 68:1274–1276
181. Li W, Mo W, Shen D, Sun L, Wang J, Lu S, Gitschier JM, Zhou B (2005) PLoS Genet 1:e36
182. Murray CK, Ellis MW, Hospenthal DR (2004) Am J Trop Med Hyg 71:685–686
183. Shoeb HA, Tawfik AF, Shibl AM, el-Feraly FS (1990) J Chemother 2:362–367

Chapter 12
Anticancer and Immunomodulatory Properties of *Tinospora*

Anil Mittal and Rana P. Singh

Abstract There has been an upsurge in the discovery of bioactive phytochemicals having a chemopreventive ability against various diseases including immunological disorders and cancer. Based on their uses in Ayurveda or herbal medicine, many plants have been the subject of experimental evaluation to provide a scientific rationale for their medicinal values. In this regard, *Tinospora* has a long history of use against various ailments including spasms, inflammation, arthritis, allergy, diabetes, cardiotoxicity, and immunosuppression. However, scientific evidences for its biological activities are limited. Recently, many studies have been carried out to support the acclaimed as well as to discover the novel potential of *Tinospora* which have also revealed its anticancer and radioprotective activities. Overall, the anticancer and immunomodulatory activities of *Tinospora* and its bioactive components could be further explored in relevant experimental model systems for its potential clinical benefits.

Keywords *Tinospora* · Anticancer · Immunomodulation · Antioxidant system · Radioprotection

Abbreviations and Symbols

AST	Average survival time
Ag	Antigen
CR-3	Complement receptor 3
CAPE	Caffeic acid phenethyl ester
CFU	Colony-forming units
CCL_4	Carbon tetra chloride
DTD	DT-diaphorase
DC	Dendritic cells

R.P. Singh (✉)
Cancer Biology Laboratory, School of Life Sciences, Jawaharlal Nehru University, New Delhi, India,
e-mail: rana_singh@mail.jnu.ac.in

K.G. Ramawat (ed.), *Herbal Drugs: Ethnomedicine to Modern Medicine*,
DOI 10.1007/978-3-540-79116-4_12,

DL	Dalton's lymphoma
EAC	Ehrlich ascites carcinoma
GSH	Reduced glutathione
GST	Glutathione S-transferase
GPX	Glutathione-peroxidase
GR	Glutathione reductase
GM-CSF	Granulocyte-monocyte colony-stimulating factor
γGT	γ-glutamyl transpeptidase
G1-4A	Arabinogalactan
HMG1	High-mobility group1
iNOS	Inducible nitric oxide synthase
IL	Interleukin
IFN	Interferon
LPx	Lipid peroxidation
LDH	Lactate dehydrogenase
LPS	Lipopolysaccharide
MST	Median survival time
MTD	Maximum tolerated level dose
MCP-1	Monocyte chemo-attractant protein-1
MPO	Myeloperoxidase
NK	Natural killer
NO	Nitric oxide
PAMP	Pathogen-associated molecular pattern
PRR	Pattern recognition receptors
RTc	Root extract of *Tinospora cordifolia*
ROS	Reactive oxygen species
SOD	Superoxide dismutase
TCE	Extract of *Tinospora cordifolia*
TIMP-1	Tissue inhibitor of metalloprotease-1
TAM	Tumor-associated macrophages
TNF	Tumor necrosis factor
TLR	Toll-like receptor
VEGF	Vascular endothelial cell growth factor

12.1 Introduction

Carcinogenesis is a multistep process induced by various types of carcinogens that ultimately lead to the development of cancer. Many biological and molecular events have been identified that are modulated by natural agents to inhibit the different stages of carcinogenesis. Epidemiological data suggest that consumption of a fiber-rich diet (low lipid content) and yellow-green vegetables can reduce the risk of various cancers [1–6]. Many laboratory studies for the chemoprevention of cancer using natural agents have revealed encouraging results [7]. In this regard, a large number of plants have been screened for their anticancer activities. These plants were found to contain specific antineoplastic phytochemicals that have been isolated from the different parts of the plants. For example, silibinin (polyphenolic flavonoid) extracted from the fruits or seeds of milk thistle has been shown to suppress prostate, skin, lung, and many other cancers [8, 9]; curcumin from turmeric inhibits colon cancer [10, 11], capsaicin from red chili inhibits prostate cancer [12], and resveratrol from red grapes inhibits breast cancer [13]. Similarly, lycopene from tomato, β-carotene from carrots, genistein from soybean, catechin from tea and ellagic acid from pomegranate etc have also been shown to possess anticancer activities [14].

Herbal medicines offer treatment for various diseases by modulation of the immune system and are classified as immunomodulators. These immunomodulators may alter the activity of the immune system by regulation of cytokine expression and secretion. Many plants have been found to have immunomodulatory activities by altering the expression of various cytokines. Such plants are *Allium sativum, Ananas cosmosus, Echinacea purpuria, Grifolia frondosa, Poria cocos, Smilax glabra, Withania somnifera, Curcuma longa*, and *Tinospora cordifolia* [15]. *Tinospora* has been shown to have various immunomodulatory properties. For example, the aqueous extract of *T. cordifolia* stem, which contains arabinogalactan, shows immunological properties [16]. Plant-derived polysaccharides are regarded as excellent immunomodulators due to their therapeutic properties, less toxicity, and fewer side effects as compared to other immunomodulators [17].

There are various species of *Tinospora*, of which *T. cordifolia* is the most extensively studied species for its biological activities. *T. cordifolia* is a glabrous, deciduous, and climbing shrub that belongs to the family Menispermaceae [18, 19] and is found throughout the tropical Indian subcontinent and China. It is popularly known as Giloy, Guduchi, or amritha and is used in Ayurvedic and veterinary medicines. In addition, it is well known for its various properties such as antispasmodic, antiperiodic, anti-inflammatory, antiarthritic, antiallergic, antidiabetic [20–22], antioxidant [23], antihepatotoxic [24], antifertility [25], and cardioprotective [26] properties including acetylcholine esterase inhibitory activity [27]. *Tinospora* root is specifically known for its antileprotic, antistress, and antimalarial activities [22, 28]. Pretreatment of animals with the aqueous extract of the stem of *T. cordifolia* shows protection against infection by *E. coli* and *Staphylococcus aureus* and also protects from mixed abdominal sepsis [29, 30].

A variety of constituents have been isolated from *T. cordifolia* including alkaloids, diterpenoids, steroids, glycosides, phenolics, sesquiterpenoids, aliphatic compounds, and polysaccharides [22]. The leaves of *T. cordifolia* contain large amounts of calcium, phosphorus, and protein [22, 31]. Various alkaloids like berberine, pulmatine, terbertarine, and magnoflorine have been isolated from the stem of *T. cordifolia*. The roots also contain other alkaloids such as choline, tinosporin, isocolumbin, and tetrahydropalmatine [32–34]. The methanolic extract of *T. cordifolia* shows the presence of norditerpene furan glycosides, phenylpropanoids, diterpine furan glycosides, and plytoecdysones [35]. The following sections focus on the anticancer and immunomodulatory activities of *Tinospora*.

12.2 Effect of *Tinospora* on Carcinogen/Drug Metabolism and Antioxidant Systems

Xenobiotic metabolism plays a critical role in delivering the active carcinogenic dose of a potential carcinogen. Generally, it consists of phase I and phase II metabolizing enzyme systems in which, due to the activity of the former, the epoxide can be formed that is an active form of carcinogen known to bind with the DNA, resulting in mutation during cell proliferation. The phase II enzyme system can make it inactive to facilitate their excretion outside the body [36]. In this regard, it has been shown that the hydroalcoholic extract of aerial roots of *T. cordifolia* modulates both phase I and phase II enzymes as well as antioxidant enzymes including cytochrome p450 reductase, cytochrome b5 reductase, glutathione transferase (GST), DT-diaphorase (DTD), superoxide dismutase (SOD), catalase, glutathione peroxidase (GPX), and glutathione reductase (GR) in mouse liver [37]. This treatment also increases reduced glutathione (GSH) content in liver and extrahepatic organs (lung, kidney, forestomach) and SOD, catalase in kidney; catalase, GST, SOD in lung; and GST, DTD, and SOD in forestomach. Inhibition of lipid peroxidation has also been shown to occur by this treatment [37]. Thus, *Tinospora* acts as a bifunctional enzyme inducer because it induces both phase I and phase II enzyme systems. Additionally, *T. cordifolia* could remove oxidative stress conditions by activating antioxidant enzymes, thereby maintaining the reducing environment, along with the removal of the reactive oxygen species and neutralization of reactive intermediate species produced from the exposure to xenobiotics including chemical carcinogens [37, 38]. Together, these reports suggest the chemopreventive and antioxidant activities of *Tinospora*.

12.3 Anticancer Activity of *Tinospora*

Different extract preparations of *Tinospora* have been tested for their anticancer activity in various cancer model systems. In Ehrlich ascites carcinoma (EAC)-bearing mice, dichloromethane extract of *T. cordifolia* (TCE) has shown anticancer activity [32]. In this study, TCE is reported to have a dose-dependent increase in the survival of EAC-bearing mice receiving 25 to 100 mg/kg doses of the extract. A 50 mg/kg body weight of dose has been found to have optimal effect on survival. Medium and average survival times for this dose were 53 and 56 d as compared to the respective non-drug-treated controls, which were 19 and 18 d, respectively. TCE has also shown a time-dependent decrease in glutathione (GSH) level but enhanced lipid peroxidation in EAC cells from mice receiving 50 mg/kg dose of TCE [32]. These observations indicate the potential cytotoxic effects of TCE on EAC cells via oxidative mechanisms.

In another study, TCE has been reported to have a cytotoxic effect on HeLa cells [39]. This cytotoxic effect of TCE on HeLa cells was found to be associated with increased lipid peroxidation, release of lactate dehydrogenase (LDH), and decrease in GST activity [40]. Another report suggests that *T. cordifolia* has been used in the successful treatment of throat cancer in humans [41]. Overall, these reports suggest the anticancer potential of *Tinospora*.

12.4 *Tinospora* Inhibits Tumor Angiogenesis

Angiogenesis is the formation of new blood vessels from preexisting ones. The physiological process of angiogenesis is complex and strictly regulated, and its deregulation can cause a number of diseases including cancer, endometriosis, diabetic retinopathy, rheumatoid arthritis, and psoriasis [42, 43]. Angiogenesis is required at the very early stages of tumor development and is important for the metastatic spread and invasiveness of tumors. Angiogenesis plays an important role in the growth and progression of solid tumors, and without angiogenesis tumors cannot grow beyond $\sim$2 mm in size [44, 45]. Therefore, to control tumor growth, invasiveness, and metastasis, antiangiogenesis strategies represent promising approaches in the treatment of various cancers [46–48].

In a study, *T. cordifolia* was found to possess antiangiogenic activity in melanoma B16F10 cell-induced capillary formation in vivo and in vitro. Many proinflammatory cytokines such as IL-1α, IL-6, TNF-α, granulocyte-monocyte colony-stimulating factor (GM-CSF), and vascular endothelial cell growth factor (VEGF) are up-regulated by melanoma B16F10 cells. After intraperitonial administration of *T. cordifolia* extract, capillary formation and the level of these cytokines are decreased while antiangiogenic agents IL-2 and tissue inhibitor of metalloprotease-1 (TIMP-1) were increased. In another experiment, employing rat aortic ring assays, the nontoxic concentrations of *T. cordifolia* extract were found to inhibit proinflammatory cytokines secreted by melanoma B16F10 cells, along with the suppression of the microvessel outgrowth from aortic rings [49]. Thus, *T. cordifolia* could inhibit angiogenesis by regulating cytokine expression and increasing the circulating level of antiangiogenic factors. This implies that *Tinospora* has the potential to inhibit tumor angiogenesis and suppress tumor growth and progression.

12.5 *Tinospora* Inhibits Cancer Metastasis

During metastasis, after getting detached from the primary cancer tissue, tumor cells penetrate the blood or lymph vessels and circulate with the body fluid and could form a secondary tumor at a distant site, which is a big obstacle in the treatment of cancer. During their journey, cancer cells encounter many immune cells, and hence any agent boosting the immune system could interfere with metastasis. Many phytochemicals have been observed to inhibit tumor cell metastasis [50]. In this regard, there are many properties associated with polysaccharide fraction isolated from the dried stem of *T. cordifolia* [16]. In vitro studies have shown that polysaccharide is a specific mitogen of B-cells but does not affect T cells. A study has also shown that intraperitonial administration of this polysaccharide fraction results in a 72% reduction in the metastatic colony formation of B16F10 melanoma cells in the lungs of syngeneic C57BL/6 mice [51]. In this study, the polysaccharide fraction also reduced neoplastic markers such as lung hexosamine, collagen hydroxyproline, uronic acid, γ–glutamyltranspeptidase (γGT), and sialic acid [51]. Overall, these observations indicate the antimetastatic effect of *Tinospora*; however, more studies on relevant animal models are needed to support this conclusion as well as to find out the associated mechanisms.

12.6 Radioprotective Potential of *Tinospora*

Radiotherapy-induced damage to normal tissues is the major limitation of the therapeutic response in antitumor therapy. On the other hand, radiation toxicity itself can cause many genetic aberrations leading to various diseases including cancer and immunosuppresion. In this regard, *T. cordifolia* has been shown to have promising radioprotective potential in terms of survival after whole-body irradiation. Additionally, it also modulates cell cycle progression, hematologic parameters, spleen-colony-forming units, and micronuclei formation. Preirradiation administration of a single dose of root extract of *T. cordifolia* (RTc, 200 mg/kg b/w) offers 76% survival in mice exposed to 10 Gy lethal gamma-irradiation, but without RTc treatment irradiated mice suffer from 100% mortality within 10 to 15 d. Most radioprotective agents show the maximum radioprotective effect at maximum tolerated level doses (MTD), but RTc shows significant protective effect at about 50% concentration of its MTD [52–54], and this ability makes *T. cordifolia* more useful for clinical applications. Most of the radiation-induced damages are due to the interaction between radiation-induced free radicals and biomolecules, and hence molecules or agents that have the ability to neutralize or scavenge these free radicals could inhibit radiation-induced damages. *T. cordifolia* possesses the ability to scavenge free radicals and prevents radiation-induced damages [55]. However, the survival ability against radiation-induced damages could be a combination of different mechanisms including scavenging free radicals, inhibition of free-radical generation, and repair of DNA and membranes [56]. Overall, these studies are indicative of the radioprotective potential of *Tinospora*; however, detailed studies are needed to explore its radioprotective mechanisms.

12.7 *Tinospora* Activates Tumor-Associated Macrophages of Dalton's Lymphoma

It has been found that the alcoholic extract of *T. cordifolia* can activate the tumor-associated macrophages (TAM) of Dalton's lymphoma (DL), which is spontaneously transplantable T-cell lymphoma. The basic functions of these macrophages are antigen presentation, phagocytosis, and secretion of IL-1, TNF, etc., which can be induced by intraperitonial administration of the *Tinospora* extract in DL-bearing mice together with the increase in life span of tumor-bearing mice and reduction in tumor growth. This extract treatment also shows the anticancer efficacy by destabilization of the membrane integrity of DL cells and activates the differentiation of TAM to dendritic cells (DC) in response to GM-CSF, tumor necrosis factor, and interleukin-4 [57, 58]. Overall, *Tinospora* could, in part, trigger its in vivo antitumor activity via activation of TAM.

12.8 Immunostimulating Properties of *Tinospora*

Many plants are known to produce immunogenic components. In this regard, various polysaccharides are known to stimulate the immune system, for example, β-glucans, which have structural identity with conserved pathogen-associated molecular patterns (PAMPs) that activate the immune system by binding the specific receptors, pattern recognition receptors. These receptors are found on various immune cells, including natural killer cells (NK cells), macrophages, monocytes, and neutrophiles [59, 60], and stimulate cytotoxic, phagocytic, and antimicrobial activities by synthesis of cytokines, chemokines, reactive-oxygen species, and N2 intermediates. α–D glucan (RR1) from *T. cordifolia* consisting of $(1 \rightarrow 4)$ linked backbone and $(1 \rightarrow 6)$ linked branches is reported to stimulate the immune system [61]. It is found to be nontoxic and nonproliferative to tumor as well as normal cells at 0 to 1000 μg/ml concentrations and proficiently activates lymphocytes of different subpopulations like NK cells, T cells, and B cells at a 100 μg/ml concentration [62]. Activation of NK cells is associated with the killing of tumor cells. RR1-mediated induction of normal lymphocytes produces IL-1β, IL-12 p70, IL-12 p40, IL-18, IL-6, IFN-γ, TNF-α, and monocyte chemoattractant proteins (MCP-1), and this cytokine profile is associated with the Th1 pathway with a self-regulatory mechanism. RR1 also increases the level of the alternative pathway component C3a des Arg of a complementary system, where C3a is a bioactive cleavage product of C3 during complementary activation in alternative pathways [63]. In another study, water extracts and/or ethanol extracts of the stems of *T. cordifolia* and *T. sinensis* were found to inhibit cyclophosphamide-caused immunosuppression and anemia in Swiss albino mice [64].

12.9 Mechanism of Macrophage Activation by $(1 \rightarrow 4)$ α-D Glucan from *Tinospora*

It has been shown that $(1 \rightarrow 4)$ α-D glucan isolated from *T. cordifolia* has many immunostimulating properties by which it modulates the macrophage functions. RR1 has been observed to inhibit phagocytosis of unopsonized zymosan bioparticles, which are yeast-derived particles containing β-glucan and mannan, in RAW 264.7 macrophages [60, 65, 66]. RR1 also inhibits the binding and internalization of opsonized zymosan A bioparticles but with lesser efficacy than that of laminarin, which is fungal-derived β-glucan [66–68]. For signaling of β-glucan, complement receptor 3 (CD11b/CD18 or CR-3) acts as a leukocyte receptor for soluble [69] and particulate glucans [70]. In RAW264.7 macrophage cells, CR3 monoclonal antibody does not inhibit RRI-induced TNF-α synthesis, suggesting that CR3 does not mediate internalization and opsonic binding of RR1; however, CR3 monoclonal antibody inhibited zymosan A-induced TNF-α synthesis [71]. RR1-induced TNF-α synthesis in macrophages is associated with NF-κB activation to elicit inflammatory response. It has also been observed that RR1 activates NF-κB by TLR-6 signaling, and this was evidenced by synthesis of IL-8 via NF-κB in TLR-6 transfected HEK293 cells [71]. However, there are many other phytochemicals that have been found to inhibit NF-κB activation in macrophages. For example, caffeic acid phenethyl ester (CAPE) and curcumin inhibit NF-κB activation as well as TNF-α synthesis in macrophages [72, 73]. Overall, NF-κB signaling appears to be an important mechanism for RR1-induced macrophage activation.

12.10 G1-4A, an Immunomodulatory Polysaccharide from *Tinospora*

Plant-derived polysaccharides are usually ideal immunomodulators because they show relatively less toxicity and fewer side effects as compared to other agents. Many studies show that they modulate the macrophage function by increasing phagocytosis, chemotaxis, and microbicidal activity and antigen presentation to T cells. For example, arabinogalactan, a polysaccharide (G1-4A) from the stem of *T. cordifolia*, modulates the macrophage function and protects against lipopolysaccharide (LPS)-induced endotoxic shock. LPS activates the cells of the immune system to release mediators of sepsis and endotoxic shock including IL-1, IL-2, IL-6, IL-8, TNF-α, endorphins, platelet-activating factor, various eicosanoids, high-mobility groups (HMG1), nitric oxide (NO), and macrophage MIF, leading to myocardial dysfunction, and renal and hepatic failure [74]. TNF-α has a crucial role in endotoxin-induced toxicity, and there are many drugs available that inhibit TNF-α production, e.g., pentoxyfilline [75], JTE-607 [76], thalidomide [77], 21-aminosteroids [78], and dexamethazone [79]. However, these drugs make the host immunocompromised [17].

G1-4A binds to macrophages and inhibits the binding of LPS to macrophages. Hence, G1-4A mimics LPS but is not toxic to mice up to a dose of 100 mg/kg body weight. Although a small amount of TNF-α is induced by G1-4A, pretreatment with G1-4A decreases the LPS-induced level of TNF-α (17). G1-4A also causes an increase in serum levels of IL-1β and IFN-γ in mice challenged with LPS. IL-1β causes a decrease in the surface expression of TLR-4, a crucial receptor for LPS, and also increases the serum level of glucocorticoids and their receptors on peritoneal macrophages [80, 81]. G1-4A also induces the NO in murine macrophages, which act as vasodilator and provide protection during endotoxic shock by inhibiting vascular thrombosis [17]. These reports provide evidence for the modulatory effects of G1-4A on cytokines and NO to induce tolerance against endotoxic shock.

12.11 Hepatoprotective Properties of *Tinospora*

Some preparations of *Tinospora* are used as liver tonic. Experimental studies also show that the extracts of *T. cordifolia* possess hepatoprotective and stimulatory properties, as has been observed in CCL_4-induced hepatotoxicity in mature albino rats [24]. In this study, administration of CCL_4 (0.7 ml/kg body weight for 7 d) causes liver damage marked by the elevation of certain specific enzymes such as serum glutamate pyruvate transminase, serum glutamate oxaloacetate transaminase, and alkaline phosphatase as well as the serum level of billirubin, which also happens in the case of jaundice. The enzyme levels return to normal after treatment with an extract of *T. cordifolia* comparable to that in a control group of rats. *Tinospora* has also been found to increase monocyte to macrophage differentiation, myelopoiesis, antigen presentation, levels of released cytokine and myeloperoxidase, and microbicidal and tumoricidal activities along with the stimulation of certain chemokines in CCL_4-intoxicated rats [24]. Overall, these observations are suggestive of protective effects of *T. cordifolia* against CCL_4-induced hepatotoxicity and immunosuppression.

12.12 Summary

The studies discussed so far suggest that *T. cordifolia* has great potential for cancer chemoprevention and immunomodulation. *Tinospora* acts as a bifunctional enzyme inducer for carcinogen/drug metabolism and induces antioxidant defense mechanisms to neutralize oxidative stress usually caused by xenobiotics including chemical carcinogens. It can inhibit angiogenesis as well as cancer metastasis. The anticancer effect of *T. cordifolia* increases the survival of EAC-bearing mice and can induce cell death in HeLa cells. It also has radioprotective potential in terms of whole-body survival, hematologic parameters, cell-cycle progression, spleen-colony-forming units (CFU), and micronuclei induction. The survival effect of *T. cordifolia* against radiation-induced damage involved various mechanisms including scavenging of free radicals, repair of DNA damage, and inhibition of free-radical generation.

T. cordifolia has many immunomodulatory functions as one of its constituents, α-D glucan polysaccharide (RR1)-activated lymphocytes of different subpopulations including NK cells, T cells, and B cells. Activation of NK cells is associated with the killing of tumor cells. RR1 induces normal lymphocytes to secrete various cytokines including interleukins, IFN-γ, TNF-α, and monocyte chemoattractant protein (MCP-1). RR1 can activate the C3a des Arg component of an alternative pathway of a complementary system. RR1 can induce TNF-α synthesis in macrophages, which is associated with NF-κB activation by TLR-6 signaling. Other polysaccharides, such as arabinogalactan (G1-4A) from the stem extract of *T. cordifolia*, modulate the macrophage function and protects against LPS-induced endotoxic shock. *T. cordifolia* also has hepatoprotective activities and can counter the immunosuppressive effects of CCL_4. Together, these studies advocate for the detailed investigation of the anticancer and immunomodulatory properties of *Tinospora*. Nevertheless, the medicinal value of this plant has immense potential in clinical applications. In the future, the identification of all biologically active components could provide mechanistic insight into their preventive and therapeutic potential against various ailments including cancer and immune diseases.

References

1. Hong WK, Sporn MB (1997) Science 78:1077
2. Chan JM, Stampfer MJ, Giovannucci E, Gann PH, Ma J, Wilkinson P, Henneken CH, Pollak M (1998) Science 279:563
3. Lippman SM, Hong WK (2002) Cancer Res 2:119
4. Singh DK, Lippman SM (1998) Oncology 12:1787
5. Sporn MB, Suh N (2000) Carcinogenesis 21:525
6. Gupta RA, Dubois RN (2000) Nat Med 6:974
7. Singh RP, Agarwal R (2006) Endocr Relat Cancer 13:751
8. Singh RP, Agarwal R (2005) Eur J Cancer 41:1969
9. Agarwal R, Agarwal C, Ichikawa H, Singh RP, Aggarwal BB (2006) Anticancer Res 26:4457
10. Shishodia S, Chaturvedi MM, Aggarwal BB (2007) Curr Probl Cancer 31:243
11. Johnson JJ, Mukhtar H (2007) Cancer Lett 255:170
12. Mori A, Lehmann S, O'Kelly J, Kumagai T, Desmond JC, Pervan M, McBride WH, Kizaki M, Koeffler HP (2006) Cancer Res 66:3222
13. Sareen D, Darjatmoko SR, Albert DM, Polans AS (2007) Mol Pharmacol 72:1466
14. Tsuda H, Ohshima Y, Nomoto H, Fujita K, Matsuda E, Iigo M, Takasuka N, Moore MA (2004) Drug Metab Pharmacokinet 19:245
15. Spelman K, Burns J, Nichols D, Winters N, Ottersberg S, Tenborg M (2006) Altern Med Rev 1:128
16. Chintalwar G, Jain A, Sipahimalani A, Banerji A, Sumariwalla P, Ramakrishnan R, Sainis K (1999) Phytochemistry 52:1089
17. Desai VR, Ramkrishnan R, Chintalwar GJ, Sainis KB (2007) Int Immunopharmacol 7:1375
18. Singh SS, Pandey SC, Srivastava S, Gupta VS, Patro B, Ghosh AC (2003) Indian J Pharmacol 35:83
19. Nadkarni KM, Nadkarni AK (1976) Indian Materia Medica, vol 1, 3rd edn. M/S Popular Prakasan, Mumbai, India
20. Chopra RN, Nayar SL, Chopra IC (1956) Glossary of Indian Medicinal Plants. Council for Scientific and Industrial Research, New Delhi, p21
21. Khosa RL, Prasad S (1971) J Res Ind Med 6:261
22. Zhao THF, Wang X, Rimando AM, Che C (1991) Planta Med 57:505
23. Subramanian M, Chintalwar GJ, Chattopadhyay S (2002) Redox Rep 7:137
24. Bishayi B, Roychowdhury S, Ghosh S, Sengupta M (2002) J Toxicol Sci 27:139
25. Gupta RS, Sharma A (2003) J Exp Biol 41:885
26. Rao PR, Kumar VK, Viswanath RK, Subbaraju GV (2005) Biol Pharm Bull 28:2319

27. Vinutha B, Prashanth D, Salma K, Sreeja SL, Pratiti D, Padmaja R, Radhika S, Amit A, Venkateshwarlu K, Deepak MJ (2007) Ethnopharmacol 109:359
28. Nayampalli S, Ainapure SS, Nadkarni PM (1982) Ind J Pharmacol 41:64
29. Thatte UM, Chhabria S, Karandikar SM, Dahanukar SA (1987) Indian Drugs 25:95
30. Thatte UM, Dahanukar SA (1989) Phytother Res 3:43
31. Kumar S, Verma NS, Pande D, Srivastava PS (2000) J Med Arom Plant Sci 22:61
32. Jagetia GC, Rao SK (2006) Biol Pharm Bull 29:460
33. Pachaly P, Schneider C (1981) Arch Pharmacol 314:251
34. Bisset NG, Nwaiwu J (1983) Planta Med 48:275
35. Gangan VD, Pradhan P, Sipahimalani AT (1997) Ind J Chem 36B:787
36. Zhang JY, Wang Y, Prakash C (2006) Curr Drug Metab 7:939
37. Singh RP, Banerjee S, Kumar PV, Raveesha KA, Rao AR (2006) Phytomedicine 13:74
38. Ketterer B (1988) Mutat Res 202:3432
39. Jagetia GC, Nayak V, Vidyasagar MS (1998) Can Lett 127:71
40. Jagetia GC, Rao SK (2006) Evid Based Complement Altern Med 3:267
41. Chauhan K (1995) Suchitra Ayurved 47:840
42. Folkman J, Shing Y (1992) J Biol Chem 267:10931
43. Carmeliet P(2003) Nat Med 9:653
44. Folkman J (1971) New Engl J Med 285:1182
45. Carmeliet P (2000) Nature 407:249
46. Dhanabal M, Jeffers M, Larochelle WJ (2005) Curr Med Chem Anti-Cancer Agents 5:115
47. El Sayed KA (2005) Mini Rev Med Chem 5:971
48. De Smet F, Carmeliet P, Autiero M (2006) Nat Chem Biol 2:228
49. Leyon PV, Kuttan G (2004) Int Immunopharmacol 4:1569
50. Deorukhkar A, Krishnan S, Sethi G, Aggarwal BB (2007) Expert Opin Investigat Drugs 16:1753
51. Leyon PV, Kuttan GJ (2004) Ethnopharmacol 90:233
52. Uma Devi P, Ganasoundari A, Rao BSS, Srinivasan KK (1999) Radiat Res 151:74
53. Riklis E, Kol R, Marco R (1990) Int J Radiat Biol 57:699
54. Goel HC, Ganguly SK, Prasad J, Jain V (1996) Ind J Exp Biol 34:1194
55. Kapil, A, Sharma S (1997) J Ethnopharmacol 58:89
56. Goel HC, Prasad J, Singh S, Sagar RK, Agrawala PK, Bala M, Sinha AK, Dogra R (2004) J Radiat Res 45:61
57. Singh N, Singh SM, Shrivastava P (2004) Immunopharmacol Immunotoxicol 26:145
58. Singh N, Singh SM, Shrivastava P (2005) Immunopharmacol Immunotoxicol 27:1
59. Allavena P, Chieppa M, Monti P, Piemonti L (2004) Crit Rev Immunol 24:179
60. Brown GD, Gordon S (2003) Immunity 19:311
61. Singh SS, Pandey SC, Srivastava S, Gupta VS, Patro B, Ghosh AC (2003) Indian J Pharmacol 35:83
62. Nair PK, Rodriguez S, Ramachandran R, Alamo A, Melnick SJ, Escalon E, Garcia Jr PI, Wnuk SF, Ramachandran C (2004) Int Immunopharmacol 4:1645
63. Ember JA, Jagels MA, Hugli TE (1998) In: Volanakis JE, Frank MM (ed) The Human Complement System in Health and Disease. Dekker, New York, p 241
64. Manjrekar PN, Jolly CI, Narayanan S (2000) Fitoterapia 71:254
65. Brown GD, Gordon S (2005) Cell Microbiol 7:471
66. Brown GD, Taylor PR, Reid DM, Willment JA, Williams DL, Martinus-Pomares L, Wong SY, Gordon S (2002) J Exp Med 196:407
67. Czop JK, Austen KF (1985) J Immunol 134:2588
68. Goldman R (1988) Exp Cell Res 174:481
69. Thornton BP, Vetvicka V, Pitman M, Goldman RC, Ross GD (1996) J Immunol 156:1235
70. Ross GD, Cain JA, Myones BL, Newman SL, Lachmann PJ (1987) Complement 4:61
71. Nair PK, Melnick SJ, Ramachandran R, Escalon E, Ramachandran C (2006) Int Immunopharmacol 6:1815
72. Fitzpatrick LR, Wang J, Le T (2001) J Pharmacol Exp Ther 299:915
73. Abe Y, Hashimoto S, Horie T (1999) Pharmacol Res 39:41

74. Das UN (2000) Crit Care 4:290
75. Krakauer T, Stiles BG (1999) Clin Diagn Lab Immunol 6:594
76. Kakutani MK, Waga TI, Iwamura H, Wakitani K (1999) Inflamm Res 48:461
77. Moreira AL, Wang J, Sarno EN, Kaplan G (1997) Braz J Med Biol Res 30:1199
78. Howe LM (2000) Expert Opin Investigat Drugs 9:1363
79. Davenpeck KL, Zagorski J, Schleimer RP, Bochner BS (1988) J Immunol 161:6861
80. Alves-Rosa F, Vulcano M, Beigier-Bompadre M, Fernandez G, Palermo M, Isturiz MA (2002) Clin Exp Immunol 128:221
81. Nomura F, Akashi S, Sakao Y, Sato S, Kawai T, Matsumoto M, Nakanishi K, Kimoto M, Miyake K, Takeda K, Akira S (2000) J Immunol 164:3476

Chapter 13
Safety and Efficacy of Phytomedicines in Cancer Prevention and Treatment

Giuseppina Benoni and Laura Cuzzolin

Abstract In this review we discuss some aspects of herbal use either to prevent cancer or to treat the disease or the side effects of chemotherapy. The most powerful reasons, for cancer patients, to use phytomedicines are related to the wish to leave no option untried and to the dissatisfaction with mainstream oncology treatments. In the review, herbs commonly used in cancer and their mechanism of action are referred. Moreover, clinical trials about the use of some herbs for treating the side effects of chemotherapy and radiation are cited. As regards the safety data of phytomedicines in cancer patients, considering the narrow therapeutic window of chemotherapic drugs, the risk of clinically relevant herb-drug interactions can increase: in the USA more than 100,000 deaths per year can be attributed to drug interactions, most of them connected to the use of herbs. Some experts believe that the potential risk of herb drug interactions is enough to recommend patients on chemotherapy not to use herbal therapies. At present, oncologists must be encouraged to discuss herbal use with their patients, and they should be aware of possible herb anticancer drug interactions. Furthermore, physicians should advise patients to refrain from using herbs, especially when their effects have not been well investigated.

Keywords Herbal drugs · Cancer · Chemoprevention

13.1 Introduction

Many people are keen to use herbal medicines either to prevent cancer or to treat the disease or side effects of chemotherapy, but most remain unaware of the limited scientific evidence regarding the efficacy of these remedies and the potential drug interactions and adverse effects that may result from their consumption. Most trials are small, open-label, uncontrolled and not randomised. Preliminary evidence suggests that some herbs may have a role in preventing cancer or adjoin therapies for

G. Benoni (✉)
Department of Medicine & Public Health-Section of Pharmacology, University of Verona, Italy, Policlinico G.B. Rossi, Piazzale L.A. Scuro, 37134 Verona, Italy
e-mail: Giuseppina.benoni@univr.it

K.G. Ramawat (ed.), *Herbal Drugs: Ethnomedicine to Modern Medicine*,
DOI 10.1007/978-3-540-79116-4_13,

its treatment, but it appears that herbal medicines are unlikely to provide a miracle cure for cancer [1], and their effects are not always predictable.

Most cancer patients combine, rather than replace, conventional therapy with herbs [2]. Studies on cancer patients sampled from the general population indicate that herb users are mostly young women high on the social ladder [3, 4] who are looking for miraculous cures [5]. In addition, they tend to be more health conscious [6], are willing to spend out of pocket [7] or are being treated with chemotherapy [8].

Several studies indicate that young patients and those who live long after receiving a cancer diagnosis are more likely to seek some forms of complementary and alternative medicine (CAM) [9]. One concern is the fact that approx. 70% of those who report using herbs do not inform their physicians of their decision [10]. Furthermore, the fact that most conventional medical practitioners have very little knowledge of herbs makes educating herb users a challenging task.

13.1.1 Reasons for Using Phytomedicines in Cancer Treatment and Prevention

Different reasons lead patients with acute, serious or life-threatening conditions and patients with chronic or benign diseases to try CAM. For cancer patients, the most powerful motivations relate to the desire to leave no option untried and to the dissatisfaction with mainstream oncology treatments [11]. In general, most cancer patients are satisfied with conventional treatment. Nevertheless, the attention paid to their symptoms and side effects is not satisfactory for them. It is currently estimated that more than half of the patients diagnosed with cancer have explored herbal medicines in an attempt to treat cancer and the adverse effects associated with conventional cancer treatments. For example, PC-SPES is sometimes used as the sole treatment for prostate cancer for its presumed action as an anticancer agent [12, 13].

According to a survey of gynaecological cancer patients on their reasons for using herbs [14], a large number of patients resorted to herbs to improve their quality of life and, if they were taking other drugs, to decrease their adverse effects, whereas 15% of patients used herbs to treat the cancer directly. Several products are under investigation in clinical trials for this reason. However, they have not been under appropriate trial development so far [15].

Another reason for using herbs is related to their cancer-prevention properties. For example, the more or less conscious use of green tea by men has become more and more popular. Indeed, in a recent case-control study, supported by in vitro research, a protective effect of green tea against prostate cancer was suggested [16, 17]. The risk was thought to decline along with the increase in frequency, duration, and quantity of green tea consumption. However, conflicting results on the use of green tea as a protective substance in relation to prostate cancer have been reported by epidemiological studies [16]. In general, there is a lack of evidence on the efficacy of most herbs in the prevention of cancer.

In addition to these reasons, cancer patients generally have the same reasons as other people for using CAM [18]. Most cancer patients want to increase their body's ability to fight the disease (50%) and improve their physical and emotional health (40% and 35% respectively) [19]. In a recent report, a study performed at a community hospital with a cancer centre found that dietary supplement users have a better quality of life than non-users [20]. In contrast, previous data showed opposite results [4–21].

13.2 Methods

Medline and Embase searches were performed between 1990 and 2007 using the terms "phytotherapic compounds", "phytomedicine", "efficacy", "safety" and "epidemiological data" combined with the terms "cancer", "patients", "chemopreventive", "chemotherapy", "side effect" and "antitumour activity".Information about the use of herbal products on the examined subpopulations derived from medical records, prescription data, personal interviews, self-completion questionnaires or a combination of these.

13.3 Epidemiological data

It is estimated that up to one third of the entire population of the USA has used CAM, and most individuals in that country have used herbal products on a routine basis [10]. Most research on the use of CAM is conducted in the USA, with multiple recent surveys suggesting that 25 to 84% of US cancer patients have used CAM therapies and 13% to 63% herbal preparations [10], with vary utilization rates according to geographic areas and types of cancer. Apparently, patients with breast cancer tend to use more herbal products than individuals with other types of malignancy, since women use CAMs more than men [10]. In particular, it seems that colorectal and breast cancer patients are more likely to use dietary supplements than lung cancer patients [22].

A recent descriptive survey conducted in 14 European countries has shown that 36% of cancer patients used some kind of CAM. Among those countries, the utilization rates range from 15% to 73% of the population in Greece and Italy, respectively [19]. Furthermore, herbs have resulted to be the most common CAM therapy in 13 countries, most of them being specific in each country, i.e. mistletoe in Switzerland, olive leaf paste in Greece and Aloe vera in Serbia and Spain. An Australian study has reported that herbal treatments and naturopathy are the most popular CAMs used by cancer patients, representative over 30% of the recorded CAM use [23].

In a nationwide, cross-sectional survey carried on in Japan, 92% of the CAM users with cancer use products such as mushrooms, herbs and shark cartilage [3].

A Canadian survey has reported that 25% of breast cancer patients use herbal products [9]. In a study by Richardson et al. [24] regarding various CAM approaches, a statistic on 453 male and female patients from 8 different cancer outpatient clinics has shown that 62% of these cancer patients reported use of herbs and vitamin products. Navo et al. [14] have reported that gynaecologic oncology patients use CAM (49.6% to 56.3%) among all herbal products. In general, most of them gathered information about herbs on their own from internet and media, but also looked for the opinion of their friends and family.

Data from the Women's Healthy Eating and Living study have shown that up to 80% of non-stage IV breast cancer patients took dietary supplements such as vitamins, antioxidants and herbs [25]. Recently, an important increasing trend in the use of herbal products rather than other supplements has been observed [26]. A recent study on patients enrolled in early-phase chemotherapy trials at the Mayo Clinic Comprehensive Cancer Centre reports a high use of herbal products. More than 80% simultaneously used vitamins, herbs and minerals in addition to their experimental chemotherapeutic agents [27], although this is often an exclusion criterion formally not allowed during this type of treatment.

13.4 Herbs Commonly Used in Cancer Treatment and Prevention

Several herbs and herbal compounds are believed to suppress the transformative, hyperproliferative and inflammatory process that initiates carcinogenesis [28]and dramatically inhibit the vascular endothelial growth factor (VEGF), whose production is considered essential for the migration of cancer cell and for angiogenesis.

Polyphenols are receiving considerable attention for the potential implications in cancer therapy, particularly resveratrol, (grapes), quercetin (broccoli, onions and tea), curcumin (turmeric), genistein (soy), epigallocatechin (green tea) and their sources. In particular, quercetin strongly inhibited in a time and dose-dependent the expression of the mutated p53 protein in breast cancer, while the anti-tumour activity of resveratrol occurs through p53-mediated apoptosis [29, 30]. Another mechanism for anticancer activity of resveratrol is the inhibition of a wide range of angiogenic growth factors including VEGF expression [31].

Curcumin, found in turmeric and its derivatives, has demonstrated a significant inhibition of VEGF and directly inhibits angiogenesis in vivo and in vitro. Moreover, curcumin is also known for interfering with the endothelial cell function by inhibiting the specific integrin engagement. However, it does not interfere with normal cell function; this being the beauty of these safe compounds [32]. Moreover, Aggarwal et al. [33] reported that the ability of curcumin stems of suppressing the proliferation of a wide variety of tumour cells entails the anticancer effect of down-regulating the Nf-kB, AP-1 and Egr-1 transcription factors and the expression of COX2, LOX, NOS, MMP-9, uPA, and TNF.

Regarding green tea and its extracts, findings have shown that they suppress chromosomal abnormalities induced by carcinogens. For this reason, interest in these compounds has grown as they also play a role in delaying the cumulative genetic damage necessary for a cell to evolve from normality to one with aggressive metastatic capabilities [34]. An appropriate dose of VEGF blockade would be 2 to 4 g of standardized green tea extract. Each gram of this extract provides 400 to 500 mg of EGCG [35].

The seed cones from magnolia trees inhibited the growth of new blood vessels in mice. Silybin, a silymarin compound in milk thistle, bound to phosphatidylcholine, was shown to inhibit VEGF when used as a single agent against human ovarian cancer [36].

Aloe-emodina, a hydroxyanthraquinone from *Aloe vera* and other plants, has been indicated as a new antiangiogenic compound with a remarkable photocytotoxic effect on tumour cells [37]. Emodin has been reported to be non-toxic for normal cells while possessing specific toxicity for neuroectodermal tumour cells [38, 39].

Edible berries (bilberry, cranberry, elderberry, raspberry seed) may have potent chemopreventive properties by inhibiting both H2O2 and TNF-α, which induced VEGF expression [40]. Ginkgo has multiple actions including antiedemic, antihypoxic, antioxidant, antiplatelet, free-radical scavenging and microcirculatory properties [10–41].

Terpenes, and especially triterpenes, due to their broad spectrum of biological activity, are studied by many researchers since there is growing interest in the evaluation of triterpenoid extracts for cancer treatment [42]. Several triterpene compounds that have recently appeared in the literature in preclinical studies suggest potential therapeutic opportunities. Betullinic acid, boswellic acid, squalene, ursolic acid, oleanolic acid and triterpenoid saponins are among the growing number of compounds evaluated in preclinical and some clinical studies [43].

Alfa-bisabolol is a sesquiterpene alcohol present in the essential oil of chamomile and other plants which exerts, according to Cavalieri et al. [44], a strong time- and dose-dependent cytotoxic action against human pancreatic cell line. However, it failed to affect the viability of human fibroblasts. Based on these results, alfa-bisabolol should be considered a novel promising compound for inhibiting the growth and survival of transformed cells with high malignancy.

The efficacy of Ginseng in the treatment of physical performance, psychomotor performance, cognitive function and immune modulation has been investigated [45]. The conclusion of this analysis is that the efficacy of ginseng is not established beyond reasonable doubt for any of these indications. However, a retrospective trial involving 4634 patients suggested a dose-response relationship between ginseng consumption and a decrease in the risk of cancer, with a 40% relative risk reduction [46].

Iscador, a derivative of mistletoe, is an example of an immunostimulant causing either an increase in cytotoxic T lymphocytes or natural killer cells or endogenous production of interferon and interleukins. Iscador is available in many mainstream European cancer clinics. European governments have approved studies on the efficacy of iscador against cancer. The results are mixed, since some studies suffer methodological shortcomings [47]. According to another systematic review [1], none of the methodologically well-made trials showed efficacy in survival.

Several mushroom-derived compounds have been approved for use as cancer treatment in Japan. Trials on polysaccharide Kureha (PSK), an extract of the mushroom *Coriolus versicolor*, showed superior survival with PSK in both gastrectomy and esophagectomy [1]. Two randomized trials of PSK given after curative resection for colorectal cancer showed that both disease-free and overall survival rates were significantly higher in the PSK group [48]. Results were less encouraging in breast cancer and leukaemia [49, 50]. The proposed mechanism was immune modulation of a beta glucan, a natural carbohydrate, present in great quantities in mushroom, bound to specific receptors on neutrophils. An interesting association between the response to PSK treatment and HLA type was reported [51].

Several studies have been published on the chemistry, pharmacology and clinical applications of *Echinacea*, in particular *E. purpurea*. Among these, many have described the use of *Echinacea* for immune system stimulation [10]. *Astragalus* has been proved to have immunologic benefits by stimulating macrophage and natural killer cell activity and inhibiting T-helper cell type2 cytokines. McCulloch et al, [52] found evidence that Astragalus-based Chinese herbal medicine may increase the effectiveness (by improving survival, tumour response and performance status) and reduce the toxicity of standard platinum-based chemotherapy for advanced non-small-cell lung cancer. Garlic has been used in a variety of conditions as a chemopreventive agent for gastrointestinal tumours [53].

13.5 Herbs for Treating the Side Effects of Chemotherapy and Radiation

Most adverse side effects of chemotherapy and radiotherapy are still a big concern for patients and their families. No provisions are made for the side effects of chemotherapy, and sometimes patients are not aware of these effects. In this respect, to treat the side effects of chemotherapy and radiotherapy, the use of "safe natural" herbs may be more appealing. Insomnia and other sleep disturbances are common in cancer patients. Insomnia is a multifactorial health concern that currently affects at least one out of three cancer patients, and yet most insomnia sufferers do not consult their physician regarding pharmaceutical options for their relief.

In a multicenter study in Baltimore, 44% of breast and lung cancer patients reported experiences of sleep disturbances [54]. In a cross-sectional survey, 31% of patients attending clinics for breast, gastrointestinal, genitourinary, gynaecologic and various skin cancers reported insomnia [55]. Up to 75% of cancer patients visiting an intensive care unit reported sleep disturbances [56]. Since long-term use of hypnotic drugs is associated with dependence and other risks, herbal remedies as sleep aids are required, especially when the disease has become chronic [57]. Many physicians have expressed the need to educate themselves and their patients to use, appropriately, these herbs and have questioned whether sufficient research data exist in order to make recommendations for insomnia relief and restoration of normal sleep quality. Some herbal sedatives have a long history of use in Europe but they are yet relatively new in the USA.

Valeriana contains constituents of the essential oil sesquiterpenes, which are of special interest for their sedative effects. Valeriana has shown sedative activity in a meta-analysis of clinical trials for the treatment of insomnia [58] and has been proved useful in sleep, since it is active against anxiety, which may interfere with sleep. Numerous studies have investigated the use of **kava** in the treatment of anxiety and sleep disorders. A meta-analysis reported that kava was effective in reducing anxiety [59].

Chamomile is very popular in the USA and Europe for its relaxing effect. Clinical trials on the sedative effects of chamomile are lacking and thus needed. In a pilot study, 10 out of 12 patients immediately entered into a deep sleep lasting an average of 90 min [58].

Lavender is used in aromatherapy as a holistic relaxant and, when inhaled, has been reported to have sedative effects in both animals and humans [60]. Patients undergoing radiation therapy were treated with lavender unscented carrier oil or other essential oils for inhalation: the carrier group experienced less anxiety than did the essential oil group. There are anecdotal and historical accounts of the sleep-enhancing benefits of hops and *Lemon balm*. Both herbs are purported to have sedative and hypnotic effects, but clinical evidence from MEDLINE is scarce [58].

Passionflower, another herb with mild sedative effects, is often combined with valerian in the herbal sleep aids found in health food stores. In a double-blind, randomized, placebo-controlled study, the effect on anxiety of passionflower did not differ from that of the oxazepam group [61].

A meta-analysis was conducted on **St. John's wort** including 23 randomized trials (*N*:1757 outpatients) with mild to moderate depressive disorders. St. John's wort was found to be significantly more effective than placebo. Other authors concluded that St. John's wort was not effective in 200 adults with major depression [58].

Regarding the efficacy of topical ***Aloe vera*** in reducing the side effects on skin of radiation therapy, some authors [62, 63] reported no improvement in clinical studies compared to placebo or aqueous cream.

Ginger has demonstrated effectiveness with chemotherapy-induced nausea in a meta-analysis of randomized clinical trials [64]. Important adaptogens to consider in oncology include *Eleutherococcus senticosus, Panax ginseng, Rhodiola rosea, Schisandra chinensis, Rhaponticum carthamoides and Aralia manchurica*. Extensive studies using various cytotoxic therapies in animal and human models with tumours have demonstrated that these adaptogens reduced chemotherapic drug toxicity, in combination with cytotoxic agents, particularly in bone marrow restoration, and they enhanced antitumour and antimetastatic effects [65, 66].

13.6 Safety Data of Herbal Compounds in Cancer Patients

There is relatively little quality information about the adverse effects of herbs and interactions between herbs and conventional medicines. This is due to a lack of knowledge by healthcare practitioners, particularly about the way these effects should be reported, and to the fact that consumers are less likely to report incidents associated with herbal products than those having to do with conventional medicines [9]. Cancer patients using herbal compounds should be warned that some of the adverse effects of these therapies are often similar to symptoms associated with their disease or treatment, thus making it difficult to discern if the disease or the "remedy" is the problem [67, 68]. In addition, given the relatively poor quality control standards of these products in many countries around the world, herbal medicines containing the wrong herbal product and/or potentially toxic adulterants have been reported [69, 70].

When adverse events and/or interactions with conventional drugs are reported in the literature, it is important, at first, to note whether the herbal product was authenticated. The quality of herbal products is highly variable with respect to the concentrations of its major and characteristic ingredients [71]. This appears to be the case both in countries where no strict regulation for these products exists (e.g. the USA) and in countries where a stricter regulation has been formulated (e.g. Germany). In addition pollution from pathogenic microorganisms, pesticides and heavy metals makes the health risks even greater [71].

As an example, regarding cannabis, sometimes used for palliative purposes by cancer patients [67], it has recently been demonstrated that the kind bought in Dutch " coffee shops" could be contaminated by bacteria and fungi that may seriously harm ill patients [72]. To avoid the risks associated with pollution, patients should be extra careful, because herbs that seem safe for healthy people under normal conditions might not be so safe for certain groups of patients such as cancer patients [73]. It has been clearly established, in any case, that a variety of herbal medicines may produce serious side effects [71].

Considering the narrow therapeutic window of chemotherapic drugs, the use of herbs, overall, increases the risk of clinically relevant herb-anticancer-drug interactions. It is estimated that the interactions are responsible for more unexpected toxicities of chemotherapic drugs in cancer patients [74].

Interactions between herbs and anticancer agents are currently not sufficiently documented. Basically, interactions are mostly thought to be the result of enzyme interactions (like those of cytochrome P450 metabolic pathways) and drug-trasporting proteins (ATP-binding transporters) which can be influenced in activity and expression by herbal medicine ingredients [74]. If given together, some herbs may indirectly induce the metabolism of cytotoxic agents, potentially leading to non-therapeutic systemic drug levels or inhibiting metabolism, in turn leading to potentially lethal toxicities. An additional consideration for cancer chemotherapy is that herb-mediated induction of various enzymes and transporters may also take place in tumour cells and subsequently result in resistance to chemotherapic drugs such as antracyclines, epipodophyllotoxins, taxanes and vinca alkaloids [75].

For St. John's wort, interactions with irinotecan, imatinib and docetaxel have been studied [76–78]. Regardless of the modulation of CYP3A4 and P-glycoprotein activity observed with St. John's wort, its crucial role in the elimination of many important anticancer drugs is of particular concern [10]. This suggests that interactions between St John's wort and such agents are likely to have clinical and toxicological implications and that rigorous tests for possible interactions are urgently needed. Currently, research is ongoing, for instance, to study the effects of medicinal cannabis on irinotecan and docetaxel metabolism [67]. Clinically significant interactions have been seen in some cases by Meijerman [74].

Some data suggest that *Echinacea* is likely to interact with anticancer drugs that are a substrate for CYP3A4 and that the interaction depends on the relative extraction of drugs at the hepatic and intestinal sites and on the route of administration [74]. Moreover, like several other herbal preparations, *Echinacea* found in retail stores often does not contain the labelled species [79]. Obviously, the high variability observed in the concentration of constituents of the herb may have implications for *Echinacea's* ability to modulate drug absorption and disposition.

Studies using in vitro and in vivo animal models have also indicated that various garlic constituents can induce the activity of CYP3A4 [80] and conjugating enzymes such as glutathione S-transferases and quinine reductase [81, 82]. Although the product composition and the designs of the clinical studies vary widely [83], there appears to be little likelihood of significant interactions between garlic and the anticancer drugs that are predominantly metabolized by CYP3A4.

Evidence indicates that some preparations of ginseng have phytoestrogenic effects, which suggests that its use with soy supplementation should be discouraged in women with breast cancer or endometrial cancer [84]. Substantial variability in ginsenoside content has been reported among commercial ginseng preparations, indicating that significant, clinical effects on the pharmacokinetics of the anticancer drugs metabolized by CYP3A4 could be brand specific [85].

The most important concerns with kava toxicity are the reports of hepatotoxicity in kava users. Some cases have been sufficiently severe such that liver transplants have been required and three deaths have resulted from hepatotoxicity associated with kava [86]: it has been suggested that the genetic deficiency of CYP2D6 might be associated with a susceptibility to kava toxicity in some of the observed cases [87]. In the South Pacific, where the genetic deficiencies of this enzyme are not known, no reports of liver failure have been reported, while genetic deficiencies occur in about 10% of Europeans [88]. Two cases of hepatic insufficiency, in the South Pacific area, were reported after kava consumption which were resolved when the patients ceased taking kava and were explained as allergic reactions [89]. Cancer patients who are undergoing treatment, have recently undergone treatment or are taking medications should avoid kava. If a cancer patient sees no other feasible alternatives than kava, he or she should consume tea rather than a concentrated extract since no cases of actual liver failure are known to be associated with the traditional aqueous extracts.

An interaction between the pharmacokinetics of anticancer drugs and grapefruit juice has been described. Grapefruit juice intake resulted in a 26.2% lower AUC of etoposide after oral intake. The median absolute bioavailability with and without pretreatment with grapefruit juice was 52.4% and 73.2%, respectively [90]. From clinical studies, it is reported that ginkgo may interfere with the pharmacokinetics of anticancer drugs metabolized by CYP2C19 or CYP3A4 [41].

Black cohosh, soy, may decrease the efficacy of tamoxifen, and therefore women taking this agent should especially avoid soy [91]. Moreover, at present it seems prudent to discourage soy-derived products in patients with estrogen-dependent tumours (e.g. breast cancer or endometrial cancer), as experimental data in mice indicate that soy can stimulate the growth of these tumours [92].

Selenium and squalamine may increase the toxicity of cisplatin, black cohosh, *Echinacea*, salicylate-containing herbs, while bilberry, cramp bark, meadowsweet, poplars, red clover, uva ursi, white willow, and wintergreen may increase the hepatoxicity of metrotexate [93].

Valerian appears to have a wide margin of safety. No studies have been found indicating any effects of valerian on the cytochrome enzymes involved in drug metabolism. For this reason there is no evidence of any potential effects on the clearance of chemotherapic drugs. The major drug interactions with valerian are those involving the potentiation of effects of sedatives, hypnotics or anaesthetics. Cancer patients should be advised to quit valerian use before surgeries for its possible withdrawal symptoms [58].

No toxicity has been reported for lavender: however, the herb potentiated the sleep-inducing activity of several agents including alcohol and hexobarbital in laboratory studies [94]. There are several reports of skin reactivity to chamomile [95]. Chamomile showed a moderate inhibitory effect on CYP3A4 in vitro with unclear clinical relevance. The formation of sister chromatid exchanges by daunorubicin has been decreased in rats pretreated with chamomile essential oil as compared to those given corn oil. Since sister chromatid exchanges may be correlated with antitumour activity of this anthracycline, this finding is of potential concern for patients undergoing anthracycline therapy [96].

Boudreau et al. [97] evaluated the biological and toxicological properties of *Aloe vera*, as a topical and oral agent. Ingestion of *Aloe vera* is associated with diarrhoea, electrolyte imbalance, kidney dysfunction and conventional drug interactions; episodes of contact dermatitis erythema and phototoxicity have been reported from topical applications. Rabe et al. [98] described as a case study the first case of acute hepatitis due to this compound.

The controversy surrounding antioxidants in combination with conventional therapy has focused on drugs that are believed to achieve their cytotoxic effects by generating free radicals, such as alkylating agents, anthracycline, platinum compounds, topoisomerase II inhibitors and radiation.

Although it has been reported that antioxidant and free-radical scavengers can actually prevent cancer [99], there is evidence that these agents could also interfere with chemotherapy (e.g. alkylating agents, anthracyclines and platinum analogues) and radiation therapy by acting as free-radical scavengers [84].

However, particular adjunctive agents, such as Mesna, produce their effects by quenching free radicals and do not seem to decrease the efficacy of chemotherapy.

Individuals treated with drugs that deplete the antioxidant status may require replenishment of antioxidants after treatment. However, the published clinical reports do not clearly define the role of antioxidant supplementation [100]. In some cases, the combinations of herbs and drugs could be considered synergic if the efficacy is greater than the effect of the single agent. Some herbs are known to enhance the cytotoxic effect of drugs through unknown mechanisms. Herbs like *Angelica sinensis* or *Zingiber officinale* might increase the toxicity of anticancer drugs [101].

13.7 Conclusions

More than 100,000 deaths per year in the USA can be attributed to drug interactions, placing this problem between the fourth and sixth leading cause of death, most of these deaths being connected to the use of herbs [10]. As a result, it is of the utmost importance to clarify their use before exposing patients to drugs with a narrow therapeutic window, which includes most anticancer agents. In clinical practice, the clinician should, in every patient contact, consider that most cancer patients will not spontaneously report herb use. A study revealed that, of the 48% of cancer patients using concomitant CAM, the vast majority had acted without informing their doctor [102].

For this reason, it is important for healthcare practitioners to discuss herbal medicine with their patients. One of the biggest challenges for healthcare professionals is to face their lack of substantive knowledge about the herbs that patients are using. Herbal products are commonly perceived as "natural" and thus as "innocent", a perception that is hard to change [103]. For this reason, the patient will answer "no" if asked about herb use [18], as proved by the St. John's wort tea case. Some herbal teas, like the St. John's wort variants, are produced by companies regarded highly reliable and trustworthy in the community and are sold in the supermarket as normal tea. Patients who drink this tea may not be as consciously aware of the risks associated with herbs as they would be if they had bought such products in their local pharmacy or drugstore. Therefore, the physician should actively ask for herbal and supplement use while explaining their potential hazards [73–104].

Finally, there is almost no information on whether the commonly applied dose range of herbal medicine is critical for anticancer drug interactions in patients. In this respect, evaluation of the data obtained in vitro and in vivo methods must be carried out and verified in well-designed clinical trials.

Some experts believe that the potential risk of herb-drug interactions is enough to recommend patients on chemotherapy to not use herbal therapies and herbal preparations to be interrupted prior to surgery or radiation therapy [105]. At present, oncologists must be encouraged to discuss herbal use with their patients and they should be aware of possible herb-anticancer drug interactions. Furthermore, physicians should advise patients to refrain from using herbs, especially when their effects are well known.

References

1. Cassileth BR, Deng G (2004) Oncologist 9:80
2. Tascilar M, de Jong FA, Verweij J, Mathijssen RHJ (2006) Oncologist 11:732
3. Hyodo I, Amano N, Eguchi K, Narabayashi M et al (2005) J Clin Oncol 23:2645
4. Paltiel O, Avitzour M, Peretz T et al(2001) J Clin Oncol 19:2439
5. Esinberg DM, Kessler RC, Foster C et al (1993) N Engl J Med 328:246
6. Cassileth BR (2002) Gastrointestinal Oncology: Principles and Practice. In: Kelsen DP, Daly JM, Levin B, Kern SE, Tepper JE (eds) Complementary and Alternative Medicine. Lippincott Williams and Wilkins, Philadelphia, p 165
7. Hess DJ (2002) Am J Public Health 92:1579
8. Richardson MA, Straus SE (2002) Semin Oncol 29:531
9. Smith M, Boon HS (1999) Patient Educ Counsel 38:109
10. Sparreboom A, Cox MC, Acharya MR, Figg WD (2004) J Clin Oncol 22:2489
11. Ernst E, Cassileth BR (1999) Eur J Cancer 35:1608
12. Kosty MP (2004) J Clin Oncol 22:3657
13. Straus SE (2002) N Engl J Med 347:1997
14. Navo MA, Phan J, Vaughan C, Palmer L et al (2004) J Clin Oncol 22:671
15. Vickers AJ, Kuo J, Cassileth BR (2006) J Clin Oncol 24:136
16. Jian L, Xie LP, Lee AH et al (2004) Int J Cancer 108:130
17. Siddiqui IA, Adhami VM, Saleem M et al (2006) Mol Nutr Food Res 50:130
18. Kronenberg F, Mindes J, Jacobson JS (2005) Cancer Invest 23: 420
19. Molassiotis A, Fernadez-Ortega P, Pud D et al (2005) Ann Oncol 16: 655
20. Lis CG, Cambron JA, Grutsch JF et al (2006) Support Care Cancer 14:193
21. Cassileth BR, Lusk EJ, Guerry D et al (1991) N Engl J Med 324:1180
22. Gupta D, Lis CG, Birdsall TC et al (2005) Support Care Cancer 13:912
23. Girgis A, Adams J, Sibbritt D (2005) Oncol Res 15:281
24. Richardson MA, Sanders T, Palmer JL et al (2000) J Clin Oncol 18:2505
25. Pierce JP, Faerber S, Wright FA et al (2002) Control Clin Trials 23:728
26. Cassileth BR, Lucarelli CD (2003) Herb-Drug Interactions in Oncology. Hamilton, Canada:B.C. Decker, 1–467.
27. Dy GK, Bekele L, Hanson LJ et al (2004) J Clin Oncol 22:4810
28. Oh GS, Pae HO, Choi BM, Seo EA, Kim DH et al (2004) Cancer Lett 205:23
29. Avila MA, Velasco JA, Cansado J, Notorio V (1994) Cancer Res 54:2424
30. Zhang S, Cao HJ, Davis FB, Tang HY, Davis PJ, Lin HY (2004) Br J Cancer 91:178
31. Cao Y, Fu ZD, Wang F, Liu HY, Han R (2005) J Asian Nat Prod Res 7:205
32. Arbiser JL, Klauber N, Rohan R, van Leeuwen R, Huang MT, Fisher C, Flynn E, Byers HR (1998) Mol Med 4:376
33. Aggarwal BB, Kumar A, Bharti AC (2003) Anticancer Res 23:363
34. Sazuka M, Murakami S, Isemura M, Satoh K, Nukiwa T (1995) Cancer Lett 98:27
35. Yance D (2005) In: Proceedings from the conferences on Medicines from the Earth. 4–6 June 2005, Brevard, NC, p 189
36. Gallo D, Giacomelli S, Ferlini C, Raspaglio G et al (2003) Eur J Cancer 39:2403
37. Cárdenas C, Quesada AR, Medina MA (2006) Cell Mol Life Sci 63:3083
38. Wasserman L,Avigad S,Beery E, Nordenberg J, Fenig E (2002) Am J Dermatopathol 24:17
39. Pecere T, Gazzola MV, Mucignat C et al (2000) Cancer Res 60:2800
40. Roy S, Khanna S, Alessio HM, Vider J, Bagchi M, Sen CK (2002) Free Radic Res 36:1023
41. DeFeudis FV, Papadopoulos V, Drieu K (2003) Fundam Clin Pharmacol 17:405
42. Melnick SJ (2006) J Pediatr Hematol Oncol 28:221
43. Spor MB, Suh N (2000) Carcinogenesis 21:525
44. Cavalieri E, Mariotto S, Fabrizi C, Carcereri de Prati A et al (2004) Biochem Biophys Res Commun 315:589
45. Vogler BK, Pittler MH, Ernst E (1999) Eur J Clin Pharmacol 55:567
46. Yun T-K, Choi S-Y (2003) Int J Epidemiol 27:359

47. Kienle GS, Berrino F, Bussing A et al (2003) Eur J Med Res 8:109
48. Mitomi T, Tsuchiya S, Iijima N et al (1992) Dis Colon Rectum 35:123
49. Morimoto T, Ogawa M, Orita K et al (1996) Eur J Cancer 32A:235
50. Iino Y, Yokoe T, Maemura M (1995) Anticancer Res 15:2907
51. Yokoe T, Iino Y, Takei H et al (1997) Anticancer Res 17:2815
52. McCulloch M, See C, Shu X-J et al (2006) J Clin Oncol 24:419
53. Levi F, Pasche C, LaVecchia C (1999) Br J Cancer 79:1283
54. Engstrom CA, Strohl RA, Rose L, Lewandowski L, Stefanek ME (1999) Cancer Nurs 22:143
55. Davidson JR, MacLean AW, Brundage MD, Schulze K (2002) Soc Sci Med 54:1309
56. Nelson JE, Meier DE, Oei EJ et al (2001) Crit Care Med 29:277
57. Savard J, Morin CM (2001) J Clin Oncol 19:895
58. Block KI, Gyllenhaal C, Mead MN (2004) Integr Cancer Ther 3:128
59. Pittler MH, Ernst E (2003) Cochrane Database Syst Rev 1:CD003383
60. Lis-Balchin M, Hart S (1999) Phytother Res 13:540
61. Akhondzadeh S, Naghavi HR, Vazirian M et al (2001) J Clin Pharm Ther 26:363
62. Heggie S, Bryant GP, Tripcony L et al (2002) Cancer Nurs 25:442
63. Su CK, Mehta V, Ravikumar L et al (2004) Int J Radiat Oncol Biol Phys 60:171
64. Boon H, Wong J (2004) Expert Opin Pharmacother 5:2485
65. Panossian A, Wilkman G, Wagner H (1999) Phytomedicine 6:287
66. Bocharova OA (1999) Vestn Ross Akad Med Nauk 5:49
67. de Jong FA, Engels FK, Mathijssen RH et al (2005) J Clin Oncol 23:2886
68. Shukla Y, Pal SK (2004) Asian Pac J Cancer Prevent 5:3
69. Shaw D, Leon C, Kolev S, Murray V (1997) Drug Safety 17:342
70. Keen R, Deacon AC, Delves HT, Moreton JA, Frost PG (1994) Postgrad Med J 70:113
71. De Smet PA (2004) Clin Pharmacol Ther 76:1
72. Hazekamp A, Sijrier R, Verpoorte R et al (2005) Pharm Weekbl 140:402
73. De Smet PA (2002) N Engl J Med 347:2046
74. Meijerman I, Beijnen JH, Schellens JHM (2006) Oncologist 11:742
75. Broxterman HJ, Lankelma J, Hoekman K (2003) Drug Resist Update 6:111
76. Mathijssen RH, Verweij J, de Bruijn P et al (2002) J Natl Cancer Inst 94:1247
77. Frye RF, Fitzgerald SM, Laguttuta TF et al (2004) Clin Pharmacol Ther 76:323
78. Koromoski BJ, Parise RA, Egorin MJ et al (2005) Clin Cancer Res 11:6972
79. Gilroy CM, Steiner JF, Byers T et al (2003) Arch Intern Med 163:699
80. Raucy JL (2003) Drug Metab Dispos 31:533
81. Guyonnet D, Belloirm C, Suschetet M et al (2000) Mutat Res 466:17
82. Munday R, Mundsay C (1999) Nutr Cancer 34:42
83. Lawson L, Ransom D, Hughes B (1992) Thromb Res 65:141
84. Weiger W, Smith MR, Boon H et al (2002) Ann Intern Med 137:889
85. Harkey MR, Henderson GL, Gershwin ME et al (2001) Am J Clin Nutr 73:1101
86. Gow PJ, Connelly NJ, Hill RL, Crowley P, Angus PW (2003) Med J Aust 178:442
87. Teschke R (2003) Z Gastroenterol 41:395
88. Moulds RFW, Malani J (2003) Med J Aust 178:451
89. Russman S, Barguil Y, Cabalion P et al (2003) Eur J Gastroenterol Hepatol 15:1
90. Reif S, Nicolson MC, Bisset D et al (2002) Eur J Clin Pharmacol 58:491
91. Ju YH, Doerge DR, Allred KF et al (2002) Cancer Res 62:2474
92. Allred CD, Allred KF, Ju YH et al (2001) Cancer Res 61:5045
93. Kelly KM (2004) Eur J Cancer 40:2041
94. Atanassova-Shopova S, Roussinov K (1970) Bull Inst Physiol 8:69
95. Pereira F,Santos R, Pereira A (1997) Contact Dermatitis 36:307
96. Herznande-Ceruelos A, Madrigal-Bujaidar E, de la Cruz C (2002) Toxicol Lett 135:103
97. Boudreau MD, Beland FA (2006) J Environ Sci Health C Environ Carcinog Ecotoxicol Rev 24:103
98. Rabe C, Musch A, Schirmacher P, Kruis W, Hoffmann R (2005) World J Gastroenterol 11:303
99. Gopalakrishna R, Gundimeda U (2002) J Nutr 132:3819S

100. Ladas EJ, Jacobson JS, Kennedy DD et al (2004) J Clin Oncol 22:517
101. Poppenga RH (2002) Clin Tech Small Anim Pract 17:6
102. Vapiwala N, Mick R, DeNittis A et al (2005) Proc Am Soc Ther Radiat Oncol 63:S451
103. Weitzman S (1998) Int J Cancer Suppl 11:69
104. Cauffield JS (2000) Pharmacotherapy 20:1289
105. Swisher EM, Cohn DE, Goff BA et al (2002) Gynecol Oncol 84:363

Chapter 14
Novel Leads from Herbal Drugs for Neurodegenerative Diseases

Maheep Bhatnagar

Abstract This article summarizes the salient features of new therapeutic leads from herbal sources for various types of neurodegenerative diseases. Efforts made in using existing knowledge on several popular medicinal plants, particularly those utilized in the Indian traditional medicinal and Ayurvedic system discussed in light of recent research. A few promising plants such as *Asparagus racemosus, Bacopa monnieri, Centella asiatica*, and *Mucuna pruriens* are worth exploring for drug development for neuroprotection.

Keywords Medicinal plants, Neurodegeneration, Neuroprotection, Neurodegenerative diseases

14.1 Introduction

Herbal medicines are being used by about 80% of the world's population, primarily in the developing countries, for primary healthcare, because of its better cultural acceptability, better compatibility with the human body, and lesser side effects. According to the World Health Organization's (WHO) definition traditional medicine comprises therapeutic practices that have been in existence, often for hundreds of years, before the development and spread of modern systems of medicine and are still in use. Herbal drugs constitute only those traditional medicines that primarily use medicinal plant preparations for therapy. The earliest recorded evidence of their use in Indian, Chinese, Egyptian, Greek, Roman, and Syrian texts, dates back about 5000 years ago. However, the last few years have witnessed a significant increase in their use in the developed world. India, China, and other neighboring countries have well founded traditional systems of medicines where plant based therapeutic agents occupy an important niche in health management.

M. Bhatnagar (✉)
Animal Biotechnology and Neuroscience Laboratory, Department of Zoology,
M.L. Sukhadia University, Udaipur-313001, India
e-mail: mbhatnagar@yahoo.com, m.maheep@gmail.com

K.G. Ramawat (ed.), *Herbal Drugs: Ethnomedicine to Modern Medicine*,
DOI 10.1007/978-3-540-79116-4_14, © Springer-Verlag Berlin Heidelberg 2009

The traditional use of herbal preparations existed in Indian medicinal systems long before their usefulness was accepted in modern scientific approaches to health-care. Even today in the majority of rural and urban areas in India, as well as in several other parts of world, traditional herbal health-care is practiced. According to one estimate, herbal products now constitute a major portion −20% of international pharmaceutical industry. About 1500 to 1800 products are sold in US markets alone [1].

Although a large number of scientific studies have been conducted in India and world wide on medicinal plants, traditionally proven herbal preparations/formulations do not yet fully conform to standards of drug testing, safety, and efficacy. Though opportunities for developing traditional preparations as drugs with international acceptance are enormous, there is hardly any intensive effort in this direction at the government or industry level in India, which has resulted in a decline in enthusiasm for studying plant preparations for their therapeutic value. But one has to believe that approx 25,000 living plant species in India contain a much greater diversity of bioactive compounds than any chemical library made by man. Although it is also true that extracts prepared from medicinal plants contain a variety of molecules with potent biological activities, unfortunately, it is often difficult to analyze their biological activities because of their complex nature and possible synergistic actions [2, 3].

14.2 Ayurveda, Herbal Drugs, and the Central Nervous System

Plant products or plant parts are the main source of medicines in Ayurveda, where they are either used as extracts or decoctions. Charak Samhita (900 BC) is the first recorded treatise of the Ayurveda. It consists of eight sections with descriptions of more than 341 Indian plants useful as medicines. In Ayurveda, the drugs that act on the central nervous system CNS are broadly classified into ten categories (Table. 14.1).

Table 14.1 Ayurvedic classification

1. **Medya** (Intelligence and memory)
2. **Samjnasthapa** (Resuscitative)
3. **Madakari** (Narcotic)
4. **Nidrajanana** (Hypnotic/sedative)
5. **Vedanastrhapana** (Analgesic)
6. **Apasmarahara** (Antiepileptic)
7. **Unmadhara**(Anti-insanity)
8. **Jawarahara** (Antipyretic)
9. **Chardi** (Anti-emetic)
10.**Vamnopza** (Emetic)

The *medhya* in Ayurveda refers to all substances that promote or are beneficial for *medha* (intellect), *dhriti* (concentration), and retention and *smiriti* (memory). In modern pharmacology, these substances are called nootropic agents. A number of Ayurvedic drugs are grouped as rasayanas (rejuvenators). According to Ayurvedic physicians, rasayanas increase the life span and delay ageing by counteracting the degenerative changes associated with ageing. Modern medicines have also recognized the rasayana group of drugs as antistress/adaptogens and could be useful in treating various CNS disorders.

Ayurvedic drugs are also garnering much attention for diseases with no or inadequate drugs for treatment in modern medicine, such as metabolic and degenerative disorders. Most of these diseases have multifactorial causation, and there is a growing realization that in such conditions, a combination of drugs acting at a number of targets simultaneously is likely to be more effective than drugs acting at one target; the one target, one drug paradigm is not likely to be satisfactory in such cases. Ayurvedic drugs, which are most often multi component have a special relevance for such conditions, and are attracting much attention. No doubt this would require a detailed study to obtain a proof of the concept, but these are the opportunities offered [2].

The multi target approach is the basic doctrine of Ayurveda, which takes a holistic view of the human system. In therapeutic terms, it implies that the treatment of disease should not be directed to a single tissue or organ but to the body system as a whole, taking into consideration the interconnectivity of the various bodily organs and their mutual dependence. For various reasons Ayurveda has not been exposed much to modern scientific study. Investigations of the biological activity of multi component Ayurvedic drugs will also bring Ayurveda into the main stream of scientific research. Encouraging research and development studies will help to develop a much needed interface between Ayurveda and modern medicine, and may further confirm how best the two might complement each other [2, 3].

India has a well recorded and well practiced knowledge of traditional herbal medicine, but, unlike China, it has not been able to capitalize on this herbal wealth by promoting its use in the developed world, despite the later's renewed interest in herbal medicine. Disseminating knowledge of traditional system of medicine can be achieved by judicious product identification based on diseases that have no medicine or only palliative therapy. Ayurvedic masters had thus developed certain dietary and therapeutic measures to arrest/delay ageing and rejuvenating the whole functional dynamics of the bodily organs. This revitalization and rejuvenation is known as the 'Rasayana chikitsa" (rejuvenation therapy). "Rasayana chikitsa" is not only a drug therapy, but a specialized discipline of Ayurveda that deals mainly with the preservation and promotion of health by revitalizing the metabolism and enhancing immunity. "Rasayana" therapy is done for a particular period of time with a strict regimen of diet and excercise. "Rasayana" drugs are rich in antioxidants, and are good hepatoprotectents and immunomodulators. Shushruta (an ancient Ayurvedic surgeon), in defining "rasayana" therapy, said that it arrests ageing (Vayasthapam) and increases life expectancy (Ayushkaram), intelligence (Medha), and strength (Bala), thereby enabling one to avoid diseases.

There has been much research on the plants used as "Rasyana drugs" in order to understand them in the modern context. Puri [4–7] gave an account of the herbs used in various rasayana preparations, while Udupa [8] studied the effects of rasayana drugs on psychosomatic stress. Rasyana drugs have been proven effective in the treatment of epilepsy [9], convulsive disorders, reduction of anxiety and apprehension, and in calming the mind [10].

14.3 Neurodegenerative Diseases:

Unlike other categories of diseases, such as infections or trauma, neurodegenerative diseases share a common etiology or some clinicopathological features. There are two major common features of the degenerative diseases:

1. They are diseases of neurons, which selectively affect one or more functional systems of neurons. For example, there is selective degeneration of striatonigral dopaminergic neurons.
2. They are marked generally by symmetric and progressive involvement of the central nervous system.

These diseases also differ among themselves quite distinctively. Some of them have a heritable pattern, while others are sporadic. Some show atrophy or loss of affected neurons, with or without specific features. Thus, specific diagnosis of such diseases can be made only by correlating the clinical as well as pathological findings. For a better understanding, these diseases are grouped according to the part or parts of the brain that are principally involved (Table. 14.2).

Table 14.2 Neurodegenerative diseases and areas of brain affected

Disease	Affected area of brain
Alzheimer's disease (AD) Pick's disease	Basal ganglia and brain stem
Huntingdon's disease Idiopathic Parkinson's disease Post encephalitic Parkinson's Striatonigral degeneration Progressive supranuclear palsy	Basal ganglia and brain stem
Shy–Drager syndrome Olivopontocerebellar degeneration Friedreich's ataxia Ataxia telengiectasia	Spinocerebeller area
Amyotropic lateral sclerosis Werding Hoffman disease Kugelberg welander Syndrome	Motor neurons

Alzheimer's disease (AD) is the most common disease of aged persons [10]; its onset is rare before the age of 50. The most frequent and initial complaints are impairment of concentration, memory, and other higher intellectual functions. Over the course of 5 to 10 years, progressive memory loss, disorientation, and language dysfunction often leads to a mute, immovable state. Death usually results from infections. Although 10% of cases appear to be clearly inherited, most are sporadic. Pathologically, the disease is characterized by a widening of the cerebral sulci. Microscopic features include the presence of neurofibrillary tangles and senile plaques. Senile plaques are found most frequently in the cerebral cortex and limbic structures. They range from 20 to 150 μm in diameter, are composed of focal collections of dilated, tortuous neuritis that often contain neurofibrillary tangles arranged around a central amygdaloid core. This Congo red positive core is composed of a β-amyloid protein. In addition to these two characteristics, AD is also characterized by a deficiency of acetylcholine and, the enzymes acetylcholinesterase (AchE) and choline acetyltransferase (ChAT), in the cerebral cortex, hippocampus, and amygdala. The gene responsible for the β-APP protein is located on chromosome 21 [11, 12].

Huntington's disease (HDD) usually appears between the ages of 20 and 50 years. It is characterized by extrapyramidal movements combined with progressive dementia. It is an autosomal dominant condition. Those who inherit it from their fathers usually manifest it much earlier in life than those who inherit it from their mothers. The defective gene is assigned to chromosome 4. In HDD, the brain is comparatively small (less than 1000 g) and demonstrate marked atrophy in Caudate and Putman, with conspicuous dilatation of the frontal poles of the lateral ventricles. There may be secondary atrophy of the cortex and Globus pallidus. Microscopically, there is severe loss of both large and small neurons in the dorsal region of these nuclei. There is also marked fibrillary gliosis.

Parkinson's is known as a disturbance of motor function, characterized by an expressionless face, slow voluntary movements, and progressively shortened accelerated step, rigidity, and, in most cases, characteristic tremors. Such motor disturbances are seen in a number of other diseases, that show damage to the nigrostriatal dopaminergic system. The disease may be developed as a result of use of dopamine antagonists, toxins, pesticides, stress etc. Ideopathic Parkinson's or paralysis agitans is a sporadic, progressive disorder and is the most common form of the disease. The onset age of the disease is 50 to 80 years. Broadly speaking the disease shows depigmentation of the substantia nigra and locus coeruleus. Microscopically, there is a loss of melanin containing neurons from these regions, with extraneuronal pigment deposition, and the presence of lewy bodies in the remaining neurons. Lewy bodies are intra cytoplasmic, eosinophillic, round, or elongated inclusions that consist of a dense core surrounded by a paler rim. The loss of the dopamine containing neurons from the substantia nigra results in dopaminergic depletion in the striatum, because this is the principal site of their axonal projections. The severity of the Parkinson's syndrome is proportional to the severity of the dopaminergic deficiency. This deficiency can be compensated at least in part by replacement therapy with L-dopa [12].

Motor neuron disease (Amyotrophic lateral sclerosis) is characterized by neuronal degeneration, concentrated mainly in the upper and lower motor neurons of the pyramidal motor system. The upper motor neurons are in the motor cortex and their axons traverse the internal capsule and corticospinal tract to synapse on the lower motor neurons in the cranial motor nerve nuclei and the motor neurons in the anterior horn of the spinal cord. Pathologically, degeneration of the upper motor neurons produces axonal loss and atrophy of the corticospinal tracts in the lateral columns of the spinal cord. The neurological symptoms exhibited by patients include spasticity, hyperflexia, muscle atrophy and weakness. The disease is also sporadic in most cases, but an approximately 2:1 male predominance has been noted. Onset is typically in late middle age, with usually fatal outcome in 2 to 6 years. There is no known treatment [12].

Increased generation of oxidative free radicals or an impaired antioxidant defense mechanism has been identified in the ageing process and neurodegenerative conditions, including Parkinson's and Alzheimer's disease, and in chronic stress induced perturbed state [13, 14]. Major oxidative free radical scavenging enzymes are superoxide dismutase, catalase, and glutathione peroxidase. Deficient functioning of these enzymes leads to an accumulation of toxic oxidative free radicals and consequent degenerative changes. Potential antioxidant therapies for CNS disorders should therefore include either natural antioxidant enzymes or agents that are capable of augmenting the function of these oxidative free radicals scavenging enzymes [15]. By virtue of their proposed properties and clinical uses, rasayana may provide potential intervention against oxidative threats, both in healthy and diseased states [16]. Bhattacharya et al. [17] have shown that the active principle of *Withania somnifera* increased cortical and striatal concentrations of antioxidant enzymes. Jain et al. [14] and Shukla et al. [18] also reported the effects of *Semicarpus anacardium* and *W. somnifera* on SOD, CAT, GPX, and LPO and demonstrated a significant increase in SOD and decrease in LPO. Panda and Kar [19] investigated the effects of *W. somnifera* root powder, administered for 15 and 30 days, and demonstrated a significant decrease in LPO and increase in both SOD and CAT.

14.4 Medicinal Plants, Neurodegenerative Diseases, and Therapeutics:

According to a WHO report (2001), approx 450 million people suffer from mental or behavioral disorders, yet only a small fraction of them receive even the most basic treatment. This amounts to 12.3% of the global burden of disease, which may increase to 15% by 2020. It is now becoming exceedingly apparent that the available psychotherapeutics does not properly meet the therapeutic demands of the vast majority of patients with mental health problems, and that herbal remedies remain the ultimate therapeutic hope for many such patients in the world. Critical analysis of our current understanding of the most popular and well studied CNS active medicinal plants reveals that many therapy related questions have not yet been properly investigated.

Drugs having CNS activity are widely used around the world. Mostly these are synthetic in nature, which are expensive, and with numerous side effects. In a systematic analysis published in 1997, it was reported that 157 of 520 drugs (30%) approved by the Food and Drug Administration (FDA) in the USA for the 11years period between 1983–1994, were natural products or their derivatives [20]. This report revealed in addition, that when focused efforts were made to discover natural products for clinical use, the success level rose dramatically. Thus, during the same period, 61% of anticancer agents approved were natural products or their derivatives. In the absence of targeted programs for natural products, there was no success. There were no analgesics, antidepressants, anxiolytics, or other CNS active drugs derived from natural products that were approved during the 11 year time period analyzed.

The situation has not changed much in recent years. Although identification of hits and leads from secondary plant metabolites continues to be a major goal of many drug discovery projects, most of the efforts do not concentrate on the search for agents potentially useful for the treatment of CNS disorders. In addition, comparatively few reports on the neuronal function modulating activities of herbal extracts and their active constituents appear regularly in various journals. So there is a dire need for a concentrated effort to define and understand the most appropriate therapeutic plants or to exploit them for identifying and developing CNS active drugs. On the other hand, it is worth noting that in India, a number of plants are available that show remarkable CNS activity, but to date, no systematic or scientific study has been carried out on such plants. In the past, several medicinal plants were identified for their effects on CNS. Satyavati [21] and Dhawan [22, 23] published a comprehensive list of various Indian plants with proven CNS activity. Properties of some of these plants are summarized in Table. 14.3.

At present, in modern medicine, therapies for many CNS diseases are not available. Drugs available for the treatments of anxiety, depression, and mental health conditions are not satisfactory, as the majority of them are of synthetic origin or derived from synthetic molecules. Chlorpromazine, reserpine, tricyclic and MAO depressants, benzodiazines, meprobamate, etc., used for the treatment of psychiactric patients; bicuculline, pentylenetetrazole, amphetamine, methylphenidates, etc., for CNS stimulation; barbiturates, hydantoin, oxazolidone, succinimides, and acetyl urea as antileptics; bromocriptine, apomorphine, levodopa, amatidine, trihexiphenydryl, and procyclidines for the treatment of Parkinsons's, are some of the drugs that have severe side effects and lead to dependency development.

Herbal remedies for such conditions have been known since time immemorial, and efforts made during the past few decades reconfirm that several herbs are indeed therapeutically useful for the treatment of various CNS disorders. Although all the active principles and mode of action of these herbal drugs/ formulations have not yet been precisely defined, but available data indicate that they could be useful sources to develop better, effective, safer, and cheaper drugs with novel modes of action. In the past few years, the number of reports on well-planned clinical trials using

Table 14.3 Composition of certain formulations in practice for treatment of CNS disorders

Name of formulation and its composition	Therapeutic implications
Mentat *Hyocyamus reticulate* *Mucuna pruriens* *Sida cordifolia* *Withania somnifera*	Anti-Parkinson's
Trasina *Banisteria cappi* *Claviceps purpurea* *Datura stramonium* *Mucuna pruriens* *Vicia faba*	Anti-Alzheimer's
BR-16A *Acorus calamus* *Bacopa monnieri* *Celestrus peiniculatus* *Centella asiatica* *Evolvulus alsenoides* *Mukta pisthi* *Nardostachys jatamanasi* *Orchis mascula* *Prunus amygadalus* *Syzygium aromaticum* *Valarana wallachii* *Withania somnifera* *Zinziber officianalis*	Anti-Parkinson's
Galanthamine *Galanthus warnori*	Anti-Alzheimer's
EGB761 *Ginkgo biloba*	Dementia, dementia of Alzheimer's type, and multi-infarct dementia
Brain vita *B. monnieri* extract	Cognitive functions
GB8 *Swiss Ginkgo biloba*	Cognitive functions

standardized herbal extracts has increased. No doubt, the consequences of these efforts will result in standardized herbal formulations acceptable to both patients and industry. A brief account of the potential herbs of Indian origin and those that are not native to India with therapeutic leads for the treatment of various CNS diseases including neurodegenerative disorders is presented below.

14.5 Plants Native to India

14.5.1 Acorus calamus

Acorus calamus, known as vacha in Ayurveda, is an aromatic marsh herb cultivated throughout India. Its medicinally useful part is its rhizome. Its essential oil consists of a range of sesquiterpene hydrocarbons, alcohols, and ketons (e.g., acorone, acoragermacrone, calamendiol) as well as eugenol, methyl isoeugenol, and phenyl propane derivatives α- and β-asaron. In Ayurveda, vacha has been acclaimed as a treatment for epilepsy and as a tranquilizer, in addition its action on bronchial diseases, indigestion, etc. The juice of the herb is also recommended in Sushruta Samhita for intellectual vigour and longevity. The ethanolic extract of the rhizome has been shown to possess neuroprotective action in acrylamide-disabled rats [24].

14.5.2 Asparagus racemosus

This is a tall, highly branched scandent under shrub, found growing wild in almost all parts of India and also in the Himalayas up to an elevation of 1500 m. It is widespread in upper Gangetic planes, especially in Bihar, where it comes up profusely after rains. Its medicinally useful part is the root, which contains glycosides (4%), as well as free sugar and oligosaccharides. The roots have been shown to contain, in addition to the ubiquitous sitosterol, undecanyl cetanoate, 4, 6-dihydroxy-2-o-(2′hydroxyisobutyl) benzaldehyde, flavonoids and pyrrolizidine alkaloid, and asparagamine A [25]. An extract was shown to possess a potent antioxidant property and was evaluated against kainic acid induced hippocampal and striatal neuronal damage in anesthetized mice. The extract supplemented mice showed reduced lipid peroxidation and an increase in glutathione peroxidase and glutathione contents, thus demonstrating its antioxidant activity [26, 27]. Recent studies have demonstrated the neuroprotective effects of asparagus root extract in an *in vivo* model of aluminum and paraquat induced neurodegeneration [28, 29].

14.5.3 Bacopa monnieri

This plant is commonly found in wet marshy and damp places throughout India. It is a succulent, creeping herb with a stem 10 to 20 cm long that produces roots at nodes. The drug contains alkaloid (brahmine) and triterpenoid saponins (bacosides A and B) [30, 31]. The plant is used as a nervine tonic, and diuretic and is commonly used to treat asthma, epilepsy, insanity, and hoarseness. It is a major constituent of medhya rasayana formulations, which facilitates learning and improves memory [32–37]. Initial studies using 50% ethanolic extract of the whole plant without roots demonstrated its effects on short and long term memory retention. Bhattacharya and his coworkers [38] studied the effects of *B. monnieri* extract on AD using a rat model. Oral administration of a 5 to 10 mg extract per kilogram of body weight markedly reduced the memory deficits as well as acetylcholine concentrations, choline acetylase activity, and muscuranic receptor binding in the hippocampus and frontal cortex [38].

14.5.4 *Celastrus peniculatus*

Kumar and Gupta [39] demonstrated that *Celastrus peniculatus*, along with *W. somnifera*, shilajit, and *Convolvulus pleuricaulis* showed CNS activity. Seed oil is known to affect the CNS [40, 41], and to be useful in psychiatric patients [41]. The beneficial effects of *C. peniculatus* include being a stimulant [42], memory enhancement [43], and improvement in IQ of mentally retarded children [44].

14.5.5 *Centella asiatica*

Locally called mandukparni, *Centella asiatica* is highly prized in Ayurveda as a medhya (promoter of memory and intellect) and as an agent that prevents ageing (vayahsthapana) from the days of Charaka and Sushruta. Mandukaparni is also used for treatment of epilepsy, mental health disorders (mania, hysteria), insomnia, etc., [45]. A double blind clinical trial in 30 mentally retarded children (age 7- to 18 years) who received the drug for 3 months showed improved cognitive functions [46]. A recent study has also demonstrated improved learning and memory in wistar rats. The extract was found to be effective in preventing cognitive deficits in an intracerebroventricular STZ model of Alzheimer's in rats. Asiatic acid has been patented as a therapeutic agent for the treatment of dementia. Asiatic acid and related synthetic analogs have been shown to protect cultured neurons from glutamate-induced excitotoxicity [47].The alcoholic (50%) extract of the plant has shown enhanced neuropeptide synthesis in the brain. It also induces an increase in brain protein content and acquisition efficiency [48].

14.5.6 *Convolvulus pleuricaulis*

Convolvulus pleuricaulis is used in traditional systems of medicine in the treatment of anxiety, neurosis, insanity, and epilepsy, and also as a brain tonic. The whole plant is one of the most important medhya rasayana drugs in Ayurveda. It improves balance and vitiation in kapha-vata-pitta doshas (physiological functions), and the herb is astringent and bitter. *C. pleuricaulis* is used traditionally to treat nervous debility, insomnia, fatigue, low energy levels, and as a brain tonic, alterative, and febrifuge. The whole herb is used medicinally in the form of a decoction along with cumin and milk in fever, nervous debility, and memory loss. The plant is reported to be a prominent memory improving drug, psychostimulant, and tranquilizer, and it reduces mental tension. The methanolic extract of *C. microphyllus* Sieb. Ex Spreng (*C. pleuricaulis* Choisy) showed enhanced release of nerve growth factor (NGF). NGF prevents experimentally induced or age related degeneration of basal forebrain cholinergic cell bodies in adult rats and can also restore lesion-induced loss of cognitive functions [49, 50].

14.5.7 Crocus sativus

Also called a kunkumam or keshara in Ayurveda. It is a small perennial cultivated in certain parts of the Jammu, Kashmir, Himachal Pradesh, and Uttarakhand states of India. The medicinally useful part is the stigma, which is dried and marketed as saffron. The important constituents of saffron are its pigments (crocin-1,2,3,4) and essential oils. Four crocetins (F, G, H, I) have also been isolated [51]. The alcoholic extract of saffron ameliorates the impairment effect on learning and memory processes. It has been also shown that crocin inhibits neuronal death induced by both internal and external apoptotic stimuli [52], thus it is considered as neuroprotector. Crocin prevents the activation of c-jun kinase phosphorylation, which is involved in the signaling cascade for neuronal death [53].

14.5.8 Curculigo orchioides

Curculigo is locally called kalimusli and is found all over India. Pharmacological investigations revealed that the 70% ethanol extract of the rhizomes are a sedative and anticonvulsant. The plant is known to contain steroids and triterpenoids and several phenolic compounds [54, 55].

14.5.9 Curcuma longa

Curcuma longa is a perennial rhizomatous plant growing all over India. Curcumin or C. *longa* extract shows strong antioxidant activity. β-amyloid-induced oxidative stress appears to be an important pathway of neuronal cell death in AD. The methanolic extract of turmeric led to the isolation of Calebin-A and the curcumins, which effectively protects neuronal cells against β-amyloid deposition. In another study, curcumin, on oral administration to alcohol-fed rats, caused a significant reversal of brain lipid peroxidation, thus indicating a neuroprotective role [56]. *In vivo* experiments showed that oral intake of curcumin significantly reduces the duration and clinical severity of demylenation in experimental allergic encephalitis [57].

14.5.10 *Cyprus rotundus*

Known as mustaka in Ayurveda, *Cyprus rotundus* is a perennial grass growing almost everywhere in India. Its tubers contain medicinally useful essential oils (sesquiterpenoids, monoterpenes, aliphatic alcohols, acetates). A receptor binding assay demonstrated that isocurcuminol, a constituent of *C. rotundus*, modulates GABAergic neurotransmission via enhancement of endogenous receptor ligand binding, and thus having a bearing in epilepsy [58].

14.5.11 *Ficus religiosa*

In Ayurveda, it is claimed that the leaves of *Ficus religiosa* possesses anticonvulsant activity. The leaf extract was evaluated for its activity against pentylenetetrazole (PTZ, 60 mg/kg i.p.) induced convulsions in albino rats. The study revealed 80 to 100% protection against PTZ induced convulsions when given 30 to 60 min prior to the induced convulsions [59].

14.5.12 *Ginkgo biloba*

An extract prepared from green leaves (EGB761) was identified as therapeutically useful for the treatment of peripheral circulatory disturbances, as it is a vasodilator. Results showed that EGB761 is not a vasodilator in the classical sense, but it is a CNS function modulating and neuroprotective agent, sustainable for the therapy of patients with cerebrovascular disorders or cerebral insufficiencies. Ginkgo is widely used in Europe for treating dementia. It improves blood flow in the brain and contains flavonoids that act as antioxidants. It is presumed that ginkgo may improve thinking, learning, and memory, and results are very encouraging in people with AD [60].

More than 40 components of ginkgo have been identified and isolated. Two of the most important groups of active chemicals are flavonoids (quercetin, kaempferol, isorhemnetin) and terpenes (lactones or terpenoids, which include bilobalide and several ginkgolides A, B, C, J and M). Individual constituents have been studied in hundreds of in vitro, animal and human experimental systems [61, 62]. Recently, efforts have been made to understand the effects of ginkgo on dementia of the Alzheimer's type and in the closely related multi-infarct dementia. These double-blind randomized trials in patients have shown that EGB761 is efficacious in delaying the clinical deterioration with dementia, or in bringing about symptomatic improvement [63–65]. The mechanism of ginkgo's therapeutic effects are not fully understood, but they are attributed to the synergistic effects of its constituents [66, 67]. These act to varying degrees as scavengers of free radicals and chemicals, implicated in the pathophysiology of AD. The dose is 40 mg, three times a day, of an extract standardized to 24% flavonoid, glycoside, and 6% terpenoid.

14.5.13 Mucuna pruriens

In Ayurveda *Mucuna pruriens* is called atmagupta or kapikacchu. It is a herbaceous creeper growing in several parts of India. Its seeds, roots and pod bristles are medicinally useful parts. An important chemical constituent of the plant is the nonprotein amino acid L-dopa, which is present in seeds. In addition, β-sitosterol, lecithin, glutathione, and gallic acid are other important constituents [45, 68–70]. Beans of this plant are used as nutritive food in some parts of India. It is also used as a therapeutic agent in various reproductive and nervous diseases [71–73]. An Ayurvedic formulation containing *M. pruriens* beans is used in the treatment of Parkinson's disease.

14.5.14 Nardostachys jatamanasi

The plant of *Nardostachys jatamanasi* is used by Santhal tribals in the treatment of madness, epilepsy, loss of unconsciousness, convulsions, etc., [74]. The decoction of the root is also reported to be useful in mental disorders, insomnia, etc. *N. jatamanasi* is reported to yield 2% volatile oil containing an ester, an alcohol and two alkaloids [75, 76]. The rhizome of jatamanasi yields jatamanashic acid [77, 78]. Various extracts of jatamanasi root showed sedative effects in rats. The ethanolic extract of *N. jatamanasi* reduced rat brain serotonin, and though it showed no effect on the CNS, but oil from the rhizome showed depressant action on the CNS [79]. A preparation comprising *N. jatamanasi, Centella asiatica, Acorus calamus, Rauwolfia serpentina, Saussurea lappa*, and *Valeriana wallichii* showed significant improvement in case of schizophrenic patients [80]. The ethanol extract showed potent inhibition of acetylcholinesterase reaction rate [45].

14.5.15 Plumbago zeylanica

This plant is called chitraka in Ayurveda. It is a perennial shrub growing wild in the hotter parts of India. Its roots and root bark are medicinally useful. The chief constituent is plumbagin. The ehanol extract of the root has shown spontaneous motility in rats with a concomitant increase in dopamine and a metabolite homovanillic acid level in striatum, indicating a dopaminergic pathway for stimulatory action on the CNS [81]. The plant has also been useful in the treatment of schizophrenia.

14.5.16 *Semecarpus anacardium*

In Ayurveda *Semecarpus anacardium* is commonly known as bhallataka. The tree commonly grows in the hotter areas of India and the foothills of the Himalayas. The fully developed nut is valued medicinally. A phenolic glycoside, anacardoside, has been isolated. Besides the phenol, several biflavonoids have been obtained from the defatted nuts of the plant. A cytological and ultrastructural study on Swiss rats from the author's laboratory has shown neuroprotective effects of the ethanol extract [82–84].

14.5.17 *Swertia chirayita*

The *Swertia chirayita* is known as kiraatatikta in Ayurveda and chiraytta is its Indian trade name. It grows in the temperate Himalayas from Kashmir to Bhutan. Though the root is considered more potent, the whole herb is used medicinally. More than 20 n-polyhydroxylated xanthones, such as swertianin, swerchirin, and mangiferin, identified. Mangiferin has been shown to be free radical scavenger [85] and a superoxide scavenger. It is an inhibitor of the expression of inducible nitric oxide synthatase and TNF genes, thus revealing its potential for the treatment of neurodegenerative disorders [86].

14.5.18 *Withania somnifera*

Withania somnifera is commonly known as ashwagandha in Ayurveda. It is an evergreen shrub growing throughout the drier and subtropical parts of India. The medicinally valuable part is its root, which is used in Ayurvedic preparations. It is used as a powder, decoction, medicated wine, etc. The main pharmacologically active constituents are alkaloids (withanine, cuscohygrine, tropane, anahygrine, somniferin, anaferine, withananine, withananinine) and steroid lactones (withanolides). The total alkaloid content of its roots varies between 0.1 and 0.3%, although in some cases a higher yield is also been reported. Karnic [87] suggested that *W. somnifera* should be considered the premier herb for all negative conditions associated with ageing. Evidences shows that the drug is useful in preventing senile dementia and AD [42]. The drug helps in slowing down the progression of AD [15]. Glycowithanolides were also found to reverse both cognitive deficits and perturbed central cholinergic markers induced as a result of neurodegeneration. Bhattacharya et al. [88], while studying foot shock induced changes in the rat brain, showed that *W. somnifera* also normalized SOD and LPO activity and enhanced CAT and GPX activity. Its inclusion as a rasayana drug is supported by several studies showing the neuroprotective activity of *W. somnifera* roots. Damage to neuron circuits in the brain leads to several diseases, such as memory deficit, AD, and Parkinson's disease. In vitro investigation using the methanolic extract of its roots demonstrated the formation of dendrites. This resulted in a significant increase in the percentage of cells with neuritis in human neuroblastoma SK-N-SH cells [89]. A study in our laboratory showed that the root extract significantly protected the neurons in the hippocampal regions of rats under stress conditions [90]. Oral administration of the root extract improved memory acquisition and retention in experimental animals [90]. The effects of the methanolic extract on cholinergic, GABAergic, and glutamatergic receptors have also been demonstrated in wistar rats, as well as activation of cholinergic signal transduction cascade in the cortical and basal forebrain region [91].

14.6 Plants not Native to India

14.6.1 Galanthus wornorii

Galanthamine is a pure unaltered extract of *Galanthus wornorii* [92]. A recent study by Willcock and coworkers [93] has shown that galanthamine appears to slow the progression of neurodegenerative conditions. It also reversibly and competitively inhibits acetylcholinesterase and enhances the response of nicotinic receptors to acetylcholine. A study in 653 Alzheimer's patients showed that galanthamine slows down the decline of the functional abilities as well as cognition.

14.6.2 Huperzine serrata

Huperzine A was identified for the first time in a Chinese medicinal herb, *H. serrata*. In a well designed placebo-controlled trial using huperzine, 58% of the patients with AD showed significant improvement in memory and cognitive and behavioral functions after taking 200 mg of the drug, twice per day, for up to 8 weeks. Another double-blind experiment using injected huperzine A also confirmed positive effects in patients with dementia [94].

14.6.3 Lavandula stoechas

Another type of neurodegenerative disease is epilepsy. Lavandula has been used for a long time in traditional medicine as anticonvulsant. Gilani et al. [95] validated its anticonvulsant effects. The study revealed that an aqueous methanolic extract (600 mg/kg) significantly reduced the severity and increased the latency of onset of convulsions induced by PTZ. Moreover, in isolated rabbit jejunum preparations, it caused a dose dependent relaxation of spontaneous contraction and inhibited K^+induced contractions, suggesting Ca^+ channel blockade [95].

14.7 Conclusion

The use of medicinal plant products in the form of household formulations, or traditional Indian medicinal systems has been in practice for a long time. Efforts made during the past few decades by systematic experimental and clinical studies have confirmed that several of these products are indeed therapeutically potent, not only in treating cough and cold, but also diseases of the CNS, including neurodegenerative diseases, for which only symptomatic treatment is available so far. Although all active principles and their mode of action are not yet precisely defined, available data on their neurophamacological activity indicate that they could be sources of better therapy for diseases for which treatment is not yet fully developed. Extracts prepared from many medicinal plants contain a variety of bioactive molecules. Unfortunately, it is extremely difficult to analyze them because of their complex synergistic activity.

Acknowledgements The author is grateful to Dr. S.D. Shukla and Mr. D. Sharma (TRF) for their help in prearing this manuscript.

References

1. Seth PK, Kakkar P (2002) Annl Neurosci 9:17
2. Bhatnagar M (2005) Mol Cellular Neuobiology. Narosa, New Delhi
3. Rajini M, Kanaki NS (2008) Phytochemical standardization of herbal drugs and polyherbal formulations. In: Ramawat KG, Merillon JM (eds) Biotechnology: Bioactive Molecules. Springer, Berlin Heidelberg New York, p 349
4. Puri HS (1970) Q J Crude Drug Res 10:1555
5. Puri HS (1970b) Indian Drugs7:15
6. Puri HS (1971) Q. J Crude Drug Res 11:1742
7. Puri HS (1972) Indian Drugs 9:11
8. Udupa KN (1973) J Res Edu Ind Med 8:1
9. Puri HS (2003) Ayurvedic Herbs of Rejuvenation and Longevity. Taylor and Francis, London
10. Walton J (1998) Oxford University Press, Oxford
11. Gonzalo-Ruiz A, San ZJM, Geula C, Gonzalo P (2005) J Chem Neuroanat 29:31
12. Houghton PJ, Howes M-J (2005) Neurosignals 14:6
13. Maxwell SRJ (1995) Drugs 49:345
14. Jain S, Shukla SD, Sharma K, Bhatnagar M (1998) Indian J Gerontol 12:60
15. Bast A, Haenen GR, Dolemea GJ (1991) Am J Medical 91:2
16. Ghosal S (1991) Indian J Indigen Med 8:1
17. Bhattacharya SK, Kumar A, Ghosal S (1995) Phytother Res 9:110
18. Shukla SD (2001) Ph D thesis, MLS University, Udaipur, India
19. Panda S, Kar A (1997) Ind J Physiol Pharmacol 41:424
20. Cragg GM, Newman DJ, Sanders KM (1997) J Nat Prod 60:52
21. Satyavati GV, Raina MK, Sharma M (1976) Medicinal Plants of India. ICMR, New Delhi
22. Dhawan BN, Shrimali RC (1995) Indian J Pharmacol 27:202
23. Singh HK, Dhawan BN (1997) Indian J Pharmacol 29:5359
24. Shukla PK, Khanna VK, Ali MM, Mayura RR, Handa SS, Srimal RC (2002) Phytother Res 16:256
25. Rastogi RP, Mehrotra BN (1998) Comp Ind Med Plants 5:107

26. Parihar MS, Hemnani T (2004) J Neural Transmiss 111:1
27. Tripathi YB, Chaurasia S, Tripathi E, Upadhyaya A, Dubey GP (1996) Indian J Exp Biol 34:521
28. Saxena, G, Singh M, Saxena A, Bhatnagar M (2007) J Herb Med Toxicol 2:33
29. Saxena G (2005) Ph D thesis, MLS University, Udaipur, India
30. Dhawan BN, Singh H (2002) Ann Neurosci 9:34
31. Rastogi RP, Dhar ML (1960) J Sci Ind Res 19b:455
32. Chatterji N, Rastogi RP, Dhar ML (1963) Indian J Chem 1:212
33. Chatterji N, Rastogi RP, Dhar ML (1965) Indian J Chem 3:24
34. Kawai K, Istaka Y, Shibata S, Kulshreshtha DK, Ratogi RP (1973) Acta Cryst 29B:2497
35. Rastogi S, Pal R, Kulshrestha DK (1993) J Heterocycl Chem 2:149
36. Rastogi S, Kulshrestha DK (1999) Indian J Chem 38b:353
37. Garai S, Mahto SB, Ohtani K, Yamasaki K (1996) Phytochemistry 42:815
38. Bhattacharya SK, Kumar A, Ghosal S (1999) Pharmacol Toxicol 4:11
39. Kumar MH, Gupta YK (2002) Phytomedicine 9:302
40. Joglekar GC, Balwani JM (1967) J Res Indian Med 1:190
41. Lewis WH, Elvin-Lewis MPF (2003) Medical Botany. Wiley, New York p 527
42. Kakrani HK, Nair GV, Kalyani GA, Satyanarayan D (1985) Fitoterapia 56:293
43. Karranth KS, Haridas KK, Gunasundarai S, Guruswami MN (1980) Arogya Health Sci VI:137
44. Nalini K, Aroor AR, Kumar KB, Rao A (1986) Alt Med 1:355
45. Dev S (2006) Prime Ayurvedic Drugs. Anamaya, New Delhi, p 105
46. Kumar MHV, Gupta YK (2002) J Ethanopharmacology 79:253
47. Lee MK, Kim SR, Sung SH, Lim D, Kim H, Choi H, Park HK, Te S, Ki YC (2000) Res Commun Mol Pathol Pharmacol 108:75
48. Khare CP (2006) Indian Medicinal Plants. Springer, Berlin Heidelberg New York, p 169
49. Pavia MR, Davis RE, Schwartz RO (1990) Ann Rep Med Chem 25:21
50. Singh RH, Mehta J (1977) Indian Med Yoga Homeopathy, 12:18
51. Li YC, Wu TS (2001) Chem Pharm Bull 50:1305
52. Soeda S, Iwata K, Hosoda Y, Shimeno H (2001) Life Sci 69:2887
53. Ochiai T, Soeda S, Ochno S, Tnka H, Shoyama Y, Shimeno H (2004) Neurochem Int 44:321
54. Tandon M, Shukla YN (1995) Curr Res Med Aromat Plants 17:42
55. Valls J, Rchard T Larronde F, Leblais V, Muller B, Delauney, JP, Ramawat KG, Merillon JM (2006) Fitoterapia 77:416
56. Aggarwal BB, Shishodia S, Young-Joon S (2007) The Molecular Targets and Therapeutic Uses of Curcumin in Health And Disease. Springer, Berlin Heidelberg New York
57. Natrajan C, Bright JJ (2002) J Immunol 168:6506
58. Ha JH, Lee KY, Choi HC, Cho J, Kong BS, Lim JC, Lee DU (2002) Biol Pharm Bull 25:128
59. Indurwale NH, Biyani KR (2001) Adv Pharmacol Toxicol 2:51
60. Watnabe CMH, Wolffram S, Ader P, Rimbach G, Pecker L, Maguire JJ, Schultz PG, Gohil KC (2001) Proc Natl Acad Sci USA 98:6577
61. Chavez ML, Chavez PI (1998) Hosp Pharm 33:658
62. Van Beek TA, Bombardelli E, Morazzoni P, Peterlongo F (1998) Fitoterapia 69:195
63. Le Bars PL, Katz MM, Berman N, Itil TM,Fredman AM, Saltzberg AFA (1997) JAMA 278:1327
64. Oken BS, Storzbach DM, Kaye JA (1998) Arch Neurol 55:1409
65. Sivarajan VV, Balchandran I (1999) Ayurvedic Drugs and Plant Sources. Oxford & IBH, New Delhi
66. Behl C, Davis J, Leslie R, Schubert D (1994) Cell 77:817
67. Maitra I, Marcocci L, Droys-Lefais M, Packer L (1995) Biochem Pharmacol 49:1649
68. Manyam BV, Parikh KM (2002) Ann Neurosci 9:40
69. Vaidya ND (1925) A Treatise on Experianced Remedies. Vaidya Kalpatru, Ahemdabad, p 100
70. Vaidya AB, Rajgopalan TG, Mankodi NA, Antarkar DS, Tatehd PS, Purohit AV, Wadia NH (1978) Neurol India 26:171
71. Maw GN (1999) Annu Rep Med Chem 34:74

72. Dutt VC (1980) Materia Medica of Hindus (Krishnades Sanskrit Studies, vol II) Varanasi 148–149
73. Damodaran M, Ramaswami R (1937) Biochem 31:2149
74. Manyam BV, Bhatt NV (2001) Ayur Int 1:6
75. Jain SK, Tarafdar CR (1970) Econ Bot 24:241
76. Bose BC, Vijayavergia R, Bhatnagar JN (1957) Indian J Med Sci 11:799
77. Bose BC, Gupta SS, Vijayavergia R, Saifi AQ, Bhatnagar JN (1957) Curr Sci 26:278
78. Chaudhry GR, Sharma VN, Siddiqui S (1951) J Sci Ind Res 10B:48
79. Chaudhry GR, Dhar MM, Nityanand P, Dhar ML (1958) J Sci Ind Res 17B:159
80. Chopra IC, Jamwal KS, Khajuria BN (1986) Indian J Med Res 42:385
81. Mahal AS, Ramu MG, Chaturvedi DD, Thomas KM, Senapati HM, Murthy NNS (1976) Indian J Psychiatr 18:283
82. Bopaiah CP, Pradhan N (2001) Phytother Res 15:153
83. Shukla SD, Jain S, Sharma K, Bhatnagar M (2000) Indian J Exp Biol 38:1007
84. Bhatnagar M, Shukla SD, Bhatnagar R (2005) Herb Pharma Ther 5:21
85. Pauletti PM, Castro-Gamba I, Silva DHS, Young MCM, Tomazela DM, Eberlin MN, Bolzani VS (2003) J Nat Prod 66:1384
86. Watanabe CMH, Wolffram S, Ader P, Rimbach G, Pecker L, Maguire JJ, Schultz PG, Gohil KC (2001) Proc Natl Acad Sci USA 98:6577
87. Karnic LR (1991) Indian Med J 3:1
88. Bhattacharya A, Ghosal S, Bhattacharya SK (2001) J Ethnopharmacol 74:1
89. Tohda C, Kuboyama K, Komatsu K (2000) Neuroreport 11:198
90. Jain S, Shukla SD, Sharma K, Bhatnagar M (2001) Phytother Res 15:544
91. Schliebs R, Leibmann A, Bhattacharya SK, Kumar A, Ghosal S, Bigl V (1997) Neurochem Int 30:181
92. Venturi VM, Piccinin GL, Taddei I (1965) Boll Soc Ital Biol Sper 11:593
93. Wilcock GK, Lilienfield S, Gaens E (2000) BMJ 4:1445
94. Grant WB (1997) Alz Dis Rev 2:42
95. Gilani AH, Aziz N, Khan MA, Shaheen F, Jableen Q, Siddiqui BS, Herzig JW (2000) J Ethnopharmacol 71:161

Chapter 15
Traditional Medicine for Memory Enhancement

Melanie-Jayne R. Howes and Peter J. Houghton

Abstract In traditional practices of medicine, numerous plants have been used to alleviate memory impairment both in healthy individuals and those with disease states which are now recognised as specific cognitive disorders such as Alzheimer's disease (AD). An ethnopharmacological approach has provided leads to identify plants and their compounds that may have potential to modulate cognitive abilities by different modes of action. A variety of therapeutic targets have been identified as relevant in the treatment of cognitive disorders, including modulation of the cholinergic system, which may be achieved by the inhibition of acetylcholinesterase (AChE), and neuroprotection against glutamate-induced overstimulation of *N*-methyl-D-aspartate (NMDA) receptors, by the use of NMDA receptor modulators. Other activities considered to be relevant in the alleviation of cognitive impairment include anti-inflammatory, antioxidant and estrogenic activities. Two of the currently licensed drugs used to treat cognitive symptoms in AD, galantamine and rivastigmine, were derived from plant sources and have been characterised as inhibitors of AChE. However, some plant extracts which occur as a complex mixture of components, such as *Ginkgo biloba* L. extract, have demonstrated relevant biological activities in relation to cognitive function, but the compounds responsible for the observed effects or the mechanisms of action have not been well characterised. Amongst the many plants reputed to enhance cognitive function in a variety of traditional medicines including Ayurvedic, Chinese, European, African and South American medicines, relatively few have been extensively studied to determine any pharmacological basis for their historical uses. Some of those plants that have generated particular interest in understanding and establishing their potential for alleviating cognitive impairment are discussed.

Keywords Acetylcholinesterase inhibitors · Alzheimer's disease · Anti-inflammatory · Antioxidant · Estrogenic · Memory · Traditional medicine

---

M.-J.R. Howes (✉)
Royal Botanic Gardens, Jodrell Laboratory, Kew, Richmond, Surrey, TW9 3AB, UK
e-mail: m.howes@kew.org

K.G. Ramawat (ed.), *Herbal Drugs: Ethnomedicine to Modern Medicine*,
DOI 10.1007/978-3-540-79116-4_15,

Abbreviations

ACh	Acetylcholine
AChE	Acetylcholinesterase
AD	Alzheimer's disease
AMPA	Amino-3-hydroxy-5-methyl-4-isoxazole propionic acid
APP	Amyloid precursor protein
Anti-ChE	Anti-cholinesterase
BACE	β-Secretase
BChE	Butyrylcholinesterase
ChAT	Choline acetyltransferase
ChE	Cholinesterase
CNS	Central nervous system
COX	Cyclo-oxygenase
DMHA	Dimethylhuperzine A
ERT	Estrogen-replacement therapy
GABA	γ-Aminobutyric acid
5-HT	5-Hydroxytryptamine
IL-1	Interleukin-1
KUT	Kami-utan-to
LOX	Lipoxygenase
LTB_4	Leukotriene B_4
mRNA	Messenger ribonucleic acid
NF-κB	Nuclear transcription factor κB
NGF	Nerve growth factor
NMDA	*N*-methyl-D-aspartate
NO	Nitric oxide
NSAID	Non-steroidal anti-inflammatory drug
PAF	Platelet-activating factor
PG	Prostaglandin
PUFA	Polyunsaturated fatty acid
TCM	Traditional Chinese medicine
TNF	Tumour necrosis factor
VIP	Vasoactive intestinal peptide

15.1 Introduction

15.1.1 Cognitive Disorders

Loss of memory may occur in healthy people, where it is often associated with the ageing process, or it may also occur in specific disease states in which cognitive function is impaired. Cognitive dysfunction can be a feature of a variety of disorders, including schizophrenia [1], multiple sclerosis [2], and the many pathologies associated with dementia. Dementia, which involves a deterioration of cognition, may present in patients with central nervous system (CNS) infections, Huntington's disease, Parkinson's disease, Lewy body disease, Pick's disease, vascular disease and Alzheimer's disease (AD) [3, 4]. AD is reported to be the most common type of dementia [4] and is estimated to account for between 50% and 60% of dementia cases in persons over 65 years of age [5]. There is a need for effective therapies that target loss of memory both in healthy individuals and those with specific disorders that involve cognitive impairment. At present there is a lack of therapeutic substances that can alleviate memory loss in both cases. There are many challenges involved in developing effective therapies for treating or preventing general memory loss associated with ageing and for treating or preventing the memory impairment which is a feature of the different pathologies associated with the variety of cognitive disorders such as dementia, although there may be some common therapeutic targets which may be investigated to alleviate cognitive impairment in different disease states and in general memory loss. The common symptoms of neurodegenerative diseases, such as loss of memory, have been recognised as a feature of increasing age for a long time, and it is these symptoms, rather than the specific disease states which are diagnosed in modern medicine, which have been acknowledged in many traditional practices of medicine.

15.1.2 Alzheimer's Disease

AD is a progressive, neurodegenerative disease that primarily affects the elderly population and it is a major public health concern. The etiology of AD is still not fully understood. The main symptoms associated with AD involve cognitive dysfunction, which primarily involves loss of memory and recognition skills [4, 6, 7]. Several other features present in the later stages of the disease, including language deficits, depression, agitation, mood disturbances and psychosis [8, 10]. Although the pathology of AD has not been fully elucidated, senile plaque and neurofibrillary tangle formation, and associated oxidative and inflammatory processes and neurotransmitter disturbances, occur in the CNS. In particular, a consistent neuropathological feature associated with memory loss is a deficit of acetylcholine (ACh) (**1**), a neurotransmitter which is associated with cognitive function, and a cholinergic deficit has been correlated with the severity of AD [11–14].

15.1.3 Cholinergic Function

The presynaptic synthesis of ACh (**1**) from choline and acetyl coenzyme A in neurons is catalysed by the enzyme choline acetyltransferase (ChAT) [15]. Once synthesised, ACh is stored in synaptic vesicles and is released upon neuronal stimulation into the synaptic cleft [15]. ACh may then stimulate postsynaptic receptors (or presynaptic receptors to regulate ACh release) before being hydrolysed at the ester bond in the ACh molecule to yield choline and acetate, under the action of the enzyme acetylcholinesterase (AChE). Thus, ACh has a short half-life, and attempts to prolong its action have been investigated as a therapeutic target to restore cholinergic function in some cognitive disorders such as dementia.

AChE consists of a complex protein of the α/β hydrolase-fold type having an overall ellipsoid shape containing a deep gorge about 20 Å in depth. At the bottom of this gorge are four main subsites known as the 'esteratic site', the 'oxyanion hole', the 'anionic subsite' and the 'acyl pocket', and it is in this region where ACh hydrolysis appears to occur, although the initial binding of ACh is thought to occur at an outer region known as the 'peripheral site' [16]. In addition to its role in hydrolysing ACh, AChE has also been associated with other functions including a role as an adhesion protein, a bone matrix protein, in neurite growth and in the production of amyloid fibrils, which are characteristically found in the brain tissue of AD patients [17], although the clinical relevance of these actions has yet to be explored more thoroughly. Another type of cholinesterase (ChE) is butyrylcholinesterase (BChE), which is found mainly in serum and glia, unlike AChE, which is mainly distributed in neurons and erythrocytes in vertebrates [18]. The physiological role of BChE is unclear, with no endogenous natural substrate having been identified. BChE is known to hydrolyse toxic esters such as cocaine, and it is considered to have a detoxifying action [18].

15.1.4 Symptoms Related to Memory Disorders in Traditional Practices of Medicine

Many disease states now known and characterised by diagnostic tests in modern orthodox 'Western' medicine, including cognitive disorders such as AD, were not recognised in the past as such in some more traditional practices of medicine, or may be described differently in other cultures. Consultation with traditional practitioners and healers or referral to historical literature is an unlikely approach to yield relevant information if modern medical terms are used. A more productive approach would involve searching for information according to the treatment of characteristic symptoms of a particular disease. AD is characterised by a loss of short-term and eventually long-term memory as a feature of general cognitive decline. In the later stages of the disease, language deficits, depression and agitation can occur. Therefore, investigations into traditional uses should be focused on those substances which have a reputation for improving memory or cause a general stimulation of learning processes or intellect, in whichever way it is viewed in a particular cultural context.

As well as those materials used for medicinal purposes, obtaining knowledge on the types and uses of poisonous substances may also provide some indication of their physiological effects. For example, a substance that produces symptoms of sweating, flushing of the skin and intestinal contractions may indicate that it contains compounds which stimulate cholinergic activity, perhaps by inhibiting AChE and thus ACh (**1**) hydrolysis. Compounds that may produce these effects could be of interest for study as potential cognitive enhancers.

It must also be emphasised that understanding the etiology of a disease in Western medicine may not be the same as that in other traditions. The approaches to treatment will reflect the symptoms which manifest and the recognised causes. Many African cultures include spiritual factors in the cause of disease, whilst Ayurvedic and Chinese medicine view the correction of imbalance as an aspect of the appropriate treatment. The interpretation of medical conditions described according to these traditional practices compared to the orthodox medical approach may present problems. Consequently, knowledge of the underlying pathologies which could be associated with the described and observed symptoms may be necessary to help characterise disease states.

15.2 Activities Relevant to the Treatment of Cognitive Disorders

15.2.1 Precursors of Acetylcholine

Adequate availability of choline has been proposed to enable sufficient ACh (**1**) synthesis for neurotransmission. Precursors of ACh (e.g. choline and lecithin) have been investigated for their effects on synthesis and release of ACh, with a view to increasing ACh release and cholinergic activity. Few clinical or animal studies have reported any significant beneficial effects on cognitive function with these compounds [8]. Therapy failure may be due to impaired uptake mechanisms of choline causing the reduction in ACh synthesis, and not due to insufficient choline supply. This is apparent as it has been reported that more choline occurs in the cerebrospinal fluid of AD patients than in patients without AD, and that choline levels increase with disease progression [8, 19]. Therapy with ACh precursors may be limited by side-effects, including gastrointestinal disturbances such as nausea, vomiting and diarrhoea.

15.2.2 Muscarinic Receptor Stimulation

Direct cholinergic receptor stimulation has been explored as one therapeutic target to enhance cognitive function. Cholinergic agonists are reported to facilitate learning and memory, but cholinergic antagonists impair learning and memory [20, 21]; thus cholinergic agonists may be useful in cognitive disorders. Direct stimulation of the M_1 muscarinic receptor with agonists such as xanomeline (**2**), which is an analogue of the natural compound arecoline, is reported to improve cognition both in animal models and in AD patients, and antagonists of central presynaptic M_2 receptors, which include analogues of the naturally derived himbacine (**3**), also enhance cognitive ability by increasing the release of ACh (**1**) [22]. To date, treatment with compounds that directly interact with muscarinic receptors has not been a major approach to alleviate cognitive disorders such as AD, perhaps due to the unpleasant cholinergic side-effects such as gastrointestinal contraction and sweating, reported to be associated with muscarinic receptor modulators.

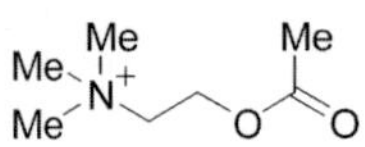

1 Acetylcholine

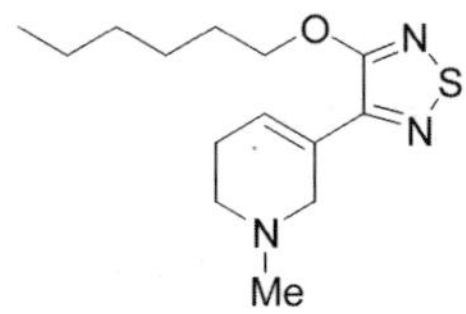

2 Xanomeline

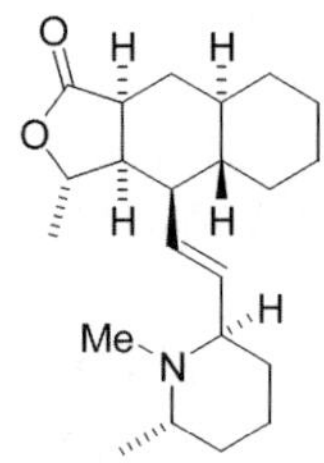

3 Himbacine

15.2.3 Nicotinic Receptor Stimulation/Nicotinic Agonists

Behavioural studies have shown that nicotinic receptors participate in cognitive functions [23]. Nicotinic receptors are reduced in cortical brain areas in AD [24–28], and nicotine (**4**) upregulates nicotinic receptors and increases ACh (**1**) release [29–31]. Nicotine treatment in various in vivo studies, including administration to rats with cholinergic brain lesions and in aged monkeys, has been shown to improve cognitive function [32, 34].

Thus, nicotinic agonists may enhance cholinergic neurotransmission and therefore cognitive function in some disorders that feature memory impairment. In support of this, some studies suggest smoking may protect against AD development, and administration of nicotine to AD patients and to healthy (non-AD) elderly people improved cognitive function [35–39]. However, some cohort studies have shown that smoking shows either no association with AD risk or moderately increases AD risk [40–42]. Although the effects of smoking on memory and AD risk are inconclusive, the effects of nicotine on cognitive ability are still of interest. In addition to modulating cholinergic activity, nicotine has shown other activities which may be relevant in some neurodegenerative diseases featuring memory loss. Nicotine inhibits β-amyloid formation in vitro [43, 44], inhibits the neurotoxic effects of glutamate [45] and also enhances the effects of nerve growth factor (NGF) [31].

A number of other alkaloids are reported to be nicotinic agonists and could therefore be investigated, or their structures modified, to develop new therapeutic compounds. Alkaloids including lobeline (**5**) from *Lobelia inflata* L. (Campanulaceae) and cytisine (**6**), found in a number of plants including species of *Sophora*

(Leguminosae), have binding affinity for nicotinic receptors [46–49], but these compounds do not appear to have been developed for any pharmaceutical purposes, perhaps due to toxicity.

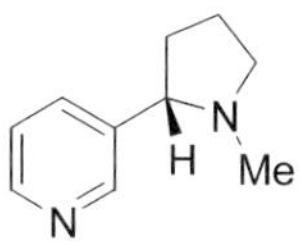

4 Nicotine

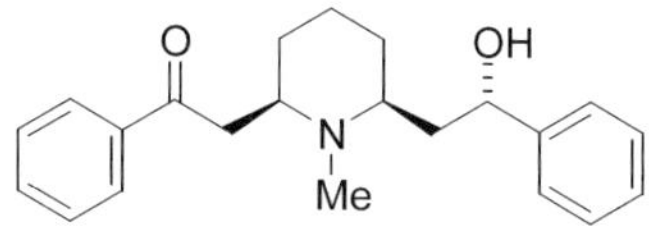

5 Lobeline

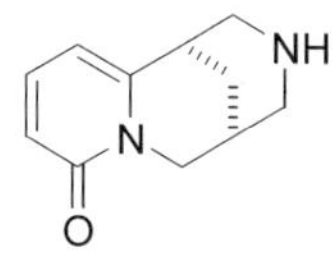

6 Cytisine

15.2.4 Cholinesterase Inhibitors

Inhibition of ACh (**1**) hydrolysis by AChE, through the use of AChE inhibitors, has been a more successful approach than attempts to use compounds which directly stimulate cholinergic receptors to modulate cholinergic function, as AChE inhibitors prolong the half-life of ACh, and therefore the availability of ACh released into the neuronal synaptic cleft.

Over the last decade, some AChE inhibitors have been licensed for clinical use for the symptomatic relief of mild to moderately severe AD. However, these drugs only alleviate some of the cognitive symptoms of the disease, rather than treat the disease, and may not be effective in some patients. The synthetic drug tacrine (**7**) (Cognex) was the first AChE inhibitor to be licensed, but its use was limited by adverse effects, including hepatotoxicity [50, 51]. The current AChE inhibitors licensed for use in AD include donepezil (**8**) (Aricept), rivastigmine (Exelon) and galantamine (Reminyl) [52], with the latter two drugs based on naturally derived compounds. In addition to modulating cholinergic function, some AChE inhibitors are reported to interfere with β-amyloid metabolism and thus could reduce senile plaque formation, one of the pathological occurrences in AD [53].

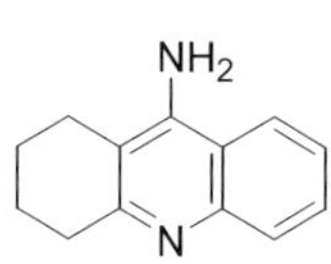

7 Tacrine

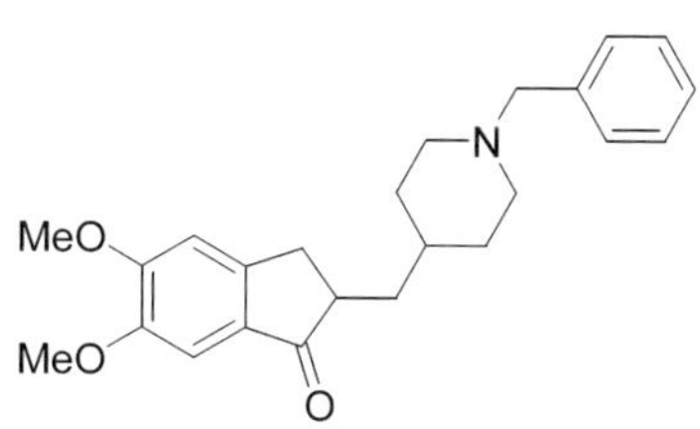

8 Donepezil

15.2.5 Anti-Inflammatory Activity

Some reports have indicated that the use of anti-inflammatory compounds may modify the progression of AD, since inflammatory processes have been linked with AD pathology [54–57]. Some studies have indicated that non-steroidal anti-inflammatory drugs (NSAIDs), which inhibit cyclo-oxygenase (COX) activity, may reduce the risk of developing AD, and patients with rheumatoid arthritis, who often use NSAIDs, are suggested to have a lower incidence of AD [58–62]. In addition to inhibition of COX, it has also been suggested that NSAIDs may act via other mechanisms such as anti-amyloidogenic effects [63]. In view of the adverse effects commonly associated with COX inhibitors currently in clinical use [63], new anti-inflammatory compounds may be developed, including those which are naturally derived, which may have potential in modifying the progression of cognitive disorders such as AD with fewer adverse effects.

There are numerous examples of plant extracts and their constituents which display anti-inflammatory effects [64–68]. Consequently, there is some potential for novel anti-inflammatory agents to be identified from plant sources, although plants as a source of new anti-inflammatory drugs have not been extensively exploited to achieve this aim. Various flavonoids have been associated with anti-inflammatory activity [65, 66, 69] and may have potential for use in some CNS disorders in which inflammatory processes are known to occur. Some structural features of flavonoid molecules required for COX inhibition are considered to be the presence of a pyrocatechol group in at least one of the flavonoid rings; however, flavones lacking these substituents can also inhibit COX [68, 69]. Other plant-derived compounds with potential for use in inflammatory disorders include ferulic acid, which is an antioxidant and anti-inflammatory compound. Ferulic acid is of interest since it ameliorated the reduction in ACh (**1**) levels in the cortex and the inflammatory responses in the hippocampus induced by β-amyloid in mice and, significantly, it also improved cognitive function [70].

15.2.6 Antioxidant Activity

Antioxidants have been suggested to reduce the risk of developing dementia, although evidence to support this hypothesis is under review [71]. Free-radical reactions, which are reported to initiate cell injury, have been implicated in the pathology of various diseases including ageing processes, atherosclerosis, ischemic heart disease and neurodegenerative diseases which involve cognitive impairment [72–74]. Antioxidants have therefore been considered as a means to modify and minimise neuronal degeneration in cognitive disorders.

A wide variety of plants have been associated with antioxidant effects [75–77]. It is therefore not surprising that many different and structurally diverse phytochemicals have also shown antioxidant activity, including some cinnamic acids, coumarins, diterpenoids, flavonoids, monoterpenoids, phenylpropanoids and tannins [78–84]. The antioxidant properties of *Camellia sinensis* Kuntze (Theaceae), commonly known as green tea, are well documented, and some studies suggest that *C. sinensis* extracts and some of the catechin components have protective mechanisms in neurodegenerative disorders [85]. For example, (-)-epigallocatechin gallate

(**9**) had protective effects against β-amyloid-induced neurotoxicity in vitro, an effect associated with its scavenging reactive oxygen species [86]. However, the ability of the catechins to cross the blood-brain barrier may be restricted due to their polarity, thus limiting any therapeutic effect in practice.

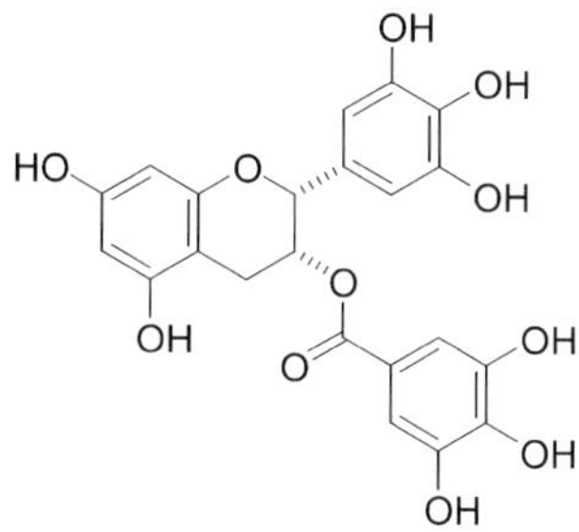

9 Epigallocatechin 3-*O*-gallate

Although many plants and their compounds have shown antioxidant effects in vitro, relatively few have been explored for their therapeutic and clinical relevance, particularly in relation to cognitive disorders. One plant which has shown favourable effects in the CNS is *Thymus vulgaris* L. (Lamiaceae). A study investigating *T. vulgaris* essential oil showed that it maintained higher polyunsaturated fatty acid (PUFA) levels in various tissues, including the brain in rats, indicating protective antioxidant effects [87]. Other examples of plants which show antioxidant activity, particularly with reference to CNS pathologies, are described later in this chapter.

15.2.7 Estrogenic Activity

For a number of years, conclusions from epidemiological evidence indicated that estrogen-replacement therapy (ERT) had a preventative role against AD development, and estrogen treatment in women with AD enhanced cognitive function [88–91]. The mechanisms by which estrogens may protect against AD are unclear but may be mediated via interaction with estrogen receptors in the CNS, or perhaps by effects on neurotransmitter systems, modulation of NGF, enhancement of cerebral blood flow or antioxidant effects [92–94] or other unknown mechanisms. However, evidence from some more recent studies [95, 96] does not support an association between high estrogen levels and a reduced incidence of AD. Nevertheless, the estrogenic activities of some plant extracts have been explored as one possible explanation for their reputed memory-enhancing effects [97, 98].

Soya beans, the seeds of *Glycine max* Merr. (Leguminosae), form an important part of the traditional diet in China and other parts of the Far East and are frequently a staple of the diet of vegetarians and vegans. Soya contains isoflavones including genistein (**10**) and daidzein (**11**), which have been characterised as phytoestrogens.

Some studies indicate that phytoestrogens may alter anxiety, learning and memory in vivo [99]. Phytoestrogens, particularly the soya isoflavones, are reported to improve cognitive function, not only in some animal studies but also in some clinical studies, and have been suggested to offer protection against AD development [100–102]. One study in student volunteers suggested that a high soya diet (100 mg total isoflavones/d for 10 weeks) may improve short- and long-term memory in both females and males [100]. Another study showed that consumption of soya isoflavones by postmenopausal women for a period of 12 weeks improved cognitive function [103]. In a double-blind, randomised, placebo-controlled trial, some cognitive benefits, particularly verbal memory, occurred in postmenopausal women taking isoflavone supplements for 6 months [104].

The mode of action of phytoestrogens to explain favourable effects on cognition is unclear, and they may act similarly to ERT and via interaction with the estrogen receptor or by other mechanisms, perhaps independently of the estrogen receptor. For example, genistein showed a neuroprotective effect against β-amyloid-induced neurotoxicity in vitro [105], and it ameliorated β-amyloid peptide-induced hippocampal neuronal apoptosis in vitro, which could be associated with an antioxidant effect [106]. It has been proposed that although phytoestrogens may exert some neuroprotective effects analogous to that of antioxidants, they are not functionally equivalent to the endogenously active estrogens [107].

Many other plants have been considered to display estrogenic effects in some studies, but their physiological significance and any potential clinical relevance in improving cognition require further investigation. *Pueraria lobata* (Willd.) Ohwi (Leguminosae), a plant used in traditional Chinese medicine (TCM), and some of its component isoflavones (e.g. puerarin (**12**)) have shown estrogenic activity in vitro [108]. Puerarin also attenuates the deficits of inhibitory avoidance performance in rats, which was associated with an increase in cholinergic activity via nicotinic, but not muscarinic, receptors, in addition to activation of *N*-methyl-D-aspartate (NMDA) receptors, and a decrease in serotonergic neuronal activity [109]. Another study in which postmenopausal women were treated with *P. lobata* (equivalent to 100 mg isoflavones) for 3 months suggested it could promote some favourable effects on cognitive function [110]. It is apparent that some phytoestrogens do show potential for use in some cognitive-related disorders, but more extensive and longer-term studies are needed.

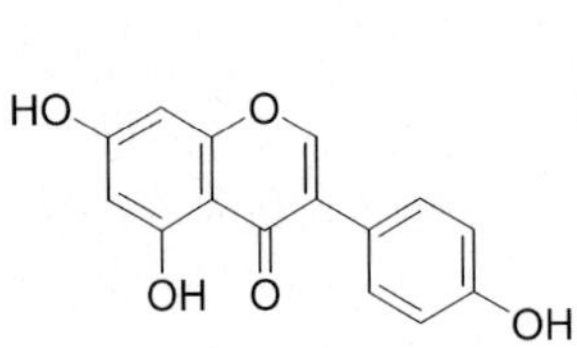

10 Genistein

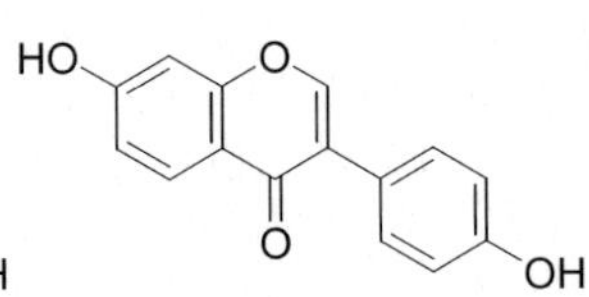

11 Daidzein

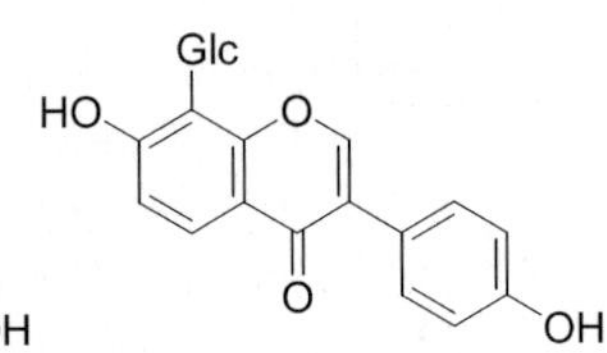

12 Puerarin

15.2.8 NMDA Antagonists

Glutamate may induce neuronal degeneration by overstimulation of NMDA receptors. NMDA receptor modulators may have potential use in some CNS disorders including schizophrenia, stroke, epilepsy, Parkinson's disease, Huntington's disease and AD [111]. Memantine (**13**) (Ebixa), an uncompetitive NMDA receptor antagonist, is reported to be neuroprotective [71], is a licensed drug for the treatment of AD symptoms and has been shown to be therapeutically effective in AD patients [112, 113].

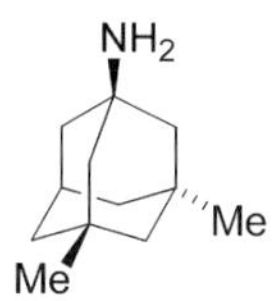

13 Memantine

15.3 Plants as a Source of Useful Therapeutic Agents in Cognitive Diseases

Numerous plants are reputed in traditional practices of medicine to alleviate the cognitive decline that can be associated with general ageing, but they may also be relevant in the treatment of specific cognitive disorders such as AD and other dementias. Thus, plants reputed to have 'anti-ageing' or 'memory-enhancing' effects could also be considered for their potential efficacy in disorders now recognised to be associated with cognitive dysfunction, including those conditions in which dementia occurs.

A mixture of plants is commonly prescribed in some practices of traditional medicine including Ayurveda and TCM. The plant constituents may not only act synergistically with other constituents from the same plant, but they may also enhance the activity of compounds from other plants in a particular remedy or herbal formula. For example, the interaction of a compound at a target receptor may affect the activity of another compound at that receptor, possibly due to allosteric effects, which can occur at some types of receptor [114]. An ethnopharmacological approach can assist with the search for plants and, eventually, potential new drugs that could be relevant for the treatment of cognitive disorders, including AD.

There are numerous examples of plants used in various traditional practices of medicine which have a reputation for influencing cognitive functions. However, relatively few of these plants have been investigated to establish any scientific basis for their reputed effects. The plants described below are mainly those which have stimulated an interest in establishing a pharmacological basis for the reputed effects. It should also be considered that there are many other traditional medicines which have yet to be investigated for a scientific basis to explain their traditional uses and many that have shown interesting biological activity in some studies but which have not been investigated extensively. Although some of these remedies have been promoted as ‘alternative’ or ‘complementary’ therapies, in many instances, there is a lack of substantial evidence for their efficacy and safety or their potential interactions with other medicines. Some of these remedies may contain compounds which are more active when combined in a mixture than when isolated and used alone. Thus, the use of a plant extract may be preferred to single isolated constituents. In addition, in a mixture such as an extract, there may be a variety of compounds with polyvalency, i.e. the different compounds present act in a number of different but relevant ways, and at different molecular targets to produce the overall pharmacological effects for treating a particular condition.

15.4 Plants Used in Traditional Ayurvedic Medicine

Ayurvedic medicine is the oldest medical system in the world with written records in Sanskrit dating back at least 5000 years. It originates from the Indian subcontinent and has also influenced the traditional medical system in Thailand. The practice of Ayurvedic medicine is now widely used throughout the world as a complementary medicine.

15.4.1 Areca catechu L.

Arecoline (**14**) is the major alkaloid of those present in betel or areca nuts, the fruit of the palm tree *Areca catechu* L. (Arecaceae), which is extensively chewed to induce salivation and euphoria throughout the Indian subcontinent and other parts of south-east Asia. It is estimated that 500 million people regularly chew betel nut (often referred to as ‘pan’ or ‘paan’ in India) in a form which is usually shredded, mixed with lime and wrapped in a leaf from the *Piper betel* Blanco (Piperaceae) plant, although chewing of betel nuts has been positively correlated with an incidence of oral cancer [115]. As a direct result of the cholinergic activity induced by this plant, excessive salivation occurs, which is associated with a muscarinic effect, and CNS stimulatory and euphoric effects develop, which is considered to be associated with a nicotinic receptor stimulant effect [116].

Arecoline has been reported as an M_1/M_3 partial agonist [117] and was shown to bind to M_2 muscarinic receptors [118]. Arecoline has been considered as a treatment for cognitive impairment since it showed improvement in scopolamine-induced cognitive impairment and passive avoidance performance in vivo, indicating a cholinergic action [119, 120]. When arecoline was administered to AD patients, it enhanced verbal memory [121] and moderately improved cognitive function and recognition skills [122, 123].

Derivatives of arecoline have been synthesised in order to improve selectivity for cortical muscarinic receptors. Examples of arecoline derivatives include xanomeline (**2**), reported to be functionally selective for the M_1 receptor [22], which delayed cognitive decline and reduced hallucinations and delusions when given to AD patients [116]. Other derivatives of arecoline are Lu 25-109-T and talsaclidine, which are also M_1 functionally selective receptor agonists. Although Lu 25-109-T showed encouraging results in vitro [124], it failed to improve cognition when tested clinically in patients with mild to moderate AD [125]. Talsaclidine has been shown to increase cholinomimetic central activation in animals and humans without some of the side-effects observed with AChE inhibitor therapy, but higher doses are linked with adverse effects including salivation and sweating, and, disappointingly, cognitive function was not significantly improved with this compound [22, 126]. Other tests on rhesus monkeys did show some improvement in memory-related tasks, but at doses which produced unacceptable adverse effects [127].

15.4.2 Bacopa monniera Wettst.

Bacopa monniera Wettst. (Scrophulariaceae), known by the common name 'brahmi', has been used in Ayurvedic medicine for almost 3000 years as a nerve tonic and to improve intellect and memory [128]. Various investigations have attempted to substantiate and identify a scientific basis for the reputed effects. A number of in vivo studies have shown *B. monniera* extracts to improve cognitive function [128–130]. The mode of action to explain these effects has yet to be fully elucidated. Some studies suggest that the antioxidant effects of *B. monniera* may protect the CNS from oxidative damage. Extracts of *B. monniera* have been reported to induce a dose-related increase in superoxide dismutase, catalase and glutathione peroxidase activities in the rat frontal cortex, striatum and hippocampus [131], to dose-dependently inhibit nitric oxide (NO)-related toxicity (DNA damage) in cultured rat astrocytes [132] and to inhibit aluminium-induced neurotoxicity in the rat brain [133]. The antioxidant effects of *B. monniera* have also been suggested to alter amyloid plaque formation [134, 135]. In addition to antioxidant effects, *B. monniera* has also shown anti-inflammatory activity in vivo [128, 136]. It is reported to modulate the cholinergic system [128] and has an anxiolytic action [137]. It is therefore possible that *B. monniera* may exert multiple effects on the CNS.

Although the majority of relevant studies which have investigated the reputed cognitive-enhancing effects of *B. monniera* have focused on extracts rather than isolated constituents, it is the triterpenoid saponins, reported to occur in the aerial parts of *B. monniera* [138–142], which have been associated with the activity. The dammarane-type triterpenoid saponins, a mixture known as bacoside A which includes bacoside A_3 (**15**), have been shown to protect rat brains from smoking-induced apoptosis [143] and from structural and functional impairment of mitochondria [144].

Clinical studies have so far only been undertaken on healthy volunteers. In one double-blind, placebo-controlled study with normal healthy subjects treated with *B. monniera* extract, no acute effects on cognitive function were observed [145]. However, a double-blind, randomised, placebo-controlled study on patients aged 40 to 65 years showed that those given a standardised extract of *B. monniera* resulted in a significant improvement in retention of new information but no difference in the rate of learning, attention, and verbal and visual short-term memory [146]. Another similar study investigated the speed of visual information processing, learning rate and memory consolidation and concluded that the *B. monniera* extract improved higher-order cognitive processes such as learning and memory [147].

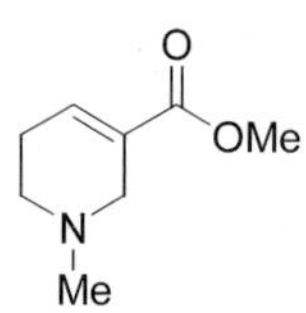

14 Arecoline

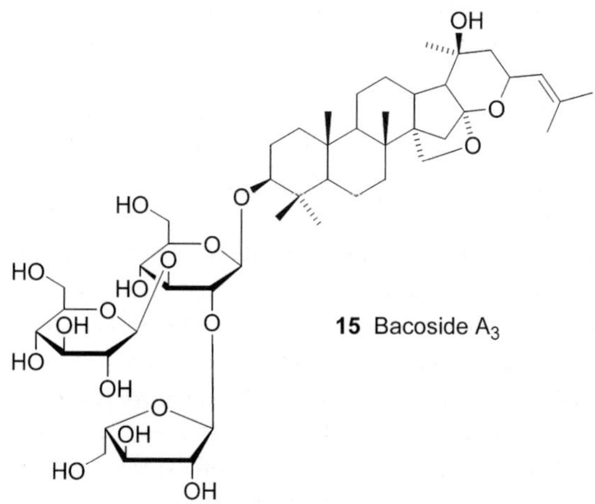

15 Bacoside A_3

15.4.3 Centella asiatica (L.) Urb.

An ancient Ayurvedic remedy, *Centella asiatica* (L.) Urb. (Apiaceae), also known by the synonym *Hydrocotyle asiatica* L., is reputed to restore youth, memory and longevity [148]. In Sanskrit, and commonly as an herbal product, it is known as 'gotu kola'. An Ayurvedic formulation composed of four herbs including *C. asiatica*, is used to retard age and prevent dementia, and the herb combined with milk is given to improve memory [149]. In TCM *C. asiatica* has been used for various disorders, such as traumatic diseases, and for combating physical and mental exhaustion [150, 151]. The essential oil from *C. asiatica* leaf contains monoterpenoids, including bornyl acetate, α-pinene, β-pinene and γ-terpinene [150, 152], all of which are reported to inhibit AChE [153–155]. However, monoterpenoid AChE inhibitors are weak compared to the anti-ChE alkaloid, physostigmine [154]. In view of the relatively weak anti-ChE activity of monoterpenoids reported to date, it is unlikely that they would be therapeutically effective in cognitive disorders.

The pharmacological basis to explain the reputed antiamnesic effects of *C. asiatica* has been explored in a number of studies. An alcoholic extract of *C. asiatica* was tranquillising in rats, an activity that was attributed to a triterpenoid, brahmoside [148, 156]. Further studies showed the extract of *C. asiatica* leaf to be sedative, antidepressant and potentially cholinomimetic in vivo [156], and asiaticoside, a triterpenoid saponin from *C. asiatica*, is a reported anxiolytic [157, 158]. These findings suggest that *C. asiatica* may be appropriate to treat symptoms of depression and anxiety in AD, and that it may also influence cholinergic activity and, thus, cognitive function. Cognitive enhancing effects have been observed in rats following oral administration of an aqueous extract of *C. asiatica*; this effect was associated with an antioxidant mechanism in the CNS [159]. Further evidence for the antioxidant benefits of *C. asiatica* extract were observed when it was administered orally to rats and was shown to protect the rat brain against age-related oxidative damage [160]. Knowledge regarding the compounds responsible for these effects is lacking, although phenolic compounds in *C. asiatica* have been correlated with antioxidative activity in vitro [161]. An aqueous extract of this plant has also been shown to decrease seizures and prevent cognitive impairment in an animal model for epilepsy [162].

Alterations in other neurotransmitter systems have been associated with AD pathology [163, 164]. Interestingly, an aqueous extract of *C. asiatica* leaf modulated dopamine, 5-hydroxytryptamine (5-HT) and noradrenaline systems in rat brain and improved learning and memory processes in vivo [165]. Glutamate may induce neuronal degeneration by overstimulation of NMDA receptors, and memantine (**13**), an NMDA receptor antagonist, is therapeutically effective in AD patients [112, 113]. The triterpenoid asiatic acid (**16**) (found in *C. asiatica*) and its derivatives have been shown to protect cortical neurons from glutamate-induced excitotoxicity in vitro [166]. A neuroprotective action may contribute to changes in the dendritic morphology of neurons in the CNS. This has been suggested by a study in which stimulation of neuronal dendritic growth in hippocampal neurons was observed following administration of leaf extracts of *C. asiatica* to neonatal rat pups [167]. Components in *C. asiatica* including asiatic acid have also been suggested to accelerate repair of damaged neurons [168]. This action, perhaps in conjunction with other observed biological activities relevant to the alleviation of cognitive dysfunction, could explain the basis for the reputed memory-enhancing effects of this plant.

15.4.4 Celastrus paniculatus Willd.

Celastrus paniculatus Willd. (Celastraceae) seeds and seed oil have been used in Ayurvedic medicine to stimulate intellect and to sharpen the memory [169, 170]. Many of the studies undertaken to establish any pharmacological basis for the reputed effects of *C. paniculatus* have focused on the seeds and seed oil. When administered orally to rats, the seed oil decreased levels of noradrenaline, dopamine and 5-HT in the brain, which was correlated with an improvement in learning and memory processes, without inducing neurotoxic effects [171]. Administration of the seed oil to rats also reversed a scopolamine-induced task deficit, but this effect was not associated with anti-ChE activity [172]. Other studies have explored more polar extracts from the seeds of *C. paniculatus* rather than the seed oil. An aqueous seed extract showed an antioxidant effect in the CNS, which may provide some explanation for the reputed benefits on memory, since this extract enhanced cognition in vivo [173]. A seed extract is also reported to increase brain phospholipid content in vivo, possibly as a consequence of increased myelination [174]. Aqueous seed extracts protected neuronal cells against glutamate-induced toxicity [175] and H_2O_2-induced toxicity [176], with methanol and ethanol extracts in addition to the seed oil also showing the latter effect [177]. Although the neuroprotective effect of the polar extracts was attributed to their antioxidant properties, the seed oil, which was the most potent neuroprotector, was suggested to act via a different mechanism [177].

Another *C. paniculatus* extract was evaluated for NMDA and γ-aminobutyric acid (GABA) receptor binding and NGF effects, but did not produce any response [178]. Studies on the flowers from *C. paniculatus* have shown a methanol extract to be anti-inflammatory [179], which may also have some relevance in the management of neurodegenerative disorders. A polyherbal formula (Abana) containing *C. paniculatus* as a component amongst other herbs is used in Ayurvedic medicine, and dose-dependently improved memory in both young and aged rodents and reversed scopolamine- and diazepam-induced amnesia and reduced brain ChE activity [180]. The contribution of each of the component herbs of this formula to the observed effects, or if any synergistic effect occurred, is unknown. Although a number of studies have attempted to elucidate the mechanisms of action to explain the reputed effects of *C. paniculatus* on cognitive function, the compounds responsible for the observed activities have yet to be established.

15.4.5 *Clitoria ternatea* L.

The roots of the Indian medicinal plant *Clitoria ternatea* L. (Leguminosae) have a reputation for promoting intellect [170, 181]. This reputed effect may be related to effects on cholinergic activity in the CNS, as some studies have shown. A study investigating both the aerial parts and roots of *C. ternatea* showed that alcoholic root extracts were more effective than extracts of the aerial parts in attenuating memory deficits in rats [182]. Enhanced memory retention following oral administration of *C. ternatea* root extract was associated with increased levels of ACh (**1**) and ChAT in rat brain, but no relationship with inhibition of AChE activity was established, and cortical AChE activity was actually found to be increased [182]. However, another study showed that the triterpenoid taraxerol (**17**) from *C. ternatea* inhibited AChE both in vitro and in the brain of rodents in vivo, but it was not as potent as physostigmine [183]. An aqueous extract of the root also increased ACh levels in rat hippocampus following oral administration, and it was hypothesised that this effect may be due to an increase in ACh synthetic enzymes [184]. Other studies have indicated that *C. ternatea* extract can act as a nootropic, an anxiolytic, an antidepressant and an anticonvulsant and has antistress [185] and anti-inflammatory activities [186].

Further studies are necessary to establish the mechanism of action to explain the observed effects of the root extract on the CNS, and also to identify the compounds responsible for activity. It has been suggested that memory enhancement in vivo could be explained by an increase in functional growth of neurons of the amygdala, since this effect was observed in rodents orally administered with an aqueous root extract of *C. ternatea* [187].

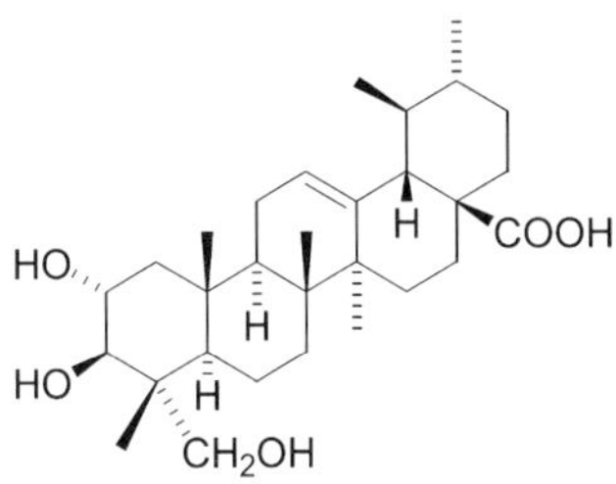

16 Asiatic acid

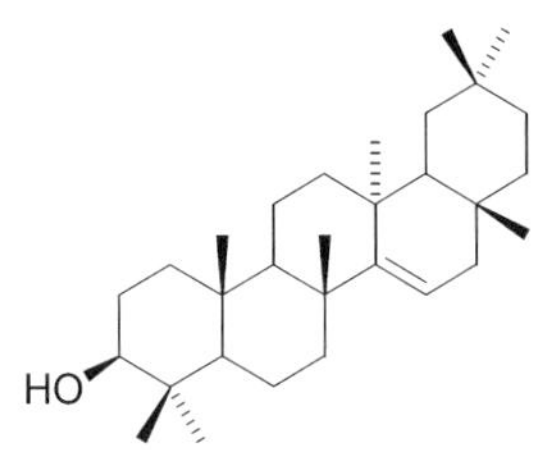

17 Taraxerol

15.4.6 Curcuma longa L.

Regarded as a 'Rasayana' herb in Ayurveda to counteract ageing processes, *Curcuma longa* L. (Zingiberaceae) has also been used for culinary purposes and in the textile industry. Much research has focused on curcumin (**18**), a curcuminoid from *C. longa* rhizomes, and it has been shown to modulate a variety of molecular targets. In particular, studies have shown that some curcuminoids are associated with antioxidant and anti-inflammatory activities, but in general, studies with particular attention to cognitive disorders and any clinical relevance are lacking. In addition, further evaluation of potentially active compounds from *C. longa*, other than the curcuminoids, may contribute to the understanding of the traditional uses of this herb.

The antioxidant activity of curcumin is well documented [188, 189], and it is suggested to be the underlying mechanism to explain a number of beneficial effects on cognition. Curcumin was shown to be neuroprotective in vitro [190] and protected against ethanol-induced brain injury in vivo following oral administration, an effect that was related to a reduction in lipid peroxide levels and enhancement of glutathione in rat brain [191]. A neuroprotective action of curcumin was also observed in an animal model of Parkinson's disease, an effect also attributed to its antioxidant properties [192]. It also dose-dependently improved motor and cognitive impairment and significantly attenuated the associated oxidative stress in the brain when orally administered to rodents [193]. Some compounds from *C. longa*, including curcumin, demethoxycurcumin (**19**), bisdemethoxycurcumin (**20**) and calebin A (**21**) (and some synthetic analogues), were shown to protect PC12 cells from β-amyloid insult in vitro [194, 195]; this activity was also suggested to be due to an antioxidant effect [196]. It is proposed that the hydrophobic bridge of the conjugated network in the curcumin structure enables penetration into the blood-brain barrier, and the more hydrophilic phenolic polar groups are important for its binding to β-amyloid [197].

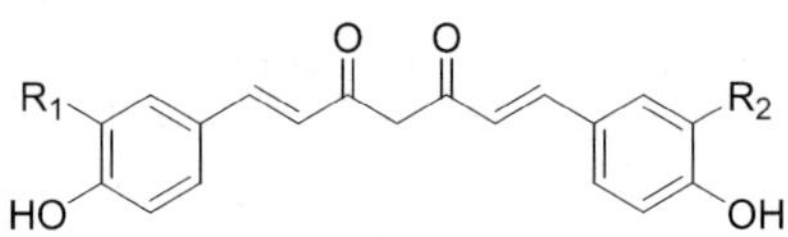

18 $R_1 = R_2$ = OMe; Curcumin
19 R_1 = OMe, R_2 = H; Demethoxycurcumin
20 $R_1 = R_2$ = H; Bisdemethoxycurcumin

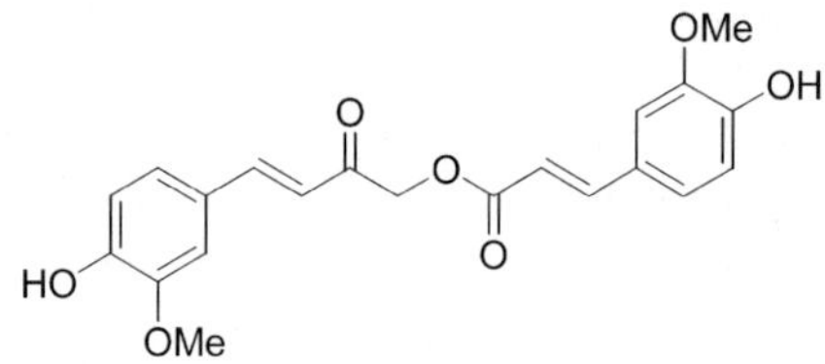

21 Calebin A

Curcumin is also reported to be anti-inflammatory [65, 198] and has been suggested to modulate eicosanoid biosynthesis and COX-1, COX-2 and lipoxygenase (LOX) activities [199–202]. It also inhibits nuclear transcription factor κB (NF-κB) activation [65], although the clinical significance of the latter action in cognitive disorders is unclear. Attempts to improve selectivity for the COX enzymes and efficacy by developing compounds based on the structures of the curcuminoids could be a route to new anti-inflammatory drugs. Some curcuminoid pyrazole and isoxazole analogues have been synthesised and are reported to inhibit COX and to be anti-inflammatory in vivo, and it was shown that replacement of the β-diketo in the curcumin structure with a pyrazole ring enhanced the COX-2/COX-1 selectivity [203].

15.4.7 Withania somnifera (L.) Dunal

Withania somnifera (L.) Dunal (Solanaceae) root, known as 'ashwagandha' in Sanskrit, is classed among 'Rasayanas', the rejuvenative tonics, and its use dates back almost 4000 years. It is considered to be one of the most highly regarded herbs in Ayurvedic medicine. The Ayurvedic scholar Charaka (10 BC) described the reputed effects associated with *W. somnifera*: 'One obtains longevity, regains youth, gets a sharp memory and intellect and freedom from diseases, gets a lustrous complexion and strength of a horse' [204]. It has also traditionally been used to treat some inflammatory conditions such as arthritis.

Numerous studies provide experimental evidence to support the traditional uses of *W. somnifera*, and many of these have associated the biological activities with various steroidal derivatives found in the root. The sitoindosides IX (**22**) and X isolated from the root augmented learning acquisition and memory in both young and old rats [205]. This observation could be explained by a modulatory effect on cholinergic function, since another study with an extract of *W. somnifera* containing the sitoindosides VII–X and withaferin A (**23**) resulted in enhanced AChE activity in the lateral septum and globus pallidus and decreased AChE activity in the vertical diagonal band, enhanced muscarinic M_1 receptor binding in the lateral and medial septum and the frontal cortices, and increased muscarinic M_2 receptor binding sites in cortical regions, when administered to rodents [206]. However, it was not shown to affect $GABA_A$, benzodiazepine receptor binding, or NMDA or amino-3-hydroxy-5-methyl-4-isoxazole propionic acid (AMPA) glutamate receptor subtypes in this study [206]. Another study showed a similar extract containing the same sitoindosides and withaferin A to reverse the ibotenic acid-induced cognitive deficit and the reduction in cholinergic markers (e.g. ACh (**1**), ChAT) in rodents [207]. Tests in vitro have shown a methanol extract of *W. somnifera* to inhibit AChE [208] and some withanolides inhibit AChE and BChE, with withaferin A, 2,3-dihydrowithaferin A (**24**) and 5β,6β-epoxy-4β-hydroxy-1-oxowitha-2,14,24-trienolide (**25**) being more active against AChE activity [209, 210]. These studies indicate that *W. somnifera* and some of its components, particularly some withanolides and sitoindosides, may improve cognition by influencing cholinergic neurotransmission, although other mechanisms of action have also been suggested.

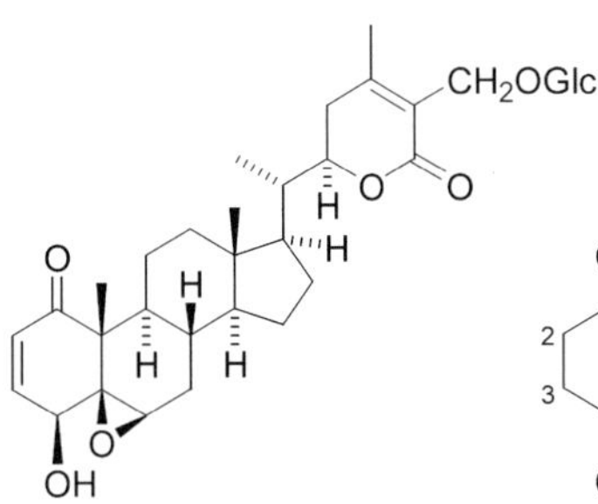

22 Sitoindoside IX

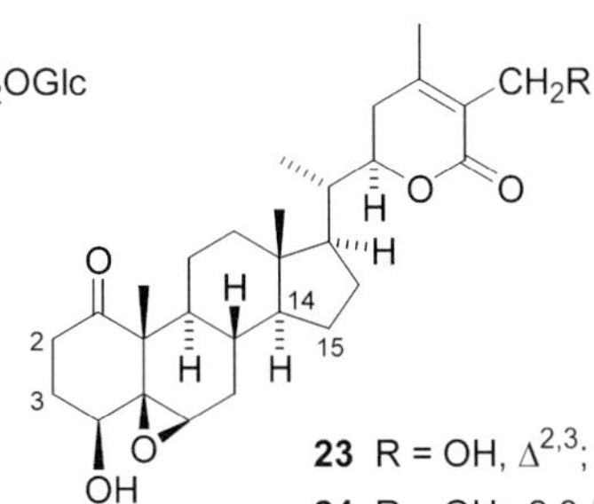

23 R = OH, $\Delta^{2,3}$; Withaferin A

24 R = OH; 2,3-Dihydrowithaferin A

25 R = H; $\Delta^{2,3}$, $\Delta^{14,15}$

A reversal of scopolamine-induced disruption of acquisition and attention and attenuation of amnesia was observed when a root extract of *W. somnifera* was administered to mice, and these effects were suggested to be due to a nootropic effect [211]. Other studies have shown that *W. somnifera* root (methanol extract), and some of the withanolide derivatives in particular, could dose-dependently promote dendrite formation in human neuroblastoma cells in vitro [212, 213]. Withanolide A and withanosides IV and VI are also reported to extend axons and dendrites, respectively, in vitro [214], and withanolide A is considered to reconstruct neuronal networks in vivo [215]. The clinical relevance of these effects is unknown, but cholinergic function could be modulated if neurite outgrowth were to occur in the CNS at therapeutic doses and with consideration of pharmacokinetic profiles of the relevant compounds in vivo. The main metabolite of withanoside IV following oral administration was identified as the aglycone sominone, which induced axonal and dendritic regeneration and synaptic reconstruction in cultured rat cortical neurons [216]. The use of *W. somnifera* to improve some memory disorders shows promise, if compounds demonstrate a suitable pharmacokinetic profile for efficacy. A neuroprotective action has also been considered as an explanation for the reputed effects of the root of this medicinal plant, since a root extract significantly reduced the number of hippocampal degenerating cells in the brains of stressed rodents [217] and was neuroprotective in an animal model of Parkinson's disease [218].

In addition to the steroidal derivatives, other compounds may contribute to the reputed and pharmacological effects of *W. somnifera* root preparations. The cognitive benefits associated with nicotine (**4**) [37–39] could also explain the effects on memory observed with *W. somnifera*, since this alkaloid is reported to occur in *W. somnifera* root [148]. However, other studies have not reported it to be present [219, 220]. Chemical variation in *W. somnifera* plants has been reported to occur [221], which emphasises the need for standardisation of plant-derived products for therapeutic use.

Other activities that may be advantageous in the alleviation of some cognitive disorders are anti-inflammatory and antioxidant effects, both of which have been associated with *W. somnifera*. Inhibition of lipid peroxidation both in vitro and in vivo has been observed with extracts of the root [222, 223], with the root extract and the glycowithanolides (containing equimolar concentrations of sitoindosides VII–X and withaferin A) protecting against lipid peroxidation due to an antioxidant action [224, 225]. In addition to the withanolides being antioxidant [189, 226–228] and decreasing lipid peroxidation in rodent brains, withanolides and sitoindosides (VII–X) also enhanced catalase and glutathione peroxidase activities in rat frontal cortex and striatum [225, 226, 229]. Phenolic compounds from *W. somnifera* root might also contribute to the overall antioxidant properties of this plant [230].

Evidence for the anti-inflammatory effects of *W. somnifera* is apparent in several studies. Root extracts were effective against arthritis and associated biochemical markers in rodents [231, 232]; they reduced serum protein levels (α2-macroglobulin, an indicator of inflammatory conditions) [233, 234] and also reduced interleukin-1 (IL-1) and tumour necrosis factor (TNF)-α levels in vivo [235], which may be of some clinical relevance as some inflammatory mediators have been linked with senile plaque formation in some cognitive disorders. The compounds responsible for these observations require further study, although a dimeric withanolide, ashwagandhanolide, is an inhibitor of COX-2 activity [228]. The leaves of *W. somnifera* leaves are also reported to have anti-inflammatory activity [236].

W. somnifera root has been extensively studied and shown to possess a variety of activities that could be relevant to improve cognitive impairment. It appears to have therapeutic potential for treating memory-related disorders, but further evidence of clinical safety and efficacy is still needed before this promising herbal drug could be considered for wider use in cognitive disorders.

15.5 Plants Used in Traditional Chinese Medicine (TCM)

The practice of TCM has been documented for thousands of years, and the medicinal preparations used include various substances of animal, fungal and plant origin. TCM has also influenced the traditional medicine practiced in neighbouring regions, such as Japan, Korea and Vietnam.

15.5.1 *Evodia rutaecarpa* (Juss.) Benth.

The plant described in the Pharmacopoeia of the People's Republic of China (2005) as *Evodia rutaecarpa* (Juss.) Benth. (Rutaceae) is used in TCM for its reputed cardiotonic, restorative and analgesic effects [237]. Extracts and alkaloids isolated from this plant have been investigated for activities that might help to explain the reputed restorative effects. An ethanol extract of this plant and four compounds present, dehydroevodiamine (**26**), evodiamine (**27**), rutaecarpine (**28**) and synephrine (**29**), have been shown to be anti-inflammatory in vitro [238], an action that has been implicated as potential therapy in some cognitive disorders.

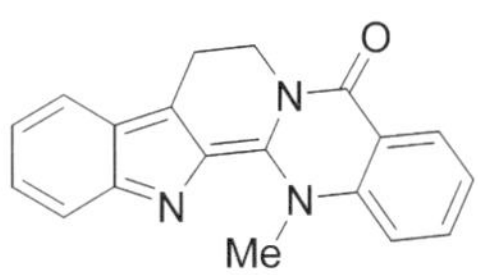

26 Dehydroevodiamine

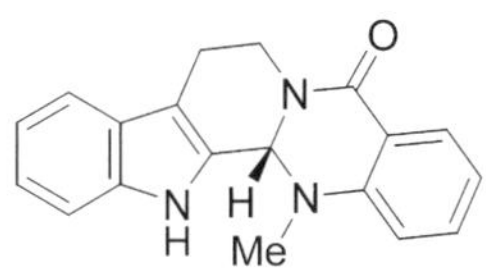

27 Evodiamine

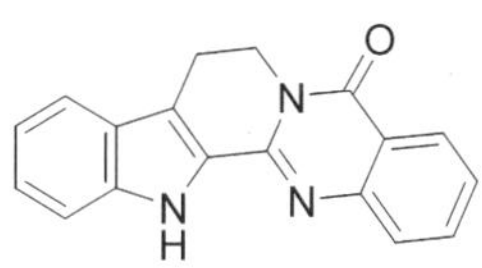

28 Rutaecarpine

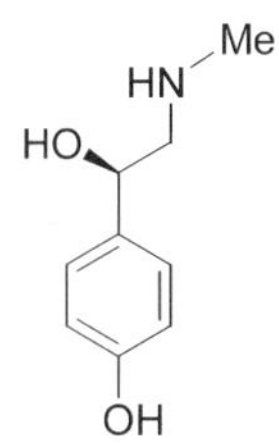

29 Synephrine

The alkaloid rutaecarpine is also reported to inhibit COX-2 activity in vitro and to be anti-inflammatory in vivo [239], although another study showed evodiamine to inhibit COX-2 induction and NF-κB activation, whilst rutaecarpine did not show these effects [240]. Evodiamine has also been shown to inhibit both constitutive and induced NF-κB activation and NF-κB-regulated gene expression [241]. Dehydroevodiamine increases cerebral blood flow in vivo [242], an action that might also improve cognitive function, and it may have a neuroprotective action, since it inhibited glutamate uptake and release in vitro [243]. Dehydroevodiamine (hydrochloride) also prevents impairment of learning and memory and neuronal loss

[244] and is not considered to be associated with any serious adverse effects [245]. *E. rutaecarpa* extract and dehydroevodiamine both inhibited AChE in vitro and reversed scopolamine-induced memory impairment in rats [246]. In addition to reversing scopolamine-induced amnesia, dehydroevodiamine is reported to be even more effective in the reversal of β-amyloid-peptide-induced amnesia in vivo [247], indicating that it may improve cognitive ability by influencing cholinergic function and by other mechanisms.

Since the alkaloids rutaecarpine and dehydroevodiamine are inhibitors of ChE, their chemical structures have been used as a basis for the development of new ChE inhibitors, including synthetic compounds which combine the structural features of these alkaloids with tacrine (**7**). Inhibition of both AChE and BChE in vitro occurred with the synthetic analogues (8*Z*)-5,6-dihydro-8*H*-isoquino[1,2-*b*]quinazolin-8-imine, 5,8-dihydro-6*H*-isoquino[1,2-*b*]quinazoline, 5,7,8,13-tetrahydroindolo [2′,3′:3,4]pyrido[2,1-*b*]quinazoline and *N*-(2-phenylethyl)-*N*-[(12*Z*)-7,8,9,10-tetrahydroazepino [2,1-*b*]quinazolin-12(6*H*)-ylidene]amine, with 5,7,8, 13-tetrahydroindolo [2′,3′:3,4]pyrido[2,1-*b*]quinazoline and *N*-(2-phenylethyl)-*N*-[(12*Z*)-7,8,9,10-tetrahydroazepino [2,1-*b*]quinazolin-12(6*H*)-ylidene]amine showing higher affinity for BChE, but the dibenzo-analogue of dehydroevodiamine (13-methyl-5,8-dihydro-6*H*-isoquino[1,2-*b*]quinazolin-13-ium chloride) showed greater selectivity for AChE compared to BChE [248].

15.5.2 *Ginkgo biloba* L.

In Europe, leaf preparations of *Ginkgo biloba* L. (Ginkgoaceae) were used for the treatment of circulatory disorders in the 1960s, and they are now a popular herbal remedy with a reputation for alleviating memory problems. In Iran, *G. biloba* has been used traditionally to improve memory associated with blood circulation abnormalities [249]. The use of *G. biloba* in TCM dates back for centuries, and the Pharmacopoeia of the People's Republic of China (2005) includes *G. biloba* seeds as a remedy for cough and asthma and to reduce leukorrhoea and urination [237].

There has been extensive research to determine any pharmacological basis which might explain the reputed effects of *G. biloba* on memory, and a number of clinical studies have also been conducted. Much of this research has used a standardised extract of *G. biloba* known as EGb 761, which contains flavonoid glycosides and terpenoid lactones amongst various other constituents. This extract has shown a variety of activities relevant to improving cognitive function, particularly neurodegenerative-related disorders such as AD, thus indicating that the extract may have a number of different modes of action. EGb 761 has shown favourable effects on cerebral circulation and neuronal cell metabolism [250, 252, 252] and on the cholinergic system [253], and it has antioxidant activity [254–256]. EGb 761 reduced apoptosis both in vitro and in vivo [257, 258], and it was neuroprotective against NO- and β-amyloid-induced toxicity in vitro [259, 260], with the latter effect being associated with the flavonoid fraction (CP 205) [261]. Ginkgolides, the main constituents of the non-flavonoid fraction of EGb 761, are neuroprotective against hypoxia-induced injury in cortical neurons [262].

EGb 761 also protected mitochondria from β-amyloid-induced toxicity [263] and modulated synaptic and mitochondrial plasticity in vitamin E-deficient rodents [264]. Other studies showed EGb 761 and the terpenoid lactone bilobalide (**30**) to protect against ischemia-induced neuronal death, and they reduced mitochondrial gene expression in vivo [265]. Furthermore, bilobalide reduced both glutamate and aspartate release in cortical slices [266], indicating a neuroprotective action. One mechanism to explain a neuroprotective action of bilobalide is that it acts as an antagonist at $GABA_A$ receptors [267].

Some *G. biloba* leaf constituents, including some flavonoids (e.g. quercetin (**31**)) and terpenoids (e.g. bilobalide and ginkgolides A (**32**), B (**33**) and C (**34**)), have been associated with a vasodilatory action [268, 269], which could also benefit memory function. Anti-inflammatory activity is also associated with *G. biloba* and its components. Ginkgolide B antagonises platelet-activating factor (PAF) [270], ginkgetin (**35**), a biflavone from *G. biloba* leaves, inhibits phospholipase A_2, and ginkgetin and the biflavone mixture downregulate COX-2 expression [271, 272]. The antioxidant properties of *G. biloba* extract have shown potential relevance in modulating AD pathology in vivo. An extract reduced oxidative stress resulting from senile plaques in vivo and progressively reversed structural changes in dystrophic neurites associated with senile plaques, thus indicating that neurotoxicity associated with the senile plaques in AD could be partially reversible with antioxidant therapies such as *G. biloba* extract [273]. Some evidence suggests that *G. biloba* and its components quercetin, kaempferol and isorhamnetin may also have estrogenic activity in some circumstances [274, 275], but any physiological relevance of this effect is unclear.

These activities, and perhaps other modes of action yet to be elucidated, might explain the effects on cognition observed when *G. biloba* extracts have been tested in vivo. Extracts have been shown to enhance cognition in both young and old rats [276, 277], to improve short-term memory in mice [278] and spatial learning and memory in rats with aluminium-induced brain dysfunction [279], to reduce cognitive impairment and hippocampal damage after ischemia in rats [280] and to attenuate scopolamine-induced amnesia in rats [281], perhaps indicating modulation of cholinergic function. This hypothesis is supported by another study in which a standardised *G. biloba* extract containing 24% flavone glycosides inhibited AChE activity both in vitro and ex vivo at doses which correlated with effects against scopolamine-induced deficits in a passive avoidance test in mice [129]. In another study, the antagonistic effect of *G. biloba* extract on spatial memory deficits in rodents was attributed to cholinergic activity but was also suggested to be partly due to a histaminergic mechanism of action [282].

The clinical efficacy of *G. biloba* extracts (including EGb 761) has been evaluated in several studies including double-blind, placebo-controlled, multicentre trials

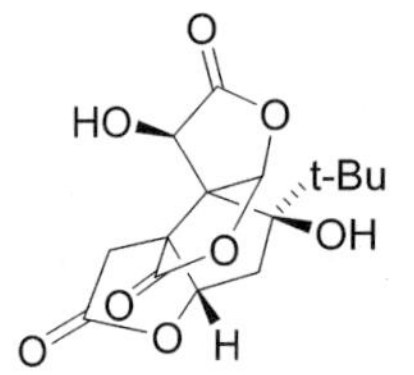

30 Bilobalide

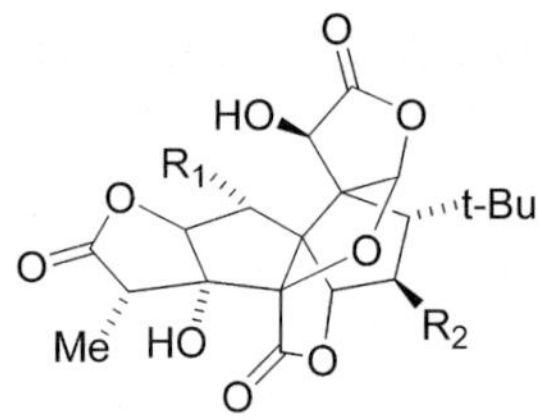

32 $R_1 = R_2 = H$; Ginkgolide A
33 $R_1 = OH, R_2 = H$; Ginkgolide B
34 $R_1 = R_2 = OH$; Ginkgolide C

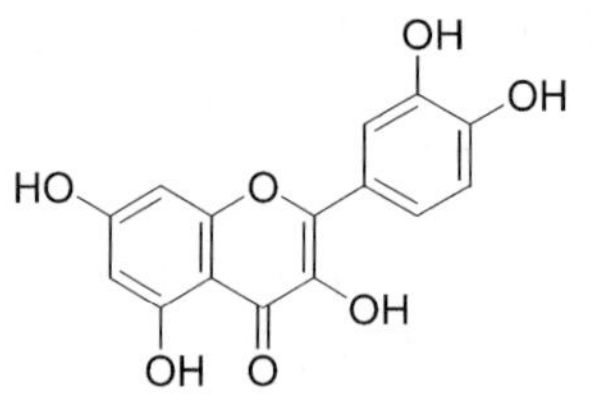

31 Quercetin

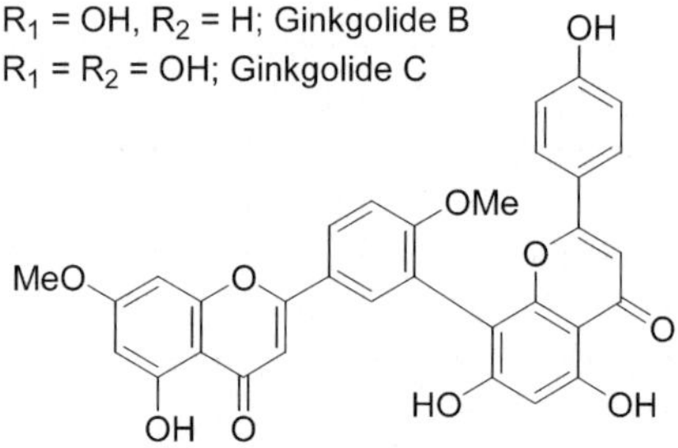

35 Ginkgetin

with administration to both AD and healthy subjects [283–292]. Although these trials have indicated that *G. biloba* can modestly improve cognitive ability, some of the data need to be interpreted with caution since some results were based on self-assessment questionnaires rather than more objective methods of analysis. *G. biloba* extract is also being investigated for any role it may have in the prevention of AD [293]. In another randomised, double-blind, placebo-controlled trial, 120 mg *G. biloba* extract was administered twice daily to assess if it could improve cognitive performance in multiple sclerosis patients. Although this extract was not found to significantly improve cognition in this study, it was suggested to influence some cognitive processes such as mental flexibility [294].

G. biloba is probably one of the most studied herbal remedies for alleviating memory problems, and there is substantial evidence, both pharmacological and clinical, to encourage further study on its potential in the treatment of some cognitive disorders. As is often typical with herbal products, the active compounds need to be characterised, appropriate doses need to be established and both short- and long-term safety needs to be evaluated. Clinical trials to date have generally shown oral administration of *G. biloba* to be well tolerated, with no serious adverse effects [287, 295]. It is also important to consider potential drug interactions, particularly as patients requiring therapies to improve cognition may be already taking a number of other medicines. One study showed that *G. biloba* supplementation in AD patients taking donepezil (**8**) (5 mg/d) did not have a major impact on the pharmacokinetics and pharmacodynamics of donepezil [296], although the use of *G. biloba* with antiplatelet or anticoagulant medicines may increase the risk of haemorrhage [297].

15.5.3 *Huperzia serrata* (Thunb.) Trevis.

In TCM a prescription prepared from *Huperzia serrata* (Thunb.) Trevis. (Lycopodiaceae) has been a treatment for memory loss [298, 299]. Of the alkaloids isolated from *H. serrata*, huperzine A (**36**) has been extensively studied for pharmacological and clinical effects in relation to treatment of cognitive disorders.

A range of studies in animals have shown this alkaloid to improve memory-retention processes in cognitively impaired aged and adult rats [300] and to attenuate cognitive deficits in chronically hypoperfused rats [301] and in gerbils following ischemia [302]. The principal mechanism of action thought to be responsible for the cognitive-enhancing effects of huperzine A is modulation of cholinergic function by inhibition of ChE; it reversibly inhibits AChE both in vitro and in vivo [303–305]. Huperzine A is more selective for AChE than BChE, was less toxic than the synthetic AChE inhibitors donepezil (**8**) and tacrine (**7**) and significantly improved memory and behaviour in AD patients in a multicentre, double-blind trial [306, 307]. In phase IV clinical trials in China, huperzine A improved memory in elderly, AD and vascular dementia patients, with limited adverse effects [308]. Pharmacokinetic studies have indicated that huperzine A is rapidly absorbed, widely distributed in the body and eliminated at a moderate rate [308].

In addition to ChE inhibition, other effects may contribute to the cognitive benefits of this remedy. Huperzine A also favourably affects other neurotransmitter systems to improve memory [309]. It was also neuroprotective against β-amyloid peptide [310], oxygen-glucose deprivation [311], free-radical-induced cytotoxicity [312] and glutamate [313] and it acts as an NMDA receptor antagonist in the cerebral cortex [314]. The enantiomers of huperzine A concentration-dependently inhibit NMDA receptor binding without stereoselectivity, although stereoselectivity is reported in the inhibition of AChE, with the (+)-isomer of huperzine A being less potent than the natural (−)-isomer [315]. Huperzine A is also suggested to attenuate apoptosis by inhibiting the mitochondria-capase pathway [316] and to have neurotrophic effects [308]. Huperzine B (**37**), also from *H. serrata*, attenuates H_2O_2- and oxygen-glucose-deprivation-induced injury in the rat pheochromocytoma cell line PC12, indicating that it has a neuroprotective action [317, 318].

Huperzine A appears to be therapeutically advantageous over some other known ChE inhibitors since it is a potent, reversible and relatively selective inhibitor of AChE, it shows other activities that may be relevant in alleviating cognitive dysfunction and it has shown efficacy in clinical trials in cognitively impaired patients with few adverse effects. It is therefore not surprising that the structures of huperzines A

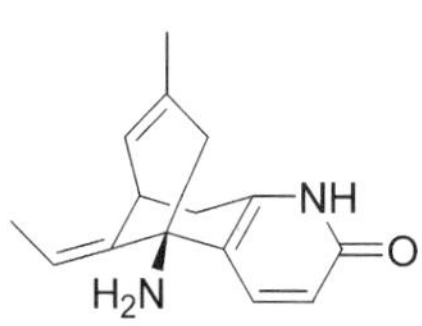

36 Huperzine A

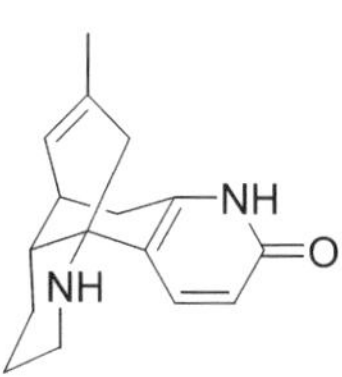

37 Huperzine B

and B have been used as templates for the synthesis of new compounds, with the aim of developing potentially new drugs with improved efficacy and safety.

One study involved an attempt to develop a compound, (±)-14-fluorohuperzine A, to enhance the H-bond between the C-14 methyl of huperzine A and the backbone carbonyl of His440 on AChE. The racemic form of 14-fluorohuperzine A inhibited AChE in vitro with a potency that was 62-times less than huperzine A [319]. Other attempts to develop fluorinated analogues of huperzine A ((±)-12,12,12-trifluorohuperzine A, (±)-14,14,14-trifluorohuperzine A, (±)-12,12,12,14,14,14-hexafluorohuperzine A and (±)-12-fluorohuperzine A) have been unsuccessful as these compounds did not inhibit AChE more potently than huperzine A [320, 321].

Analogues of huperzine A synthesised to achieve 5-substitution with either a hydroxyl group, a fluoro group or an acetoxyl group were assessed for their anti-AChE activity in vitro. The AChE inhibitory activities of these 5-substituted huperzine A analogues were also less potent than huperzine A when tested in vitro, indicating that the C-5 amino group in huperzine A (which can form a quaternary ammonium under physiological conditions to imitate ACh (**1**)) is an important structural feature for AChE inhibition [322]. Other compounds synthesised are the (*E*)- and (*Z*)-5-desamino huperzine A derivatives, which, although more potent than the 5-fluoro and 5-hydroxyl derivatives in the inhibition of AChE in vitro, were still less potent than huperzine A [322, 323]. Another synthetic derivative of huperzine A, (−)-dimethylhuperzine A (DMHA), showed AChE inhibitory activity comparable to (−)-huperzine A, and although the enantiomer (+)-DMHA was inactive against AChE activity, both enantiomers were equally effective in protecting against glutamate-induced neurotoxicity [324]. Analogues of huperzine B have also been synthesised with the aim of improving AChE inhibitory potency, but although some of these analogues are reported to be up to four-fold more potent than huperzine B, they were not as potent as huperzine A in the inhibition of AChE [325].

Other studies have involved synthesis of hybrid compounds, containing structural features of both huperzine A and other known ChE inhibitors. Pharmacomodulation of huperzine A and tacrine has produced compounds that include a combination of the carbobicyclic substructure of huperzine A and the 4-aminoquinoline substructure of tacrine [326, 327]. One of these compounds, *rac*-(*E*)-12-amino-13-ethylidene-6,7,10,11-tetrahydro-9-methyl-7-11-methanocycloocta[*b*]quinoline hydrochloride, was less potent than tacrine in the inhibition of AChE, but it was more active than the (*Z*)-stereoisomer [326]. Other derivatives synthesised which lack the ethylidene substituent (*rac*-12-amino-6,7,10,11-tetrahydro-9-methyl-7,11-methanocycloocta[*b*]quinoline hydrochloride and *rac*-12-amino-9-ethyl-6,7,10,11-tetrahydro-7,11-methanocycloocta[*b*]quinoline hydrochloride) inhibited AChE more potently than tacrine [326].

Other synthetic compounds are the huprines. Huprine X ((−)-12-amino-3-chloro-9-ethyl-6,7,10,11-tetrahydro-7,11-methanocycloocta[*b*]quinoline hydrochloride) (**38**) was found to bind to AChE in a similar manner to huperzine A and tacrine at the acylation site in the active site gorge; it also interferes with the binding of peripheral-site ligands [328]. Huprine X also had an affinity for AChE that was 180 times that of huperzine A, 1200 times that of tacrine and 40 times that of donepezil [328], and it may act as an agonist at muscarinic M_1 and M_2 and at nicotinic receptors [329]. Both (±)-huprine X and (±)-huprine Y ((±)-12-amino-3-chloro-9-methyl-6,7,10,11-tetrahydro-7,11-methanocycloocta[*b*]quinoline hydrochloride) (**39**) increased the level of ACh in the synaptic cleft more effectively than tacrine, and the interaction of (±)-huprine X with nicotinic receptors was weaker than that of (±)-huprine Y [330], although the nicotinic receptor binding effect of huprine X was not shown in a separate study [331]. (±)-Huprines Y and Z (**40**), which differ in structure by the halogen at position 3 (chlorine and fluorine, respectively), are more potent inhibitors of AChE than BChE and both compounds inhibited brain AChE (ex vivo), with (±)-huprine Y being approximately five times more potent than (±)-huprine Z [332].

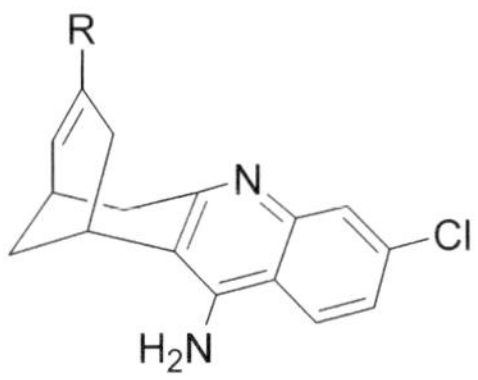

38 R = Et; Huprine X
39 R = Me; Huprine Y

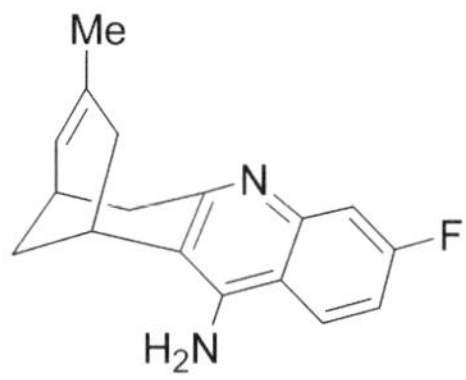

40 Huprine Z

Other huperzine A hybrids that have been synthesised include structural features of both huperzine A and E2020 (donepezil). These hybrid compounds were synthesised with the aim of enabling an interaction between the 5,6,7,8-tetrahydroquinolinone of huperzine A and the active site of AChE, and an interaction between the benzyl piperidine of E2020 and the peripheral binding site of AChE, but these derivatives were less potent than E2020 in the inhibition of AChE *in vitro* [333].

15.5.4 Magnolia officinalis Rehder & E.H.Wilson

In TCM, the bark of the root and stem of *Magnolia officinalis* Rehder & E.H.Wilson (Magnoliaceae) has been used as a remedy for alleviating anxiety and nervous disturbances. Many of the more recent studies investigating any basis for the CNS effects of *M. officinalis* have focused on the biphenolic lignans isolated from *M. officinalis*, honokiol (**41**) and magnolol (**42**), and although a variety of in vitro tests and some *in vivo* studies have been undertaken that may explain the reputed effects, there is a lack of clinical evidence for efficacy.

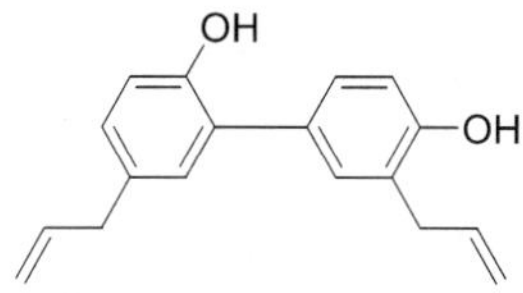

41 Honokiol

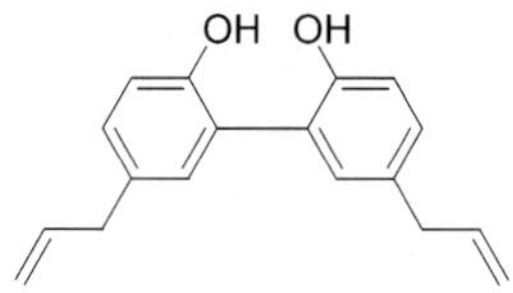

42 Magnolol

Modulation of cholinergic function could explain any favourable effects of this remedy on memory, as both honokiol and magnolol increase ChAT activity, they inhibit AChE activity in vitro and they increase hippocampal ACh (**1**) release in vivo [334]. Other activities that may preserve cognitive function have also been associated with *M. officinalis* and the component lignans. An extract [335], magnolol [336, 337] and honokiol [337, 338] showed antioxidant activity, and magnolol was neuroprotective in vitro [339]. In another study, both magnolol and honokiol, the latter being the most potent, were neuroprotective against glutamate-, NMDA- and H_2O_2-induced mitochondrial dysfunction in vitro, effects associated with an antioxidant action and antagonism of excitatory amino-acid-induced toxicity; these two compounds are suggested to differ in some mechanisms by which they are neuroprotective [340, 341]. The anti-inflammatory activity of magnolol both in vitro and in vivo could be explained by its ability to inhibit COX and 5-LOX [342, 343] or by regulating the NF-κB pathway [344]. Modulation of the NF-κB pathway has also been shown to occur with honokiol [345] and it protected against cerebral ischemia-reperfusion injury in rodents, which was attributed to its antioxidant, anti-inflammatory and antiplatelet aggregation properties [346, 347]. Also from the bark of *M. officinalis* are the (+)- and (−)-enantiomers of syringaresinol and a mixture of their glucosides; these compounds promoted dose-dependent neuritogenesis in vitro [348]. These potential neurotrophins and the lignans honokiol and magnolol appear to show some activities that could be relevant in disorders in which cognition is impaired, but more in-depth studies, particularly those assessing their clinical relevance and safety, are needed.

15.5.5 *Polygala tenuifolia* Willd.

Polygala tenuifolia Willd. (Polygalaceae) root is a remedy used in TCM for cardiotonic and cerebrotonic effects and is considered to act as a sedative and tranquilliser and to alleviate amnesia, neuritis and insomnia [349, 350]. According to the Chinese Materia Medica, the root is supposed to produce an effect upon the will and mental powers, improving understanding and strengthening memory, and the Pharmacopoeia of the People's Republic of China (2005) includes *P. tenuifolia* root as a remedy to anchor the mind and for forgetfulness [237].

Since in TCM a mixture of herbs, rather than one single herb or other substance, is commonly prescribed, a number of studies have investigated a traditional Chinese prescription, known as DX-9386, for any activities that could have relevance in improving cognitive processes. DX-9386 is composed of *P. tenuifolia* in addition to *Panax ginseng* C.A. Mey. (Araliaceae), *Acorus gramineus* [Soland.] (Acoraceae) and *Poria cocos* (Schwein.) F.A. Wolf (Fomitopsidaceae) (in the ratio 1:1:25:50 dry weight). Although this prescription has shown a number of favourable biological activities in relation to treating cognitive dysfunction, the contribution of each of the components in the prescription to the observed effects is unclear. Studies conducted have shown that DX-9386 may slow the ageing process as it ameliorated memory impairment and reduced lipid peroxide levels [351] and prolonged lifespan [352] in senescence-accelerated mice, and it ameliorated ethanol-induced memory impairment in rodents [353]. The enhancement of hippocampal long-term potentiation of synaptic transmission by DX-9386 was attributed to *P. ginseng* and *P. cocos*, with *P. tenuifolia* only producing a minor effect [354].

Other studies have also investigated *P. tenuifolia* amongst a mixture of 12 prescription components in a formula known as kami-utan-to (KUT), a remedy used in traditional Japanese medicine to treat psychoneurological diseases. Studies have shown KUT to dose-dependently upregulate ChAT activity and increase NGF secretion in vitro, and to improve passive avoidance behaviour and induce ChAT activity in the cerebral cortex of aged rats and in scopolamine-induced memory impaired rats [355, 356]. These activities were mainly attributed to the *P. tenuifolia* content of the prescription, since the effects on ChAT activity and NGF secretion were not as pronounced when treated with KUT in the absence of *P. tenuifolia* root, and *P. tenuifolia* root extract alone was shown to upregulate ChAT activity and increase NGF secretion in vitro [355, 356]. The clinical potential of KUT in cognitive disorders has shown some promise, as KUT treatment improved memory-related behaviour in AD patients [356].

Some studies on *P. tenuifolia* root extract alone in the absence of other traditional remedies have provided more evidence to explain the use of this plant in TCM for CNS effects. Extracts reversed scopolamine-induced cognitive impairment and, to some extent, improved memory and behavioural disorders induced by CNS lesions in rodents; in addition, they showed a neuroprotective action against glutamate and amyloid precursor protein (APP) in vitro and dose-dependently inhibited AChE activity in vitro [357–359], indicating an effect on cholinergic function in addition to other modes of action. Although extracts of *P. tenuifolia* have been associated with a neuroprotective action and an aqueous extract enhanced axonal length in cortical neurons treated with amyloid in vitro, the aqueous extract was not effective in recovering dendritic atrophy and synaptic loss [360]. Anti-inflammatory activity could also contribute to the favourable CNS effects associated with this plant. An aqueous extract of *P. tenuifolia* root inhibited IL-1 mediated TNF secretion by astrocytes [361] and dose-dependently inhibited ethanol-induced IL-1 secretion in vitro [362].

The compounds responsible for the effects of *P. tenuifolia* are unknown, although some cinnamic acid derivatives may explain the suggested cholinergic action as sinapic acid, a cinnamic acid derivative from *P. tenuifolia* root, increased ChAT activity in the frontal cortex in brain-lesioned rats [355]. Also from *P. tenuifolia* root is the acylated oligosaccharide tenuifoliside B, which has a sinapoyl moiety in its structure. This compound showed a cerebral protective effect on KCN-induced anoxia, and it ameliorated the scopolamine-induced impairment of performance in a passive avoidance task in rodents [363]. It is suggested that sinapic acid, or a sinapoyl moiety, could be an important structural feature for the molecular interactions that modulate cognitve function. To support this theory, a further study was conducted which showed that sinapic acid also inhibited KCN-induced hypoxia and scopolamine-induced memory impairment and it inhibited basal-forebrain-lesion-induced cerebral cholinergic dysfunction in rodents [364]. Tenuigenin, extracted from *P. tenuifolia*, inhibited β-amyloid secretion via β-secretase (BACE) inhibition in vitro [365]. Other compounds that have been associated with some of the in vitro effects observed with *P. tenuifolia* extracts include saponins with the aglycone presenegenin (onjisaponins A, B, E, F and G), which increased NGF levels in astrocyte cultures, with onjisaponin F also inducing the ChAT mRNA level in rat basal forebrain cells [366].

15.5.6 Salvia miltiorhiza Bunge

The dried root of *Salvia miltiorhiza* Bunge (Lamiaceae), also known as Chinese sage or 'dan shen', is red in colour and was therefore believed to be a treatment for blood disorders in folk medicine. In TCM the root has been used as a remedy to stabilise the heart and calm the nerves and to treat circulatory disorders, insomnia and neurasthenia, and to alleviate inflammation [151, 367]. The Pharmacopoeia of the People's Republic of China (2005) includes *S. miltiorhiza* root as a remedy for fidgets and insomnia, amongst other indications [237]. *S. miltiorhiza* root extracts have been investigated for a wide range of activities in relation to effects on the cardiovascular system but also for effects on the CNS, and in cerebral ischemia in particular. Some of the biological activities associated with *S. miltiorhiza* could also be relevant to the modulation of cognitive function and may provide some explanation for the traditional uses of this plant in some neural disorders.

S. miltiorhiza extracts have been shown to modulate the action of some neuropeptides, although studies on this subject are relatively limited and the pharmacological activities reported often provide theoretical explanations for their clinical relevance in CNS disorders. *S. miltiorhiza* has been suggested to modify the actions of vasoactive intestinal peptide (VIP), a neuropeptide distributed within the gastrointestinal tract and CNS [368], and somatostatin [369], a CNS neuropeptide that has been implicated in learning and memory [370, 371]; both neuropeptides are considered to play a role in changes involved in cerebral ischemia. *S. miltiorhiza* has also been suggested to influence the action of substance P, which is associated with neuronal damage following cerebral ischemia [372] and which is decreased in the brains of AD patients [373]. The compounds in *S. miltiorhiza* that protect against ischemic damage in the CNS appear to be the tanshinones [374], including tanshinone IIb, which was effective in reducing stroke-induced brain damage [375].

S. miltiorhiza may modify ischemic damage to the CNS by antioxidant effects. Since the metabolism of free fatty acids from the breakdown of lipid membranes and the generation of oxygen free radicals occur in ischemia, further brain injury occurs [376] and cognitive ability may be affected. *S. miltiorhiza* reduces lipid peroxidation and could therefore protect against these effects on the CNS during ischemia [377–379]. The herbal mixture known as '*Salvia* compositus', composed of *Dalbergia odorifera* T.C.Chen (Leguminosae) and *S. miltiorhiza*, is reported to modulate electrical activities of the cerebral cortex [380], to ameliorate cerebral oedema [381] and to inhibit oxidation of lipids [382].

Several compounds isolated from *S. miltiorhiza* root have shown antioxidant effects, which may explain any protective effects this plant may have against ischemic damage, and could also be relevant in modifying the progression of some cognitive disorders. The antioxidant modes of action of compounds from *S. miltiorhiza* are well documented [383]. Caffeic acid dimers, trimers and tetramers are some of the antioxidant compounds from the root of this plant and include rosmarinic acid (**43**) and salvianolic acids A (**44**) and B (**45**) [378, 384], with the latter two compounds also protecting against memory impairment induced by cerebral ischemia in rodents [385, 386]. Salvianolic acid B also protects against amyloid-induced neurotoxicity in vitro, which was associated with an antioxidant effect [387]. A series of quinones from the root of *S. miltiorhiza* have shown antioxidant activity in vitro [388, 389]. The antioxidant potency of these compounds was rosmariquinone, dihydrotanshinone (**46**), miltirone 1 > dehydrorosmariquinone > cryptotanshinone (**47**) > tanshinone IIa (**48**), which suggested that the structural features important

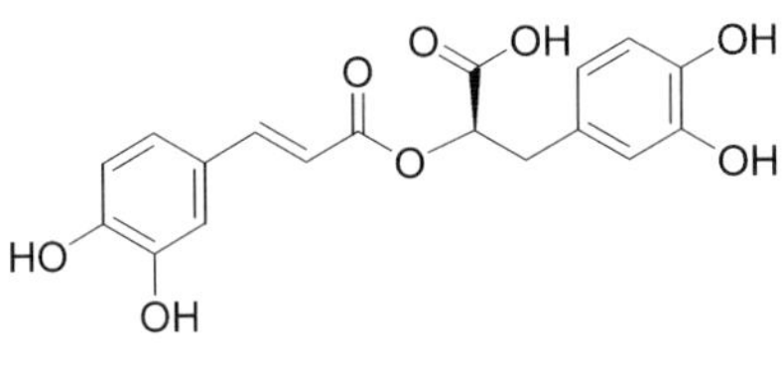

43 Rosmarinic acid

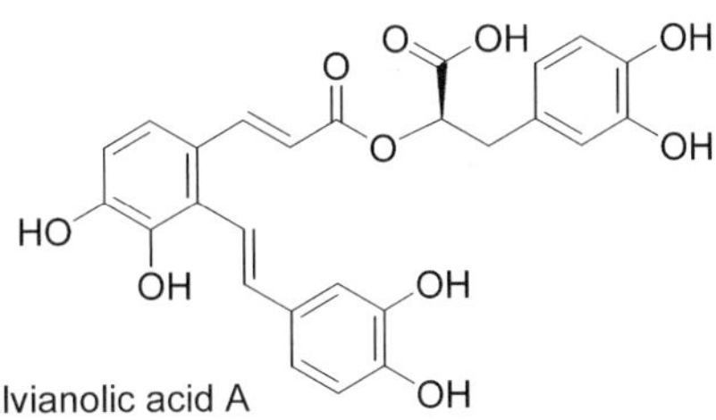

44 Salvianolic acid A

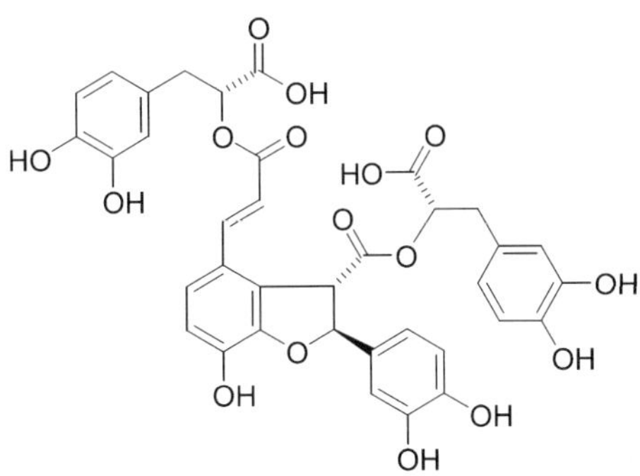

45 Salvianolic acid B

for the antioxidant activity were additional conjugated double bonds in the A ring, a dihydrofuran ring rather than a furan ring and an isopropyl substituent *ortho* to a quinone carbonyl rather than a dihydrofuran ring [388].

The anti-inflammatory effects of some *S. miltiorhiza* components could perhaps be exploited for use in cognitive disorders. Tanshinones from *S. miltiorhiza* root were anti-inflammatory in vivo and were active against 5-LOX in vitro, although they were not as active as the crude extracts [349, 390], suggesting that other compounds from the root may be more potent against LOX activity, or perhaps a synergistic effect might occur with the crude extract. In addition to an anti-inflammatory action in vivo, tanshinone I (**49**) inhibited phospholipase A_2 and prostaglandin (PG) E_2 formation in vitro, but did not affect COX-2 activity or expression [391], but cryptotanshinone does inhibit COX-2 activity [392]. These studies suggest that cryptotanshinone, but not tanshinone I, may have potential in modulating inflammatory processes in cognitive disorders such as AD, since many studies suggest that COX-inhibiting drugs prevent or delay the onset of AD [58–62].

Other, perhaps more relevant, activities of *S. miltiorhiza* root are inhibition of neuronal cell death by inhibition of presynaptic glutamate release, currently a therapeutic target in AD, and protection against β-amyloid-induced neuropathological changes in the hippocampus in vivo, effects observed with an extract and tanshinone, respectively [393, 394]. Also significant is that the first diterpenoids to show AChE inhibitory activity were isolated from *S. miltiorhiza* root, with dihydrotanshinone and cryptotanshinone being the most active [395]. These dihydrofurans were more active than the furans tanshinones I and IIa [395], suggesting that the more flexible dihydrofuran improves the binding affinity to the active site of the enzyme.

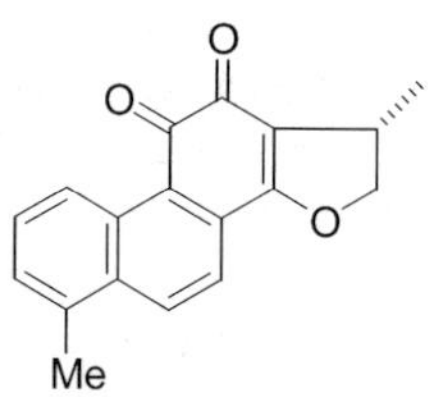

46 Dihydrotanshinone

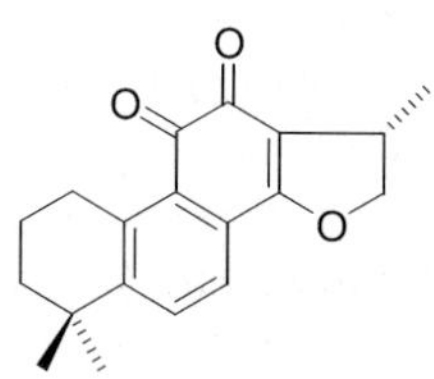

47 Cryptotanshinone

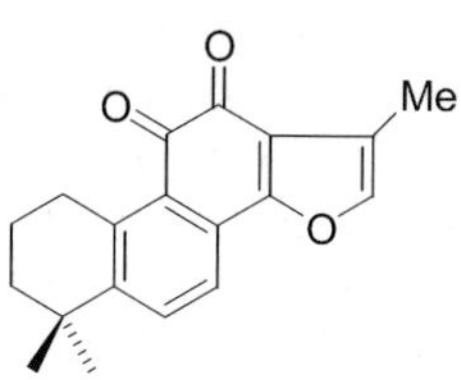

48 Tanshinone IIa

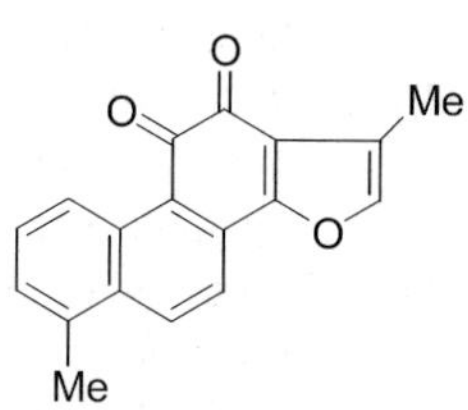

49 Tanshinone I

Although many activities associated with the treatment of cognitive disorders have been associated with numerous traditional remedies, studies to determine the bioavailability and if active compounds reach the site of action in the CNS are very limited. Studies on *S. miltiorhiza* compounds with regard to these parameters have shown that the bioavailability of salvianolic acid B [396] and tanshinone IIa [397] is low, and that penetration of cryptotanshinone across the blood-brain barrier may be limited in vivo [398], effects which may affect efficacy. In addition, an extract of *S. miltiorhiza* induced cytochrome P_{450} [399], which raises the possibility of drug interactions, and treatment with a preparation of *S. miltiorhiza* in patients on long-term warfarin therapy increased haemorrhage risk [400].

15.6 Plants Used in Traditional European Medicine

The practice of herbal medicine in Europe has been influenced by the remedies used and described by the ancient Greeks and by the traditions of Middle Eastern countries. Also often used are those herbal remedies used in North American traditional medicine, whose uses were learnt from the native Americans by the European settlers largely in the period between the 17th and 19th centuries.

15.6.1 Galanthus and *Narcissus* Species

Galanthus species (Amaryllidaceae) were used traditionally in Bulgaria and Turkey for neurological conditions [401]. The alkaloid galantamine (**50**) was originally isolated in the mid-20th century from *G. woronowii* Losinsk., commonly known as the ‘snowdrop’, but has now also been isolated from some species of *Narcissus* (Amaryllidaceae) and *Leucojum aestivum* L. (Amaryllidaceae) [401]. Galantamine is one of the few drugs of natural origin used to alleviate symptoms in AD, and it is now a licensed drug in Europe for this purpose.

The mode of action of galantamine is principally by inhibition of AChE. It is reported to bind at the base of the active site gorge of the enzyme (*Torpedo californica* AChE), interacting with both the choline-binding site and the acyl-binding pocket, and the tertiary amine is suggested to form an H-bond (via the *N*-methyl group) near the top of the gorge (to Asp-72) [402]. Galantamine is more selective for AChE than BChE and provides complete oral bioavailability [403, 404].

In clinical, multicentre, randomised, controlled trials, galantamine was well tolerated and significantly improved cognitive function when administered to AD patients [405, 406]. The cognitive benefits produced with galantamine treatment appear to be sustained for at least 3 years, which is a much longer time than for other drugs of this type [407]. Galantamine could have advantages over some other known AChE inhibitors in clinical use, as it is also a positive allosteric modulator of nicotinic receptors [408, 409], another activity that could modulate cholinergic and therefore cognitive function. There is also evidence to suggest that galantamine may be of some therapeutic value in vascular dementia and in dementia with Lewy bodies, as well as AD patients [410, 411], and it also improved memory and attention in patients with schizophrenia who were stabilised on risperidone [412].

In view of the success of galantamine as a naturally derived drug, it is not surprising that other alkaloids from species of *Narcissus* have been investigated for their anti-ChE activity. In one study, 26 extracts from different species of *Narcissus* and 23 Amaryllidaceae alkaloids were evaluated, but only 7 of these alkaloids, those with galantamine and lycorine (**51**) structural skeleton types, inhibited AChE [413]. In this study, 11-hydroxygalantamine (**52**) had similar potency to galantamine in the inhibition of AChE, epinorgalantamine (**53**) was less potent and sanguinine (**54**) was even more potent than galantamine, an effect attributed to the additional hydroxyl group in this structure, for interaction with AChE [413]. Among the lycorine type alkaloids, assoanine (**55**) was the most active AChE inhibitor in this study, but it was still four-fold less potent than galantamine [413]. Another study identified the alkaloid ungiminorine as an AChE inhibitory component, although the activity was relatively weak [414].

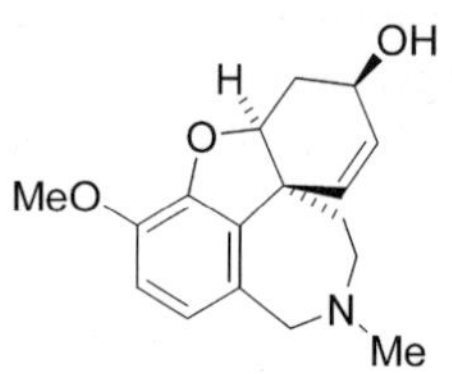

50 Galantamine

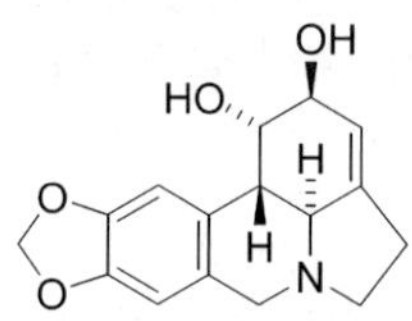

51 Lycorine

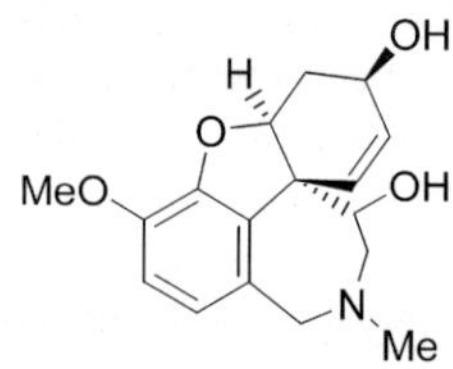

52 11-Hydroxygalantamine

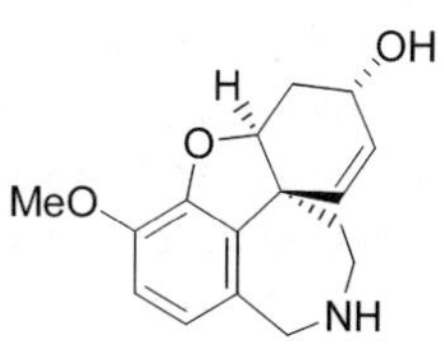

53 Epinorgalantamine

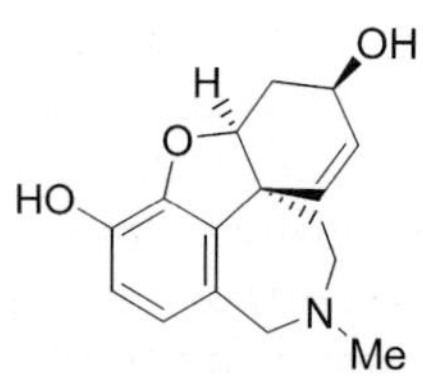

54 Sanguinine

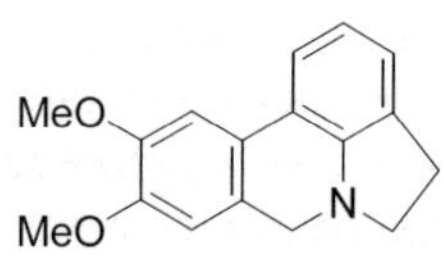

55 Assoanine

15.6.2 Melissa officinalis L.

In traditional European medicine *Melissa officinalis* L. (Lamiaceae) has been used as a remedy for over 2000 years, and it is reputed to treat melancholia, neuroses and hysteria, and the plant has been acclaimed for promoting long life and for restoring memory [415–417]. John Hill (1751) reported that *M. officinalis* was 'Good for disorders of the head and stomach' [418]. *M. officinalis* has also been used in other traditional practices of medicine and was considered as a treatment for depression in Arabic medicine [417] and used to treat hysteria in traditional Greek medicine [419].

Some clinical studies have investigated the reputed cognitive effects of *M. officinalis*. Some improvement in cognitive performance has been reported in healthy (non-AD) participants in randomised, placebo-controlled, double-blind, crossover studies, following treatment with cholinergically active (determined using in vitro studies) *M. officinalis* dried leaf [420] or a standardised extract [421], and cognitive improvements and a positive effect on agitation were also reported in AD patients administered an extract of *M. officinalis* for 4 months in a double-blind, randomised, placebo-controlled trial [422]. The results of a double-blind, placebo-controlled trial to investigate the effects of aromatherapy with *M. officinalis* oil in patients with clinically significant agitation associated with dementia indicated that *M. officinalis* oil may be a safe and effective treatment for agitation in people with AD [423].

Some investigations have been conducted to ascertain a pharmacological basis for the reputed cognitive enhancing effects of *M. officinalis* and to explain the favourable effects observed in some clinical trials. The essential oil and an ethanolic extract from *M. officinalis* weakly inhibit AChE [424] and some monoterpenoids that have been identified in the essential oil of *M. officinalis* including citral (a mixture of the isomers geranial and neral) [415, 425, 426] are also weak inhibitors of AChE [155]. Another study showed no AChE inhibitory effect to be associated with aqueous and methanolic extracts of *M. officinalis* [427].

Other activities of *M. officinalis* that may be relevant in AD therapy and that may explain the reported cognitive improvements include antioxidant effects [425, 428, 429], possible estrogenic effects [97] and binding to muscarinic and nicotinic receptors in vitro [430, 431]. However, the randomised, placebo-controlled, double-blind, balanced-crossover study which investigated the acute effects on cognition and mood of a standardised extract of *M. officinalis* did not correlate the modulation of cognitive function observed with an effect on cholinergic neurotransmission [421]. It was proposed that the low cholinergic binding properties shown in this study could be a result of a loss of volatile components in this particular extract [421]. Other studies indicate that different *M. officinalis* extracts may vary in their cholinergic receptor binding properties [431]. The differences observed in the cholinergic binding effects of *M. officinalis* preparations could be due to natural chemical variation in the *M. officinalis* used or different extraction methods; chemical variation can occur in *M. officinalis* essential oils due to various factors [426, 432–434]. These possibilities emphasise the need for quality control in the production of herbal remedies for therapeutic purposes. Continued investigation on the modes of action, including modulation of neurotransmitter systems in the CNS, would be useful to further assess the potential of compounds from *M. officinalis* extracts and essential oil for use in cognitive disorders.

15.6.3 Salvia officinalis L. and *S. lavandulifolia* Vahl

Research into the historical literature has identified several quotes in 16th- and 17th-century English herbals, describing sage (species of *Salvia* (Lamiaceae)) to improve memory [98]. In his late-16th-century English herbal, Gerard writes about sage: 'It is singularly good for the head and brain and quickenethe the nerves and memory', and Culpeper, writing about 50 years later, says that 'It also heals the memory, warming and quickening the senses', whilst Hill in 1756 poignantly describes the tragic effects associated with ageing by stating, 'Sage will retard that rapid progress of decay that treads upon our heels so fast in latter years of life, will preserve faculty and memory more valuable to the rational mind than life itself' [98]. Effects on the CNS have been reported for a number of species of *Salvia*, including sedative and hypnotic, hallucinogenic, memory-enhancing, anticonvulsant, neuroprotective and anti-Parkinsonian activities [435, 436].

In recent years a variety of studies have been conducted to investigate if there is any scientific evidence to explain the traditional uses of sage for improving memory, with many studies focusing on extracts and essential oils from *S. officinalis* L. and *S. lavandulifolia* Vahl. An ethanolic extract and the steam-distilled oil of *S. officinalis*, and the oil of *S. lavandulifolia* gave inhibition of AChE at relatively low concentrations in vitro [430]. The cyclic monoterpenoids 1,8-cineole (**56**) and α-pinene (**57**) were shown to inhibit AChE in vitro and were considered to explain the anti-ChE activity of the *S. lavandulifolia* oil, although other constituents of the oil may also have contributed, perhaps synergistically [154, 437].

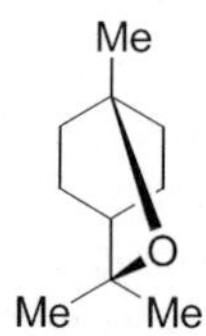

56 1,8-Cineole

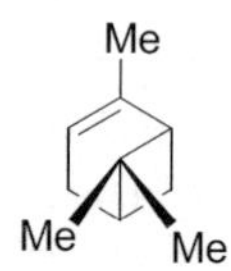

57 α-Pinene

Although the observed anti-ChE activity of the monoterpenoids is particularly interesting, since many of the previously reported AChE inhibitors of natural origin were amines, the monoterpenoids were considerably less potent by a factor of at least 10^3 than the alkaloid inhibitors such as physostigmine [154]. The relevance of in vitro test results with sage extracts and oils has been explored by further tests in vivo. One study showed that oral administration of *S. lavandulifolia* oil to rats decreased AChE activity in both the striatum and the hippocampus, compared to the control rats, suggesting that one or more oil constituents or their metabolites reach the brain and inhibit AChE in select brain areas, which is consistent with evidence of inhibition of AChE in vitro [430, 438]. Another study assessed the effect of an ethanolic extract of *S. officinalis* on memory retention of passive avoidance learning in rodents; the potentiation of memory retention observed with this treatment may have been associated with an interaction with the muscarinic and nicotinic cholinergic systems [439].

Extracts from some species of *Salvia*, including from both *S. officinalis* and *S. lavandulifolia*, have also shown antioxidant effects [98, 428, 440]. An aqueous methanolic extract of *S. officinalis* dose-dependently inhibited lipid peroxidation [428], and antioxidant effects were also shown with an ethanolic extract of *S. lavandulifolia*; both the water-soluble and chloroform-soluble fractions of the latter extract gave similar activity [98]. Compounds isolated from species of *Salvia* that have shown antioxidant effects include salvianolic acids I, K and L and various other phenolic compounds [441, 442].

Species of *Salvia* have also been investigated for activities relevant to producing an anti-inflammatory action. Inhibition of eicosanoid synthesis was observed with an ethanol extract of *S. lavandulifolia*, although this effect was relatively weak [98]. Some essential oil constituents from *S. lavandulifolia* have also been assessed for their effects on eicosanoid synthesis, but only the monoterpenoid α-pinene (comprising 5% of the essential oil) produced significant activity, although it did show some weak selectivity for inhibition of leukotriene B_4 (LTB_4) generation [98]. Ursolic acid, from *S. officinalis*, has anti-inflammatory activity in vivo [443].

Some other, perhaps relevant, activities have also been reported for sage. A standardised extract of *S. officinalis* and one of its components rosmarinic acid (**43**) were neuroprotective against β-amyloid-induced toxicity in vitro [444]. In addition, dose-dependent estrogenic activity was observed in vitro with an ethanolic extract from *S. lavandulifolia* [98], and although the possible benefits of estrogenic compounds on cognition are still unclear, an estrogenic effect of sage might also provide some explanation for the reputed effects in traditional medicine.

Some clinical studies with human volunteers, including AD patients, have also been reported. In a placebo-controlled, double-blind, balanced, crossover study, subjects (healthy young adults) received a standardised oil extract of *S. lavandulifolia* and vehicle (sunflower oil) alone, with a 7-d washout period between each treatment [445]. The sage treatment was associated with significant effects on cognitive ability, including improvements in immediate word-recall scores [445]. A similar study also showed a positive modulation of mood and cognition in healthy young adults when given doses of a standardised essential oil of *S. lavandulifolia* in a placebo-controlled, double-blind, balanced, crossover study [446]. In a small pilot trial, 11 patients showing mild to moderate symptoms of AD were orally administered *S. lavandulifolia* oil, which significantly improved cognitive function; a reduction in neuropsychiatric symptoms and an improvement in attention were observed [447]. Further evidence for the cognitive enhancing effects of sage was shown in a multicentre, double-blind, randomised, placebo-controlled trial. In this study, patients showing some typical AD symptoms were treated with an extract of *S. officinalis* and significantly better outcomes in measurements of cognitive function were observed; there was no significant difference in side-effects between the treated and placebo groups, but a greater incidence of agitation in the placebo group was observed [448], suggesting the sage treatment may have also alleviated agitation.

15.7 Plants Used in African and South American Traditional Medicine

African traditional medicine, which is diverse because of the vast range of habitats, languages and cultural groups, has a long history of use, and in some countries up to 90% of the population relies on plants as the only source of medicines [449]. Countries north of the Sahara have a similar ethnopharmacology because of an influence from Islamic cultures over many centuries. Sub-Saharan cultures are more diverse but do share some common features, such as the consideration of spiritual influences and beings in the disease and healing process. In sub-Saharan Africa, the influence of European culture came quite late and was diversified because of the different colonial powers and the climate being generally not very amenable to growing some of the traditional European plants. Therefore, the endemic medical systems were arguably more preserved than in many parts of South America, where European domination occurred two or three centuries earlier. However, in the more remote parts of the continent, especially where there was also considerable biodiversity, extensive knowledge about the medicinal uses of the local plants remained, and consequently, several drugs have been included in European medicine over the years.

15.7.1 Physostigma venenosum Balf.

The calabar bean, the seeds of *Physostigma venenosum* Balf. (Leguminosae), was used traditionally in Africa, particularly south-eastern Nigeria, for ritual deaths associated with the funeral of a chief and as an ordeal poison, claimed to determine the guilt or innocence of persons accused of a crime [450–452]. Rapid death was an indication that the suspect was guilty and innocence was shown by survival. This logic does appear to have some scientific basis, as differences in absorption might arise due to psychosomatic influences, with nervous sipping by guilty suspects enabling greater absorption [453]. The toxic effects of the calabar bean extract were found to be due to excessive cholinergic stimulation resulting in increased salivation, nausea, bradycardia, muscle cramps and respiratory failure, as well as CNS effects. This effect was attributed to the presence of an alkaloid with an unusual pyrroloindole skeleton, physostigmine (**58**), also known as eserine [454], which potently inhibits AChE [455]. Physostigmine has been shown to inhibit both G_1 and G_4 AChE forms, the major AChE isoenzymes present in mammalian CNS [456]. Physostigmine also inhibits with similar potency BChE, an enzyme that has been implicated in the etiology and progression of AD [457].

Physostigmine protects mice against cognitive impairment caused by oxygen deficit, and it improves learning [452] and antagonises scopolamine-induced impairment of cognitive function in rats [458]. Physostigmine, a short-acting reversible AChE inhibitor, is also reported to have shown significant cognitive benefits in both normal and AD patients [306, 459], but clinical use may be limited by its short half-life, which would require frequent dosing. To develop compounds which are lipophilic, and so able to cross the blood-brain barrier, and which can improve on the pharmacokinetic profile of physostigmine, a number of compounds structurally related to physostigmine that inhibit AChE have been synthesised with the aim of developing new drugs to alleviate cognitive dysfunction in AD.

The most therapeutically successful of these compounds to date is rivastigmine (**59**), an AChE inhibitor that is now licensed for use in Europe for the symptomatic treatment of mild to moderate dementia in AD or Parkinson's disease. Rivastigmine inhibits AChE in the cortex and hippocampus and preferentially inhibits the G_1 form of AChE [460], and it improves cognition in AD patients, an effect that has been correlated with the level of AChE inhibition [461–465]. It also potently inhibits AChE and BChE in Alzheimer's plaques and tangles [466]. Other synthetic compounds being investigated include cymserine (**60**), which has a structure based on the backbone of physostigmine and which potently and concentration-dependently binds with human BChE [467], and a structural analogue of cymserine, bisnorcymserine, which also potently inhibits BChE in vitro [468]. Another derivative of physostigmine is eptastigmine (heptyl-physostigmine tartrate) (**61**), in which the carbamoylmethyl group in position 5 of the side chain has been substituted with a carbamoylheptyl group [469]. Although it is an effective inhibitor of both AChE and BChE and it improves cognitive performance in AD patients, this compound caused haematologic adverse effects, resulting in the suspension of any further clinical trials [469].

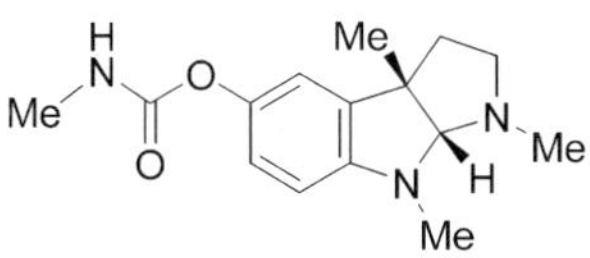

58 Physostigmine

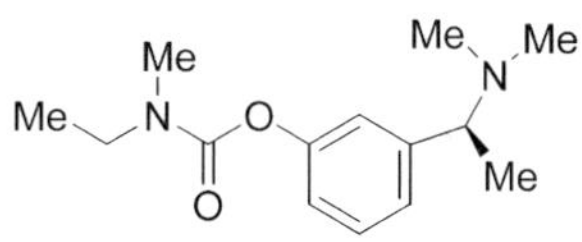

59 Rivastigmine

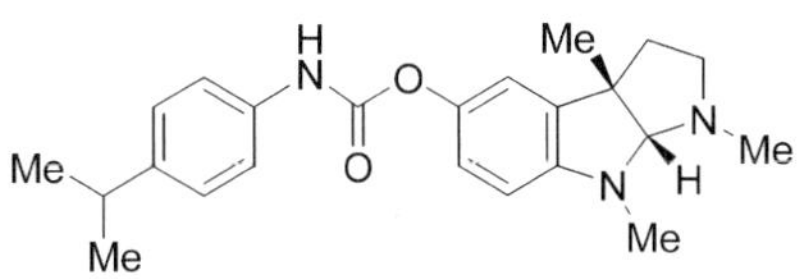

60 Cymserine

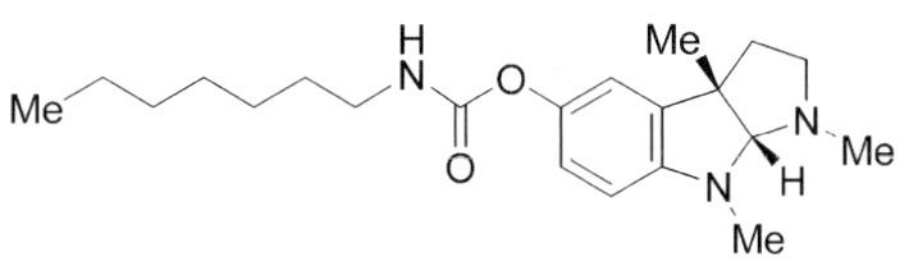

61 Eptastigmine

15.7.2 *Pilocarpus* Species

Jaborandi leaves are obtained from various species of *Pilocarpus* (Rutaceae), found in South America, but members of this genus are also found in the West Indies and Central America, although to a lesser extent. The leaves from species of

Pilocarpus, including *P. microphyllus* Stapf, have been used traditionally in South America to induce sweating and urination, considered to eliminate toxins from the body. The leaves contain a number of imidazole alkaloids including pilocarpine (**62**) [450], which has structural similarities to ACh (**1**) and is a muscarinic receptor agonist [116].

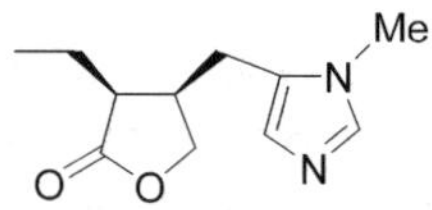

62 Pilocarpine

Chewing the leaves of Jaborandi therefore produces cholinergic effects such as contraction of the pupils and excessive salivation. It is for this action that salts of pilocarpine have been used as topical treatments to reduce the raised intra-ocular pressure that occurs in glaucoma. Since modulation of the cholinergic system is considered to be involved in learning and memory processes, muscarinic receptor modulators could be used clinically to achieve cognitive improvements. Pilocarpine has been shown to enhance cognitive performance in rodents [470–472], although studies to investigate cognitive effects in humans are lacking, probably due to its poor pharmacokinetic profile as it poorly penetrates the blood-brain barrier, in addition to having undesirable side-effects, such as nausea and vomiting, sweating and bradycardia. Although, it has been suggested that co-administration of a compound that blocks peripheral muscarinic receptors with pilocarpine, or with other muscarinic receptor agonists, may reduce the potential for adverse effects [473]. It should also be noted that CNS symptoms may be induced or exacerbated with the use of pilocarpine eye drops in patients with AD [474, 475].

15.7.3 Ptychopetalum olacoides Benth.

Ptychopetalum olacoides Benth. (Olacaceae) originates from the Amazon and the roots have been used as a traditional remedy for a variety of ailments of the CNS and stress, particularly age-related conditions. The roots are known commonly as 'mara-puama', 'muirapuama' or 'miranta' and are now internationally available in general health-food stores. An ethanol extract of the roots of this plant improved memory retrieval when administered to both young and ageing rodents [476, 477]. The mechanisms of action to explain these effects and the active compounds are unknown. Some studies show root extracts of *P. olacoides* to have antioxidant [478], AChE inhibitory [479] and neuroprotective activities [480], effects that could explain the experimental and reputed effects on memory function. More studies are necessary to investigate further the underlying modes of action that the compounds from the roots of this plant have on the CNS and any clinical relevance this remedy may have in cognitive disorders.

15.8 Conclusions

Many plants have a reputation in a variety of traditional practices of medicine for alleviating symptoms of cognitive disorders such as memory decline; such plants have been used for medicinal purposes for a long time and in some cases continue to be used. There have been numerous scientific studies on a number of these plants to establish any pharmacological basis for their historical uses, although the extent to which each species has been investigated varies considerably. Very few of the plants that have a reputation for modifying cognitive abilities have been extensively studied, and there is frequently a lack of knowledge about the compounds responsible for the observed activities, and reliable clinical studies are also uncommon. Amongst those plants which have been subjected to more thorough investigations is *Ginkgo biloba*, which has shown biological activities relevant to modulation of cognitive function both in vitro and in vivo, and there is some evidence of efficacy in both healthy and AD subjects.

It is particularly interesting that of the four main currently licensed drugs (donepezil (**8**), galantamine (**50**), memantine (**13**) and rivastigmine (**59**)) used to treat cognitive symptoms in AD, and which have been investigated for their potential for use in other cognitive disorders, two of these (galantamine and rivastigmine) were derived from plant sources. Some of the therapeutic single chemical entities that have been developed for clinical use have arisen from toxicological investigations, very often combined with chemical modification of the active toxic compound once its structure has been determined; rivastigmine, which is based on the structure of physostigmine (**58**), is an example of this. In other cases, a compound useful in the treatment of a cognitive disorder has been discovered by studies on a plant species which has attracted interest because of its traditional use for some other purposes, such as galantamine.

It should also be considered that, although the development of new and effective 'orthodox' drugs as single chemical entities for some cognitive disorders is one aim, the use of plant extracts can still be valuable, particularly as in many cases more than one constituent in a plant is responsible for the overall effect, and it is too simplistic to conceive that a single compound has the same effect or efficacy. Herbal preparations consisting of complex mixtures for therapeutic purposes are generally not accepted in modern Western medicine, and consequently many potentially useful plants will not be accepted for clinical use, although they may be extensively used in complementary and alternative therapies. In these circumstances, the role of scientific research is to provide evidence for their therapeutic applications in conventional Western medicine and to identify compounds responsible for the activity so that herbal products with the appropriate qualitative and quantitative chemical profile can be made available, with greater assurance of safety and efficacy.

In addition to concerns regarding the quality control issues associated with herbal medicinal products, the potential for their interactions with other medicines has not been widely studied and these issues need to be addressed.

In conclusion, it is apparent that the pharmacological activities of plants and their compounds often appear to reflect their uses in traditional medicine, although traditional medicines used to treat other disorders and poisons can also provide leads to develop herbal medicines and drugs to alleviate symptoms in cognitive disorders.

Acknowledgements We would like to thank Dr N. C. Veitch (Royal Botanic Gardens, Kew) for drawing the chemical structures.

References

1. Piskulic D, Olver JS, Norman TR, Maruff P (2007) Psychiat Res 150:111
2. Amato MP, Zipoli V, Portaccio E (2006) J Neurol Sci 245:41
3. Berman SB, Greenamyre JT (2006) Curr Neurol Neurosci Rep 6:281
4. Grossman H, Bergmann C, Parker S (2006) Mt Sinai J Med 73:985
5. Francis PT, Palmer AM, Snape M, Wilcock GK (1999) J Neurol Neurosurg Psychiat 66:137
6. Desgranges B, Baron J-C, de la Sayette V, Petit-Taboué M-C, Benali K, Landeau B, Lechevalier B, Eustache F (1998) Brain 121:611
7. Förstl H, Hentschel F, Sattel H, Geiger-Kabisch C, Besthorn C, Czech C, Mönning U, Beyreuther K (1995) Arzneimittelforschung 45:394
8. Kumar V, Durai NB, Jobe T (1998) Clin Geriatr Med 14:129
9. McGuffey EC (1997) J Am Med Assoc NS37:347
10. Wragg RE, Jeste DV (1989) Am J Psychiat 146:577
11. Kanof P, Davis KL (1995) J Neurochem 64:749
12. Giacobini E (1990) Prog Brain Res 84:321
13. Perry E, Tomlinson E, Blessed G, Bergmann K, Gibson P, Perry R (1978) Br Med J 2:1457
14. Perry E (1986) Br Med Bull 42:63
15. Marshall IG, Parsons SM (1987) Trends Neurosci 10:174
16. Houghton PJ, Ren YH, Howes MJ (2006) Nat Prod Rep 23:181
17. Zhang X (2004) Curr Drug Target CNS Neurol Disord 3:137
18. Giacobini E (2004) Pharmacol Res 50:433
19. Giacobini E (1993) Prog Brain Res 98:447
20. Blokland A (1996) Brain Res Rev 21:285
21. Karczmar AG (1995) Prog Neuropsychopharmacol Biol Psychiat 19:187
22. Clader JW, Wang YG (2005) Curr Pharm Des 11:3353
23. Mihailescu S, Drucker-Collin R (2000) Arch Med Res 31:131
24. London ED, Ball MJ, Walker SB (1989) Neurochem Res 14:745
25. Martin-Ruiz C, Court J, Lee M, Piggott M, Johnson M, Ballard C, Kalaria R, Perry R, Perry E (2000) Acta Neurol Scand Suppl 176:34
26. Nordberg A (1993) J Reprod Fertil Suppl 46:145
27. Perry EK, Smith CJ, Court JA, Perry RH, Fairbairn AF (1990) J Neural Transm 2:149
28. Shimohama S, Taniguchi T, Fujiwara M, Kameyama M (1986) J Neurochem 46:288
29. Balfour DJK, Fagerström KO (1996) Pharmacol Ther 72:51
30. Flores CM, Rogers SW, Pabreza LA, Wolfe B, Kellar KJ (1992) Mol Pharmacol 41:31
31. Whitehouse PJ, Kalaria RN (1995) Alzheimer Dis Assoc Dis 9:3
32. Buccafusco JJ, Jackson WJ (1991) Neurobiol Ag 12:233
33. Decker MW, Majchrzak MJ, Anderson DJ (1992) Brain Res 572:281
34. Rezvani AH, Levin ED (2001) Biol Psychiat 49:258

35. Graves AB, Mortimer JA (1994) J Smoking-Related Disord 5:79
36. Min SK, Moon IW, Ko RW, Shin HS (2001) Psychopharmacology 159:83
37. Newhouse PA, Kelton M (2000) Pharm Acta Helv 74:91
38. Sahakian B, Jones G, Levy R, Gray J, Warburton D (1989) Br J Psychiat 154:797
39. van Duijn CM, Hofman A (1991) Br Med J 302:1491
40. Kukull WA (2001) Biol Psychiat 49:194
41. Merchant C, Tang MX, Albert S. Manly J, Stern Y, Mayeux R (1999) Neurology 52:1408
42. Tyas SL, Pederson LL, Koval JJ (2000) Ann Epidemiol 10:409
43. Salomon AR, Marcinowski KJ, Friedland RP, Zagorski MG (1996) Biochemistry 35:13568
44. Zamani MR, Allen YS (2001) Biol Psychiat 49:221
45. Akaike A, Tamura Y, Yokaota T, Shimohama S, Kimura J (1994) Brain Res 644:181
46. Decker MW, Majchrzak MJ, Arneric SP (1993) Pharmacol Biochem Behav 45:571
47. Miller DK, Crooks PA, Dwoskin LP (2000) Neuropharmacology 39:2654
48. Xiao P, Kubo H, Komiya H, Higashiyama K, Yan YN, Li JS, Ohmiya S (1999) Phytochemistry 50:189
49. Mineur YS, Somenzi O, Picciotto MR (2007) Neuropharmacology 52:1256
50. Hammel P, Larrey D, Bernuau J, Kalafat M, Fréneaux E, Babany G, Degott C, Feldmann G, Passayre D, Benhamou JP (1990) J Clin Gastroenterol 12:329
51. Watkins PB, Zimmerman HJ, Knapp MJ, Gracon SI, Lewis KW (1994) J Am Med Assoc 271:992
52. Evans J (2001) Chem Br 37:47
53. Castro A, Martinez A (2006) Curr Pharm Des 12:4377
54. Heneka MT (2006) Clin Neurosci Res 6:247
55. McGeer EG, McGeer PL (1997) CNS Drugs 7:214
56. McGeer EG, McGeer PL (1999) Curr Pharm Des 5:821
57. Moore AH, O'Banion MK (2002) Adv Drug Deliv Rev 54:1627
58. Breitner JCS (1996) Annu Rev Med 47:401
59. Breitner JCS, Welsh KA, Helms MJ, Gaskell PC, Gau BA, Roses AD, Pericak-Vance MA, Saunders AM (1995) Neurobiol Ag 16:523
60. Jenkinson ML, Bliss MR, Brain AT, Scott DL (1989) Br J Rheumatol 28:86
61. McGeer PL, McGeer E, Rogers J, Sibley J (1990) Lancet 335:1037
62. McGeer PL, Schulzer M, McGeer EG (1996) Neurology 47:425
63. Ho L, Qin, WP, Stetka BS, Pasinetti GM (2006) CNS Drugs 20:85
64. Bingöl F, Şener B (1995) Int J Pharmacognosy 33:81
65. Calixto JB, Otuki MF, Santos ARS (2003) Planta Med 69:973
66. Calixto JB, Campos MM, Otuki MF, Santos ARS (2004) Planta Med 70:93
67. Handa SS, Chawla AS, Sharma AK (1992) Fitoterapia 63:3
68. Welton AF, Tobias LD, Fiedler-Nagy C, Anderson W, Hope W, Meyers K, Coffey JW (1986) Prog Clin Biol Res 213:231
69. Alcaraz MJ, Ferrándiz ML (1987) J Ethnopharmacol 21:209
70. Yan JJ, Cho JY, Kim HS, Kim KL, Jung JS, Huh SO, Suh HW, Kim YH, Song DK (2001) Br J Pharmacol 133:89
71. Doraiswamy PM, Xiong G (2006) Expert Opin Pharmacother 7:1
72. Maxwell SJ (1995) Drugs 49:345
73. Slater TF (1984) Biochem J 222:1
74. Spiteller G (1993) J Lipid Mediat 7:199
75. Deans SG, Noble RC, Pénzes L, Beregi E (1993) Natural antioxidants from aromatic and medicinal plants. In: Fehér J, Blázovics A, Matkovics B, Mézes M (eds) Role of Free Radicals in Biological Systems. Akademiai Kiado, Budapest, p 159
76. Masaki H, Sakaki S, Atsumi T, Sakurai H (1995) Biol Pharm Bull 18:162
77. Yanishlieva NV, Marinova E, Pokorny J (2006) Eur J Lipid Sci Technol 108:776
78. Aesbach R, Löliger J, Scott, BC, Murcia A, Butler J, Halliwell B, Aruoma OI (1994) Food Chem Toxicol 46:1739
79. Foti M, Piattelli M, Baratta MT, Ruberto G (1996) J Agric Food Chem 44:497

80. Okuda T, Kimura Y, Yoshida T, Hatano T, Okuda H, Arichi S (1983) Chem Pharm Bull 31:1625
81. Pietta PG (2000) J Nat Prod 63:1035
82. Ruberto G, Baratta MT (2000) Food Chem 69:167
83. Schwarz K, Ternes W (1992) Z Lebensm Unters Forsch 195:99
84. Seidel V, Verholle M, Malard Y, Tillequin F, Fruchart J, Duriez P, Bailleul F, Teissier E (2000) Phytother Res 14:93
85. Ramassamy C (2006) Eur J Pharmacol 545:51
86. Choi YT, Jung CH, Lee SR, Bae JH, Baek WK, Suh MH, Park J, Park CW, Suh SI (2001) Life Sci 70:603
87. Youdim KA, Deans SG (1999) Biochim Biophys Acta 1438:140
88. Birge SJ (1997) Am J Med 103:1S
89. Fillet H, Weinreb H, Cholst I, Luine V, McEwan B, Amador R, Zabriskie J (1986) Psychoneuroendocrinology 11:337
90. Honjo H, Ogino Y, Naitoh K, Urabe M, Kitawaki J, Yasuda J, Yamamoto T, Ishihara S, Okada H, Yonezawa T, Hayashi K, Nambara T (1989) J Steroid Biochem 34:521
91. Schneider LS, Farlow MR, Henderson VW, Pogoda JM (1996) Neurology 46:1580
92. Mooradian AD (1993) J Steroid Biochem Mol Biol 45:509
93. Norbury R, Cutter WJ, Compton J, Robertson DM, Craig M, Whitehead M, Murphy DG (2003) Exp Gerontol 38:109
94. Sohrabji F, Miranda RC, Toran-Allerand D (1994) J Neurosci 14:459
95. Hogervorst E, Williams J, Budge M, Riedel W, Jolles J (2000) Neuroscience 101:485
96. Mulnard RA, Cotman CW, Kawas C, van Dyck CH, Sano M, Doody R, Koss E, Pfeiffer E, Jin S, Gamst A, Grundman M, Thomas R, Thal LJ (2000) J Am Med Assoc 283:1007
97. Howes MJR, Houghton PJ, Barlow DJ, Pocock VJ, Milligan SR (2002) J Pharm Pharmacol 54:1521
98. Perry NSL, Houghton PJ, Sampson J, Hart S, Lis-Balchin M, Hoult JRS, Evans P, Jenner P, Milligan S, Perry EK (2001) J Pharm Pharmacol 53:1347
99. Lephart ED, West TW, Weber KS, Rhees RW, Setchell KDR, Adlercreutze H, Lund TD (2002) Neurotoxicol Teratol 24:5
100. File SE, Jarrett N, Fluck E, Duffy R, Casey K, Wiseman H (2001) Psychopharmacology 157:430
101. Kim H, Xia H, Li L, Gewin J (2001) Biochem Soc Trans 29:216
102. Pan Y, Anthony M, Watson S, Clarkson TB (2000) Menopause 7:209
103. Duffy R, Wiseman H, File SE (2003) Pharmacol Biochem Behav 75:721
104. Kritz-Silverstein D, Von Muhlen D, Barrett-Connor E, Bressel MAB (2003) Menopause 10:196
105. Bang OY, Hong HS, Kim DH, Kim H, Boo JH, Huh K, Mook-Jung I (2004) Neurobiol Dis 16:21
106. Zeng HY, Chen Q, Zhao BL (2004) Free Radical Biol Med 36:180
107. Zhao LX, Chen Q, Brinton RD (2002) Exp Biol Med 227:509
108. Boue SM, Wiese TE, Nehls S, Burow ME, Elliott S, Carter-Wientjes CH, Shih BY, McLachlan JA, Cleveland TE (2003) J Agric Food Chem 51:2193
109. Hsieh MT, Kuo LH, Tsai FH, Wang WH, Wu CR (2002) Planta Med 68:901
110. Woo J, Lau E, Ho SC, Cheng F, Chan C, Chan ASY, Haines CJ, Chan TYK, Li M, Sham A (2003) Menopause 10:352
111. Ritzen A, Mathiesen JM, Thomsen C (2005) Basic Clin Pharmacol Toxicol 97:202
112. Reisberg B, Ferris S, Mobius HJ, Schmitt F, Doody R (2002) Neurobiol Ag 23:2039
113. Winblad B, Poritis N (1999) Int J Geriatr Psychiat 14:135
114. Sieghart W (2006) Adv Pharmacol 54:231
115. Warnakulasuriya S, Trivedy C, Peters TJ (2002) Br Med J 324:799
116. Broadley KJ, Kelly DR (2001) Molecules 6:142
117. Lim DY, Kim IS (2006) Acta Pharm Sin 27:71
118. Yang YR, Chang KC, Chen CL, Chiu TH (2000) Chin J Physiol 43:23

119. Carey GJ, Costall B, Domeney AM, Gerrard PA, Jones DNC, Naylor RJ, Tyers MB (1992) Pharmacol Biochem Behav 42:75
120. Reikkinen P, Reikkinen M, Sirvio J (1993) J Pharmacol Exp Ther 267:1484
121. Soncrant TT, Raffaele KC, Asthana S, Berardi A, Morris PP, Haxby JV (1993) Psychopharmacology 112:421
122. Raffaele KC, Asthana S, Beradi A, Haxby JV, Morris PP, Schapiro MB, Soncrant TT (1996) Neuropsychopharmacology 15:163
123. Tariot PN, Cohen RM, Welkowictz JA, Sunderland T, Newhouse PA, Murphey DL, Weingartner H (1988) Arch Gen Psychiat 45:901
124. Meier E, Frederiksen K, Nielsen M, Lembol HL, Pedersen H, Hyttel J (1997) Drug Dev Res 40:1
125. Thal LJ, Forrest M, Loft H, Mengel H (2000) Neurology 54:421
126. Wienrich M, Meier D, Ensinger HA, Gaida W, Raschig A, Walland A, Hammer R (2001) Life Sci 68:2593
127. Terry AV, Buccafusco JJ, Borsini F, Leusch A (2002) Psychopharmacology 162:292
128. Russo A, Borrelli F (2005) Phytomedicine 12:305
129. Das A, Shanker G, Nath C, Pal R, Singh S, Singh HK (2002) Pharmcol Biochem Behav 73:893
130. Vohora D, Pal SN, Pillai KK (2000) J Ethnopharmacol 71:383
131. Bhattacharya SK, Bhattacharya A, Kumar A, Ghosal S (2000) Phytother Res 14:174
132. Russo A, Borrelli F, Campisi A, Acquaviva R, Raciti G, Vanella A (2003) Life Sci 73:1517
133. Jyoti A, Sethi P, Sharma D (2007) J Ethnopharmacol 111:56
134. Holcomb LA, Dhanasekaran M, Young KA, Manyam BV (2004) Neurobiol Ag 25:S244
135. Holcomb LA, Dhanasekaran M, Hitt AR, Young KA, Riggs M, Manyam BV (2006) J Alzheimers Dis 9:243
136. Channa S, Dar A, Anjum S, Yaqoob M, Atta-ur-Rahman (2006) J Ethnopharmacol 104:286
137. Bhattacharya SK, Ghosal S (1998) Phytomedicine 5:77
138. Chakravarty AK, Sarkar T, Masuda K, Shiojima K, Nakane T, Kawahara N (2001) Phytochemistry 58:553
139. Chakravarty AK, Garai S, Masuda K, Nakane T, Kawahara N (2003) Chem Pharm Bull 51:215
140. Garai S, Mahato SB, Ohtani K, Yamasaki K (1996) Phytochemistry 42:815
141. Hou CC, Lin SJ, Cheng JT, Hsu FL (2002) J Nat Prod 65:1759
142. Pawar RS, Khan SI, Khan IA (2007) Planta Med 73:380
143. Anbarasi K, Kathirvel G, Vani G, Jayaraman G, Devi CSS (2006) Neuroscience 138:1127
144. Anbarasi K, Vani G, Devi CSS (2005) J Environ Pathol Toxicol Oncol 24:225
145. Nathan PJ, Clarke J, Lloyd J, Hutchison CW, Downey L, Stough C (2001) Hum Psychopharmacol Clin Exp 16:345
146. Roodenrys S, Booth D, Bulzomi S, Phipps A, Micallef C, Smoker J (2002) Neuropsychopharmacology 27:279
147. Stough C, Lloyd J, Clarke J, Downey LA, Hutchison CW, Rodgers T, Nathan PJ (2001) Psychopharmacology 156:481
148. Kapoor LD (1990) Handbook of Ayurvedic Medicinal Plants. CRC, Boca Raton, FL
149. Manyam BV (1999) J Altern Complement Med 5:81
150. Brinkhaus B, Lindner M, Schuppan D, Hahn EG (2000) Phytomedicine 7:427
151. Tang W, Eisenbrand G (1992) Chinese drugs of plant origin. Springer-Verlag, Berlin Heidelberg, Germany
152. Asakawa Y, Matsuda R, Takemoto T (1982) Phytochemistry 21:2590
153. Miyazawa M Watanabe H, Kameoka H (1997) J Agric Food Chem 45:677
154. Perry NSL, Houghton PJ, Theobald A, Jenner P, Perry EK (2000) J Pharm Pharmacol 52:895
155. Ryan MF, Byrne O (1988) J Chem Ecol 14:1965
156. Sakina MR, Dandiya PC (1990) Fitoterapia 61:291
157. Chen SW, Wang WJ, Li WJ, Wang R, Li YL, Huang YN, Liang X (2006) Pharmacol Biochem Behav 85:339
158. Wijeweera P, Arnason JT, Koszycki D, Merali Z (2006) Phytomedicine 13:668

159. Kumar MHV, Gupta YK (2002) J Ethnopharmacol 79:253
160. Subathra M, Shila S, Devi MA, Panneerselvam C (2005) Exp Gerentol 40:707
161. Zainol MK, Abd-Hamid A, Yusof S, Muse R (2003) Food Chem 81:575
162. Gupta YK, Kumar MHV, Srivastava AK (2003) Pharmacol Biochem Behav 74:579
163. Fowler CJ, O'Neill C, Winblad B, Cowburn RT (1992) Acta Neurol Scand 85:59
164. Uchihara T, Kondo H, Kosaka K, Tsukagoshi H (1992) Acta Neuropathol 83:271
165. Nalini K, Aroor AR, Karanth KS, Rao A (1992) Fitoterapia 63:232
166. Lee MK, Kim SR, Sung SH, Lim DY, Kim H, Choi H, Park HK, Jew SS, Kim YC (2000) Res Commun Mol Pathol Pharmacol 108:75
167. Rao KGM, Rao SM, Rao SG (2006) Evid-Based Comp Altern Med 3:349
168. Soumyanath A, Zhong YP, Gold SA, Yu XL, Koop DR, Bourdette D, Gold BG (2005) J Pharm Pharmacol 57:1221
169. Nadkarni KM (1976) Indian Materia Medica, 3rd edn. Popular Prakashan, Bombay
170. Warrier PK, Nambiar VPK, Ramankutty C (1995) Indian Medicinal Plants, vol 2. Orient Longman, India
171. Nalini K, Karanth KS, Rao A, Aroot AR (1995) J Ethnopharmacol 47:101
172. Gattu M, Boss KL, Terry AV, Buccafusco JJ (1997) Pharmacol Biochem Behav 57:793
173. Kumar MHV, Gupta YK (2002) Phytomedicine 9:302
174. Bidwai PP, Wangoo D, Bhullar NK (1987) J Ethnopharmacol 21:307
175. Godkar PB, Gordon RK, Ravindran A, Doctor BP (2004) J Ethnopharmacol 93:213
176. Godkar P, Gordon RK, Ravindran A, Doctor BP (2003) Fitoterapia 74:658
177. Godkar PB, Gordon RK, Ravindran A, Doctor BP (2006) Phytomedicine 13:29
178. Dev S (1997) Curr Sci 73:909
179. Ahmad F, Khan RA, Rasheed S (1994) J Ethnopharmacol 42:193
180. Parle M, Vasudevan M (2007) J Health Sci 53:43
181. Misra R (1998) Med Res Rev 18:383
182. Taranalli AD, Cheeramkuzhy TC (2000) Pharm Biol 38:51
183. Kumar V, Mukherjee K, Pal BC, Houghton PJ, Mukherjee PK (2007) Planta Med 73:10.1055
184. Rai KS, Murthy KD, Karanth KS, Nalini K, Rao MS, Srinivasan KK (2002) Fitoterapia 73:685
185. Jain NN, Ohal CC, Shroff SK, Bhutada RH, Somani RS, Kasture VS, Kasture SB (2003) Pharmacol Biochem Behav 75:529
186. Devi BP, Boominathan R, Mandal SC (2003) Fitoterapia 74:345
187. Rai KS, Murthy KD, Rao MS, Karanth KS (2005) Phytother Res 19:592
188. Miquel J, Bernd A, Sempere JM, Diaz-Alperi J, Ramirez A (2002) Arch Gerentol Geriatr 34:37
189. Scartezzini P, Speroni E (2000) J Ethnopharmacol 71:23
190. Koo BS, Lee WC, Chung KH, Ko JH, Kim CH (2004) Life Sci 75:2363
191. Rajakrishnan V, Viswanathan P, Rajasekharan KN, Menon VP (1999) Phytother Res 13:571
192. Zbarsky V, Datla KP, Parkar S, Rai DK, Aruoma OI, Dexter DT (2005) Free Radical Res 39:1119
193. Kumar P, Padi SSV, Naidu PS, Kumar A (2007) Methods Find Exp Clin Pharmacol 29:19
194. Kim DSHL, Kim JY (2001) Bioorg Med Chem Lett 11:2541
195. Park SY, Kim DSHL (2002) J Nat Prod 65:1227
196. Kim DSHL, Park SY, Kim JY (2001) Neurosci Lett 303:57
197. Balasubramanian K (2006) J Agric Food Chem 54:3512
198. Maheshwari RK, Singh AK, Gaddipati J, Srimal RC (2006) Life Sci 78:2081
199. Ramsewak RS, DeWitt DL, Nair MG (2000) Phytomedicine 7:303
200. Rao CV (2007) Adv Exp Med Biol 595:213
201. Skrzypczak-Jankun E, McCabe NP, Selman SH, Jankun J (2000) Int J Mol Med 6:521
202. Srivastava KC, Bordia A, Verma SK (1995) Prostaglandins Leukotrienes Essent Fatty Acids 52:223
203. Selvam C, Jachak SM, Thilagavathi R, Chakraborti AK (2005) Bioorg Med Chem Lett 15:1793

204. Upton R (2000) (ed) Ashwaganda Root. American Herbal Pharmacopoeia and Therapeutic Compendium, AHP, US
205. Ghosal S, Lal J, Srivastava R, Bhattacharya SK, Upadhyay SN, Jaiswal AK, Chattopadhyay U (1989) Phytother Res 3:201
206. Schliebs R, Liebmann A, Bhattacharya SK, Kumar A, Ghosal S, Bigl V (1997) Neurochem Int 30:181
207. Bhattacharya SK, Kumar A, Ghosal S (1995) Phytother Res 9:110
208. Vinutha B, Prashanth D, Salma K, Sreeja SL, Pratiti D, Padmaja R, Radhika S, Amit A, Venkateshwarlu K, Deepak M (2007) J Ethnopharmacol 109:359
209. Choudhary MI, Yousuf S, Nawaz SA, Ahmed S, Atta-ur-Rahman (2004) Chem Pharm Bull 52:1358
210. Choudhary MI, Nawaz SA, Zaheer-ul-Haq, Lodhi MA, Ghayur MN, Jalil S, Riaz N, Yousuf S, Malik A, Gilani AH, Atta-ur-Rahman (2005) Biochem Biophys Res Commun 334:276
211. Dhuley JN (2001) Phytother Res 15:524
212. Tohda C, Kuboyama T, Komatsu K (2000) NeuroReport 11:1981
213. Zhao J, Nakamura N, Hattori M, Kuboyama T, Tohda C, Komatsu K (2002) Chem Pharm Bull 50:760
214. Kuboyama T, Tohda C, Zhao J, Nakamura N, Hattori M, Komatsu K (2002) NeuroReport 13:1715
215. Kuboyama T, Tohda C, Komatsu K (2005) Br J Pharmacol 144:961
216. Kuboyama T, Tohda C, Komatsu (2006) Eur J Neurosci 23:1417
217. Jain S, Shukla SD, Sharma K, Bhatnagar M (2001) Phytother Res 15:544
218. Ahmad M, Saleem S, Ahmad AS, Ansari MA, Yousuf S, Hoda MN, Islam F (2005) Human Exp Toxicol 24:137
219. Das PK, Malhotra CL, Prasad K (1964) Arch Int Pharmacodyn Ther 150:356
220. Schwarting AE, Bobbitt JM, Rother A, Atal K, Khanna KL, Leary JD, Walter WG (1963) Lloydia 26:258
221. Dhar RS, Verma V, Suri KA, Sangwan RS, Satti NK, Kumar A, Tuli R, Qazi GN (2006) Phytochemistry 67:2269
222. Dhuley JN (1998) J Ethnopharmacol 60:173
223. Panda S, Gupta P, Kar A (1997) Curr Sci 72:546
224. Bhattacharya A, Ramanathan M, Ghosal S, Bhattacharya SK (2000) Phytother Res 14:568
225. Chaurasia SS, Panda S, Kar A (2000) Pharm Res 41:663
226. Bhattacharya A, Ghosal S, Bhattacharya SK (2001) J Ethnopharmacol 74:1
227. Mishra L, Singh BB, Dagenais S (2000) Altern Med Rev 5:334
228. Subbaraju GV, Vanisree M, Rao CV, Sivaramakrishna C, Sridhar P, Jayaprakasam B, Nair MG (2006) J Nat Prod 69:1790
229. Bhattacharya SK, Satyan KS, Ghosal S (1997) Ind J Exp Biol 35:236
230. Prakash D, Suri S, Upadhyay G, Singh BN (2007) Int J Food Sci Nutr 58:18
231. Begum V, Sadique J (1988) Ind J Exp Biol 26:887
232. Rasool M, Varalakshmi P (2007) Fundam Clin Pharmacol 21:157
233. Anbalagan K, Sadique J (1981) Curr Sci 50:88
234. Anbalagan K, Sadique J (1981) Ind J Exp Biol 19:245
235. Dhuley JN (1997) J Ethnopharmacol 58:15
236. Sudhir S, Budhiraja RD, Miglani GP, Arora B, Gupta LC, Garg KN (1986) Planta Med 1:61
237. Chinese Pharmacopoeia Commission (2005) Pharmacopoeia of the People's Republic of China. People's Medical Publishing House, China
238. Ko HC, Wang YH, Liou KT, Chen CM, Chen CH, Wang WY, Chang S, Hou YC, Chen KT, Chen CF, Shen YC (2007) Eur J Pharmacol 555:211
239. Moon TC, Murakami M, Kudo I, Son KH, Kim HP, Kang SS, Chang HW (1999) Inflammat Res 48:621
240. Choi YH, Shin EM, Kim YS, Cai XF, Lee JJ, Kim HP (2006) Arch Pharmacal Res 29:293
241. Takada Y, Kobayashi Y, Aggarwal BB (2005) J Biol Chem 280:17203
242. Haji A, Momose Y, Takeda R, Nakanishi S (1994) J Nat Prod 57:387
243. Lim DK, Lee YB, Kim HS (2004) Neurochem Res 29:407

244. Shin KY, Park CH, Choi SH, Lee SH, Kim HS, Suh YH (2004) Neurobiol Ag 25:S589
245. Park CH, Choi SH, Kim CH, Seo JH, Koo JW, Rah JC, Kim HS, Jeong SJ, Joo WS, Lee YJ, Kim YS, Kim MS, Suh YH (2003) Arzneimittelforschung 53:393
246. Park CH, Kim S, Choi W, Lee Y, Kim J, Kang SS, Suh YH (1996) Planta Med 62:405
247. Wang HH, Chou CJ, Liao JF, Chen CF (2001) Eur J Pharmacol 413:221
248. Decker M (2005) Eur J Med Chem 40:305
249. Ross IA (2001) Medicinal plants of the world, volume 2. Humana, Totowa, NJ
250. Heiss WD, Zeiler K (1978) Pharmakotherapie 1:137
251. Loffler T, Lee SK, Noldner M, Chatterjee SS, Hoyer S, Schliebs R (2001) J Neural Transm 108:1457
252. Tea S, Celsis P, Clanet M, Marc-Vergnes J-P, Boeters U (1987) Therapiewoche 37:2655
253. Kristofiková Z, Benesová O, Tejkalová H (1992) Dementia 3:304
254. Barth SA, Inselmann G, Engemann R, Heidemann HT (1991) Biochem Pharmacol 41:1521
255. Marcocci L, Packer L, Droy-Lefaix M-T, Sekaki A, Gardès-Albert M (1994) Methods Enzymol 234:462
256. Rojas P, Garduno B, Rojas C, Vigueras RM, Rojas-Castaneda J, Rios C, Serrano-Garcia N (2001) Neurochem Res 26:1245
257. Schindowski K, Leutner S, Kressmann S, Eckert A, Muller WE (2001) J Neural Transm 108:969
258. Yao Z, Drieu K, Papadopoulos V (2001) Brain Res 889:181
259. Bastianetto S, Ramassamy C, Dore S, Christen Y, Poirier J, Quirion R (2000) Eur J Neurosci 12:1882
260. Bastianetto S, Zheng WH, Quiron R (2000) J Neurochem 74:2268
261. Bastianetto S, Quirion R (2002) Cell Mol Biol 48:693
262. Zhu L, Wu XM, Yang L, Du F, Qian ZM (2007) Brain Res 1166:1
263. Eckert A, Keil U, Kressmann S, Schindowski K, Leutner S, Leutz S, Muller WE (2003) Pharmacopsychiatry 36:S15
264. Bertoni-Freddari C, Fattoretti P, Caselli U, Paoloni R, Solazzi M (2002) Cell Mol Biol 48:709
265. Chandrasekaran K, Mehrabian Z, Spinnewyn B, Drieu K, Kiskum G (2001) Brain Res 922:282
266. Jones FA, Chatterjee SS, Davies JA (2002) Amino Acids 22:369
267. Kiewert C, Kumar V, Hildmann O, Rueda M, Hartmann J, Naik RS, Klein J (2007) Brain Res 1128:70
268. Kuboto Y, Tanaka N, Umegaki K, Takenaka H, Mizuno H, Nakamura K, Shinozuka K, Kunitomo M (2001) Life Sci 69:2327
269. Nishida S, Satoh H (2004) Clin Chim Acta 339:129
270. Braquet PG, Hosford D, Koltai M (1994) Neuropsychopharmacology 10:693S
271. Baek SH, Yun SS, Kwon TK, Kim JR, Chang HW, Kwak JY, Kim JH, Kwun KB (1999) Shock 12:473
272. Kwak WJ, Han CK, Son KH, Chang HW, Kang SS, Park BK, Kim HP (2002) Planta Med 68:316
273. Garcia-Alloza M, Dodwell SA, Meyer-Luehmann M, Hyman BT, Bacskai BJ (2006) J Neuropathol Exp Neurol 65:1082
274. Oh SM, Chung KH (2004) Life Sci 74:1325
275. Oh SM, Chung KH (2006) J Steroid Biochem Mol Biol 100:167
276. Petkov VD, Kehayov R, Belcheva S, Konstantinova E, Petkov VV, Getova D, Markovska V (1993) Planta Med 59:106
277. Wang YF, Wang L, Wu J, Cai JX (2006) Br J Pharmacol 148:147
278. Stoll S, Scheuer K, Pohl O, Müller WE (1996) Pharmacopsychiatry 29:144
279. Gong QH, Wu Q, Huang XN, Sun AS, Shi JS (2005) Life Sci 77:140
280. Paganelli RA, Benetoli A, Milani H (2006) Behav Brain Res 174:70
281. Chopin P, Briley M (1992) Psychopharmacology 106:26
282. Yamamoto Y, Adachi Y, Fujii Y, Kamei C (2007) Brain Res 1129:161
283. Hartley DE, Heinze L, Elsabagh S, File SE (2003) Pharmacol Biochem Behav 75:711

284. Hofferberth B (1994) Hum Psychopharmacol Clin Exp 9:215
285. Kanowski S, Herrmann WM, Stephan K, Wierch W, Hörr R (1997) Phytomedicine 4:3
286. Konowski S, Hoerr R (2003) Pharmacopsychiatry 36:297
287. Le Bars PL, Katz MM, Berman N, Itil TM, Freedman AM, Schatzberg AF (1997) J Am Med Assoc 278:1327
288. Le Bars PL, Velasco FM, Ferguson JM, Dessain EC, Kieser M, Hoerr R (2002) Neuropsychobiology 45:19
289. Mazza M, Capuano A, Bria P, Mazza S (2006) Eur J Neurol 13:981
290. Napryeyenko O, Borzenko I (2007) Arzneimittelforschung 57:4
291. Rai GS, Shovlin C, Wesnes KA (1991) Curr Med Res Opin 12:350
292. Rigney U, Kimber S, Hindmarch I (1999) Phytother Res 13:408
293. Vellas B, Andrieu S, Ousset PJ, Ouzid M, Mathiex-Fortunet H (2006) Neurology 67:S6
294. Lovera J, Bagert B, Smoot K, Morris CD, Frank R, Bogardus K, Wild K, Oken B, Whitham R, Bourdette D (2007) Mult Scler 13:376
295. Cieza A, Maier P, Poppel E (2003) Arch Med Res 34:373
296. Yasui-Furukori N, Furukori H, Kaneda A, Kaneko S, Tateishi T (2004) J Clin Pharmacol 44:538
297. Chiappelli F, Navarro AM, Moradi DR, Manfrini E, Prolo P (2006) Evid-Based Comp Altern Med 3:411
298. Foster S (1989) Herbalgram 18/19:21
299. Skolnick AA (1997) J Am Med Assoc 277:776
300. Lu W-H, Shou J, Tang X-C (1988) Acta Pharm Sin 9:11
301. Wang LM, Han YF, Tang XC (2000) Eur J Pharmacol 398:65
302. Zhou J, Zhang HY, Tang XC (2001) Neurosci Lett 313:137
303. Ashani Y, Peggins JO, Doctor BP (1992) Biochem Biophys Res Commun 184:719
304. Laganière S, Corey J, Tang X-C, Wülfert E, Hanin I (1991) Neuropharmacology 30:763
305. McKinney M, Miller JH, Yamada F, Tuckmantel W, Kozikowski AP (1991) Eur J Pharmacol 203:303
306. Shu Y-Z (1998) J Nat Prod 61:1053
307. Small GW, Rabins RV, Barry PP, Buckholtz NS, DeKosky ST, Ferros SH, Finkel SI, Gwyther LP, Khachaturian ZS, Lebowitz BD, McRae TD, Morris JC, Oakley F, Schneider LS, Streim JE, Sunderland T, Teri LA, Tune LE (1997) J Am Med Assoc 278:1363
308. Wang R, Yan H, Tang XC (2006) Acta Pharm Sin 27:1
309. Ou LY, Tang XC, Cai JX (2001) Eur J Pharmacol 433:151
310. Xiao XQ, Zhang HY, Tang XC (2002) J Neurosci Res 67:30
311. Zhou J, Fu Y, Tang XC (2001) Neurosci Lett 306:53
312. Xiao XQ, Yang JW, Tang XC (1999) Neurosci Lett 275:73
313. Ved HS, Koenig ML, Dave JR, Doctor BP (1997) NeuroReport 8:963
314. Wang XD, Zhang JM, Yang HH, Hu GY (1999) Acta Pharm Sin 20:31
315. Zhang YH, Chen XQ, Yang HH, Jin GY, Bai DL, Hu GY (2000) Neurosci Lett 295:116
316. Zhou J, Tang XC (2002) FEBS Lett 536:21
317. Wang ZF, Zhou J, Tang XC (2002) Acta Pharm Sin 23:1193
318. Zhang HY, Tang XC (2000) Neurosci Lett 292:41
319. Zeng FX, Jiang HL, Tang XC, Chen KX, Ji RY (1998) Bioorg Med Chem Lett 8:1661
320. Kaneko S, Nakajima N, Shikano M, Katoh T, Terashima S (1996) Bioorg Med Chem Lett 6:1927
321. Kaneko S, Shikano M, Katoh T, Terashima S (1997) Synlett SI:447
322. Zhou GC, Zhu DY (2000) Bioorg Med Chem Lett 10:2055
323. Högenauer K, Baumann K, Enz A, Mulzer J (2001) Bioorg Med Chem Lett 11:2627
324. Rajendran V, Prakash KRC, Ved HS, Saxena A, Doctor BP, Kozikowski AP (2000) Bioorg Med Chem Lett 10:2467
325. Rajendran V, Saxena A, Doctor BP, Kozikowski AP (2002) Bioorg Med Chem Lett 12:1521
326. Badia A, Banos JE, Camps P, Contreras J, Gorbig DM, Munoz-Torrero D, Simon M, Vivas NM (1998) Bioorg Med Chem 6:427
327. Carlier PR, Du DM, Han YF, Liu J, Pang YP (1999) Bioorg Med Chem Lett (1999) 9:2335

328. Camps P, Cusack B, Mallender WD, El Achab R, Morral J, Muñoz-Torrero D, Rosenberry TL (2000) Mol Pharmacol 57:409
329. Roman S, Vivas NM, Badia A, Clos MV (2002) Neurosci Lett 325:103
330. Ros E, Aleu J, de Aranda IG, Munoz-Torrero D, Camps P, Badia A, Marsal J, Solsona C (2001) Eur J Pharmacol 421:77
331. Roman S, Badia A, Camps P, Munoz-Torrero D, Clos MV (2005) Brain Res 1061:73
332. Alcalá MDM, Vivas NM, Hospital S, Camps P, Muñoz-Torrero D, Badia A (2003) Neuropharmacology 44:749
333. Zeng FX, Jiang HL, Zhai YF, Zhang HY, Chen KX, Ji RY (1999) Bioorg Med Chem Lett 9:3279
334. Hou YC, Chao PD, Chen SY (2000) Am J Chin Med 28:379
335. Zhou Y, Xu R (1992) Zhongguo Zhongyao Zazhi 17:368
336. Chen YL, Lin KF, Shiao MS, Chen YT, Hong CY, Lin SJ (2001) Basic Res Cardiol 96:353
337. Lo YC, Teng CM, Chen CF, Chen CC, Hong CY (1994) Biochem Pharmacol 47:549
338. Chiu JH, Ho CT, Wei YH, Lui WY, Hong CY (1997) Life Sci 61:1961
339. Lee MM, Hseih MT, Kuo JS, Yeh FT, Huang HM (1998) NeuroReport 9:3451
340. Lin YR, Chen HH, Ko CH, Chan MH (2005) Neuropharmacology 49:542
341. Lin YR, Chen HH, Ko CH, Chan MH (2006) Eur J Pharmacol 537:64
342. Wang JP, Hsu MF, Raung SL, Chen CC, Kuo JS, Teng CM (1992) Naunyn-Schmiedeberg's Arch Pharmacol 346:707
343. Wang JP, Ho TF, Chang LC, Chen CC (1995) J Pharm Pharmacol 47:857
344. Tse AKW, Wan CK, Zhu GY, Shen XL, Cheung HY, Yang MS, Fong WF (2007) Mol Immunol 44:2647
345. Tse AKW, Wan CK, Shen XL, Yang MS, Fong WF (2005) Biochem Pharmacol 70:1443
346. Liou KT, Shen YC, Chen CF, Tsao CM, Tsai SK (2003) Brain Res 992:159
347. Liou KT, Lin SM, Huang SS, Chih CL, Tsai SK (2003) Planta Med 69:130
348. Chiba K, Yamazaki M, Umegaki E, Li MR, Xu ZW, Terada S, Taka M, Naoi N, Mohri T (2002) Biol Pharm Bull 25:791
349. Chang H, But PP (2001) Pharmacology and applications of Chinese material medica, vols 1–2. World Scientific, Singapore
350. Duke JA, Ayensu ES (1985) Medicinal Plants of China, vols 1–2, 2nd ed. Reference Publications, Algonac, MI
351. Nishiyama N, Zhou Y, Saito H (1994) Biol Pharm Bull 17:1481
352. Zhang Y, Takashina K, Saito H, Nishiyama N (1994) Biol Pharm Bull 17:866
353. Nishiyama N, Zhou Y, Takashina K, Saito H (1994) Biol Pharm Bull 17:1472
354. Smriga M, Saito H, Nishiyama N (1995) Biol Pharm Bull 18:518
355. Yabe T, Iizuka S, Komatsu Y, Yamada H (1997) Phytomedicine 4:199
356. Yamada H, Yabe T (1997) Curr Top Phytochem 1:157
357. Chen YL, Hsieh CL, Wu PHB, Lin JG (2004) J Ethnopharmacol 95:47
358. Lee HJ, Ban JY, Koh SB, Seong NS, Song KS, Bae KW, Seong YH (2004) Am J Chin Med 32:599
359. Park CH, Choi SH, Koo JW, Seo JH, Kim HS, Jeong SJ, Suh YH (2002) J Neurosci Res 70:484
360. Naito R, Tohda C (2006) Biol Pharm Bull 29:1892
361. Kim HM, Lee EH, Na HJ, Lee SB, Shin TY, Lyu YS, Kim NS, Nomira S (1998) J Ethnopharmacol 61:201
362. Koo HN, Jeong HJ, Kim KR, Kim JC, Kim KS, Kang BK, Kim HM, Kim JJ (2000) Imunopharmacol Immunotoxicol 22:531
363. Ikeya Y, Takeda S, Tunakawa M, Karakida H, Toda K, Yamaguchi T, Aburada M (2004) Biol Pharm Bull 27:1081
364. Karakida F, Ikeya Y, Tsunakawa M, Yamaguchi T, Ikarashi Y, Takeda S, Aburada M (2007) Biol Pharm Bull 30:514
365. Jia HX, Jiang Y, Ruan Y, Zhang YB, Ma X, Zhang JZ, Beyreuther K, Tu PF, Zhang D (2004) Neurosci Lett 367:123

366. Yabe T, Tuchida H, Kiyohara H, Takeda T, Yamada H (2003) 10:106
367. Huang KC (1993) The pharmacology of Chinese herbs. CRC Press Ltd., Boca Raton, USA
368. Kuang P, Wu W, Zhang F, Liu J, Pu C (1989) J Tradit Chin Med 9:203
369. Kuang P, Wu W, Zhu K (1993) J Tradit Chin Med 13:38
370. Dutar P, Vaillend C, Viollet C, Billard JM, Potier B, Carlo AS, Ungerer A, Epelbaum J (2002) Neuroscience 112:455
371. Lamirault L, Guillou JL, Micheau J, Jaffard R (2001) Eur J Neurosci 14:567
372. Kuang P, Wu W, Liu J, Zhang F, Pu C (1991) J Tradit Chin Med 11:123
373. Quigley BJ, Kowall NW (1991) Neuroscience 41:41
374. Lam BYH, Lo ACY, Sun X, Luo HW, Chung SK, Sucher NJ (2003) Phytomedicine 10:286
375. Yu XY, Lin SG, Zhou ZW, Chen X, Liang J, Duan W, Yu XQ, Wen JY, Chowbay B, Li CG, Sheu FS, Chan E, Zhou SF (2007) Neurosci Lett 417:261
376. Traystman RJ, Kirsch JR, Koehler RC (1991) J Appl Physiol 71:1185
377. Kuang P, Tao Y, Tian Y (1996) J Tradit Chin Med 16:138
378. Liu G-T, Zhang T-M, Wang B-E, Wang Y-W (1992) Biochem Pharmacol 43:147
379. Zhao B, Jiang W, Zhao Y, Hou J, Xin W (1996) Biochem Mol Biol Int 38:1171
380. Fan S, Sun L, Wang Z, Wang W, Xu S (1979) Acta Pharm Sin 14:199
381. Kuang P, Li Z, Zhang F, Tao Y, Liu J, Wu W (1995) J Tradit Chin Med 15:135
382. Zhang X, Kuang P, Wu W, Yin X, Kanazawa T, Onodera K, Metoki H, Oike Y (1994) J Tradit Chin Med 14:195
383. Zhu YZ, Huang SH, Tan BKH, Sun J, Whiteman M, Zhu YC (2004) Nat Prod Rep 21:478
384. Huang YS, Zhang JT (1992) Acta Pharm Sin 27:96
385. Du GH, Qiu Y, Zhang JT (2000) J Asian Nat Prod Res 2:145
386. Guanhua D, Juntian Z (1997) Chin Med J 110:65
387. Lin YH, Liu AH, Wu, HL, Westenbroek C, Song QL, Yu HM, Ter Horst GJ, Li XJ (2006) Biochem Biophys Res Commun 348:593
388. Weng XC, Gordon MH (1992) J Agric Food Chem 40:1331
389. Zhang K, Bao Y, Wu P, Rosen RT, Ho C (1990) J Agric Food Chem 38:1194
390. Paulus K, Bauer R (2000) Natural products research in the new millenium abstract book, 48th annual meeting of the Society for Medicinal Plant Research. Zurich, Switzerland, P2A/73
391. Kim SY, Moon TC, Chang HW, Son KH, Kang SS, Kim HP (2002) Phytother Res 16:616
392. Jin DZ, Yin LL, Ji XQ, Zhu XZ (2006) Eur J Pharmacol 549:166
393. Kuang P, Xiang J (1994) J Tradit Chin Med 14:45
394. Li LX, Dai JP, Ru LQ, Yin GF, Zhao B (2004) Acta Pharm Sin 25:861
395. Ren YH, Houghton PJ, Hider RC, Howes MJR (2004) Planta Med 70:201
396. Wu YT, Chen YF, Hsieh YJ, Jaw I, Shiao MS, Tsai TH (2006) Int J Pharm 326:25
397. Hao HP, Wang GJ, Cui N, Li J, Xie L, Ding ZQ (2006) Planta Med 72:1311
398. Yu XY, Lin SG, Chen X, Zhou ZW, Liang J, Duan W, Chowbay B, Wen JY, Chan E, Cao J, Li CG, Zhou SF (2007) Curr Drug Metab 8:365
399. Kuo YF, Lin YL, Don MJ, Chen RM Ueng YF (2006) J Pharm Pharmacol 58:521
400. Hu ZP, Yang XX, Ho PCL, Chan SY, Heng PWS, Chan E, Duan W, Koh HL, Zhou SF (2005) Drugs 65:1239
401. Heinrich M, Teoh HL (2004) J Ethnopharmacol 92:147
402. Greenblatt HM, Kryger G, Lewis T, Silman I, Sussman JL (1999) FEBS Lett 463:321
403. Bickel U, Thomsen T, Weber W, Fischer JP, Bachus R, Nitz M, Kewitz H (1991) Clin Pharmacol Ther 50:420
404. Fulton B, Benfield P (1996) Drugs Aging 1:60
405. Wilcock GK, Lilienfeld S, Gaens E (2000) Br Med J 321:1445
406. Wilkinson D, Murray J (2001) Int J Geriatr Psychiat 16:852
407. Raskind MA, Peskind ER, Truyen L, Kershaw P, Damaraju CV (2004) Arch Neurol 61:252
408. Orhan G, Orhan I, Sener B (2006) Lett Drug Des Discov 3:268
409. Woodruff-Pak DS, Vogel RW, Wenk GL (2001) Proc Natl Acad Sci USA 98:2089
410. Edwards K, Royall D, Hershey L, Lichter D, Hake A, Farlow M, Pasquier F, Johnson S (2007) Dement Geriatr Cogn Disord 23:401
411. Small G, Erkinjuntti T, Kurz A, Lilienfeld S (2003) CNS Drugs 17:905

412. Schubert MH, Young KA, Hicks PB (2006) Biol Psychiat 60:530
413. Lopez S, Bastida J, Viladomat F, Codina C (2002) Life Sci 71:2521
414. Ingkaninan K, Hazekamp A, de Best CM, Irth H, Tjaden UR, van der Heijden R, van der Greef R, Verpoorte R (2000) J Nat Prod 63:803
415. Wichtl M (2004) (ed) Herbal Drugs and Phytopharmaceuticals. Medpharm GmbH Scientific Publishers, Stuttgart, Germany
416. Kenner D, Requena Y (1996) Botanical Medicine: A European Professional Perspective. Paradigm Publications, Brookline, MA
417. McVicar J (1994) Jekka's Complete Herb Book. Kyle Cathie, London
418. Crellin JK, Philpott J (1990) Herbal Medicine Past and Present, vol 2: A Reference Guide to Medicinal Plants. Duke University Press, Durham, NC
419. Malamas M, Marselos M (1992) J Ethnopharmacol 37:197
420. Kennedy DO, Wake G, Savelev S, Tildesley NTJ, Perry EK, Wesnes KA, Scholey AB (2003) Neuropsychopharmacology 28:1871
421. Kennedy DO, Scholey AB, Tildesley NTJ, Perry EK, Wesnes KA (2002) Pharmacol Biochem Behav 72:953
422. Akhondzadeh S, Noroozian M, Mohammadi M, Ohadinia S, Jamshidi AH, Khani M (2003) J Neurol Neurosurg Psychiat 74:863
423. Ballard CG, O'Brien JT, Reichelt K, Perry EK (2002) J Clin Psychiat 63:553
424. Ferreira A, Proenca C, Serralheiro MLM, Araujo MEM (2006) J Ethnopharmacol 108:31
425. Mimica-Dukic N, Bozin B, Sokovic M, Simin N (2004) J Agric Food Chem 52:2485
426. van den Berg T, Freundl E, Czygan F-C (1997) Pharmazie 52:802
427. Adsersen A, Gauguin B, Gudiksen L, Jager AK (2006) J Ethnopharmacol 104:418
428. Hohmann J, Zupkó I, Rédei D, Csányi M, Falkay G, Máthé I, Janicsák G (1999) Planta Med 65:576
429. Ivanova D, Gerova D, Chervenkov T, Yankova T (2005) J Ethnopharmacol 96:145
430. Perry N, Court G, Bidet N, Court J, Perry E (1996) Int J Geriatr Psychiatr 11:1063
431. Wake G, Court J, Pickering A, Lewis R, Wilkins R, Perry E (2000) J Ethnopharmacol 69:105
432. Adzet T, Ponz R, Wolf E, Schulte E (1992) Planta Med 58:562
433. Meyer W, Spiteller G (1996) Z Naturforsch, C: Biosci 51:651
434. Mrlianova M, Tekel'ova D, Felklova M, Reinöhl V, Toth J (2001) Planta Med 67:178
435. Imanshahidi M, Hosseinzadeh H (2006) Phytother Res 20:427
436. Perry NSL, Howes M-J R, Houghton PJ, Perry E (2000) Why sage may be a wise remedy: effects of *Salvia* on the nervous system. In: Kintzios SE (ed) Sage: the Genus *Salvia*. Harwood, Amsterdam, p 207
437. Savelev S, Okello E, Perry NSL, Wilkins RM, Perry EK (2003) Pharmacol Biochem Behav 75:661
438. Perry NSL, Houghton PJ, Jenner P, Keith A, Perry EK (2002) Phytomedicine 9:48
439. Eidi M, Eidi A, Bahar M (2006) Nutrition 22:321
440. Matsingou TC, Petrakis N, Kapsokefalou M, Salifoglou (2003) J Agric Food Chem 51:6696
441. Lu YR, Foo LY (2001) Tetrahedron Lett 42:8223
442. Lu YR, Foo LY (2001) Food Chem 75:197
443. Baricevic D, Sosa S, Della Loggia R, Tubaro A, Simonovska B, Krasna A, Zupancic A (2001) J Ethnopharmacol 75:125
444. Iuvone T, De Filippis D, Esposito G, D'Amico A, Izzo AA (2006) J Pharmacol Exp Ther 317:1143
445. Tildesley NTJ, Kennedy DO, Perry EK, Ballard CG, Savelev S, Wesnes KA, Scholey AB (2003) Pharmacol Biochem Behav 75:669
446. Tildesley NTJ, Kennedy DO, Perry EK, Ballard CG, Wesnes KA, Scholey AB (2005) Physiol Bevah 83:699
447. Perry NSL, Bollen C, Perry EK, Ballard C (2003) Pharmacol Biochem Behav 75:651
448. Akhondzadeh S, Noroozian M, Mohammadi M, Ohadinia S, Jamshidi AH, Khani M (2003) J Clin Pharm Ther 28:53
449. Hostettmann K, Marston A, Ndjoko K, Wolfender JL (2000) Curr Org Chem 4:973

450. Evans WC (2002) Trease and Evans Pharmacognosy, 15th edn. Saunders, London
451. Holmstedt B (1972) In: Swain T (ed) Plants in the Development of Modern Medicine. Harvard University Press, Cambridge, MA, p 303
452. McCaleb R (1990) Herbalgram 23:15
453. Mann J (1992) Murder, Magic and Medicine. Oxford University Press, Oxford
454. Julian PL, Pikl J (1935) J Am Chem Soc 51:755
455. Kamal MA, Greig NH, Alhomida AS, Al-Jafari AA (2000) Biochem Pharmacol 60:561
456. Ogane N, Giacobini E, Struble R (1992) Brain Res 589:307
457. Greig NH, Utsuki T, Yu QS, Zhu XX, Holloway HW, Perry T, Lee B, Ingram DK, Lahiri DK (2001) Curr Med Res Opin 17:159
458. Yoshida S, Suzuki N (1993) Eur J Pharmacol 250:117
459. Sitaram N, Weingartner H, Gillin JC (1978) Science 201:272
460. Polinsky RJ (1998) Clin Ther 20:634
461. Agid Y, Dubois B, Anand R, Gharabawi G (1998) Curr Ther Res Clin Exp 59:837
462. Cutler NR, Veroff AE, Anand R, Hartman R, Mancione L (1999) Neurology 52:A173
463. Grossberg G, Desai A (2001) Expert Opin Pharmacother 2:653
464. Kumar V, Sugaya K, Messina J, Veach J (1999) Neurology 52:A395
465. Spencer CM, Noble S (1998) Drugs Ag 13:391
466. Eskander MF, Nagykery NG, Leung EY, Khelghati B, Geula C (2005) Brain Res 1060:144
467. Kamal MA, Al-Jafari AA, Yu QS, Greig NH (2006) Biochim Biophys Acta 1760:200
468. Kamal MA, Klein P, Yu QS, Tweedie D, Li YZ, Holloway HW, Greig NH (2006) J Alzheimers Dis 10:43
469. Braida D, Sala M (2001) CNS Drug Rev 7:369
470. De-Mello N, Souza-Junior IQ, Carobrez AP (2005) Behav Brain Res 158:263
471. Levin ED, Torry D (1996) Psychopharmacology 123:88
472. Prediger RDS, De-Mello N, Takahashi RN (2006) Eur J Pharmacol 531:176
473. Wang H, Lu Y, Chen HZ (2003) Neurosci Lett 344:173
474. Fraunfelder FT, Morgan R (1994) J Am Med Assoc 271:1742
475. Reyes PF, Dwyer BA, Schwartzman RJ, Sacchetti T (1987) J Neurol Neurosurg Psychiat 50:113
476. da Silva AL, Piato ALS, Bardini S, Netto CA, Nunes DS, Elisabetsky E (2004) J Ethnopharmacol 95:199
477. da Silva AL, Piato AL, Ferreira JG, Martins BS, Nunes DS, Elisabetsky E (2007) J Ethnopharmacol 109:449
478. Siqueira IR, Cordova CAS, Creczynski-Pasa TB, Elisabetsky E, Nunes DS, Netto CA (2002) Pharm Biol 40:374
479. Siqueira IR, Fochesatto C, da Silva AL, Nunes DS, Battastini AM, Netto CA, Elisabetsky E (2003) Pharmacol Biochem Behav 75:645
480. Siqueira IR, Cimarosti H, Fochesatto C, Nunes DS, Salbego C, Elisabetsky E, Netto CA (2004) Life Sci 75:1897

Chapter 16
Neuroprotective Herbs for Stroke

Hocheol Kim

Abstract In recent years, many attempts have been made to document research data from extracts of composite formulas, single herbs, or single compounds from traditional Eastern medicine (TEM) herbs, according to orthodox pharmacological actions. Basic and clinical research in TEM constitutes an abundant source of new drug discovery and development with the integration of TEM and Western pharmacology. This article reviews herbs that have been documented to have a neuroprotective effect in *in vitro* and *in vivo* ischemic model systems and the neuroprotective compounds isolated from them. The neuroprotective mechanisms of herbs and single compounds relevant to the treatment of brain ischemia, including antioxidant, antiexcitotoxic, and anti-inflammatory effects were also discussed.

Keywords Cerebral ischemia · Herbs · Neuroprotection · Traditional medicine · Stroke

Abbreviations

4-VO	4-vessel occlusion
BCAo	Bilateral common carotid artery occlusion
CAT	Catalase
CCA	Common carotid artery
CNS	Central nervous system
COX-2	Cyclooxygenase-2
DPPH	2,2-diphenyl-1-picrylhydrazyl,
(-)-EGCG	(-)-epigallocatechin gallate
GABA	Gamma-aminobutyric acid
GCRP	Calcitonin gene-related peptide
GPX	Glutathione peroxidase

H. Kim (✉)
Department of Herbal Pharmacology, College of Oriental Medicine, Kyung Hee University, Seoul 130-701, Korea
e-mail: hckim@khu.ac.kr

K.G. Ramawat (ed.), *Herbal Drugs: Ethnomedicine to Modern Medicine*,
DOI 10.1007/978-3-540-79116-4_16,

GSH-Px	Glutathione peroxidase
GSH	Glutathione
IFN-γ	Interferon γ
IL-1β	Interleukin β
iNOS	Inducible nitric oxcide synthase
KA	Kainic acid
LPS	Lipopolysaccharide
MDA	Malondialdehyde
NMDA	N-methyl-D-aspartate
NO	Nitric oxide
NR2B	NMDA receptor 2B
OGD	Deprivation of oxygen and glucose
PGE_2	Prostaglandin E_2
PGI_2	Prostaglandin I_2
PTK	Protein tyrosine kinase
rt-PA	Recombinant tissue-plasminogen activator
SD	Sprague–Dawley
SOD	Superoxide dismutase
TBARS	Thiobarbiturate reactive substances
TCM	Traditional Chinese medicine
TEM	Traditional Eastern medicine
tMCAo	Transient middle carotid artery occlusion
TNF-α	Tumor necrosis factor α
TXA_2	Tromboxane A_2
VCAM-1	Vascular cell adhesion molecule-1
VEGF	Vascular endothelial growth factor
VIP	Vasoactive intestinal peptide

16.1 Introduction

Stroke is the third leading cause of death in industrialized countries, followed by cancer and heart attack. It is a major cause of permanent disability for which there is currently no effective treatment. Most strokes (80%) are ischemic and the majority of ischemic strokes result from occlusion of a major cerebral artery by thrombus or embolism, which leads to loss of blood flow in a specific region. The remaining strokes are hemorrhagic, where a blood vessel bursts either in the brain or on its surface [1].

Two major approaches have been developed in ischemic stroke. The first is to establish reperfusion by dissolution of the clot using thrombolytic drugs. At present, rt-PA is the only thrombolytic drug approved for the treatment of acute ischemic stroke; rt-PA administration is restricted to within 3 h of stroke and its use increases the risk of hemorrhagic transformation [2].

The second approach is to develop neuroprotective agents that interfere with the biochemical cascade of events that leads to cell death in the penumbra area that surrounds the core. This protection would attenuate lots of the clinical problems of stroke, including motor disability and spatial hemiplegia. However, although more than 37 potential neuroprotective agents have been studied in more than 114 clinical trials [3], none of them is clinically efficacious and in use in the Western world [4].

In Far Eastern countries such as Korea, China, and Japan, stroke has been treated by traditional Eastern medicine (TEM) for thousands of years. TEM is also known as traditional Chinese medicine (TCM), traditional Korean medicine, Sino-Japanese medicine, oriental medicine, traditional herbal medicine, and traditional Asian medicine. In China, traditional medicine is prevalent; approximately one third of patients are treated with traditional medicine [5], and in Korea, 25% of stroke patients also visit traditional medicine doctors [6].

Extensive experience and abundant clinical data on stroke treatment have been accumulated on TEM. Basic and clinical research in TEM constitutes a potentially rich source of drug discovery and development with the integration of TEM and western pharmacology. In recent years, many attempts have been made to document research data about extracts of composite formulas, single herbs, or single compounds from TEM herbs according to orthodox pharmacological actions. Groups of TEM herbs have been identified as potential sources for compounds with predominant effects on the circulation, thrombogenesis, inflammatory processes, and neuroprotection.

This chapter reviews herbs and prescriptions that have been screened for neuroprotective effects in *in vitro* and *in vivo* ischemic model systems and the neuroprotective compounds isolated from them. Neuroprotective mechanisms of prescriptions, herbs, and single compounds relevant to the treatment of brain ischemia, including antioxidant, antiexcitotoxic, and anti-inflammatory effects are also discussed.

16.2 Stroke Therapy in Traditional Medicine

Stroke is the first of the four major serious syndromes and the most acute disease in TEM. Stroke in TEM is called 'wind stroke' because it happens abruptly like the wind. The concept of stroke in TEM is quite different in many ways from that held by western medicine. The syndrome is characterized by the sudden appearance of hemiplegia, deviated eyes and mouth, and impeded speech that may or may not start with sudden loss of consciousness.

The theoretical systems of TEM are based on the doctrines of yin and yang, the five elements, viscera, and meridian systems. Gong and Sucher nicely reviewed the basic principles and classification of wind stroke in TEM [7]. Generally, the disease state is considered mainly as a destruction of the harmonious components of yin and yang. Wind stroke is considered to be caused either by weak internal strength (so-called '*qi*') invaded by strong external 'bad wind' or by excessive internal 'fire', such as anger, fatigue, heavy drinking, or dietary problems. Both can violate the harmonious negative-positive balance of the self, which eventually leads to stroke.

The treatment of wind stroke in TEM is aimed at creating equilibrium between the relative strength of the patient's body resistance and the intensity of endogenous and exogenous pathogenic factors. Because wind stroke in TEM is caused by hyperactivity of liver yang, obstruction of the heart orifices by phlegm, excessive heat, or blood stasis. The treatments in TEM include heat-clearing drugs, antirheumatics, drugs for dispersing exterior wind, drugs for promoting blood circulation, drugs for relieving phlegm, drugs for subduing interior wind, drugs for resuscitation, or tonics for deficiency syndromes, in accordance with the cause. Recently, these drugs have been demonstrated to have antioxidant, anti-inflammatory, and antiglutamate effects. Usually, drugs for clearing away heat, inducing resuscitation, expelling wind, activating blood and removing stasis are used in the early acute stage, and drugs for invigorating and treating deficiencies are used in the later stage.

In the book *Tongeuibokam*, one of the famous classics in TEM in Korea, 123 different prescriptions for stroke are recorded in the wind stroke section [8]. The herbs used in these prescriptions include, Saposhnikoviae Radix, Ligustichi Radix, Ginseng Radix, Angelicae Sinensis Radix, Paeoniae Radix, Arisaematis Rhizoma, Atylactylodis Rhizoma, Notopterygii Rhizoma seu Radix, Ephedrae Herba, and Scutellariae Radix, in decreasing frequency. Herbs listed in Table 16.1 are used to treat neurological symptoms of strokes.

16.3 Neuroprotective Herbs for Stroke

Cerebral ischemic injury is the result of an obstruction of blood flow in a major cerebral vessel, which will lead to a core of severely ischemic brain tissue that may not be salvaged. However, the ultimate size of the brain infarct also depends on the penumbra, a zone of tissue around the core of the infarct where blood flow is maintained above a neuronal disabling level or the critical 20 to 25% of normal blood flow. Decreased blood flow leads to severe impairment of cellular function by disruption of ATP-dependent processes [9].

Ischemia and subsequent reperfusion provide circumstances that produce oxygen radical production. Several studies have suggested a relationship between cerebral ischemia and oxidative stress in humans [10, 11]. Therefore, antioxidants have been evaluated as neuroprotective agents in stroke [12]. Many herbs, especially those that contain flavonoids, are suggested to have antioxidant effects and possibly might be protective against brain injury caused by ischemia and reperfusion.

Many herbs have also been shown to have anti-inflammatory properties, and thus there is potential for novel anti-inflammatory agents to be identified from plant sources. For example, numerous flavonoid compounds have been associated with anti-inflammatory activity and may have the potential for use in the management of inflammatory disorders [13].

Glutamate toxicity and glucose deprivation is one mechanism of neuronal injury following ischemia. For a thousand years in TEM, some herbs have been used as tranquilizers for their central nervous system (CNS) inhibitory effects. Some of

Table 16.1 Herbs used in stroke therapy in TEM with their main component compounds

Medicine	Scientific name	Chinese	Main components
Drugs for clearing away heat			
Scutellariae Radix	*Scutellaria baicalensis*	黃芩	Baicalein, biacalin, wogonin, wogonoside
Coptidis Rhizoma	*Coptis japonica* *C. chinensis*	黃連	Berberine, coptisine, worenine, palmatine, jatrorrhizine
Gardeniae Fructus	*Gardenia jasminoides*	梔子	Gardenoside, geniposide, shanzhiside, gardoside
Rhei Radix et Rhizoma	*Rheum palmatum*	大黃	Rhein, aloe-emodin, sennoside A,B,C,D,E,F
Bovi Calculus	*Bos taurus*	牛黃	Cholic acid, deoxycholic acid, bilirubin, taurine
Antirheumatics			
Saposhnikoviae Radix	*Saposhnikovia divaricata*	防風	Bisabolene, imperatorin, hellopterin, ledebourielol
Notopterygii Rhizoma seu Radix	*Notopterygium incisum*	羌活	α-thujene, α-pinene, β-pinene, β-ocimine,
Angelicae pubescens Radix	*Angelica pubescens*	獨活	Othol, columbiamotin, xanthotoxin, angelol
Gentianae Macrophyllae Radix	*Gentiana macrophylla*	秦艽	Gentianine, gentianidine, gentiopieroside
Aconiti Radix	*Aconitum carmichaeli*	附子	Aconitine, hypaconitine, mesaconitine
Stephaniae Tetrandrae Radix	*Sinomenium acutum* *Stephania tetrandra*	防己	Hanfangchin A, fangchinoline
Drugs for promoting blood circulation			
Ligustichi Radix	*Ligusticum chuanxiong*	川芎	Chuanxiongzine, ligustilide
Salviae Miltorrhizae Radix	*Salvia miltiorrhiza*	丹蔘	Tanshinone I,IIA, cryptotanshinone, dihydrotanshinone
Carthami Flos	*Carthamus tinctorius*	紅花	Carthamin, saffloryellow A, B, safflomin A

Table 16.2 (continued)

Medicine	Scientific name	Chinese	Main components
Tonics for deficiency syndromes			
Gingseng Radix	*Panax ginseng*	人蔘	Ginsenoside-Rb1, Rc, Rg1, Re
Aconthopanacis Senticosi Radix	*Acanthopanax senticosus*	刺五加	Eleutheroside A-E
Astragali Radix	*Astragalus mumbranaceus*	黃芪	Astragaloside I, II,IV, soyasaponin I
Atylactylodis Rhizoma	*Atractylodes japonica*	白朮	Atractylol, atractylon, junipercamphor, atractylolide
Poria	*Poria cocos*	茯苓	β-pachyman, pachymic acid
Schizandriae Fructus	*Schizandra chinensis*	五味子	Schizandrin, deoxyschizandrin, schizandrol
Dioscoreae Rhizoma	*Dioscorea batatas*	山藥	Choline, allantoin, batasine, abscine
	D. japonica		
Rehmanniae Radix	*Rehmannia glutinosa*	地黃	Catalpol, digydrocatalpol, remannioside
Angelicae Sinensis Radix	*Angelicae gigas*	當歸	Ligustilide, N-butylidenece phthalide,
	A. sinensis		
Paeoniae Radix	*Paeoniae lactiflora, P. veitchii*	白芍藥	Paeoniflorine, paeonol
Drugs for releaving phlegm			
Pinelliae Tuber	*Pinellia termata*	半夏	Choline, l-ephedrine, β-sitosterol
Arisaematis Rhizoma	*Arisaema amrnese*	南星	Saponin, alkaloids, benzoic acid, d-mannitol
Platycody Radix	*Platycodon grandiflorum*	桔梗	Platycodoside, platycodin D
Pruni Mandshuricae Semen	*Prunus mandshurica*	杏仁	Amygdalin, prunase
Drugs for dispersing exterior wind			
Ephedrae Herba	*Ephedra sinica,*	麻黃	Ephedrine, d-pseudoephedrine, l-N-methylephedrine
Puerariae Radix	*Pueraria thunbergiana*	葛根	Daidzin, daidzein, puerarin
Angelicae Dahuricae Radix	*Angelica dahurica*	白芷	l-tetradecene, lauric acid, agidol,
Mentae Herba	*Mentha arvensis*	薄荷	Menthol, menthacamphor
Cinnamomi Cortex	*Cinnamomum cassia*	肉桂	Cinnamon oil, cinnamic aldehyde, cinnamic acid
Schizonepetae Herba	*Schizonepeta tenuifolia*	荊芥	Pulegone, d-menthone, isomenthone, d-limonene
Asari Radix	*Asarum sieboldii*	細辛	α-pinene, β-pinene, safrole, asarone
Zingiberis Rhizoma	*Zingiberis officinale*	生薑	Zingiberol, borneol, 6-gingerol, 6-shogaol
Viticis Fructus	*Vitex rotundifolia*	蔓荊子	Rotundifuron, prerotundifuran, vitexilactone

Table 16.3 (continued)

Medicine	Scientific name	Chinese	Main components
Drugs for subduing interior wind			
Gastrodiae Rhizoma	*Gastrodia elata*	天麻	Gastrodin, gastrodioside, vanillyl alcohol
Uncariae Ramulus cum Uncis	*Uncaria.rhynchophylla*	釣鉤藤	Rhynchophylline, isorhynchophylline
Bombyx Corpus	*Bombyx mori*	白僵蠶	Ammonium sxalate, pyridine-2,6-dicarboxyl acid
Scorpion	*Buthus martensi*	全蝎	Buthatoxin, betaine
Drugs for promoting *qi* circulation			
Citri Percarpium	*Citrus unshiu, C. reticulate*	陳皮	Limonene, α,β-pinene, hesperidin
Aurantii Fructus	*Citus aurantium*	枳殼	Hesperidin, neohesperidin, naringin
Saussurae Radix	*Saussurea lappa*	木香	Aplotaxene, α,β-costene
Linderae Radix	*Lindera strychnifolia*	烏藥	Lindenenol, lindenene, lineroxide
Magnoliae Cortex	*Magnolia obovata, M. officinalis*	厚朴	β-eudesmol, magnolol, honokiol
Drugs for resuscitation			
Moschus	*Moschus moschiferus*	麝香	Muscone, muscopyridine
Acori Graminei Rhizoma	*Acorus gramineus*	石菖蒲	β-asarone, α-asarone, caryophyllene, α-humulene

these herbs are known to have neuroprotective effects by the antagonism of excitatory amino acids, particularly glutamate, which is increased in the early postcerebral ischemia period and activates NMDA receptors.

Individual herbs are comprised of many compounds, and it is therefore difficult to investigate the exact neuroprotective mechanism of many herbs, even though some show high efficacy in *in vitro* and *in vivo* ischemia models. The effective mechanisms of herbs may include antioxidant, anti-inflammatory, and antiglutamate actions.

16.4 Single Herb Extracts and Their Active Compounds

More than 100 herbs have been used for stroke prevention and therapy in TEM. Described below are single herbs and their individual active compounds that have been demonstrated to have neuroprotective effects *in vitro* and *in vivo*.

16.4.1 Panax ginseng

The root of *Panax ginseng* C. A. Meyer (Araliaceae), usually 4 to 6 years after cultivation, has been used as a general tonic or adaptogen and is frequently featured in TEM prescriptions. Many studies have validated the empirical usage of ginseng over thousands of years for *qi*-invigorating and cardiovascular effects in TEM. The main active components of *P. ginseng* are ginsenosides, which have been shown to have a variety of beneficial effects, including anti-inflammatory, antioxidant, and anticancer effects [14]. *P. notoginseng*, from the same genus as *P. ginseng*, and its saponins exert a protective effect on ischemic brain damage *in vivo* [15, 16] and focal cerebral ischemia [17, 18].

Ginseng has neuroprotective effects in transient focal or global cerebral ischemia [19–21]. Red ginseng powder, which is steamed ginseng under pressure, also prevents delayed neuronal death in gerbils. Ginsenoside Rb1 (Fig. 16.1) protects the brain from reversible focal brain ischemia in rats [22] and protects hippocampal CA1 neurons by scavenging free radicals [23]. Ginsenoside Rb1 stimulates the expression of the mitochondrion-associated antiapoptotic factor Bcl-x(L) in vitro and *in vivo* [24]. Ginsenoside Rg1 increases ischemia-induced cell proliferation and survival in the dentate gyrus of adult gerbils [25].

With reference to the studies conducted, the neuroprotective effects of ginseng on brain ischemia can be explained by multiple mechanisms, including scavenging free radicals and inhibiting the CNS. The compounds responsible for these neuroprotective effects require further investigation, but ginsenoside Rb_1 could be one of the main neuroprotective compounds within the ginseng root.

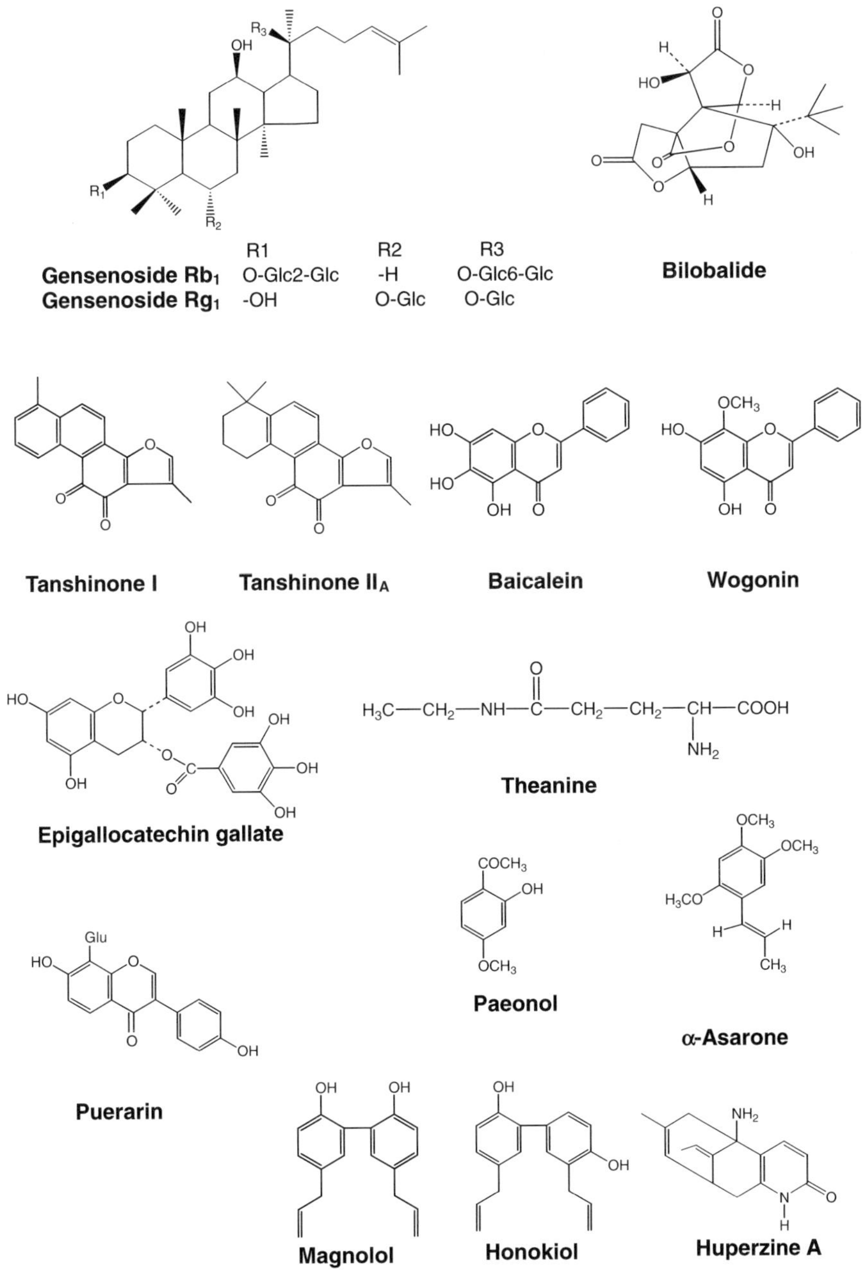

Fig. 16.1 Chemical structures of active compounds of neuroprotective herbs

16.4.2 *Salvia miltiorrhiza*

The root of *Salvia miltiorrhiza* Bunge (Labiatae) is red in color and was therefore used in TEM to treat related disorders of blood stasis with an action of quickening the blood and dispelling stasis. *S. miltiorrhiza* and its active ingredients, tanshinones and salvianolic acids, have anticoagulant, vasodilatory, anti-inflammatory, free-radical scavenging, mitochondrial protective, and other activities. Experimental studies have shown that *S. miltiorrhiza* dilated coronary arteries and scavenged free radicals in ischemic diseases. Clinical trials also indicated that *S. miltiorrhiza* was an effective medicine for angina pectoris, myocardial infarction, and stroke [26].

Several studies have investigated possible mechanisms for the protective effect of *S. miltiorrhiza* against cerebral ischemia. *S. miltiorrhiza* attenuated dysfunction of VIP [27] and modified ischemic cell changes by modulating somatostatin [28] in cerebral ischemia. *S. miltiorrhiza* also decreased the size of the infarcted area after CCA ligation in gerbils by inhibiting presynaptic glutamate release and stimulating GABA release [29]. Inhibition of nitric oxide (NO) formation could explain the CNS protective effects observed with *S. miltiorrhiza* [30].

S. miltiorrhiza may offer an additional therapeutic approach to the management of stroke and ischemia. *S. miltiorrhiza* has been shown to offer protection against brain ischemia by reducing lipid peroxidation [31]. Pretreatment with S. *miltiorrhiza* reduced the infarct size in tMCAo-injured SD rats [32]. Tanshinones (Fig. 16.1) are the major lipid-soluble pharmacological constituents of *S. miltiorrhiza*. Brain infarct volume was reduced following treatment with tanshinone II_A and tanshinone II_B in MCAo mice [33, 34].

The therapeutic effect of *S. miltiorrhiza* may be partly due to its free-radical-scavenging activities, Tanshinones or other structurally related compounds may have potential for further development as neuroprotective drugs. However, systematic review on randomized control trials comparing *S. miltiorrhiza* with other medicines does not support the notion that *S. miltiorrhiza* may be beneficial to disability improvement after acute ischemic stroke [35].

16.4.3 *Ginkgo biloba*

The leaves of *Ginkgo biloba* L. (Ginkgoaceae) have been used in TEM for respiratory and circulatory disorders, and its extract has been therapeutically used for several decades to increase peripheral and cerebral blood flow as well as for the treatment of dementia. *G. biloba* extract EGb761, a standardized extract of *G. biloba* leaves, contains about 24% flavonoid glycosides, primarily quercetin, kaempferol, isorhamnetin, 6% terpene lactones, 2.8 to 3.4% ginkgolides A, B, and C, and 2.6 to 3.2% bilobalide. EGb761 has shown favorable effects on cerebral circulation and neuronal cell metabolism [36, 37] and also exhibited antioxidant activity [38, 40].

EGb761 is neuroprotective against amyloid- and NO-induced toxicity in vitro. *G. biloba* extracts attenuate scopolamine-induced amnesia in rats [41], enhance memory retention in young and old rats [42], and improve short-term memory in mice [43]. *G. biloba* attenuates delayed neuronal death in the CA1 of the hippocampus in Mongolian gerbils [44] and is also associated with reduced stroke infarct volume in mice subjected to 45 min of tMCAo [45].

Bilobalide (Fig. 16.1), a sesquiterpene trilactone constituent of *G. biloba*, reduces cerebral edema produced by triethyltin through preventing the uncoupling of oxidative phosphorylation. Pretreatment with bilobalide reduced the cerebral infarct area in MCAo mice [46]. Bilobalide protected the slices against hypoxia-induced phospholipid breakdown [47]. Bilobalide inhibited NMDA-induced phospholipase A_2 activation and phospholipid breakdown in rat hippocampal slices [48]. Ginkgolides also had protective effects on focal cerebral ischemia, and its mechanism may be relative to its inhibition of platelet-dependent thrombosis and amelioration of hemarheological partments [49].

Cumulative evidence indicates that *G. biloba* may have neuroprotective effects in brain ischemia rodent models, and bilobalide may be one of the main compounds responsible for this effect. There is no convincing evidence from trials of sufficient methodological quality to support the routine use of *G. biloba* extract to promote recovery after stroke [50]. High-quality and large-scale randomized controlled trials are needed to test its efficacy.

16.4.4 Acanthopanax senticosus

The root and stem bark of *Acanthopanax senticosus* Harms (Araliaceae), also called Siberian ginseng, have been used as a tonic and adaptogen to strengthen *qi* in TEM and to treat rheumatic arthritis and stress-induced disease [51]. *A. senticosus* includes eleutheroside, chiisanoside, senticoside, triterpenic saponin, syringin, and flavones in its compounds [52], and it has sedative, antioxidant, antihistamine, hypolipidemic, antistress, and immunomodulatory effects [14].

A. senticosus has a protective effect on CCl_4-induced liver toxicity via its antioxidative effect [53, 54]. The saponin isolated from the leaves of *A. senticosus* reduced myocardial infarct size in acutely ischemic dogs [54, 55]. The liriodendrin, syringaresinol, the hydrolysate of liriodendrin, inhibited the LPS-induced production of NO, PGE_2, and TNF-α production by macrophages and decreased expression levels of iNOS and COX-2 enzymes [56]. My group investigated whether water extracts of *A. senticosus* (100 mg/kg, i.p.) reduced infarct volume in tMCAo of SD rats. In this model, *A. senticosus* inhibited both COX-2 and OX-42 expression in the penumbral region at 24 h after MCAo [57]. These results suggest that *A. senticosus* has a neuroprotective effect via inhibition of inflammation, microglial activation, and antioxidative stress in brain ischemia.

16.4.5 *Scutellaria baicalensis*

The roots of *Scutellaria baicalensis* Georgi (Labiatae) have been used in TEM to treat inflammatory and cardiovascular disease. *S. baicalensis* contains three major polyphenolic components, namely wogonin, baicalin, and baicalein (Fig. 16.1) These three polyphenols are as free-radical scavengers of hydroxyl, DPPH, and alkyl radicals [58]. Baicalein is the most effective compound of the three polyphenols tested in preventing glutamate toxicity [59] and is also known as a selective inhibitor of 12-lipoxygenase [60]. Baicalin also acts as a neuroprotectant during cerebral ischemia [61].

S. baicalensis and baicalein reduce ischemia-reperfusion brain injury and neutrophil infiltration in MCAo rats [62, 63]. *S. baicalensis* also protects CA1 neurons against transient 4-VO in rats and inhibits microglial TNF-α and NO production *in vitro* [64]. Wogonin, 5,7-dihydroxy-8-methoxyflavone, inhibits ischemic brain injury and improves behavioral dysfunction caused by permanent MCAo [65]. Wogonin has anti-inflammatory activities in various cell types and inhibits NO production by suppressing iNOS induction and NF-kB activation in microglia [66].

16.4.6 *Camellia sinensis* (Green tea)

Green tea contains antioxidant polyphenols such as catechins and flavonols. Most of the experimental and epidemiological studies concerning green tea effects have been targeted at its possible cardiovascular, anti-inflammatory, and anticarcinogenic effects, which have been linked to the antioxidant/pro-oxidant properties of its polyphenol constituents [67, 69]. Daily ingestion of tea as an antioxidant has also been reported to prevent stroke. Green tea extract orally administered to Wistar rats for 3 weeks before induction of ischemia by occlusion of middle cerebral arteries and reperfusion minimized the eicosanoid accumulation and oxidative damage in addition to the reduction of neuronal cell death [70]. Green tea extract prevented cerebral ischemia damage caused by global ischemia-reperfusion in Mongolian gerbils [71]. (-)-EGCG has potent antioxidant properties in a green tea polyphenol and had a neuroprotective effect against neuronal damage following global ischemia in Mongolian gerbils [72]. Theanine (Fig. 16.1), a flavorous component of green tea, has a neuroprotective effect against neuronal death in transient brain ischemia. The mechanism of the neuroprotective effect of theanine is related not only to the glutamate receptor but also to other mechanisms such as the glutamate transporter [73].

16.4.7 *Pueraria thunbergiana*, *P. lobata*

The roots of *Pueraria thunbergiana* Benth. (Legumnosae) are widely used in TEM for moderating alcohol abuse, for hypotensive, antipyretic, and analgesic effects, and for treatment of the common cold. Puerarin (Fig. 16.1), daidzin, and daidzein are three of the major isoflavonoid compounds isolated from the extract of *P. lobata. P. lobata* flavonoids increase the cerebral blood flow of anesthetized mice [74] and reduce the infarct volume in MCAo by increasing the activities of SOD [75]. Their isoflavonoids have potent inhibitory effects on PGE_2 production [76], and the antioxidant effect, partly dependent on free-radical scavenging, antilipid peroxidation, and enhancement of SOD activity [77].

Isoflavones in plants are known to have an estrogenic action. Puerarin has protective effects on cultured mouse cerebral cortical neurons damaged by Glu, NMDA, or KA [78]. Puerarin also shows neuroprotective effects in MCAo rats [79]. Puerarin clearance in normal rats was much faster than that in cerebral ischemia-reperfusion rats induced by MCAo and BCAo [80]. Genistein, a PTK inhibitor, prevented the increase of p-STAT3 and DNA binding activity in ischemic reperfusion injury at 4-VO [81]. It also prevented gerbil transient ischemia via a decrease in tyrosine phosphorylation of NR2B [82].

16.4.8 Cnidium officinale, Ligusticum chuanxiong

The rhizome of *Cnidium officinale* Makino (umbelliferae) is one of the important traditional medicines used for the treatment of female genital inflammatory diseases. *C. officinale* contains a variety of volatile phthalide derivatives that have been shown to have pharmacological activities including sedative, antianemia, antifungal, smooth muscle relaxing, anti-inflammatory, analgesic, and anticomplement activities. Falcarindiol (1,9-Heptadecadiene-4, 6-diyne-3, 8-diol), which was isolated from *C. officinale*, and the ethyl acetate-soluble fraction of *C. officinale* reduced NO production and suppressed iNOS expression in BV-2 cells and primary microglia cells. The inhibition of excessive NO production played an important role in neuronal cell death in LPS-treated rat hippocampal slice cultures [83, 84].

The rhizome of *Ligusticum chuanxiong* Hort. has also been used in TCM for the same applications as *C. officnale. L. chuanxiong* inhibited platelet activation in bilateral common carotid artery occlusion (BCAo) in rabbits and corrected the TXA_2-PGI_2 imbalance in plasma after cerebral ischemia [85]. *L. chuanxiong* reduced cell damage-formation of peroxidation products after bilateral ligation of the common carotid arteries in rats [86]. Tetramethylpyrazine, a drug originally isolated from the rhizome of *L. chuanxiong*, has been used routinely in China for the treatment of stroke and angina pectoris. Tetramethylpyrazine has therapeutic potential for the treatment of dementia caused by cholinergic dysfunction and/or decrease of cerebral blood flow. Tetramethylpyrazine pretreatment showed a neuroprotective effect on cerebral ischemia in gerbils [87].

16.4.9 *Magnolia officinalis*

The cortex of *Magnolia officinalis* Rehder et Wilson (Magnoliaceae) has been used for the treatment of acute pain, diarrhea, coughs, and urinary problems in TEM. Honokiol (Fig. 16.1) and magnolol (Fig. 16.1) are the main constituents of the bark of this herb and have a variety of pharmacological activities. Honokiol has been demonstrated to be an effective antioxidant [88]. It can protect animal tissues against lipid peroxidation [89, 90] serve as an antiplatelet drug [91], and it displays an anti-inflammatory effect in activated macrophages [92, 93]. Honokiol is a potent neuroprotective agent against focal cerebral ischemic injury by its antioxidant and antiplatelet aggregation effects [94, 96].

Magnolol (5,5′-diallyl-2,2′-dihydroxydiphenyl) inhibits intracellular calcium mobilization in platelets [97], relaxes vascular smooth muscle cells [98], and has antihemostatic, antithrombotic [99], anti-inflammatory, and analgesic effects [100]. A number of other effects of magnolol have also been found, including inhibition of prostaglandin D_2 formation [100], suppression of nonselective vascular hyporeactivity to mediators [101], reduction of the formation of eicosanoid mediators [102], inhibition of neutrophil adherence [103], prevention of ischemic-reperfusion injury [104], and, most importantly, strong antioxidant activity [105]. Magnolol treatment appears to have a marked effect against heatstroke-induced cerebral ischemic insults [106].

16.4.10 *Angelica gigas, A. sinensis*

Angelicae Radix has been used as a sedative or a tonic and to treat disorders of menstruation in women, anemia, and menopause syndrome. *Angelica gigas* Nakai (Umbellaceae) is used in Korea, and *A. sinensis* (Oliv.) Diels is used in China. *A. gigas* includes decursin, decursinol angelate, angelan, and decursinol. *A. gigas* protects mice against Aβ-induced memory impairment [107]. *A. gigas* has antinociceptive effects on pain responses induced by TNF-α, IFN-γ, IL-1β, glutamate, NMDA, or kainic acid [108]. Decursin ameliorates scopolamine-induced memory impairment in mice [109]. Decursinol and decursin protect primary cultured rat cortical cells against glutamate-induced oxidative stress by both reducing calcium influx and acting on the cellular antioxidative defence system [110].

A. sinensis contains ferulic acid, ligustilide, angelicide, brefeldin A, butylidenephthalide, butyphthalide, succinic acid, nicotinic acid, uracil, and adenine. Extracts of *A. sinensis* showed a scavenging effect on peroxide and hydroxyl radicals and inhibited lipid peroxidation of the liver [111]. *A. sinensis* protects the brain from damage induced by transient forebrain ischemia in mice [112]. *A. sinensis* extract also has attenuating effects on amnesia induced by various drugs related to memory processes [113].

16.4.11 *Rhodiola rosea*

The root of *Rhodiola rosea* L. (Crassulaceae) is used as a hemostatic and tonic and for contusions. Rhodiola plants demonstrated antifatigue, antianoxia, and memory-enhancing effects. The major compositions of rhizomes of *R. rosea* are phenols such as salidroside and its aglycon tyrosol, and cinnamic glycosides such as rosin, rosavin, and rosarin. Other important constituents are flavonoids, tannins, gallic acid and its esters and essential oils. Administration of *R. rosea* extract for 10 d yielded protection against impairment in memory, as assessed by step-down passive avoidance, induced by electroshock in rats [114]. Several constituents of *R. sacra* and *R. sachalinensis* showed protective effects against beta-amyloid toxicity, oxidative stress, and apoptosis [115]. Phenolic compounds exhibited significant scavenging effects against DPPH free radicals [116]. *R. sachalinensis* treatment reduced infarct volume and attenuated COX-2 induction and microglial activation after tMCAo in rats [117].

16.4.12 *Paeonia suffruticosa, P. lactiflora*

The root bark of *Paeonia suffruticosa* Andrews (Ranunculaceae) is a drug used in TEM as both an analgesic and an anti-inflammatory agent [118], and it is prescribed in various TEM preparations for the treatment of blood stagnation. It has been reported to have strong superoxide and hydroxyl radical scavenging activity [119]. Its antioxidative effects are due to enhancing activities of SOD, CAT, and GPX [120]. Paeonol (Fig. 16.1) inhibits cerebral ischemic injury by blocking increases in Ca^{2+}, decreases in SOD activity and the content of MDA, and improved Ca^{2+}-ATPase activity in ischemic brain tissue [121, 122]. α-benzoyloxypaeoniflorin, an antioxidant monoterpene glycoside, from *P. suffruticosa* has potent radical-scavenging activity on the DPPH radical [123].

The root of *P. lactiflora* Palla has long been used to treat abdominal pain and blood deficiencies in TEM. The oral administration of *P. lactiflora* extract and paeoniflorin, a major constituent of peony root, attenuated spatial cognitive deficits caused by scopolamine in rats [124]. Paeoniflorin reduces the infarct volume as well as ameliorates the deficits in neurological symptoms caused by tMCAo in rats [125]. Paeoniflorin ameliorates memory disruption mediated by adenosine A1 receptor in rodents [126].

16.4.13 *Bombycis Corpus*

Bombycis Corpus (BC) is a *Bombyx mori* larva (silk moth larva, Bombycidae) killed by infecting with the fungus *B. bassiana*. It has been used in TEM to treat palsy, headache, convulsion, and speech problems induced by stroke and tremor.

Several sterols have been isolated from BC. BC has a protective effect against A β-induced cytotoxicity in cultured astrocyte cells through the inhibition of lipid peroxidation and protection of antioxidative enzymes such as catalase, SOD, GSH-Px, and glutathione-S transferase [127]. Sphingolipids from BC also have neurotrophic effects as shown by examining PC12 cell neurite outgrowth [128]. Pretreatment with BC protected primary hippocampal cultures from embryonic day 18 embryos against NMDA-induced toxicity [129]. BC contributes to protect human brain by inhibiting the release of glutamate.

16.4.14 Corydalis yanhusuo

The tubers of *Corydalis yanhusuo* W. T. Wang (Papaveraceae) are used in TEM mainly as an analgesic in the treatment of gastric and duodenal ulcer, rheumatism, and dysmenorrhea. *C. yanhusuo* is one of the medicinally important species of *Corydalis*. The tubers are a source of pharmacologically important alkaloids having analgesic [130], antithrombotic [131], antihypertensive [132], and anti-inflammatory effects [133]. Protopine, a component of *C. yanhusuo*, has an inhibitory activity on platelet aggregation [134], and DL-tetra-hydropalmatine has a neuroprotective effect in heatstroke-affected rats [135]. It also inhibits calcium anion entry into cells to prevent neuronal death in ischemia-reperfusion rats [136]. It reduced cerebral infarct lesion in focal ischemia-reperfusion injured rats [137].

16.4.15 Acorus gramineus

The rhizomes of *Acorus gramineus* Solander (Araceae) have been used for the improvement of learning and memory and is often included in the TEM prescriptions for stroke [138, 139]. Water extract or volatile oil from *A. gramineus* induced sedation, decreased spontaneous activity, potentiated pentobarbital-induced sleeping time, and antagonized pentylenetetrazole-induced convulsion in mice [140, 141]. The interactions of *A. gramineus* with the central dopamine (D1 and D2) receptors and the GABA binding site of $GABA_A$ receptors were thought to mediate these central inhibitory actions. The methanol extract and the essential oil from *A. gramineus* inhibited excitotoxic neuronal cell death in primary cultured rat cortical cells [142, 143]. One active principle was identified as asarone (Fig. 16.1), a major essential-oil component in the rhizomes of *A. gramineus* [144].

16.4.16 *Coptis japonica*

The rhizomes of *Coptis japonica* Makino (Ranunculaceae), or *C. chinensis* Franch., have long been prescribed in TEM for the treatment of inflammation-related diseases such as gastrointestinal disorders and infectious or inflammatory diseases. *C. japonica* contains mainly alkaloids, including berberine, magnofiorine, sanguinarine, and phenolic compounds [145]. It has been considered to have antiphlogistic, sedative, antidotal, hemostatic, and antitumor properties.

Berberine has been reported to exhibit several types of biological activities, and interest has been focused on its antioxidative potential [146, 147]. *C. japonica* extract and its active alkaloids were effective in an *in vivo* LPS plus ischemia-reperfusion model that generated $ONOO^-$ [148]. *C. chinensis* administered orally for 1 week improved scopolamine-induced learning and memory deficit in rats [149]. Although the active components of *C. japonica* that exert these bioactivities have not been fully elucidated, it has generally been considered that its alkaloids, such as berberine, palmatine, and coptisine, contribute to these activities.

16.4.17 *Phellodendri cortex*

The stem bark of *Phellodendron amurense* Ruprecht (Rutaceae) has long been used in TEM for the treatment of inflammation and fever as a traditional herb medicine having anti-inflammatory, immunostimulatory, and antitumor activities. It contains some alkaloid components such as berberine and worenine. The antioxidant, anti-inflammatory, and antityrosinase activities of *P. amurense* have been studied [150]. Phellodendri can scavenge superoxide radical (O2-·) generated through the hypoxanthine-oxidase system and hydroxyl radical OH generated through the Fenton reaction and can inhibit lipid peroxidation induced by the hydroxyl radical generation system [151].

16.4.18 *Huperzia serrata*

Huperzia serrata, a source of huperzine A (Fig. 16.1), has been used for centuries in TEM to treat fever, inflammation, blood disorders, and schizophrenia. Huperzine A acts as a potent, highly specific and reversible inhibitor of acetylcholinesterase that crosses the blood–brain barrier. Its potency of acetylcholinesterase inhibition is similar or superior to that of physostigmine, galanthamine, donepezil, and tacrine [152, 153].

Huperzine A protected PC12 cells against OGD-induced toxicity, most likely by alleviating disturbances of oxidative and energy metabolism [154]. Huperzine A treatment is protective against both brain injury and spatial memory impairment in a hypoxic ischemic brain injury of a neonatal rat model [155, 156]. Huperzine A protects against diverse neurodegenerative states observed during ischemia or

Alzheimer's disease by blocking NMDA ion channels [157]. Subchronic oral administration of huperzine A after global ischemia in gerbils significantly reduced memory impairment, reduced neuronal degeneration in the CA1 region, and partially restored hippocampal choline acetyltransferase activity [158].

16.4.19 Menispermum dauricum

The roots of *Menispermum dauricum* DC. (Menispermaceae) are used for treating sore throats, colitis, dysentery, and rheumatic arthralgia in TEM. This herb contains some alkaloids belonging to various classes such as bisbenzylisoquinoline, aporphine, proporphine, protoberberine, and oxoisoaporphine.

Phenolic alkaloids from *M. dauricum* could attenuate injury caused by left anterior descending coronary artery and BCAo in rabbit by lipid peroxidation and enhance the activity of SOD [159]. Daurisoline, which has been isolated from *M. dauricum*, crosses the blood–brain barrier and will, therefore, facilitate the functional characterization of brain calcium channels in granule cells freshly isolated from rat cerebellum as well as the exploration of P-type calcium channels as possible drug targets [160].

16.4.20 Gardenia jasminoides

The fruit of *Gardenia jasminoides* Ellis (Rubiaceae) has been used in TEM for the treatment of inflammation, jaundice, headache, edema, fever, hepatic disorders, and hypertension. Its pharmacological actions, such as protective activity against oxidative damage, cytotoxic effects, anti-inflammatory effect and fibrolytic activities, have already been demonstrated [161, 162]. Crocetin, a major component of gardenia fruits, was found to be a potent inhibitor of tumor promotion via antioxidant activity [163]. Another active component, crocin, exhibited a variety of pharmacological effects in mice, including inhibition of skin tumor growth, improvement of learning behavior previously impaired by ethanol [164], and prevention of long-term potentiation caused by ethanol in rats [165]. It could be useful as a treatment for neurodegenerative disorders accompanied by memory impairment. Crocin also combats the serum/glucose deprivation-induced ceramide formation in PC12 cells caused by increasing GSH levels and prevents the activation of the JNK pathway, which is reported to have a role in the signaling cascade downstream of ceramide for neuronal cell death [166].

16.4.21 Uncaria rhynchophylla

The branch of *Uncaria rhychophylla* (Miq.) Jacks. (Rubiaceae) has been used in TEM for relief of dizziness and treatment of tremors and convulsions [167]. It has antispasmodic effects on smooth muscle and lowers blood pressure [168, 169]. The anticonvulsive effects of this herb have been experimentally demonstrated in KA-treated rats [170]. It inhibited an increase of lipid peroxide levels evoked by ferric-chloride-induced epileptic seizures in rats [171]. Recent studies demonstrated that the extract of *U. rhynchophylla* has a neuroprotective effect on global cerebral ischemia-induced neuronal damage in rats by reduction of COX-2 mRNA and protein level *in vivo* [172]. Rhynchophylline, a major tetraacyclic oxyindole alkaloid isolated from Uncaria species, is known to have a protective effect against glutamate-induced neuronal cell death [173]. Both the extract of *U. rhynchophylla* and rhychophylline ameliorated transient cerebral ischemia-induced spatial memory deficit in mice [174].

16.4.22 Schisandra chinensis

The fruit of *Schisandra chinensis* Baillon (Schisandraceae) is used in TEM to improve liver and kidney function as an antitussive, tonic, and sedative agent. *S. chinensis* inhibited TBARS formation *in vivo* [175]. The active components of *S. chinensis* are schisandrin derivatives and lignans schisandrol A, B and schisandrin A, B, C, protect primary cultures of rat cortical cells from glutamate-induced toxicity [176]. Pretreatment with schisandrin B also protected against cerebral toxicity induced by tetrabutylhydroperoxide [177].

16.4.23 Carthamus tinctorius

Carthamus tinctorius, the flower of the safflower plant, has been used extensively in TEM for its purported ability to improve cerebral blood flow. *C. tinctorius* protected against excitotoxicity of glutamate, NMDA, kainate, quisqualate and against neuronal degeneration caused by simulated ischemia [178]. *C. tinctorius* exerted significant neuroprotective effects on rats with focal cerebral ischemic injury [179]. *C. tinctorius* reduced cell damage and the formation of peroxidation products after bilateral ligation of the common carotid arteries in rats [180].

16.4.24 Other Herbs

The root of *Dioscorea batatas* Decne or *D. japonica* Thunb, called Yam, has been used in TEM for the treatment of diarrhea, cough, dyspnea, leucorrhagia, frequent urination, and diabetes. It is composed mainly of starch with small amounts of mucilage, dioscin, and dopamine [14]. *Dioscoreae Rhizoma* was reported to have antisenility [181] and antioxidant activities [182]. My group found that water extracts of *D. japonica* protected against hippocampal cell death in 4-VO of Wistar rats.

Curcumin, an active constituent of the rhizome *Curcuma longa*, was demonstrated to have antioxidant potential in many *in vitro* and *in vivo* studies. Curcumin has a neuroprotective effect in tMCAo [183] and bilateral common carotid artery occlusion [184] that is mediated through its antioxidant activity.

Withania somnifera, referred to as *Aswagandha* in the Indian system of medicine, is a central nervous system active herb that has been used for various neurological disorders. Studies with *W. somnifera* have indicated that it exerts an antiaging effect anxiolytic and antidepressant activity [185]. The other pharmacological actions exerted by *W. somnifera* include anti-inflammatory, antistress, hemopoietic immunomodulatory, and antioxidant effects [186–188]. *W. somnifera* decreased MDA levels and hemispheric lesion area in focal cerebral ischemia induced by MCAo [189].

Spiramine T is an atisine-type diterpene alkaloid isolated from the *Spiraea japonica* var. acuta (Rosaceae). It was shown to have neuroprotective effects on cerebral ischemia-reperfusion injury produced by bilateral occlusion of the common carotid arteries in gerbils, and its mechanism might be related to reducing calcium accumulation and lipid peroxidation [190]. Spiramine T reduced the content of lipid peroxide, increased glutathione peroxidase activity, and inhibited the increase of nitric oxidase activity and NO production in the cortex during global forebrain ischemia-reperfusion in gerbils [191].

Gypenosides, saponins isolated from *Gynostemma pentaphyllum*, are widely used as they are thought to have a wide range of health benefits, including inhibition of inflammation and prevention of cardiovascular disease, due to antioxidant and lipid-lowering properties [192, 193]. Gypenosides decreased injury of DNA and RNA in rat neurons in the 4-VO cerebral ischemia-reperfusion model [194]. Gypenosides suppressed NO synthesis in murine macrophages by inhibiting iNOS enzymatic activity and attenuating NF-κB-mediated iNOS protein expression, thereby invoking a mechanism by which gypenosides may exert their therapeutic effects [195].

16.5 Multiherb Extracts

In TEM, medicines are used in the form of prescriptions to treat stroke. The prescriptions are usually composed of 4 to 15 kinds of herbs, which have main or accessory roles. More than 100 prescriptions have been generally used to treat stroke in TEM. Complex prescriptions that have been demonstrated to have neuroprotective effects *in vitro* and *in vivo* are described below.

16.5.1 Huanglianjiedu Decoction

Huanglianjiedu decoction (HLD), which consists of four herbs, Coptidis Rhizoma, Scutellariae Radix, Phellodendri Cortex, and Gardeniae Fructus, has been used in TEM as a therapy for various clinical symptoms associated with gastrointestinal disorders, inflammation, and cardiovascular diseases. These four herbs have been the subject of many investigations on the neuroprotective effects of HLD.

Chronic oral pre- and post-HLD prevented cholinergic synaptic dysfunction and serotonergic presynaptic hyperactivity induced by transient ischemia [196]. Pretreatment with oral administration of HLD protected against the impairment of learning and memory induced by transient cerebral ischemia and prevented a decrease in the ACh content of the mouse brain [197]. HLD also protected against transient forebrain ischemia induced by occlusion of both common carotid arteries in C57BL/6 mice by increasing the expression of Cu/Zn-SOD and reducing the exposure of hippocampal neurons to oxidative stress [198]. HLD extract reduced ischemia-reperfusion brain injury and neutrophil infiltration in MCAo rats [62]. The neuroprotective effect of HLD can be explained by multiple mechanisms, the anti-inflammatory and antioxidant activity of Scutellariae Radix, Coptidis Rhizoma, and Gardiniae Fructus, and inhibition of ACh release by Coptidis Rhizoma and Phellodendri Cortex.

16.5.2 *Buyanghuanwu* Decoction

Buyanghuanwu decoction (BYHWD), a traditional Chinese herbal prescription, has been used clinically for hundreds of years to treat stroke with a yang-tonic effect. The component herbs of BYHWD are Astragalis Radix, Lumbricus, Angelicae Sinensis Radix, Ligustici Chuanxiong Rhizoma, Paeoniae Radix Rubra, Persicae Semen, and Carthami Flos. BYHWD shows an antioxidant effect by raising the decline of superoxide dismutase and GPX activities in a rat model induced by pertussis vaccine [199]. BYHWD improves recovery of neurological function, reduces infarction volume, stimulates neural proliferation, and modulates VEGF and Flk1 expression in transient focal cerebral ischaemic rat brains [200]. BYHWD protects neurons from hypoxia-induced apoptosis, the mechanism of which may lie in the elimination of NO and oxygen free radicals produced during hypoxia, and also by up-regulation of Bcl-2 expression [201], and down-regulation of p53 and p21 gene expression [202].

A number of studies have demonstrated that BYHWD improves the outcome of ischemic stroke in clinical trials. BYHWD improves the metabolic imbalance of endothelin and calcitonin GCRP in patients with early cerebral infarction [203]. BYHWD promotes the proliferation of rat cortical neurons cultured in both normal and hypoxia conditions [204]. Pretreatment of BYHWD showed neuroprotective effects in the 4-VO rat model and suppressed the expression of caspase-3 p20 in the CA1 region [205].

16.5.3 *Woohwangchungsim-won*

Woohwangchungsim-won (WCW) is one of the most popular prescriptions in TEM for treating the symptoms after stroke such as semicoma, hemiplegia, deviation of the eye and mouth, retardation of language, headache, and vertigo. WCW was reported to have analgesic, anticonvulsive, antihypoxic CNS damage, antihypertensive, and anti-inflammatory effects.

Administration of WCW at 30 min before ischemia reduced the oxidative brain damage and lactate release and elevated the ATP content in global ischemia of gerbils. WCW stimulated the expression of the positive cell cycle regulators, c-Myc, c-Fos, and Cyclin D_1 and the expression of apoptosis inhibitor proteins Bcl-2 and Bcl-XL. It inhibited the expression of an apoptosis promoter protein, Bax, in human neuronal cell damage induced by nutrient depletion or cold shock [206]. WCW stimulated eNOS gene expression and inhibited VCAM-1 gene expression in the human endothelial cell line ECV304 [207].

16.5.4 Shengmaisan

Shengmai San (SMS) is a composite formula comprising three component herbs, Ginseng Radix, Ophiopogon Radix, and Schisandrae Fructus. It has been extensively used for treating cardiac disorders. Pretreatment with SMS reduced hydrogen peroxide-induced PC12 cell death [208]. SMS inhibits cerebral oxidative damage after forebrain ischemia-reperfusion in the rat [175]. The injection of SMS directly into duodenum 2 h before cerebral ischemia reduced TBARS in the bilateral carotid artery occlusion rat model. SMS also suppressed TBARS formation even when it was administered after 45 min of reperfusion following ischemia [209, 211]. These findings suggest that SMS improves the oxidative damage and thus protects against cell death in the brain ischemia rat model.

16.5.5 Qizhu Decoction

Qizhu decoction (QZD), comprising the four herbal constituents *Rhizoma Atractylodis Macrocephalae, Poria cocos, Kadix Notoginseng*, and *Radix Astragali*, is effective in preventing cerebral oxidative injury in rats [212]. The QZD was shown to have strong hydroxyl radical (·OH) scavenging activity. When the QZD was injected into rat duodenum 2 h before cerebral ischemia, the oxidative brain damage after 45 min reperfusion was strongly inhibited [213].

16.6 Conclusion

A variety of herbs and prescriptions have been demonstrated to have neuroprotective effects *in vivo* and *in vitro* that may be relevant to the treatment of stroke. The majority of *in vivo* studies have been performed in rodent ischemia models, MCAo as focal ischemia, and 4-VO and 2-VO as global ischemia. The mechanisms of neuroprotective herbs in TEM are suggested to be antioxidant, anti-inflammatory, and antiglutamate effects; however, it is difficult to be precise about mechanisms as the herbs have so many active compounds with disparate mechanisms.

In conclusion, the pharmacological activities of herbs often appear to reflect their traditional uses. The approach for screening herbs to investigate for treatment of stroke is a relatively successful method for the identification of herbs and single compounds.

Acknowledgements I am grateful to my graduate students, Jihye Sagong and Dong-hun Lee, for their assistance and to Prof. Youngmin Bu and Dr. Eun Bang Lee in NeuMed for their advice. This work was supported by a grant (PF002102-00) from the Plant Diversity Research Center of the 21st Century Frontier Research Program funded by the Ministry of Science and Technology of the Korean government.

References

1. Richard GA, Odergren T, Ashwood T (2003) Trends Pharmacol Sci 24:402
2. Wardlaw JM, Warlow CP, Counsell C (1997) Lancet 350:607
3. Kidwell CS, Liebeskind DS, Starkman S, Saver JL (2001) Stroke 32:1349
4. Madhavan R, Jacobs BS, Levine SR (2002) Neurol Res 24 Suppl 1:S27
5. Shi FL, Hart RG, Sherman DG, Tegeler CH (1989) Stroke 20:1581
6. Kim JS, Yoon SS (1997) Stroke 28:1165
7. Gong X, Sucher NJ (1999) Trends Pharmacol Sci 20:191
8. Hur J (1983) Tongeuibokam. Namsandang, Seoul
9. Barone FC, Tuma RF, Legos JJ, Erhardt JA, Parsons AA (2002) Brain inflammation, cytokines, and p38 MAP kinase signaling in stroke. In: Lick CSL (ed) New Concepts in Cerebral Ischemia. CRC, New York, pp 201–244
10. Spranger M, Krempien S, Schwab S, Donneberg S, Hacke W (1997) Stroke 28:2425
11. Ikeda K, Negishi H, Yamori Y (2003) Toxicology 189:55
12. Cherubini A, Polidori MC, Bregnocchi M, Pezzuto S, Cecchetti R, Ingegni T, di Iorio A, Senin U, Mecocci P (2000) Stroke 31:2295
13. Howes MJ, Perry NS, Houghton PJ (2003) Phytother Res 17:1
14. Kim H (2001) Herbal Pharmacology. Zipmoondang, Seoul
15. Han JA, Hu WY (1996) Zhongguo Zhong Xi Yi Jie He Za Zhi 16:506
16. Li LX (1987) Zhonghua Shen Jing Jing Shen Ke Za Zhi 20:109
17. He W, Zhu Z, Liu J, Ye H, Zeng J, Huang X, Lai F (2004) Zhong Yao Cai 27:25
18. He W, Zhu Z (2005) Zhong Yao Cai 28:403
19. Choi SR, Saji H, Iida Y, Magata Y, Yokoyama A(1996) Biol Pharm Bull 19:644
20. Wen TC, Yoshimura H, Matsuda S, Lim JH, Sakanaka M (1996) Acta Neuropathol (Berl) 91:15
21. Shah ZA, Gilani RA, Sharma P, Vohora SB (2005) J Ethnopharmacol 101:299
22. Zhang YG, Liu TP (1996) Zhongguo Yao Li Xue Bao 17:44
23. Lim JH, Wen TC, Matsuda S, Tanaka J, Maeda N, Peng H, Aburaya J, Ishihara K, Sakanaka M (1997) Neurosci Res 28:191
24. Zhang B, Hata R, Zhu P, Sato K, Wen TC, Yang L, Fujita H, Mitsuda N, Tanaka J, Samukawa K, Maeda N, Sakanaka M (2006) J Cereb Blood Flow Metab 26:708
25. Shen L, Zhang J (2003) Neurosci Lett 344:1
26. Ji XY, Tan BK, Zhu YZ (2000) Acta Pharmacol Sin 21:1089

27. Kuang P, Wu W, Liu J, Zhang F, Pu C (1991) J Tradit Chin Med 11:123
28. Kuang P, Wu W, Zhu K (1993) J Tradit Chin Med 13:38
29. Kuang P, Xiang J (1994) J Tradit Chin Med 14:45
30. Kuang P, Tao Y, Tian Y (1996) J Tradit Chin Med 16:224
31. Kuang P, Tao Y, Tian Y (1996) J Tradit Chin Med 16:138
32. Lo CJ, Lin JG, Kuo JS, Chiang SY, Chen SC, Liao ET, Hsieh CL (2003) Am J Chin Med 31:191
33. Lam BY, Lo AC, Sun X, Luo HW, Chung SK, Sucher NJ (2003) Phytomedicine 10:286
34. Yu XY, Lin SG, Zhou ZW, Chen X, Liang J, Duan W, Yu XQ, Wen JY, Chowbay B, Li CG, Sheu FS, Chan E, Zhou SF (2007) Neurosci Lett 417:261
35. Sze FK, Yeung FF, Wong E, Lau J (2005) Acta Neurol Scand 111:118
36. Heiss WD, Zeiler K, Havelec L (1978) Dtsch Med Wochenschr 103:597
37. Loffler T, Lee SK, Noldner M, Chatterjee SS, Hoyer S, Schliebs R (2001) J Neural Transm 108:1457
38. Barth SA, Inselmann G, Engemann R, Heidemann HT (1991) Biochem Pharmacol 41:1521
39. Marcocci L, Packer L, Droy-Lefaix MT, Sekaki A, Gardes-Albert M (1994) Methods Enzymol 234:462
40. Topic B, Tani E, Tsiakitzis K, Kourounakis PN, Dere E, Hasenohrl RU, Hacker R, Mattern CM, Huston JP (2002) Neurobiol Ag 23:135
41. Chopin P, Briley M (1992) Psychopharmacology 106:26
42. Petkov VD, Kehayov R, Belcheva S, Konstantinova E, Petkov VV, Getova D, Markovska V (1993) Planta Med 59:106
43. Stoll S, Scheuer K, Pohl O, Muller WE (1996) Pharmacopsychiatry 29:144
44. Calapai G, Crupi A, Firenzuoli F, Marciano MC, Squadrito F, Inferrera G, Parisi A, Rizzo A, Crisafulli C, Fiore A, Caputi AP (2000) Life Sci 67:2673
45. Clark WM, Rinker LG, Lessov NS, Lowery SL, Cipolla MJ (2001) Stroke 32:1000
46. Krieglstein J, Ausmeier F, El-Abhar H, Lippert K, Welsch M, Rupalla K, Henrich-Noack P (1995) Eur J Pharm Sci 3:39
47. Klein J, Chatterjee SS, Loffelholz K (1997) Brain Res 755:347
48. Weichel O, Hilgert M, Chatterjee SS, Lehr M, Klein J (1999) Naunyn Schmiedebergs Arch Pharmacol 360:609
49. Rui J, Li YJ, Zhang YL, Liu Y (2006) Zhongguo Zhong Yao Za Zhi 31:769
50. Liu J (2006) Explore 2:262
51. Fujikawa T, Soya H, Hibasami H, Kawashima H, Takeda H, Nishibe S, Nakashima K (2002) Phytother Res 16:474
52. Deyama T, Nishibe S, Nakazawa Y (2001) Acta Pharmacol Sin 22:1057
53. Lin CC, Huang PC (2000) Phytother Res 14:489
54. Lee S, Son D, Ryu J, Lee YS, Jung SH, Kang J, Lee SY, Kim HS, Shin KH (2004) Arch Pharm Res 27:106
55. Sui DY, Lu ZZ, Ma LN, Fan ZG (1994) Zhongguo Zhong Yao Za Zhi 19:746
56. Jung HJ, Park HJ, Kim RG, Shin KM, Ha J, Choi JW, Kim HJ, Lee YS, Lee KT (2003) Planta Med 69:610
57. Bu Y, Jin ZH, Park SY, Baek S, Rho S, Ha N, Park SK, Kim H (2005) Phytother Res 19:167
58. Gao Z, Huang K, Xu H (2001) Pharmacol Res 43:173
59. Lee HH, Yang LL, Wang CC, Hu SY, Chang SF, Lee YH (2003) Brain Res 986:103
60. Hamada H, Hiramatsu M, Edamatsu R, Mori A (1993) Arch Biochem Biophys 306:261
61. Li H, Wang H, Chen JH, Wang LH, Zhang HS, Fan Y (2003) J Chromatogr B Analyt Technol Biomed Life Sci 788:93
62. Hwang YS, Shin CY, Huh Y, Ryu JH (2002) Life Sci 71:2105
63. Zhang Y, Wang X, Wang X, Xu Z, Liu Z, Ni Q, Chu X, Qiu M, Zhao A, Jia W (2006) J Ethnopharmacol 108:355
64. Kim YO, Leem K, Park J, Lee P, Ahn DK, Lee BC, Park HK, Suk K, Kim SY, Kim H (2001) J Ethnopharmacol 77:183
65. Cho J, Lee HK (2004) Biol Pharm Bull 27:1561.
66. Kim H, Kim YS, Kim SY, Suk K (2001) Neurosci Lett 309:67

67. Lin JK, Liang YC, Lin-Shiau SY (1999) Biochem Pharmacol 58:911
68. Pan T, Jankovic J, Le W (2003) Drugs Ag 20:711
69. Wang C, Wang X, Li Y, Jiang Y, Deng S, Zhao M (1996) Hua Xi Yi Ke Da Xue Xue Bao 27:206
70. Hong JT, Ryu SR, Kim HJ, Lee JK, Lee SH, Kim DB, Yun YP, Ryu JH, Lee BM, Kim PY (2000) Brain Res Bull 53:743
71. Hong JT, Ryu SR, Kim HJ, Lee JK, Lee SH, Yun YP, Lee BM, Kim PY (2001) Brain Res 888:11
72. Lee S, Suh S, Kim S (2000) Neurosci Lett 287:191
73. Kakuda T (2002) Biol Pharm Bull 25:1513
74. Ma L, Xiao P, Guo B, Wu J, Liang F, Dong S (1999)Zhongguo Zhong Yao Za Zhi 24:238
75. Wang PY, Wang HP, Li GW (2006) Zhongguo Zhong Yao Za Zhi 31:577
76. Yamaki K, Kim DH, Ryu N, Kim YP, Shin KH, Ohuchi K (2002) Planta Med 68:97
77. Zhang G, Fang S (1997) Zhong Yao Cai 20:358
78. Dong LP, Wang TY (1998) Zhongguo Yao Li Xue Bao 19:339
79. Xu X, Zhang S, Zhang L, Yan W, Zheng X (2005) Planta Med 71:585
80. Wang L, Zhao A, Wang F, Chai Q, Chai X (1997) Zhongguo Zhong Yao Za Zhi 22:752
81. Li HC, Zhang GY (2003) Acta Pharmacol Sin 24:1131
82. Pei L, Li Y, Zhang GY, Cui ZC, Zhu ZM (2000) Acta Pharmacol Sin 21:695
83. Min KJ, Lee P, Son D, Kim H, Yeou KS (2003) Neuroreport 14:1941
84. Kim JM, Son D, Lee P, Lee KJ, Kim H, Kim SY (2003) J Pharmacol Sci 92:74
85. Liu Z (1991) Zhong Xi Yi Jie He Za Zhi 11:711
86. Leung AW, Mo ZX, Zheng YS (1991) Neurochem Res 16:687
87. Ho WK, Wen HL, Lee CM (1989) Stroke 20:96
88. Taira J, Ikemoto T, Mimura K, Hagi A, Murakami A, Makino K (1993) Free Radic Res Commun 19(Suppl 1):S71
89. Chiu JH, Ho CT, Wei YH, Lui WY, Hong CY (1997) Life Sci 61:1961
90. Liou KT, Shen YC, Chen CF, Tsao CM, Tsai SK (2003) Brain Res 992:159
91. Teng CM, Chen CC, Ko FN, Lee LG, Huang TF, Chen YP, Hsu HY (1988) Thromb Res 50:757
92. Matsuda H, Kageura T, Oda M, Morikawa T, Sakamoto Y, Yoshikawa M (2001) Chem Pharm Bull (Tokyo) 49:716
93. Son HJ, Lee HJ, Yun-Choi HS, Ryu JH (2000) Planta Med 66:469
94. Liou KT, Lin SM, Huang SS, Chih CL, Tsai SK (2003) Planta Med 69:130
95. Chen CM, Liu SH, Lin-Shiau SY (2007) Basic Clin Pharmacol Toxicol 101:108
96. Cui HS, Huang LS, Sok DE, Shin J, Kwon BM, Youn UJ, Bae K (2007) Phytomedicine 14:696
97. Teng CM, Chen CC, Ko FN, Lee LG, Huang TF, Chen YP, Hsu HY (1988) Thromb Res 50:757
98. Teng CM, Yu SM, Chen CC, Huang YL, Huang TF (1990) Life Sci 47:1153
99. Teng CM, Ko FN, Wang JP, Lin CN, Wu TS, Chen CC, Huang TF (1991) J Pharm Pharmacol 43:667
100. Wang JP, Hsu MF, Raung SL, Chen CC, Kuo JS, Teng CM (1992) Naunyn Schmiedebergs Arch Pharmacol 346:707
101. Wang JP, Raung SL, Chen CC, Kuo JS, Teng CM (1993) Naunyn Schmiedebergs Arch Pharmacol 348:663
102. Wang JP, Ho TF, Chang LC, Chen CC (1995) J Pharm Pharmacol 47:857
103. Shen YC, Sung YJ, Chen CF Eur (1998) J Pharmacol 343:79
104. Hong CY, Huang SS, Tsai SK (1996) Clin Exp Pharmacol Physiol 23:660
105. Lo YC, Teng CM, Chen CF, Chen CC, Hong CY (1994) Biochem Pharmacol 47:549
106. Chang CP, Hsu YC, Lin MT (2003) Clin Exp Pharmacol Physiol 30:387
107. Yan JJ, Kim DH, Moon YS, Jung JS, Ahn EM, Baek NI, Song DK (2004) Prog Neuropsychopharmacol Biol Psychiatry 28:25
108. Choi SS, Han KJ, Lee HK, Han EJ, Suh HW (2003) Biol Pharm Bull 26:1283

109. Kang SY, Lee KY, Park MJ, Kim YC, Markelonis GJ, Oh TH, Kim YC (2003) Neurobiol Learn Mem 79:11
110. Kang SY, Kim YC (2007) J Pharm Pharmacol 59:863
111. Wu H, Kong L, Wu M, Xi P (1996) Zhongguo Zhong Yao Za Zhi 21:599
112. Kuang X, Yao Y, Du JR, Liu YX, Wang CY, Qian ZM (2006) Brain Res 1102:145
113. Hsieh MT, Lin YT, Lin YH, Wu CR (2000) Am J Chin Med 28:263
114. Lazarova MB, Petkov VD, Markovska VL, Petkov VV, Mosharrof A (1986) Methods Find Exp Clin Pharmacol 8:547
115. Mook-Jung I, Kim H, Fan W, Tezuka Y, Kadota S, Nishijo H, Jung MW (2002) Biol Pharm Bull 25:1101
116. Lee MW, Lee YA, Park HM, Toh SH, Lee EJ, Kim YH (2004) Arch Pharm Res 23:455
117. Bu Y, Baek S, Jin ZH, Kim H (2004) Korea J Herbol 19
118. Lin HC, Ding HY, Wu YC (1998) J Nat Prod 61:343
119. Liu F, Ng TB (2000) Life Sci 66:725
120. Lee SE, Hwang HJ, Ha JS, Jeong HS, Kim JH (2003) Life Sci 73:167
121. Zhang G, Yu Z, Zhao H (1997) Zhong Yao Cai 20:626
122. Hsieh CL, Cheng CY, Tsai TH, Lin IH, Liu CH, Chiang SY, Lin JG, Lao CJ, Tang NY (2006) J Ethnopharmacol 106:208
123. Ryu G, Park EK, Joo JH, Lee BH, Choi BW, Jung DS, Lee NH (2001) Arch Pharm Res 24:105
124. Ohta H, Ni JW, Matsumoto K, Watanabe H, Shimizu M (1993) Pharmacol Biochem Behav 45:719
125. Xiao L, Wang YZ, Liu J, Luo XT, Ye Y, Zhu XZ (2005) Life Sci 78:413
126. Tabata K, Matsumoto K, Murakami Y, Watanabe H (2001) Biol Pharm Bull 24:496
127. Kim HJ, Lee WH, Yoon CH, Jeong JC, Nam KS, Kim HM, Choo YK, Lee MC, Kim CH (2001) Pharmacol Res 43:11
128. Kwon HC, Lee KC, Cho OR, Jung IY, Cho SY, Kim SY, Lee KR (2003) J Nat Prod 66:466
129. Koo BS, An HG, Moon SK, Lee YC, Kim HM, Ko JH, Kim CH (2003) Immunopharmacol Immunotoxicol 25:191
130. Hsu B, Kin KC (1962) Arch Int Pharmacodyn Ther 139:318
131. Matsuda H, Shiomoto H, Naruto S, Namba K, Kubo M (1988) Planta Med 54:27
132. Chueh FY, Hsieh MT, Chen CF, Lin MT (1995) Pharmacology 51:237
133. Kubo M, Matsuda H, Tokuoka K, Kobayashi Y, Ma S, Tanaka T (1994) Biol Pharm Bull 17:458
134. Shiomoto H, Matsuda H, Kubo M (1990) Chem Pharm Bull Tokyo 38:2320
135. Chang CK, Chueh FY, Hsieh MT, Lin MT (1999) Neurosci Lett 267:109
136. Liang C, Cheng PH, Liang CS (1998) Chin Pharm Bull 14:413
137. Liao ET, Chiang SY, Lao CJ, Tang NY, Hsieh CT, Lin JG (2001) Acta Pharmacol Sin 22:1143
138. Nishiyama N, Zhou Y, Takashina K, Saito H (1994) Biol Pharm Bull 17:1472
139. Nishiyama N, Zhou Y, Saito H (1994) Biol Pharm Bull 17:1481
140. Vohora SB, Shah SA, Dandiya PC (1990) J Ethnopharmacol 28:53
141. Liao JF, Huang SY, Jan YM, Yu LL, Chen CF (1998) J Ethnopharmacol 61:185
142. Cho J, Joo NE, Kong JY, Jeong DY, Lee KD, Kang BS (2000) J Ethnopharmacol 73:31
143. Cho J, Kong JY, Jeong DY, Lee KD, Lee DU, Kang BS (2001) Life Sci 68:1567
144. Cho J, Kim YH, Kong JY, Yang CH, Park CG (2002) Life Sci 71:591
145. Otsuka H, Fujimura H, Sawada T, Goto M (1981) Yakugaku Zasshi 101:883
146. Chun YT, Yip TT, Lau KL, Kong YC, Sankawa U (1979) Gen Pharmacol 10:177
147. Rabbani GH, Butler T, Knight J, Sanyal SC, Alam K (1987) J Infect Dis 155:979
148. Yokozawa T, Ishida A, Kashiwada Y, Cho EJ, Kim HY, Ikeshiro Y (2004) J Pharm Pharmacol 56:547
149. Hsieh MT, Peng WH, Wu CR, Wang WH (2000) Phytother Res 14:375
150. Wu TS, Hsu MY, Kuo PC, Sreenivasulu B, Damu AG, Su CR, Li CY, Chang HC (2003) J Nat Prod 66:1207
151. Kong LD, Yang C, QuiXi, Wu HP, Ye DJ (2001) Zhongguo Zhong Yao Za Zhi 26:245
152. Wang H, Tang XC (1998) Zhongguo Yao Li Xue Bao 19:27

153. Wang XD, Zhang JM, Yang HH, Hu GY (1999) Zhongguo Yao Li Xue Bao 20:31
154. Zhou J, Fu Y, Tang XC (2001) Neurosci Lett 306:53
155. Wang LS, Zhou J, Shao XM, Tang XC (2002) Brain Res 949:162
156. Wang LS, Zhou J, Shao XM, Tang XC (2003) Zhonghua Er Ke Za Zhi 41:42
157. Gordon RK, Nigam SV, Weitz JA, Dave JR, Doctor BP, Ved HS (2001) J Appl Toxicol 1:S47
158. Zhou J, Zhang HY, Tang XC (2001) Neurosci Lett 313:137
159. Wang F, Qu L, Lv Q, Guo LJ (2001) Acta Pharmacol Sin 22:1130
160. Lu YM, Frostl W, Dreessen J, Knopfel T (1994) Neuroreport 5:1489
161. Jagadeeswaran R, Thirunavukkarasu C, Gunasekaran P, Ramamurty N, Sakthisekaran D (2000) Fitoterapia 71:395
162. Koo HJ, Lee S, Shin KH, Kim BC, Lim CJ, Park EH (2004) Planta Med 70:467
163. Hsu JD, Chou FP, Lee MJ, Chiang HC, Lin YL, Shiow SJ, Wang CJ (1999) Anticancer Res 19:4221
164. Soeda S, Ochiai T, Paopong L, Tanaka H, Shoyama Y, Shimeno H (2001) Life Sci 69:2887
165. Sugiura M, Shoyama Y, Saito H, Abe K (1994) J Pharmacol Exp Ther 271:703
166. Ochiai T, Soeda S, Ohno S, Tanaka H, Shoyama Y, Shimeno H (2004) Neurochem Int 44:321
167. Huang KC (1993) The Pharmacology of Chinese Herbs. CRC, Boca Raton, FL
168. Masumiya H, Saitoh T, Tanaka Y, Horie S, Aimi N, Takayama H, Tanaka H, Shigenobu K (1999) Life Sci 65:2333
169. Kuramochi T, Chu J, Suga T (1994) Life Sci 54:2061
170. Hsieh CL, Chen MF, Li TC, Li SC, Tang NY, Hsieh CT, Pon CZ, Lin JG (1999) Am J Chin Med 27:257
171. Liu J, Mori A (1992) Neuropharmacology 31:1287
172. Suk K, Kim SY, Leem K, Kim YO, Park SY, Hur J, Baek J, Lee KJ, Zheng HZ, Kim H (2002) Life Sci 70:2467
173. Shimada Y, Goto H, Kogure T, Shibahara N, Sakakibara I, Sasaki H, Terasawa K (2001) Am J Chin Med 29:173
174. Zhang S, Murakami Y, Zheng HZ, Matsumoto K, Sakakibara I, Takayama H, Aimi N, Watanabe H (2002) J Tradit Med 19:28
175. Ichikawa H, Wang X, Konishi T (2003) Am J Chin Med 31:509
176. Kim SR, Lee MK, Koo KA, Kim SH, Sung SH, Lee NG, Markelonis GJ, Oh TH, Yang JH, Kim YC (2004) J Neurosci Res 76:397
177. Ko KM, Yiu HY (2001) Mol Cell Biochem 220:141
178. Romano C, Price M, Bai HY, Olney JW (1993) Invest Ophthalmol Vis Sci 34:72
179. Zhu H, Wang Z, Ma C, Tian J, Fu F, Li C, Guo D, Roeder E, Liu K (2003) Planta Med 69:429
180. Leung AW, Mo ZX, Zheng YS (1991) Neurochem Res 16:687
181. Li XP (1991) Zhong Xi Yi Jie He Za Zhi 11:486
182. Hou WC, Lee MH, Chen HJ, Liang WL, Han CH, Liu YW, Lin YH (2001) J Agric Food Chem 49:4956
183. Thiyagarajan M, Sharma SS (2004) Life Sci 74:969
184. Ghoneim AI, Abdel-Naim AB, Khalifa AE, El Denshary ES (2002) Pharmacol Res 46:273
185. Bhattacharya SK, iram K, osal S, attacharya A (2000) Phytomedicine 7:463
186. Bhattacharya A, Ghosal S, Bhattacharya SK (2001) J Ethnopharmacol 74:1
187. Head KA (2001) Altern Med Rev 6:141
188. Ziauddin M, Phansalkar N, Patki P, Diwanay S, Patwardhan B (1996) J Ethnopharmacol 50:69
189. Chaudhary G, Sharma U, Jagannathan NR, Gupta YK (2003) Clin Exp Pharmacol Physiol 30:399
190. Li L, Nie J, Shen Z, Wu W, Chen Z, Hao X (2001) Planta Med 67:142
191. Li L, Shen YM, Yang XS, Wu WL, Wang BG, Chen ZH, Hao XJ (2002) Brain Res 944:205
192. Li L, Jiao L, Lau BH (1993) Cancer Biother 8:263
193. Lin JM, Lin CC, Chiu HF, Yang JJ, Lee SG (1993) Am J Chin Med 21:59
194. Qi G, Zhang L, Xie WL, Chen XY, Li JS (2000) Acta Pharmacol Sin 21:1193
195. Aktan F, Henness S, Roufogalis BD, Ammit AJ (2003) Nitric Oxide 8:235
196. Kabuto H, Asanuma M, Nishibayashi S, Iida M, Ogawa N (1997) Neurochem Res 22:33

197. Xu J, Murakami Y, Matsumoto K, Tohda M, Watanabe H, Zhang S, Yu Q, Shen J (2000) J Ethnopharmacol 73:405
198. Kondo Y, Kondo F, Asanuma M, Tanaka K, Ogawa N (2000) Neurochem Res 25:205
199. Tan F (1992) Zhongguo Zhong Xi Yi Jie He Za Zhi 12:538
200. Cai G, Liu B, Liu W, Tan X, Rong J, Chen X, Tong L, Shen J (2007) J Ethnopharmacol 113:292
201. Qu HD, Tong L, Shen JG, Chen YY (2002) Di Yi Jun Yi Da Xue Xue Bao 22:35
202. Qu HD, Tong L, Shen JG (2004) Zhongguo Zhong Xi Yi Jie He Za Zhi 24:133
203. Wu Y, Jiang L (1998) Zhongguo Zhong Xi Yi Jie He Za Zhi 18:396
204. Tong L, Shen JG, Qiu XS (2002) Zhongguo Zhong Xi Yi Jie He Za Zhi 22:522
205. Li XM, Bai XC, Qin LN, Huang H, Xiao ZJ, Gao TM (2003) Neurosci Lett 346:29
206. Song K, Kim YS, Moon SK, Ko CN, Cho KH, Bae HS, Lee KS (2001) Am J Chin Med 29:445
207. Moon SK, Kim YS, Ko CN, Bae HS, Cho KH, Lee KS (2003) Am J Chin Med 31:389
208. Wang L, Nishida H, Ogawa Y, Konishi T (2003) Biol Pharm Bull 26:1000
209. Ichikawa H, Konishi T (2002) Biol Pharm Bull 25:898
210. Xuejiang W, Magara T, Konishi T (1999) Free Radic Res 31:449
211. Ichikawa H, Wang L, Konishi T (2006) Am J Chin Med 34:591
212. Xuejiang W, Ichikawa H, Konishi T (2001) Biol Pharm Bull 24:558
213. Wang XJ, Feng P (2000) Acta Pharmacol Sin 21:1141

Chapter 17
Biological Activity of Ginkgo

Ján Lehotský, Peter Kaplan, Martina Pavlikova, Peter Urban and Beata Saniova

Abstract The extract of *Ginkgo biloba* (Egb) contains more than 60 active substances with a high bioavailability. EGb has extensive protective effects on the central nervous system and cerebrovascular system. As neuroprotective agent it eliminates free radicals, acts as an antioxidant and a free-radical scavenger, reduces lipid peroxidation, and is a membrane stabilizer and an inhibitor of platelet-activating factor via the terpene ginkgolide. The anti-ischemic neuroprotective effects of Egb have been proved in cerebral and spinal cord ischemic models in animals, the action possibly being related to a functional state of mitochondria and apoptosis. However, high-quality and large-scale randomized controlled trials are needed to test its efficacy for the treatment of human stroke.

Keywords *Ginkgo biloba* extract · *EGb* · Biological activity · Neuroprotection · Ischemia

17.1 Introduction

The leaves of *Ginkgo biloba* L. have been used in ancient and modern Chinese herbal pharmacopoeia as a treatment for dysfunctions of the heart and lung and as a promoter of longevity [1]. More than 60 different biologically active substances are present in *G. biloba* leaves (Fig. 17.1). Of these, terpenoids, flavonoids and organic acids are the most important. The standardized *G. biloba* leaf extract (EGb761, Tanakan) was developed in the early 1970s and contains 24% flavonol glycosides (the flavonoid fraction), 6% terpene lactones (terpenoid fraction), 70% other substances [such as rutin, proanthocyanidines ($> 0.5\%$), organic acids (5 to 10%), glucose, rhamnose, etc.] and no more than 5 ppm of ginkgolic acids. The flavonoid fraction is primarily composed of flavanals: quercetin, kaempferol, amentoflavon,

J. Lehotský (✉)
Comenius University, Jessenius Faculty of Medicine, Department of Medical Biochemistry, Malá Hora 4, 036 01 Martin, Slovak Republic,
e-mail: Lehotsky@jfmed.uniba.sk

K.G. Ramawat (ed.), *Herbal Drugs: Ethnomedicine to Modern Medicine*,
DOI 10.1007/978-3-540-79116-4_17,

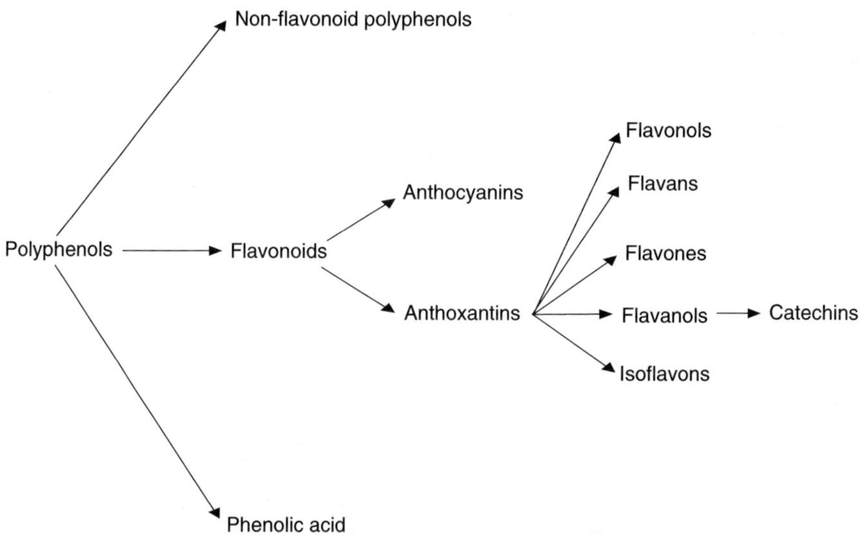

Fig. 17.1 Classification of natural plyphenols found in *EGb 761*

bilobetin, ginkgetin, idoginkgetin, 5′-methoxybilobetin and isorhamnetin. The terpenoid fraction primarily contains ginkgolides A, B, C, J and M (2.8 to 3.4%), as well as seskviterpenoic polyketone bilobalide (2.6 to 3.4%) with unique chemical structures of polycyclic ethers containing a characteristic terc-butyl group, which is very rare in natural compounds [2].

17.2 Phamacokinetics

The in vivo dynamic distribution of *Ginkgo* extract (pharmacokinetics) was investigated by several methods including positron emission tomography (PET) and LC-MS/MS [3, 4]. In general, the extent of bioavailability of ginkgolids is high, and food intake does not change it [5]. Intravenous administration of *Ginkgo* extract leads to its biphasic elimination with a prominent initial phase (half-life of 0.3 h, with peak at 10 min), followed by a slower terminal phase (half-life of 1.5 h). Tissue concentration declines by several-fold during the first 6 h and the metabolites are mainly excreted in the urine (40 to 50%) and feces (less than 30%) and traces by bile.

Interestingly, as shown by PET imaging in vivo, at least the ginkgolid B exists in two forms in the body: the original with its lactone rings closed and a second form with one of the rings open. The original form is taken up rapidly by various organs including the liver, the intestine and possibly the stomach, and consequently in plasma, the proportion of open form increases dramatically with time to shift slowly towards equilibrium. Polyphenols are usually non-lipophilic compounds and cannot cross easily through the plasma membrane. As suggested by Ramasamy [6], they might interact with plasma membrane transporters or receptors triggering intracellular signaling pathways. It is interesting that the first receptor for a dietary polyphenol has been identified for EGCG, the 67 kDa laminin receptor on vascular cells [7]. The presence of polyphenol receptors or transporters in the brain remains to be established. The beneficial effects of polyphenols appear to be a promising class of compounds for neuroprotection.

17.3 Treatment of Cerebral Ischemia

Numerous studies on the effects of EGb on cerebral ischemia have been conducted over the course of many years. EGb has extensive protective effects on the central nervous system (CNS) and cerebrovascular system. It functions as a neuroprotective agent in its ability to eliminate free radicals, acts as an antioxidant and free-radical scavenger, reduces lipid peroxidation, and is a membrane stabilizer and inhibitor of platelet-activating factor via the terpene ginkgolide B [8]. Different pathological models of cerebral ischemia were used to evaluate its effects, and measurements were taken at both the cellular and molecular levels to determine its mechanisms of action, which depend on a direct effect on neuronal cells and an indirect effect on brain circulation. In experimental models with different paradigms such as focal and global ischemia, administrated before or after the insult and given oraly or intraperitoneally [9–12], in oedema and hypoxia, it was mostly shown that EGb reduced vascular, tissue and metabolic disturbances as well as their neurological behavioural consequences. In general, in experimentally induced cerebral ischemia (MCAO), a 7-d treatment with EGb reduced the volume of CNS damage [13, 14], protected against neuronal death in the hippocampal CA1 area of the gerbil brain and extended protection to cells in the frontal lobe [15, 16].

17.3.1 Antioxidant/Scavanger Action

The chemical structure of EGb constituents is responsible for its remarkable antioxidant/reactive oxygen/nitrogen species (RONS) scavenging activity. Its neuroprotective function is based on its ability to eliminate free radicals, it acts as an antioxidant and a free-radical scavenger, and it reduces lipid peroxidation, which decreases tissue levels of ROSs and inhibits membrane lipid peroxidation. Both flavonoid and ginkgolide constituents are involved in the antioxidan/free-radical-scavenging effects of EGb. The flavonoids preferentially react with hydroxyl radicals [17] and chelate pro-oxidant transition heavy metal ions [18], which consequently inhibits the formation of new hydroxyl radicals. Significant antioxidant activity is consequently one of the most analysed protective effects of EGb on the CNS and the circulatory system.

An extract can significantly eliminate free oxygen radicals (superoxide, hydrogen peroxide and hydroxyl radicals) and has also been found to be an inhibitor of NADPH-oxidase. Scavenging activity against peroxide radicals of EGb has been proved also on liposomes and human lipoprotein, which are believed to be involved in lipid peroxidation and human atherogenesis [19]. In addition, several studies showed that EGb is able to reduce endogenous lipoperoxidation [20, 21] and can alter the activity of endogenous antioxidant enzymes such as catalase (CAT) and superoxide dismutase (SOD) in the hippocampus, striatum and substantia nigra (SN) and decrease lipoperoxidation in rat hippocampus [22]. On synaptosomal preparations from the striatum of mice, EGb761 has been shown to prevent the alteration of the neuronal dopamine uptake system and the modifications of the membrane fluidity induced by a pro-oxidant system ascorbic acid/Fe2+ [23, 24]. The flavonoid fraction of EGb761 was implicated in these protections. This efficacy was also observed in vivo. Thus dopaminergic neurons were protected against the neurotoxin MPTP when mice received EGb761 2 weeks before the neurotoxin infused peripherally through an osmotic mini-pump for 7 d [24]. The inhibition of monoaminooxidase (MAO) activity by EGb761 may be involved in this protection [25]. In an experimental model of hypoxia-induced phospholipid breakdown, it was found that bilobalide, a sesquiterpene lactone constituent of EGb761, inhibited activation of NMDA-receptor-induced phospolipase A2 and choline release in hippocampal slices [26]. EGb may play a protective role in the homeostasis of inflammation and oxidative stress, in the prevention of cell membrane damage caused by free radicals, and in neurotransmission modulation [1].

As shown by Ni et al. [27], EGb761 can prevent hydroxyl-radical-induced apoptosis in cultured neurons. Numerous studies have shown that EGb761 alone or administrated with other compounds with antioxidative properties such as vitamin E provides significant protection of membrane integrity, depresses lipoperoxidation and positively affects immune function [20]. This is one of the reasons why EGb is used for the early-stage treatment of Alzheimer's disease to interrupt the proposed oxidative stress associated with the disease [28]. In addition, it has been established by Bastianetto [29] that EGb has a protective effect against NO-induced toxicity in hippocampal cells and causes significant depression of platelet aggregation induced by the ginkgolide b fraction of EGb with a lower incidence of venous thrombosis [30].

The treatment of lumbar spinal cord ischemia may have very important clinical implications. Mechírova and Domorakova [31] have shown reasonable neuroprotective effects of EGb by histochemical analysis of NADPH-diaphorase activity (NOS-like activity) on segments of the lumbar spinal cord in rabbit subjected to 30 min ischemia and 24 h reperfusion. Reduction of reperfusion damage has been observed and was based on the scavenging activity of EGb towards free radicals produced during ischemia/reperfusion (I/R) of the spinal cord. The number of NADPH-d-positive neurons in sections of the L5 segment of the spinal cord was elevated in the ischemic spinal cord and rabbits expressed complete paraplegia. Pretreatment by EGb for 7 d caused disappearance of paraplegia, and NADPH-d activity in blood vessels and neurons was observed, similar to the controls. The authors proposed that it is the free-radical-scavenging property of EGb which decreases free radicals produced during I/R and reduces reperfusion damage [31].

Cerebral I/R insult induces inhibition of the general translation of proteins in the neocortex as well as in the highly sensitive hippocampus where, without protection, inhibition of protein synthesis persists up to the death of neurons [32]. Interestingly, the mechanism of ischemic damage protection by EGb was studied by measuring the extent to which translation inhibition occurs in rats. Thus, due to its antioxidant and antiradical ability, it can significantly reduce the I/R-induced inhibition of translation in the neocortex as well as in the highly sensitive hippocampus. EGb, with its ability to protect the translation mechanism, permits the newly synthesized mRNAs to be translated into functional proteins, thus allowing the altered gene expression to be effective. Rats pretretated with EGb at a dose of 40 mg/kg/d for 7 d were subjected to a 4-vessel occlusion model of transient forebrain ischemia for 20 min followed by 30 min, 4 h or 7 d of reperfusion. Ischemia induced inhibition of the protein synthesis rate. The ability to reinitiate proteosynthesis was significantly reduced in the frontal cortex and hippocampus in EGb-treated animals. Moreover, as shown by Fluoro-Jade B staining, it alleviates neurodegeneration and increases number of surviving neurons in highly sensitive hippocampus [12].

In the same paradigm of global ischemic damage, prophylactic oral administration of EGb at a dose of 40 mg/kg/d for 7 d showed efficient protection of surviving neurons, particularly in the most vulnerable CA1 pyramidal cells after 20 min of ischemia. Although ischemia is lethal for most populations of the CA1 pyramidal cell layer, pretreatment with EGb increased the number of NeuN immunoreactive cells (surviving neurons) in spite of a similar number of reactive astrocytes immunolabelled for GFAP (glial fibrilary acidic protein) in the stratum oriens and stratum lacunosum and moleculare [33]. As reported by the authors, the inhibition of free-radical formation afforded by EGb might explain the protection of the most vulnerable CA1 pyramidal cells against ischemic damage.

In our own laboratory, we assessed the antioxidant activity of EGb761 in in vitro experiments. The formation of TBA reactive substances (TBARS) was used as an index of lipoperoxidation (LPO) in all forebrain membranes [34]. The degree of peroxidation (lipoperoxidation index LPO) increased gradually with increasing time (30 to 60 min) of incubation in the medium containing 0.1 mmol/l $FeSO_4$/EDTA plus 1 mmol/l H_2O_2/mg of protein. The addition of 50 or 100 μl/ml EGb761 significantly protects the generation of LPO products in cerebral membranes. In addition, oxidative modifications in protein measured by the content of total sulfhydryl group level was protected up to 75 to 81% by the addition of 50 or 100 μl/ml EGb.

Ischemia creates several conditions that could account for the increased net production of free radicals or an impairment of cellular defenses that normally protect against such damage. The massive production of free radicals during I/R was observed in plenty of studies by indirect spin-trapping measurements and chemiluminiscence [35]. In our laboratory, we proved that the transient global forebrain I/R induced considerable lipoperoxidation and formation of lipoperoxidation-dependent products and direct oxidative products of neural proteins. The formation of products was time dependent and culminated in a later reperfusion period where delayed neuronal death could be observed [36–38]. In the study in our laboratory conducted on all tissue membranes, forebrain ischemia caused a small but significant increase of LPO assessed by the level of thiobarbuturic reactive substances (TBARS) and by measurement of lysine conjugate fluorescence. The TBARS level from sham-operated animals treated with EGb expressed a lower level of TBARS and in the reperfusion period reached slightly depressed values (by about 11%) compared to non-treated animals. In addition, ischemic insult caused a significant increase in levels of Lys fluorescence, which was pronounced after 24 h of reperfusion. Remarkably, in vivo treatment with EGb depressed fluorescence intensity, which suggests an anti-LPO activity after ischemic insult [34].

Ischemic insult induces oxidative modifications in protein structure, as was assayed by measurement of changes in levels of total SH groups, and alterations in tryptophane (Trp) and bityrosine fluorescence. Although ischemia significantly changed the level of SH groups during reperfusion, the EGb treatment caused a remarkable reversion of the SH-group level to non-ischemic conditions. Similarly, as shown by an analysis of fluorescence intensity of Trp and bityrosine ischemia-induced changes increased with time after ischemic insult. Pretreatment with EGb induced partial but significant recovery of Trp and bityrosine fluorescence intensity in comparison with non-treated animals. These results suggest that EGb761 has potent antioxidant activity and could act as an important factor which attenuates I/R-induced protein oxidation. The protective properties of EGb761 support its potential beneficial actions against ischemia-induced brain-related pathologies that are likely associated with the deleterious effects of reactive oxygen/nitrogen species imbalance [34].

Interestingly, the oxidative changes of neural cells induced by a brief global ischemic episode appeared mainly in the later reperfusion period, which is very similar to the time course of histopathologically assessed neuronal vulnerability [39]. In addition, oxidative protein alterations also follow disturbances in the oxidant/antioxidant balance and depression of enzymatic activities of the main antioxidant enzymes such as superoxide dismutase detected at later stages after ischemic insult in both gerbils and rats [40, 41]. Thus, oxidative alterations of neural proteins after IRI may at least partially explain functional postischemic disturbances of neuronal ion transport mechanisms [42] and ischemia-induced inhibition of global proteosynthesis [32, 43], both of which are implicated in neuronal cell damage and/or recovery from ischemic insult.

Several lines of evidence suggest that EGb allows mitochondria to maintain their respiratory activity under ischemic conditions [44] and may play a protective role in the homeostasis of inflammation and oxidative stress [22]. The antioxidant and antiradical ability is at least partially responsible for a significant reduction in the IRI-induced inhibition of translation in the neocortex and hippocampus, where, without protection, inhibition of proteosynthesis persists up to the death of neurons [32]. The scavenging-free-radical activity of EGb has been proved in different in vitro and in vivo studies [21, 45, 46].

17.3.2 Other Anti-Ischemic Effects

Recent studies using gerbils, mice and rats have provided some insight into the beneficial action of EGb. In experimentally induced cerebral ischemia, a 7-d treatment reduced the degree of CNS damage induced by middle cerebral artery ligation [46, 47]. In a similar ischemia paradigm, other investigators found that EGb protected against neuronal death in the hippocampal CA1 area of the gerbil brain [15]; a follow-up study revealed that the protection extended to cells in the frontal lobe [47]. In addition, using an Alzheimer's mutant mouse model (Tg2576), EGb treatment for 6 months markedly improved spatial cognitive performance, without affecting central β-amyloid concentrations [48]. It was found that EGb could increase the rat cerebral blood flow [49] and improved ischemic memory impairment in mice [50]. Several lines of evidence suggest that EGb alleviates the subcellular damage of cerebral ischemia [51] and allows mitochondria to maintain their respiratory activity under ischemic conditions [44]. EGb may play a protective role in the homeostasis of inflammation and oxidative stress and in the prevention of cell membrane damage caused by free radicals and neurotransmission modulation [1]. As shown by Ni et al. [27], EGb761 can prevent hydroxyl-radical-induced apoptosis in cultured neurons. EGb could prevent and treat acute cerebral ischemia damage. Regarding Egb-induced regulation of cerebral glucose utilization, bilobalide increases the respiratory control ratio of mitochondria by protecting against uncoupling of oxidative phosphorylation, thereby increasing ATP levels, a result that is supported by the finding that bilobalide increases the expression of the mitochondrial DNA-encoded COX III subunit of cytochrome oxidase [1]. It is clear that an irreversible block of protein synthesis in the selectively vulnerable CA1 field of the hippocampus necessarily leads to the death of neurons. However, the prevention of persistent inhibition of translation does not assure survival of CA1 neurons [32]. Mechanisms allowing neurons to survive, obviously including remodulation of gene expression, have not been clear until now. EGb, with its ability to protect translational machinery, permits newly synthesized mRNAs to be translated into functional proteins, thus allowing the altered gene expression to be effective.

The antiedema effect of EGb is one aspect of its potential therapeutical effects that has not been widely investigated. The first indication of inhibition of toxic edema formation in white matter induced by neurotoxic triethyltin was presented as early as 1986 by Otani [52], and later works proved the beneficial effect of EGb on cerebral edema induced also by bromethalin, a toxin [53], and by hyperthermia [54]. From all constituents of EGb, Ginkgolide B was suggested as an active factor. In a very recent study by Mdzinarishvili et al. [55], the antiedema effect of bilobalide, another EGb constituent, was tested in both in vitro (oxygen-glucose deprivation on hippocampal slices) and in vivo (middle cerebral artery occlusion in mice) conditions. Pretreatment of mice with bilobalide (10 mg/kg i.p.) not only reduced the infarct area by 43% [as judged by 2,3,5-triphenyltetrazolium chloride (TTC) staining] and edema formation by 70%, but it also reduced forebrain water content in the ischemic hemisphere by 57%.

However, as the authors showed by measurement of the water content, bilobalide does not seem to block water transport and its effect is selective for edema formation induced by ischemia. The molecular mechanism of the antiedema effect of bilobalide might involve the described protection of mitochondrial energetics and Na^+, K^+-ATPase activity under ischemic conditions [56, 57] or via bilobalide's interference with chloride fluxes [58]. Additionally, bilobalide increases the expression of glial growth factors in astroglial cultures, and thus astrocytes, with their contribution to tissue swelling, may be the potential targets of bilobalide [59].

Although the mechanism of EGb761 is unclear, it is possible that its actions are related to mitochondria and apoptosis because earlier studies found that caspase expression was altered [46, 60]. This is potentially important since mitochondria play a pivotal role in apoptosis for both intrinsic and extrinsic pathways [61]. The intrinsic and extrinsic pathways do not occur independently in vivo but are linked at different points, one of which is the bax/bcl-2 complex – an apoptotic to anti-apoptotic index [62]. Oligomerization of bax facilitates its insertion into the outer mitochondrial membrane, triggering cytochrome c release, which promotes apoptosis. Conversely, bcl-2 forms complexes with bax in such a way that the release of cytochrome c is inhibited to prevent apoptosis. Therefore, the ratio of bax/bcl-2 is crucial in determining the progress of cell apoptosis for both the intrinsic and extrinsic pathways [62]. At least two studies, by Lu et al. [63] and Loh et al. [64], corroborate the assumption of a mitochondria-based antiapoptotic protective effect of EGb. Lu et al. [63] investigated the levels of apoptotic markers in six brain regions following induced global ischemia in senescence-accelerated mice (SAMP8). A 4-d treatment with EGb significantly decreased bax/bcl-2 ratios in all brain regions in both young (1-month-old) and aged (16-month-old) mice, and the authors suggest that the bax/bcl-2 ratio provides a suitable index of apoptosis, and modulation of these markers may explain the neuroprotective action of EGb761. Interestingly, the beneficial action of EGb was efficacious despite the accelerated aging process in the animals, and future studies of the bax/bcl-2 ratio in the brain of animals and humans should be conducted. Since the protective effect of EGb761 was regional as well as global in both aged (16-month-old) and young (1-month-old) mice, this could explain why EGb761 showed protective effects for different neurological diseases such as Parkinson's [65] and Alzheimer's [66] diseases, which involves different brain areas. The remarkable cerebral protection of EGb was comapared with the antihypertensive Losartan on stroked rats. Both agents were administrated orally ($EGb = 50 mg/kg/d$) 1 week before stroke induction by middle cerebral artery occlusion. Both the mRNA and protein levels of pro-apoptotic genes (AT2 receptor, Fas, Bax and Bcl-xS) showed significant reduction in both pretreatments comparing to the vehicle group. In addition, the decrease in the number of apoptotic cells identified by TUNEL assay indicates a potent and promising therapeutic effect of EGb for stroke treatment, at least in animal experiments [64].

Moreover, the neuroprotective effect of EGb may be correlated with its effect on glucocorticoid synthesis – ginkgolides A and B inhibit corticosteroid synthesis and restore the ability to adapt to stress by reducing LPO and phospholipid content in the brain [67]. Another mechanism of EGb-evoked neuroprotection involves maintaining a balance between inhibitory/excitatory amino acids [68], platelet-activating factor receptor antagonism [69], the ability to inhibit NO-stimulated protein kinase C activity [29], and protection against ischemia-induced changes of Na^+, K^+-ATPase activity [56]. Finally, EGb protects neurons against glutamate excitotoxicity [70] and against apoptosis (cell death) induced by β-amyloid protein, a known pathogenetic factor in pathological brain ageing [71].

Bilobalide, a constituent of EGb, increases the respiratory control ratio of mitochondria by protecting against uncoupling of oxidative phosphorylation, thereby increasing ATP levels. This metabolic result is supported by the finding that bilobalide increases the expression of the mitochondrial DNA-encoded COX III subunit of cytochrome oxidase [1]. It is clear that an irreversible block of protein synthesis in the selectively vulnerable CA1 field of the hippocampus necessarily leads to the death of neurons. However, the prevention of persistent inhibition of translation does not assure survival of CA1 neurons [32]. The additional effect of EGb on MCAO-induced gap junction communication was studied on the mRNA and protein levels of connexin 43 and astrocyte gap junction intercellular communication (GJIC) induced by hypoxia-reoxygenation. Pretreatment with EGb (100 mg/l) for 7 d significantly prevented the hypoxia-reoxygenation inhibition of GJIC followed by improved expression of Cx43, leading to improved neurological deficit [72]. Changes in energy-related metabolites in the striatum of gerbils subjected to focal cerebral ischemia for 60 min after pretreatment with EGb761 and FK506, a calcium-dependent phosphatase calcineurin inhibitor, were investigated using microdialysis. The observed decreases in glucose (10% of the baseline) and pyruvate (20% of the baseline) and increase in lactate (60% of the baseline) during ischemia was significantly preserved by both EGb761 treatment and the combination (EGb761 and FK506) therapy, which suggets that preservation of energy metabolism during cerebral I/R may contribute to the neuroprotective effects of EGb [11].

A very important precaution with respect to the neuroprotective activity of EGb is communicated by the study of da Lima et al. [73]. These authors were not able to show a reduction in brain infarct size in rats after transient MCAO in conditions of unprevented, ischemia-induced fever. Acute (200 mg/kg) or chronic (100 mg/kg, once daily, for 14 d) treatment with EGb in combination or not with antipyretic dipyrone before ischemic insult failed to reduce the infarct size. The authors warn against having high expectations about being able to treat stroke with EGb. There is much biochemical evidence, but very few studies in animal models in vivo, which demonstrate EGb-induced neuroprotection against regional ichemic damage [74]. In this context, Liu [75], by searching relevant clinical trials (COCHRANE, PubMed, EMbase) and research registers, failed to find any convincing evidence from trials of sufficient methodological quality to support the routine use of *Ginkgo biloba* extract to promote recovery after human stroke. High-quality and large-scale randomized controlled trials are needed to test its efficacy.

Acknowledgements This paper was supported by grants: VEGA 3380/06, MVTS COST B30 and WCE 0064/07.

References

1. DeFeudis FV, Drieu K (2000) Curr Drug Targets1:25
2. Peng H, Li YF, Sun SG (2003) Acta Pharmacol Sin 24:467
3. Suehiro M, Simpson NR, Underwood MD, Castrillon J, Nakanishi K, van Heertum R (2005) Planta Med 71:622
4. Chen WD, Liang Y, Xie L, Lu T, Liu XD, Wang GJ(2007) Biol Pharm Bull 30:1
5. Fourtillan JB, Brisson AM, Girault J, Ingrand I, Decourt JP, Drieu K, Jouenne P, Biber A (1995) Therapie 50:137
6. Ramassamy C (2006) Eur J Pharmacol 545:51
7. Tamatani M, Che YH, Matsuzaki H, Ogawa S, Okado H, Miyake S, Mizuno T, Tohyama M (1999) J Biol Chem 274: 8531
8. Sastre J, Millán A, García de la Asunción J, Plá R, Juan G, Pallardó, O'Connor E, Martin JA, Droy-Lefaix MT, Viña J (1998) Free Radic Biol Med 24:298
9. Donato G, Volpentesta G, Lavano A, Ferraro G, Signorelli F, Amorosi A, Maltese L, De Sarro GB, Signorelli CD (2003) J Neurosurg Sci 47:149
10. Zhang YY, Li PF, Li D (2003) Acta Pharmacol Sin 24:157
11. Lin JY, Cheng FC, Chung SY, Lin MC (2004) J Biomed Sci 11:611
12. Hrehorovská M, Burda J, Domoráková I, Mechírová E (2004) Gen Physiol Biophys 23:457
13. Clark WM, Rinker LG, Lessov NS, Lowery SL, Cipolla MJ (2001) Stroke 32:1000
14. Zhang WR, Hayashi T, Kitagawa H, Sasaki C, Sakai K, Warita H, Wang JM, Shiro Y, Uchida M, Abe K (2000) Neurol Res 22:517
15. Chandrasekaran K, Mehrabian Z, Spinnewyn B, Drieu K, Fiskum G (2001) Brain Res 922:282
16. Chung SY, Wang MF, Lin JY, Lin MC, Liu HM, Cheng FC (2003) Am J Chin Med 31:533
17. Zimmermann M, Colciaghi F, Cattabeni F, Di Luca M (2002) Cell Mol Biol (Noisy-le-grand) 48:613
18. Gohil K, Packer L (2002) Cell Mol Biol (Noisy-le-grand) 48:625
19. Maitra I, Marcocci L, Droy-Lefaix MT, Packer L (1995) Biochem Pharmacol 49:1649
20. Casadesus G, Shukitt-Hale B, Joseph JA (2002) Neurobiol Ag 23:747
21. Chen LX, Chen WZ, Huang HL, Chen LX, Xie ZZ (1998) Acta Pharmacol Sin 19:359
22. Bridi R, Crossetti FP, Steffen VM, Henriques AT (2001) Phytother Res 15:449
23. Ramassamy C, Girbe F, Christen Y, Costentin J (1993) Free Radic Res Commun 19:341
24. Ramassamy C, Clostre F, Christen Y, Costentin J (1990) J Pharm Pharmacol 42:785
25. Pardon MC, Joubert C, Perez-Diaz F, Christen Y, Launay JM, Cohen-Salmon C (2000) Mech Ag Dev 113:157
26. Weichel O, Hilgert M, Chatterjee SS, Lehr M, Klein J (1999) Naunyn Schmiedebergs Arch Pharmacol 360:609
27. Ni Y, Zhao B, Hou J, Xin W (1996) Neurosci Lett 214:115
28. Bastianetto S, Ramassamy C, Doré S, Christen Y, Poirier J, Quirion R (2000) Eur J Neurosci 12:1882
29. Bastianetto S, Zheng WH, Quirion R (2000) J Neurochem 74:2268
30. Xu JP, Rui YC, Li TJ (1999) Acta Pharmacol Sin 20:423
31. Mechírová E, Domoráková I (2002) Acta Histochem 104:427
32. Burda J, Hrehorovská M, García BL, Danielisová V, Cížková D, Burda R, Némethová M, Fando JL, Salinas M (2003) Neurochem Res 28: 1237
33. Domoráková I, Burda J, Mechírová E, Feriková M (2006) Cell Mol Neurobiol 26:1193
34. Uríková A, Babusíková E, Dobrota D, Drgová A, Kaplán P, Tatarková Z, Lehotský J (2006) Cell Mol Neurobiol 26:1343
35. Lipton P (1999) Physiol Rev 79:1431
36. Murín R, Drgová A, Kaplán P, Dobrota D, Lehotský J (2001) Gen Physiol Biophys 20:431
37. Lehotský J, Murín R, Strapková A, Uríková A, Tatarková Z, Kaplán P (2004) Gen Physiol Biophys 23:401

38. Candelario-Jalil E, González-Falcón A, García-Cabrera M, Alvarez D, Al-Dalain S, Martínez G, León OS, Springer JE (2003) J Neurochem 86:545
39. Nitatori T, Sato N, Waguri S, Karasawa Y, Araki H, Shibanai K, Kominami E, Uchiyama Y (1995) J Neurosci 15:1001
40. Candelario-Jalil E, Mhadu NH, Al-Dalain SM, Martínez G, León OS (2001) Neurosci Res 41:233
41. Danielisová V, Gottlieb M, Némethová M, Burda J (2007) J Neurol Sci 253:61
42. Lehotský J, Kaplán P, Matejovicová M, Murín R, Racay P, Raeymaekers L (2002) Gen Physiol Biophys 21:31
43. Hayashi T, Saito A, Okuno S, Ferrand-Drake M, Dodd RL, Nishi T, Maier CM, Kinouchi H, Chan PH (2003) J Cereb Blood Flow Metab 23:1117
44. Janssens D, Remacle J, Drieu K, Michiels C (1999) Biochem Pharmacol 58:109
45. Dumont E, D'Arbigny P, Nouvelot A (1995) Methods Find Exp Clin Pharmacol 17:83
46. Clark WM, Rinker LG, Lessov NS, Lowery SL, Cipolla MJ (2001) Stroke 32:1000
47. Chung SY, Wang MF, Lin JY, Lin MC, Liu HM, Cheng FC (2003) Am J Chin Med 31:533
48. Stackman RW, Eckenstein F, Frei B, Kulhanek D, Nowlin J, Quinn JF (2003) Exp Neurol 184:510
49. Krieglstein J, Beck T, Seibert A (1986) Life Sci 39:2327
50. Tadano T, Nakagawasai O, Tan-no K, Morikawa Y, Takahashi N, Kisara K (1998) Am J Chin Med 26:127
51. Sun BL, Yuan H, Yang MF, Xia ZL, Zhang SM, Wang LX (2007) Int J Neurosci 117:655
52. Otani M, Chatterjee SS, Gabard B, Kreutzberg GW (1986) Acta Neuropathol 69:54
53. Dorman DC, Côté LM, Buck WB (1992) Am J Vet Res 53:138
54. Sharma HS, Drieu K, Westman J (2003) Acta Neurochir Suppl. 86:313
55. Mdzinarishvili A, Kiewert C, Kumar V, Hillert M, Klein J (2007) Neuroscience 144:217
56. Pierre S, Jamme I, Droy-Lefaix MT, Nouvelot A, Maixent JM (1999) Neuroreport 10:47
57. Chandrasekaran K, Mehrabian Z, Spinnewyn B, Drieu K, Fiskum G (2001) Brain Res 922:282
58. Klein J, Weichel O, Hilgert M, Rupp J, Chatterjee SS, Nawrath H (2003) Pharmacopsychiatry 36(Suppl 1):S78
59. Zheng SX, Zhou LJ, Chen ZL, Yin ML, Zhu XZ (2000) Acta Pharmacol Sin 21:151
60. Smith JV, Burdick AJ, Golik P, Khan I, Wallace D, Luo Y (2002) Cell Mol Biol (Noisy-le-grand) 48:699
61. Sastre J, Pallardó FV, Viña J (2000) IUBMB Life 49:427
62. Woo M, Hakem R, Mak TW (2000) Cell Res 10:267
63. Lu G, Wu Y, Mak YT, Wai SM, Feng ZT, Rudd JA, Yew DT (2006) Brain Res 1090:23
64. Loh KP, Low LS, Wong WH, Zhou S, Huang SH, De Silva R, Duan W, Chou WH, Zhu Z (2006) Neurosci Lett 398:28
65. Yang SF, Wu Q, Sun AS, Huang XN, Shi JS (2001) Acta Pharmacol Sin 22:1089
66. Stackman RW, Eckenstein F, Frei B, Kulhanek D, Nowlin J, Quinn JF (2003) Exp Neurol 184:510
67. Amri H, Drieu K, Papadopoulos V (2002) Cell Mol Biol (Noisy-le-grand) 48:633
68. Hu B, Sun S, Mei G, Chen L, Tong E (2002) Chin Med J Engl 115:1316
69. Haines DD, Bak I, Ferdinandy P, Mahmoud FF, Al-Harbi SA, Blasig IE, Tosaki A (2000) J Cardiovasc Pharmacol 35:37
70. Chandrasekaran K, Mehrabian Z, Spinnewyn B, Chinopoulos C, Drieu K, Fiskum G (2002) Cell Mol Biol (Noisy-le-grand) 48:663
71. Yao Z, Drieu K, Papadopoulos V (2001) Brain Res 889:181
72. Li Z, Lin XM, Gong PL, Zeng FD, Du GH (2005) Am J Chin Med 33:923
73. de Lima KC, Schilichting CL, Junior LA, da Silva FM, Benetoli A, Milani H (2006) Phytother Res 20:438
74. Saniova B (1999) Bratisl Lek Listy 100:156
75. Liu J (2006) Explore 2:262

Chapter 18
Vitamin E Metabolic Modulation in Plants

Guo Juan[1], Gongshe Liu[1], Chen Shuangyan[1] and Amina A. Aly[2]

Abstract Tocopherols, the lipid-soluble antioxidants known collectively as vitamin E, are synthesized only by photosynthetic organisms. These compounds play vital roles in human nutrition and health. During the past decade, the genes involved in the vitamin E synthesis pathway in plants have been isolated. Therefore, it is becoming important to manipulate these vitamin E synthesis genes to improve vitamin E content for human needs by metabolic engineering in plants. In this chapter, we summarize recent progress in the metabolic engineering of vitamin E in plants and put forward the prospect of altering vitamin E metabolic flux in the future.

Keywords Metabolic engineering · Vitamin E · Plants

Abbreviations

HPPD	ρ-hydroxyphenolpyruvate dioxygenase
HPT	Homogentisate phytyltransferase
HGGT	Homogentisic acid geranylgeranyl transferase
TC	Tocopherol cyclase
γ-TMT	γ-tocopherol methyltransferase
VTE1	Encoding of a tocopherol cyclase
VTE3	Encoding of a 2-methyl-6-phytylbenzoquinol Methyltransferase
VTE4	Encoding of a γ-tocopherol methyltransferase
ZFP-TFs	Synthetic zinc finger transcription factors
TF	Transcription factor
α-T	α-tocopherol
γ-T	γ-tocopherol
PDH	Prephenate dehydrogenase

G. Liu (✉)
Key Laboratory of Photosynthesis and Environmental Molecular Physiology, Institute of Botany, Chinese Academy of Sciences. 20 Nanxincun, Xiangshan, Haidian district, 100093 Beijing, P.R. China
e-mail: liugs@ibcas.ac.cn

K.G. Ramawat (ed.), *Herbal Drugs: Ethnomedicine to Modern Medicine*,
DOI 10.1007/978-3-540-79116-4_18, © Springer-Verlag Berlin Heidelberg 2009

tyrA Bacterial bifunctional prephenate dehydrogenase
VTE2 Homogentisate phytyltransferase
GGH Geranylgeranyl diphosphate hydratase

18.1 Introduction

The concept of metabolic pathway manipulation for the purpose of endowing microorganisms with desirable properties is very old indeed. Metabolic engineering emerged with DNA recombination as the enabling technology. Initially attention was focused, almost exclusively, on the synthetic side of this field: expression of new genes in various host cells, amplification of endogenous enzymes, deletion of genes or modulation of enzymatic activity, transcriptional or enzymatic deregulation, etc. As such, metabolic engineering was, to a significant extent, the technological manifestation of applied molecular biology, with very little engineering content [1]. Stephanopoulos et al. [1] defined metabolic engineering as the directed improvement of product formation or cellular properties through the modification of specific biochemical reaction(s) or the introduction of new one(s) with the use of recombinant DNA technology.

Metabolic engineering in plants involves the modification of endogenous pathways to increase flux toward particular desirable molecules. Giving an example, Ye et al. [2] produced transgenic plants overexpressing the phytoene synthase gene PSY, the daffodil lycopene β-cyclase gene LVY-B, and the *Erwinia* phytoene desaturase gene crtI to produce a high level of β-carotene accumulation in transgenic rice seeds. "Golden rice" is another good example of the use of metabolic engineering to improve the content of plant vitamin A, which has been hailed as a significant advance in plant nutrient quality.

Vitamin E is important for human and animal health and only synthesized in higher plants and other oxygenic photosynthetic organisms [3–5]. For human and animal health, α-tocopherol has the highest vitamin E activity [6], partly because it is retained in the human body in preference to other tocopherols and tocotrienols [7]. The function of vitamin E in mammals is to act as a free-radical scavenger for inhibiting lipid oxidation. A high intake of vitamin E is associated with a decreased risk of certain cancers and neurodegenerative and cardiovascular diseases [8]. Furthermore, new functions of vitamin E as an antihypercholesterolemic and immunostimulatory agent have been proposed [9].

However, species and types of plant tissues vary greatly in their total tocochromanol content and composition. Oilseeds are the richest source of vitamin E, with total tocochromanol levels ranging from 330 to 2000 μg per gram of oil [12]. The major form of vitamin E in oilseeds is γ-tocopherol. Unfortunately, the vitamin E activity of γ-tocopherol is only one tenth that for α-tocopherol. However, in green leaves of higher plants, the total tocochromanol content is very low, but the proportion of α-tocopherol is high [10].

Table 18.1 Important milestones in vitamin E history

Years	Work	Reference
1922	Existence of vitamin E recognized by Evans and Bishop when it became clear that this fat-soluble factor (named factor X) prevented fetal death in animals fed a diet containing rancid lard	[19]
1938	Fernholz elucidates structure of vitamin E	[20]
1938	Synthesis of vitamin E by Karrer	[21]
1955	Revelation by Gordon and colleagues that mature infants had low levels of blood tocopherol and abnormal hemolysis of erythrocytes, incubated in presence of H2O2	[22]
1967	Study by Bunyan and colleagues on antioxidant impact of vitamin E on polyunsaturated fatty acids	[23]
1991	Evidence presented by Boscoboinik that smooth muscle cell proliferation is inhibited by α-tocopherol through protein kinase C modulation	[24,25]
1997	Hydroxyphenylpyruvate dioxygenase is cloned from carrot by Garcia et al.	[26]
1998	Discovery by Fechner et al. that the expression of α-tocopherol transfer protein in the liver is induced by α- and β-tocopherol	[27]
1998	Norris et al. isolate the gene encoding ρ-hydroxyphenylpyruvate dioxygenase from Arabidopsis	[28]
1998	Shintani D and DellaPenna D clone γ-TMT from *Arabidopsis thaliana* and *Synechocystis* PCC6803 by genomics-based approach	[29]
1999	Evidence presented by Aratri et al. that increased transcription level of α-tropomyosin is caused by α-tocopherol	[30]
2001	Discovery of α-tocopherol as a transcriptional regulator of gene expression via association with a transcription factor tocopherol-associated protein	[31]
2001	Membrane-bound homogentisate phytyltransferase (HPTs) identified from *Synechocystis* sp. PCC6803 and *Arabidopsis* by Eva Collakova and Dean DellaPenna	[32]
2002	Peter Dörmann et al. identify tocopherol cyclase essential for all tocopherol biosynthesis from *Arabidopsis*	[33]
2003	Dean DellaPenna and colleagues isolate and characterized the 2-methyl-6-phytyl-1,4-benzoquinone/2-methyl-6-solanyl-1,4-benzoquinone methyltransferase (MPBQ/MSBQ MT) from *Arabidopsis*	[34]
2004	Alpha-tocopherol modulates two major signal transduction pathways centered on protein kinase C and phosphatidylinositol 3-kinase	[35]
2004	DellaPenna et al. prove that vitamin E is essential for seed longevity and for preventing lipid peroxidation during germination	[36]
1998&2005	Genetic engineering is used to elevate the tocopherol content in leaves and seeds	[28,68]
2006	Hiroshi Maeda and Dean DellaPenna reveal that tocopherols are required for proper adaptation of phloem loading at low temperatures	[37]
2005&2007	It is discovered that vitamin E modulates signal transduction in plants by influencing jasmonic acid levels	[38,39]

There is a need to increase vitamin E production through plant engineering in order to meet the demand for human consumption. Numerous studies have been carried out in this field leading to many successful examples of increased vitamin E production. Elevated tocopherol intake has also been reported to reduce the occurrence and severity of several diseases, including heart disease, some cancers, neurodegenerative diseases, and cataracts [11].

The Recommended Daily Allowance (RDA) for vitamin E was originally set at 8 and 10 mg tocopherol for adult women and men, respectively. The new RDA has been raised to 15 to 30 mg for men and women [12]. These amounts can be obtained by eating a wide variety of foods rich in vitamin E. However, daily intake of vitamin E in excess of the RDA (100 to 1000 international units) prevents cardiovascular disease and some cancers, improves immune function, and slows the progression of a number of degenerative human conditions [9]. Because of these health benefits, there is a considerable interest in increasing vitamin E content and altering its composition in favor of α-tocopherol by metabolic engineering in plants [12–15, 18].

18.2 Vitamin E History

Vitamin E was discovered at the University of California at Berkeley in 1922 (Table 18.1). Since its discovery, the constituent tocopherols and tocotrienols have been studied intensively.

18.3 Vitamin E Structure

Vitamin E consists of a mixture of tocopherols and tocotrienols (Fig. 18.1), and all are derivatives of 6-chromanol with an aliphatic side chain. The four tocopherol

A. Tocopherols

B. Tocotrienols

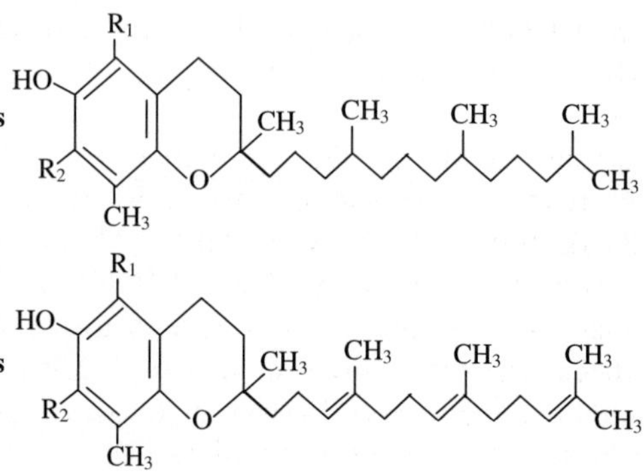

Fig. 18.1 Tocopherol and tocotrienol structures

homologs (α-, β-, γ-, and δ-) have a fully saturated C_{16} phytol side chain [15], whereas tocotrienols (α-, β-, γ-, and δ-) have a similar isoprenoid chain containing three double bonds [16]. Individual tocopherols are named according to the position and number of the methyl groups on the phenol ring, with the α-, β-, γ-, and δ-tocopherols containing three, two, two, and one methyl groups, respectively. These structural differences determine biological activity, α-homologues being the most biologically active. α-Tocopherol predominates in leaves of higher plants, whereas γ-tocopherol is often the major form in seeds [17].

18.4 Vitamin E Biosynthetic Pathway in Plants

Vitamin E biosynthetic pathway in plants was elucidated over 30 years ago. The tocopherols are exclusively synthesized in photosynthetic organisms including higher plants; significant amounts are found in all green tissues but predominantly occur in seeds [114]. In plants, tocopherol biosynthesis takes place in the plastid, and the enzymes are associated with the plastidial envelop [40, 41]. Figure 18.2 shows the vitamin E biosynthetic pathway in plants [111].

Homogentisic acid (HGA) and phytyldiphosphate (PDP), derived from cytosolic aromatic amino acid metabolism and plastidic deoxyxylulose 5-phosphate pathway, respectively (phytyl-PP for tocopherols and GGDP for tocotrienols) [42], serve as precursors for vitamin E biosynthesis.

The first step involves the production of the aromatic head group, HGA, from ρ-hydroxyphenylpyruvic acid (HPP) by the enzyme ρ-hydroxyphenylpyruvic acid dioxygenase (HPPD) [26, 28, 43–46]. The following committed reaction, the prenylation of HGA with PDP, resulting in the formation of 2-methyl-6-phytylbenzoquinol (MPBQ), is catalyzed by the homogentisate phytyltransferase (HPT), encoded by slr1736 in *Synechocystis* sp. PCC 6803 (*Syn-vte2*), or in *Arabidopsis* by the vitamin E2 gene (*At-VTE2*). Some plant species, such as oil palm, corn, and tobacco, accumulate substantial amounts of tocotrienols. These plants may harbor a VTE2 isoform exhibiting an altered substrate specificity, with preference toward geranylgeranyl diphosphate (GGDP) rather than PDP [47]. This results in the prenylation of HGA with GGDP to form 2-methyl-6-geranylgeranylbenzoquinol (MGGBQ), a tocotrienol precursor.

In *Arabidopsis* the metabolic fate of the VTE2 reaction product, MPBQ or MGGBQ, is determined by the relative enzymatic activities of the tocopherol cyclase (TC) and the MPBQ (MGGBQ) methyltransferase (MPBQ MT) encoded by the VTE1 and VTE3 loci, respectively. After methylation by MPBQ MT, TC converts MPBQ and DMPBQ to δ- and γ-tocopherol, respectively (and also cyclizes the corresponding geranylgeranylated intermediates to form δ- and γ-tocotrienols).

The VTE3 reaction product, 2,3-dimethyl-5-benzoquinol, serves as a substrate for VTE1 to form γ-tocopherol. This enzyme also utilizes MPBQ as substrate, resulting in the formation of δ-tocopherol. However, the link between the phenotype and tocopherol biosynthesis was not established at that time and continues to be elusive.

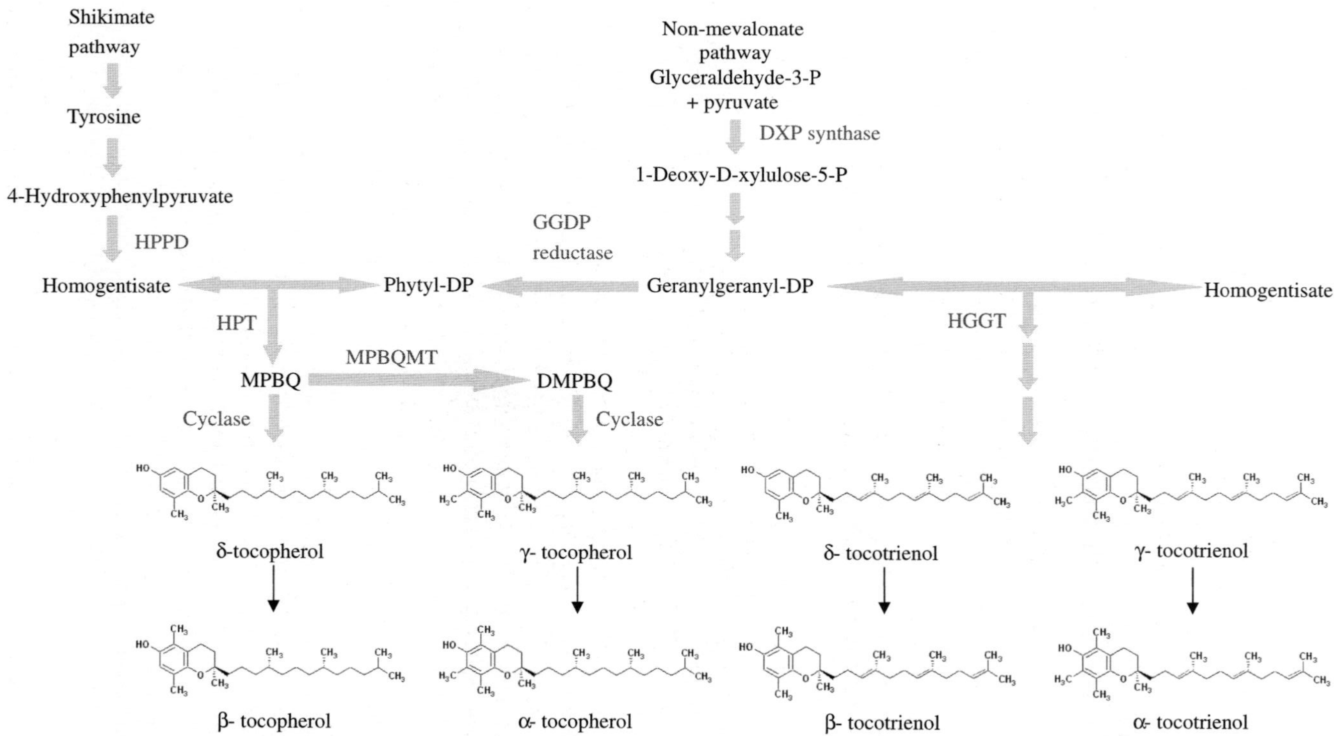

Fig. 18.2 Vitamin E biosynthetic pathway in plants (from Ref. [103] with permission of the authors). **a** R1 = R2 = CH_3, known as α-tocopherol, is designated α-tocopherol or 5,7,8-trimethyltocol; R1 = CH_3; R2 = H,known as β-tocopherol, is designated β-tocopherol or 5,8-dimethyltocol; R1 = H; R2 = CH_3, known as γ-tocopherol, is designated γ-tocopherol or 7,8-dimethyltocol; R = R = H, known as δ-tocopherol, is designated δ-tocopherol or 8-methyltocol. **b** R1 = R2 = H, 2-methyl-2-(4,8,12-trimethyltrideca-3,7,11-trienyl)chroman-6-ol, is designated tocotrienol; R1 = R2 = CH_3, formerly known as ζ1 or ζ2-tocopherol, is designated 5,7,8-trimethyltocotrienol or α-tocotrienol. The name tocochromanol-3 is also used; R1 = CH_3; R2 = H, formerly known as ε-tocopherol, is designated 5,8-dimethyltocotrienol or β-tocotrienol; R1 = H; R2 = CH_3, formerly known as γ-tocopherol, is designated 7,8-dimethyltocotrienol or γ-tocotrienol. The name plastochromanol-3 is also used; R1 = R2 = H is designated 8-methyltocotrienol or δ-tocotrienol

The final reaction to convert δ- and γ-tocopherol to β- and α-tocopherol is catalyzed by the γ-methyltransferase (γ-TMT), encoded by VTE4, and δ- and γ-tocotrienols to β- and α-tocotrienols [17, 34, 40, 48, 49].

18.5 Strategies for Vitamin E Metabolic Engineering in Plants

Metabolic engineering in plants is generally defined as the redirection of one or more enzymatic reactions to produce new compounds in an organism, improvement of existing compounds, or mediation of the degradation of compounds. Although progress in discovering biosynthetic pathway genes and in our ability to manipulate gene expression in transgenic plants has been most impressive during the past two decades, attempts to use these tools to engineer plant metabolism has met with more limited success. Though there are notable exceptions, most attempts at metabolic engineering have focused on modifying the expression of single genes affecting pathways.

There are three basic goals of metabolic engineering in plants: high production of a specific desired compound, low production of a specific unwanted compound, and the production of a novel compound, for example, a molecule that is produced in nature, but not usually in the host plant, or a completely novel compound. This includes two complementary strategies. (1) Introducing the genes that encode enzymes can increase the flux through the tocopherol biosynthetic pathway to produce elevated levels of total tocols. (2) Introducing the genes that encode enzymes can affect the composition of tocols and make α-tocopherol the predominant form of vitamin E, so as to elevate the activity of vitamin E.

There are two basic goals of vitamin E metabolic engineering in plants: (1) increasing the flux through the vitamin E biosynthetic pathway to enhance the levels of total tocochromanols and (2) altering the tocochromanol composition in favor of α-tocopherol. The approaches to achieving these goals include overexpression of single-gene, multiple-gene combinations or a transcription factor to establish single-gene or multigene control in the biosynthetic pathway for vitamin E, or use of RNAi/antisense knockout of a gene controlling the flux of some other metabolic pathway in order to increase the content or change the composition of vitamin E (Fig. 18.3).

Though progress in the discovery of genes of the biosynthetic pathway and our ability to manipulate gene expression in transgenic plants have been most impressive during the past two decades, attempts to use these tools to engineer plant metabolism has met with more limited success. Though there are notable exceptions, most attempts at metabolic engineering have focused on modifying the expression of single genes affecting pathways.

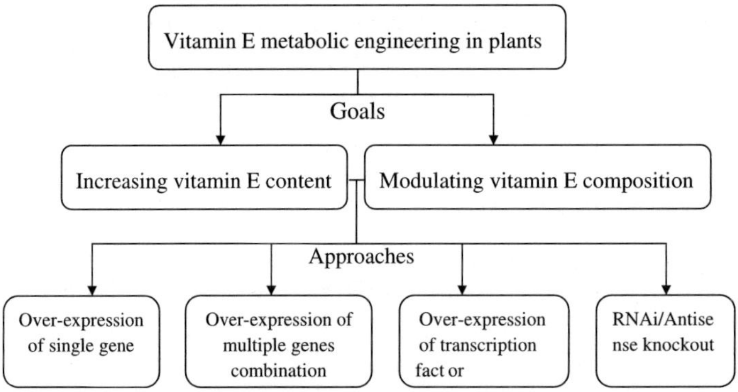

Fig. 18.3 Strategies of vitamin E metabolic engineering in plants

18.6 Isolation and Identification of Genes in the Vitamin E Biosynthetic Pathway

The biosynthetic pathway of vitamin E was elucidated several years ago, but the genes encoding the enzymes of the pathway have been identified only very recently (Table 18.2). Vitamin E biosynthetic enzymes are found in chloroplasts [40], [50–53] and chromoplasts [54].

The first potential enzymes involved in tocopherol synthesis are the hydroxyphenyl pyruvate dioxygenase (HPPD) and geranylgeranyl diphosphate reductase (GGDPR). A single copy of the gene encoding the HPPD was identified in *Daucus carota, Arabidopsis thaliana*, barley (*Hordeum vulgare*), and *Synechocystis* sp. PCC 6803 and is essential for both tocopherol and plastoquinone biosynthesis [26, 28, 45, 55–57]. HPPD is a cytosolic enzyme [64]. HPPD and GGDPR were selected as candidate regulatory enzymes based on in vivo observations showing that HGA and PDP might be limiting factors in tocopherol biosynthesis. Tissue-culture studies have shown that exogenously supplied HGA and PDP both caused significant increases in tocopherol biosynthesis [58]. The expression of gene-encoding HPPD, which is overexpressed in the leaves and seeds of both *Arabidopsis* and tobacco, proves that HGA levels limit tocopherol biosynthesis [59, 60].

In a recent patent, Grimm and Tanaka [62] identified a geranylgeranyl reductase cDNA (CHL P) from a Lambda ZAP II cDNA library of tobacco. The enzyme of geranylgeranyl reductase is involved in isoprenoid metabolism and functions in two metabolic pathways: tocopherol biosynthesis and chlorophyll biosynthesis.

Another important flux-regulating enzyme is homogentisate phytyltransferase (HPT), encoded by the *Arabidopsis* vitamin E2 (VTE2) locus. The genes encoding HPT have been independently identified using bioinformatics in three different groups in *Synechocystis* sp. PCC 6803 and *Arabidopsis* based on the similarity of their sequence to chlorophyll synthases [32, 63, 64].

Table 18.2 Enzymes and genes involved in vitamin E biosynthesis from different species

Enzyme activity	Species	Gene	Locus	Substrate	Reference
Homogentisate phytytransferase (HPT)	*Synechocystis*	Slr1736		HGA, PDP,	[32, 63, 64]
	Arabidopsis	At2g18950	VTE2, HPT1	GGDP	[32, 64]
Homogentisate geranylgeranyl	*Arabidopsis*			HGA, PDP,	[36]
transferase (HGGT)	Barley, rice, wheat			GGDP	[47]
Homogentisate prenyldiphosphate transferase	*Arabiopsis*	At3g11950	VTE2-paralog		[84, 85]
2-Methyl-6-prenylbenzoquinol	*Synechocystis*	Sll0418		MPBQ,	[73]
methyltransferase (PrBQMT)	*Arabidopsis*	At3g63410	VTE3	MGGBQ	[34, 74]
Tocopherol/tocotrienol cyclase	Sunflower			(D)MPBQ, (D)MGGBQ	[86]
(TC)	Synechocystis	Slr1737			[77]
	Arabidopsis	At4g32770	VTE1		[33, 77]
Tocopherol/tocotrienol	Maize	SXD1		δ-,γ-	[78]
methyltransferase (TMT)	Potato	StSXD1		Tocopherol,	[51]
	Synrchocystis	Slr0089		δ-,γ-	[29]
	Arabidopsis	At1g64970	VTE4	Tocotrienol	[29, 87]
Phytol kinase (PK)	*Perilla*	Slr1652		Phytol,	[88]
	Sunflower	At5g04490	VTE5	PMP	[89]
	Synechocystis				[90]
	Arabidopsis				[90]
P-Hydroxypherylpyruvic acid	*Synechocystis*	Slr0090		HPP	[28, 45, 56]
dioxygenase (HPPD)	*Arabidopsis*	At1g06590	PDS1		

Homogentisic acid geranylgeranyl transferase (HGGT) is a functionally divergent form of HPT that displays substrate specificity for GGDP in preference to PDP. cDNAs encoding HGGT, with 40 to 50% identity to *Arabidopsis* HPT, were isolated from seeds of monocot species, such as barley, wheat (*Triticum aestivum*), and rice (*Oryza sativa*) [47]. Tocotrienols are not always synthesized by dicot species, as observed by Cahoon et al. [47]. When the barley HGGT gene was overexpressed in *Arabidopsis* leaves, tocotrienols accumulated in very high levels, whereas tocopherol levels were not affected. This indicates that HGGT and HPT are highly specific for their prenyl substrates, GGDP and PDP respectively, and must compete with one another for HGA. Although overproduction of tocotrienols clearly shows that GGDP and HGA are abundantly available, the fact that HGGT overexpression has no effect on tocopherol production indicates that PDP must be limiting. Therefore, overexpression of GGDP reductase could increase PDP availability significantly and, thus, increase pathway flux.

The role of GGH in catalyzing the conversion of GGDP to PDP in tocopherol and chlorophyll biosynthesis has been suggested [10–13]. Studies on *Synechocystis* GGH deletion mutants and antisense expression of GGH in tobacco plants clearly demonstrate that GGH is essential for tocopherol biosynthesis in bacteria and plants [65–67]. However, transgenic expression of At-GGH alone, or co-overexpression of At-GGH with At-HPPD, Eh-TYRA, and At-VTE2 in soybean seeds, did not reduce tocotrienol content [68], suggesting that an additional independent pathway may govern the synthesis of PDP in plants. This was confirmed by early reports of the presence of strong phytol kinase activity in spinach leaf chloroplasts and a recent *Arabidopsis* mutant encoding a phytol kinase gene [40, 69]. The gene of phytol kinase is essential for the biosynthesis of at least 80% of the seed tocopherols in *Arabidopsis* [70], suggesting that 80% of PDP for tocopherol biosynthesis is catalyzed by phytol kinase and up to 20% of PDP is formed directly via reducing GGDP.

The activity of MPBQ/MSBQ MT has been detected in spinach (*Spinacia oleracea*) chloroplasts [41], and maize (*Zea mays*) and sunflower (*Helianthus annuus*) mutants have been identified with phenotypes that disrupt MPBQ/MSBQ MT activity [71, 72]. MPBQ/MSBQ MT was cloned from *Synechocystis* PCC6803 (SLL0418) based on the similarity of its sequence to γ-TMT [34, 73]. Because the plant ortholog of SLL0418 was not functional, MPBQ/MSBQ MT was cloned from *Arabidopsis* (VTE3) via the map-based cloning approach using partial loss-of-function alleles from EMS mutagenized *Arabidopsis* plants [34, 74]. Interestingly, the two proteins of VTE3 and SLL0418 shared less than 20% amino acid identity but displayed similar activities in the tocopherol and plastoquinone pathways [34, 74], suggesting that the convergent evolution occurred in this step of the pathway in cyanobacteria and plants. VTE3 is the only enzyme of the pathway that is also involved in plastoquinone synthesis, which catalyzes a key methylation step in both tocopherol and plastoquinone (PQ) synthesis [34, 73–75]. The essential function of VTE3 in two independent metabolic pathways can increase the challenge of tocopherol pathway engineering; as altered tocopherol intermediate pools, resulting from enhanced tocopherol flux, can lead to substrate competition between MPBQ or MGGBQ and the plastoquinol precursor MSBQ.

The function of tocopherol cyclase (TC) in tocopherol biosynthesis was discovered with the help of mutational analysis in *Synechocystis* [76, 77]. TC, encoded by the *Arabidopsis* VTE1 loci, catalyzes the formation of the chromanol headgroup of the various tocopherol isoforms, DMPBQ or MPBQ. So far, TC has been purified and characterized only from the cyanobacterium *Anabaena variabilis* [79]. Recently, genes coding for the TC from *Synechocystis* sp. PCC6803 (slr1737), *Arabidopsis thaliana* (VTE1), and maize (SXD1) have been cloned [33, 77, 78, 80]. The mutation of genes VTE1, SXD1, or slr1737 resulted in both a tocopherol deficiency and an accumulation of 2,3-dimethyl-6-phytyl-1,4-benzoquinone (DMPBQ), which suggests that TC activity is evolutionarily conserved between plants and cyanobacteria.

The γ-tocopherol methyltransferase (γ-TMT), encoded by a VTE2 locus in the *Arabidopsis* gene, was also discovered through bioinformatics analysis and confirmed by mutational analysis in *Synechocystis* [29]. A genomics-based approach was used to clone γ-TMT, which is the final enzyme in α-tocopherol synthesis.

The gene for γ-TMT in the *Synechocystis* PCC6803 genomic database was identified using *Arabidopsis* HPPDase to initiate a search. To identify the γ-TMT gene from *Arabidopsis*, the *Synechocystis* γ-TMT protein sequence was used to search the *Arabidopsis* expressed sequence tag (EST) database, resulting in the discovery of a cDNA clone with a 66% amino-acid-sequence similarity with *Synechocystis* [29, 81].

The isolation and identification of the enzymes in the vitamin E biosynthetic pathway has driven researchers to bioengineer plants with improved vitamin E content and alterations in its composition. Meanwhile, quantitative vitamin E loci in *Arabidopsis* provide insight into the regulation and/or metabolism of vitamin E in plants and has clear ramifications for improving the nutritional content of crops through marker-assisted selection and metabolic engineering [82].

18.7 Engineering Plants to Improve Vitamin E Content

Various groups have reported engineering tocopherol content composition in *Arabidopsis* leaves and seeds by overexpression of various pathway enzymes (Table 18.3) [29, 32, 47, 59, 64, 74, 91, 92].

Seed-specific expression of *Arabidopsis* VTE3 and VTE4 alone or together did not significantly alter the total level of tocopherols in transgenic soybean seed but had a dramatic impact on tocopherol composition [92]. Overexpression of VTE3 alone increases soybean seed γ- and α-tocopherol levels and correspondingly reduces the levels of δ- and β-tocopherols. Like overexpression of VTE4 in *Arabidopsis* seed [29], VTE4 overexpression in soybean seed converts γ-tocopherol almost completely to α-tocopherol, a sevenfold increase to 75% of the total, and δ-tocopherol almost completely to β-tocopherol, a tenfold increase to 25% of the total. Overexpression of VTE3 and VTE4 together shifted the tocopherol composition of soybean seeds from only 10% of α-tocopherol to about 90% α-tocopherol.

Sense expression of the barley HPD gene, under control of the 35S promoter, resulted in an up to twofold increase in the tocochromanol content of tobacco seeds. A similar increase was obtained by sense expression of the gene encoding HPT from *Synechocystis* PCC6803 under the control of the seed-specific napin promoter [64]. Dufourmantel et al. have expressed a sensitive bacterial HPPD gene from *Pseudomonas fluorescens* in plastid transformants of tobacco and soybean. HPPD accumulates to approx. 5% of the total soluble protein in transgenic chloroplasts of both species. As a result, the soybean and tobacco plastid transformants acquire a strong herbicide tolerance, performing better than nuclear transformants. In contrast, the overexpression of HPPD had no significant impact on the vitamin E content of leaves or seeds, quantitatively or qualitatively.

Overexpression of CHL P in tobacco plants resulted in a four- to sixfold and a two- to threefold increase in tocopherol level in the leaves and seeds, respectively, compared to the wild-type plants. Interestingly, the transgenic progeny plants had higher tocopherol content than the nontransformed control [62].

Table 18.3 Metabolic engineering and the potential level of vitamin E in plants

Gene source	Promoter used	Gene	Plant species engineered	Product level	Reference
A. thaliana	35S	*VTE1*	*A. thaliana* leaf	7	[102]
A. thaliana, Zea mays	Napin	*VTE1*	*B.napus* seed	0.28	[80]
				0.18	
A. thaliana	35S	*VTE1*	*N.tabacum* leaf	10.2	[96]
A. thaliana	35S	*HPPD*	*Synechocystis*	7	[68]
A. thaliana, E.herbicola	Lac	*At-HPPD, Eh-tyrA*	*A. thaliana*	1.78	[93]
			B.campestris	2.44	[68]
			G.max	2.58	
A. thaliana, Saccharomyces cerevisiae	35S	*At-HPPD & Sc-PDH*	*N.tabacum* leaf	10	[91]
Synechocystis sp. PCC6803	35S	*At-HPPD & Eh-tyrA*	*Synechocystis* sp. PCC6803	16.5	[68]
		At-VTE2 & At-GGH	*G. max* seed	14.7	
A. thaliana	35S	*HPPD, tyrA VTE2*	*B.campestris* seed	3.7	[68]
A. thaliana	35S	*HPPD, tyrA, VTE2*	*A. thaliana*	5	[68]
	35S	*HPPD & tyrA*	*A. thaliana* seed	1.8	[68]
			G. max seed	2.6	
A. thaliana	Napin	*HPPD*	—	1.0–1.2	[60]
A. thaliana	35S	*HPPD*	*A. thaliana*	0.28	[59]
				0.96–1.1	[68]
A. thaliana	35S	*HPPD*	*Synechocystis* sp. PCC6803 seed	7	[68]
A. thaliana	DC3	*HPPD*	*A. thaliana* seed	1.28	[59]
H.vulgare	35S	*HPPD*	*N.tabacum* seed	2	[60]
H.vulgare	35S	*HGGT*	*A. thaliana* leaf	10–15	[68]
			Z.mays seed	6	
Zea mays	35S	*HGGT*	Zea mays seed	8	[68]
A. thaliana Synechocystis sp. PCC6803	Napin	*HPT*	*A. thaliana* seed	2	[22]
A. thaliana	35S	*HPT*	*A. thaliana* leaf	3–4.4	[93]
			A. thaliana seed	0.4	
A. thaliana	35S	*HPT & γ-TMT*	*A. thaliana* seed	12	[93]
A. thaliana	35S	γ-TMT	*A. thaliana* seed	10	[93]
E.herbicola	35S	*TYRA*	*Synechocystis*	1.6	[68]
	Napin		*A. thaliana*	1.53	
			B.campestris	2.37	
			G.max	1.11	
E.herbicola	Napin	*TYRA*	—	1–1.4	[60]
E. herbicola	35S	*TYRA*	—	0.72–1.14	[60]
A. thaliana	Napin	*GGH*	—	0.94–1.12	[68]
A. thaliana	35S	*GGH*	—	0.9–1.06	[68]

Table 18.3 (continued)

Gene source	Promoter used	Gene	Plant species engineered	Product level	Reference
A. thaliana	35S	*VTE2*	*A. thaliana leaf*	4.4	[93]
A. thaliana	35S	*VTE2*	*G. max* seed	1.4	[93]
			A. thaliana seed	1.8	[64]
					[68]
A. thaliana	Napin	*VTE2*	—	1.29–1.53	[64]
Synechocystis sp. PCC 6803	Napin	*VTE2*	—	1.2–1.52	[64]
Synechocystis sp. PCC 6803	35S	*VTE2*	—	0.9–1.05	[68]
A. thaliana	35S	γ-TMT	*B.campestris*	2	[98]
			canola	2	
A. thaliana	35S	γ-TMT	*Latuca sativa* L.	2	[99]
A. thaliana	Napin	MPBQ-MT&γ-TMT	*G. max* seed	5	[74]
Perilla frutrscens	35S	*γ-TMT*	*G. max* seed	4.8	[88]
Solanum tuberosum	35S	Zeaxanthin epoxidase	*Solanum tuberosum*	2- to 3-fold α-T	[113]
Synthetic	35S	ZFP-TFs	*A. thaliana*	20-fold α-T	[106]
N.tabacum	35S	*CHL P*	*N.tabacum* leaf	4–6	[62]
			N.tabacum seed	2–3	

In overexpression studies, a tenfold increase in HPT activity translated into a 4.4-fold increase in leaf tocopherol levels, relative to wild-type plants [93]. Similar results were obtained in seeds, but the magnitude of the tocopherol increase was lower, ranging from 0.4- to twofold compared with wild-type levels [93, 94]. Both HPT1 and γ-TMT overexpressed in seeds resulted in a total tocopherol content 12-fold higher than wild types.

The maize HGGT gene was overexpressed in maize seeds, leading to a 20-fold increase in tocotrienol levels, which translated into an 8-fold increase in total tocols (tocopherols and tocotrienols) [47]. This result is the largest increase in tocol production ever observed in plants and significantly increases the antioxidant potential of corn. Unfortunately, because dietary tocotrienols are not absorbed as well as α-tocopherol, the large increase in tocotrienol levels observed in the HGGT overexpressing maize seeds did not add much to the vitamin E nutritional value of these plants. However, because tocotrienols have superior in vitro antioxidant activity [3], transgenic plants with elevated tocotrienol levels could be used as sources of chemical antioxidants for industrial applications, such as oxidative stabilizers for paints, coatings, and other lipophilic products. Furthermore, it has been reported that tocotrienols might have a therapeutic role in decreasing the cholesterol level in humans [95].

Overexpression of genes for TCs from *Arabidopsis* in maize and canola plants led to an 18% and 28% increase of the total tocochromanol content in the seed oil, respectively. The average δ-tocopherol content increased up to 1.6-fold and

2.7-fold, respectively [80]. Overexpression of tocopherol cyclase (ATPT2 sequence) in *Arabidopsis* resulted in a 50% increase in total tocopherol levels and over a threefold increase in δ-tocopherol levels in the seed (Table 18.3). The δ-tocopherol content increased by the conversion of 2-methyl-6-phytyl-1,4-hydroquinol (MPQ) to δ-tocopherol in overexpressing lines. Our results showed that overexpression of At-VTE1 in tobacco increased 10.2-fold, compared with that in the control [96].

In a recent report by Van Eenennaam et al. [74], the genes encoding γ-TMT and MPBQMT were overexpressed in soybean seeds to improve this important dietary source of vitamin E. The overexpression of the two tocopherol methyltransferases resulted in a 95% conversion of these lesser forms of vitamin E to α-tocopherol, which translated into a fivefold increase in vitamin E content. To put this into a real-world perspective, although four tablespoons of soybean oil from wild-type plants contains only 13 international units (IU) of vitamin E, the same volume of oil from the methyltransferase-overexpressing lines contain 65 IU of vitamin E. Because 100 IU is the recommended minimum therapeutic dose to decrease the risk of heart disease, the work of Van Eenennaam et al. [74] has done much to increase the nutraceutical potential of plant-derived vitamin E.

The enzymes catalyzing the reactions at the flux control points have been overexpressed with the aim of increasing the total tocopherol levels. The studies include the overexpression of hydroxyphenyl pyruvate dioxygenase by Tsegaye et al. [59] and Falk et al. [60], of deoxyxylulose phosphate synthase by Estevez et al. [97], and of homogentisate phytyl transferase by Collakova and DellaPenna [93] and Savidge et al. [64]. The success rates in these studies have varied, but not very drastically. Methylphytylbenzoquinone methyl transferase, tocopherol cyclase, and γ-TMT are the enzymes important in determining the tocopherol composition [14]. The overexpression of MBPQMT and γ-TMT in soybean seeds resulted in an increase of α-tocopherol by greater than 8-fold, at the expense of δ-, β-, and γ-tocopherols [98], while the overexpression of γ-TMT in the model plant *A. thaliana* increased the seed α-tocopherol levels by 80-fold [29]. In lettuce it led to a more than twofold elevation [99]. The overexpression of γ-TMT gene increased the α-tocopherol levels by more than sixfold [100]. The γ-TMT gene isolated from *Perilla frutescens* was overexpressed in soybean using a seed-specific promoter, and vitamin E content in T2 seeds was 4.8-fold higher than that in the wild type [88].

Single-gene engineering strategies for more complex approaches involved the simultaneous overexpression and/or suppression of multiple genes. The use of regulatory factors to control the abundance or activity of several enzymes is also becoming more widespread. In combination with emerging methods to model metabolic pathways, this should facilitate the enhanced production of natural products and the synthesis of novel materials in a predictable and useful manner [101].

18.8 Engineered Plants to Modulate Vitamin E Composition

Soybean seeds contain approx. 20 to 30% δ-tocopherol, 2 to 5% β-tocopherol, 60 to 70% γ-tocopherol, and 10 to 20% α-tocopherol. The overexpression of At-VTE3 (2-methyl-6-phytylbenzoquinol methyltransferase) in soybean seeds resulted in nearly a complete conversion of δ- and β-tocopherols to γ- and α-tocopherols, which indicated that VTE3 controls the flux from MPBQ to γ- and α-tocopherol [74]. When At-VTE3 was co-overexpressed with At-VTE4 (γ-tocopherol methyltransferase) in soybean, almost all β-, γ-, and δ-tocopherols were converted to α-tocopherol, with a seed accumulation of $>$ 95% α-tocopherol and up to a fivefold increase in vitamin E activity (Table 18.3) [74, 103]. These combined experiments could be extended to other agronomically important crops.

It has been demonstrated recently that overexpression of VTE3 in soybean seeds is sufficient to funnel nearly all tocopherols into the pool of α- and γ-tocopherols [74, 98]. A nontransgenic soybean contains on average 30 to 35% β- and δ-tocopherol.

Overexpression of VTE1 in *Arabidopsis* leaves resulted in up to a sevenfold increase in tocopherols and a dramatic shift in tocopherol composition from α-tocopherol (16.5%) to γ-tocopherol (80.5%), which indicated that both VTE1 and γ-TMT are limiting factors for tocopherol biosynthesis [102]. This differs from previous work, in which overexpression of γ-TMT and HPT, singly or in combination, clearly demonstrated that γ-TMT was not a limiting factor for tocopherol biosynthesis in *Arabidopsis* leaves in the absence of stress but became a limiting factor under abiotic stress [29, 93, 104]. Accumulation of tocopherol in VTE1 overexpressing plants led to a 60% and 40% decrease in ascorbate and glutathione, two key water-soluble antioxidants, respectively [102]. It is unclear why increased VTE1 activity would lead to a reduction in ascorbate and glutathione, indicating that there is still much to be learned on the regulation of tocochromanol synthesis in plants.

Arabidopsis, Perilla, and canola seeds contain predominantly γ-tocopherol. Overexpression of the gene γ-TMT led to the efficient conversion of γ-tocopherol to α-tocopherol and resulted in dramatic increase in seed α-tocopherol content [29, 81, 98, 105]. The gene encoding *Arabidopsis* γ-TMT was overexpressed in lettuce (*Latuca sativa*), which resulted in an increase in the ratio of α-/γ-tocopherol content (TR) by up to 0.8 to 320 from 0.6 to 1.2 in nontransformed plants (Table 18.3) [99]. Among the transformed plants, the total tocopherol content did not change. Therefore, γ-TMT may play an important role in determining the composition of tocopherols, but not in the total tocopherol content in transgenic plants.

γ-Tocopherol methyl transferase cDNA from *Arabidopsis thaliana*, coding for the enzyme catalyzing the conversion of the large γ-tocopherol pool to α-tocopherol, was overexpressed in *Brassica juncea* plants. T1 transgenic lines showed a shift in tocopherol profile, having α-tocopherol levels as high as sixfold over the nontransgenic controls [100].

In a recent report, designer transcription factors were used to regulate vitamin E synthesis [106]. Five three-finger zinc finger proteins (ZFPs) were designed to bind to a target 9-bp-long sequence in the promoter or coding regions of the *Arabidopsis*

GMT gene (γ-tocopherol methyltransferase). These ZFPs were separately fused to the maize opaque-2 nuclear localization signal and the maize C1 activation ED to make synthetic zinc finger transcription factors (ZFP-TFs). Overexpression of these ZFP-TFs in *Arabidopsis* seeds under the control of an embryo-specific promoter resulted in a heritable 20-fold increase in α-tocopherol compared to the control seed (Table 18.3) [106]. The dramatic increase of α-tocopherol was derived from the increased expression of γ-TMT modulated by ZFP-TFs. This provides a graphic example of designer transcription factor regulating the expression of vitamin E pathway enzymes.

18.9 Conclusions and Perspectives

It is clear from some reports that in many cases a single gene will be insufficient to achieve high-level production of desired chemicals. For example, although transgenic lines overexpressing the gene encoding HPPD showed large increases in enzyme activity compared with control lines, only small increases in tocopherol content were observed (10% in leaves and 30% in seeds) [59, 60], suggesting that HPPD alone is not sufficient to increase tocopherol biosynthetic flux. Therefore, multipoint metabolic engineering is now beginning to supersede single-point engineering as the best way to manipulate metabolic flux in transgenic plants. Several points in a given metabolic pathway can be controlled simultaneously either by overexpressing and/or suppressing several enzymes, or through the use of transcriptional regulators to control several endogenous genes. Many researchers are fast approaching a time when sophisticated strategies for the multipoint manipulation of metabolic pathways can be modeled and implemented by multiple-gene transfer, thereby facilitating the production of desirable molecules in transgenic plants [101, 107].

Combined experiments on the co-overexpression of TyrA, HPPD and VTE2, HPT and γ-TMT, VTE3 and VTE4 significantly elevated tocochromanol content or vitamin E activity to a level much higher than that derived from single-gene expression. Transcription factors offer great potential for the manipulation of metabolic pathways because of their ability to control both multiple-pathway steps and cellular processes that are necessary for metabolite accumulation [108], highlighting the potential benefit of using transcription factors to modify complex metabolic pathways in plants [109, 110]. The use of transcription factors to regulate vitamin E metabolic pathways in plants is still limited. No endogenous transcription regulators controlling the vitamin E pathway enzymes have been isolated. However, the potential of transcription factors to manipulate multiple fluxes of metabolic pathways will facilitate gene discovery and crop improvement.

Zhu et al. [112] overexpressed bacterial dihydrodipicohinate synthase in an *Arabidopsis* knockout mutant in the Lys catabolism pathway and affected a dramatic increase in free Lys in mutant. Their work elicitate us to design the overexpression of an enzyme resistant to tocopherol inhibition, together with the knock-out of the tocopherol catabolism pathway, and in order to produce much higher tocopherol levels than is produced by either strategy alone. On the other hand, we can break down the other metabolism pathways that consume common intermediates that are related to tocopherol synthesis.

Zeaxanthin is an important dietary carotenoid and can be converted to violaxanthin by zeaxanthin epoxidase. The gene for zeaxanthin epoxidase, with sense and antisense constructs, was transformed in potato (*Solanum tuberosum*). Both approaches (antisense and cosuppression) resulted in higher levels of zeaxanthin accumulation in potato tubers. In addition, α-tocopherol was elevated up to two- to threefold in the genetically transformed lines [113]. This provides an example of the use of an antisense approach to knocking out one metabolic pathway gene to modulate the content of vitamin E.

The demonstration that data obtained from engineering tocopherol synthesis in model systems can be readily transferred to crop plants indicates that we are on the cusp of an exciting era in which plant metabolic engineering can be used to have a positive impact on human nutrition and health on a global scale.

The rationale is that, although most people can obtain sufficient amounts of vitamin E from a typical diet, current foods do not provide the therapeutic levels of vitamin E that would allow the public to enjoy the added health benefits of this vitamin. Biofortified plants would provide a sustainable alternative to a prescribed regimen of vitamin E supplementation that would be available to everyone the world over.

References

1. Stephanopoulos GN, Aristidou AA, Nielsen J (1998) (eds) Metabolic Engineering Principles and Methodologies. Elsevier, New York
2. Ye X, Al Babili S, Kloti A, Zhang J, Lucca P, Bayer P, Potrykus I (2000) Science 287:303
3. Kamal-Eldin A, Appelqvist LA (1996) Lipids 31:671
4. Vidi PA, Kanwischer M, Baginsky S, Austin JR, Csucs G, Dörmann P, Kessler F, Bréhélin C (2006) J Biol Chem 281:11225
5. Dörmann P (2007) Planta
6. KamalEldin A, Appelqvist LA (1996) Lipids 31:671
7. Traber MG, Sies H (1996) Annu Rev Nutr 16:321
8. Bramley PM, Elmadfa I, Kafatos A, Kelly FJ, Manios Y, Rexborough HE, Schuch W, Sheehy PJA, Wagner KH (2000) J Sci Food Agric 80:913
9. Brigelius-Flohe R, Kelly FJ, Salonen JT, Neuzil J, Zingg JM, Azzi A (2002) Am J Clin Nutr 76:703
10. Grusak MA, DellaPenna D (1999) Annu Rev Plant Physiol Plant Mol Biol 50:133
11. Winklhofer-Roob BM, Rock E, Ribalta J, Shmerling DH, Roob JM (2003) Mol Aspects Med 24:391
12. DellaPenna D (2005a) J Plant Physiol 162:729
13. Hofius D, Sonnewald U (2003) Trends Plant Sci 8:6
14. Ajjawi I, Shintani D (2004) Trends Biotechnol 22:104
15. DellaPenna D (2005b) Trends Plant Sci 10:574
16. Münne-Bosch S, Falk J (2004) Planta 218:323
17. Münne-Bosch, Alegre L (2002) Crit Rev Plant 21:31
18. Valentin HE, Qi Q (2005) Appl Microbiol Biotechnol 68:436

19. Evans HM, Bishop KS (1922) Science 56:650
20. Fernholz E (1938) J Am Chem Soc 60:700
21. Karrer P, Fritzsche H, Ringer BH, Salomen NJ (1938) Nature 141:1057
22. Gordon HH. Nitowsky HM, Cornblath M (1955) AMA Am J Dis Child 90:669
23. Bunyan J, Murrell EA, Green J, Diplock AT (1967) Br J Nutr 21:475
24. Boscoboinik D, Szewczyk A, Azzi A (1991a) Arch Biochem Biophys 286:264
25. Boscoboinik D, Szewczyk A, Hensey C, Azzi A (1991b) J Biol Chem 266:6188
26. Garcia I, Rodgers M, Lenne C, Rolland A, Sailland A, Matringe M (1997) Biochem J 325:761
27. Fechner H, Schlame M, Guthmann F, Stevens PA, Rustow B (1998) Biochem J 331:577
28. Norris SR, Shen X, DellaPenna D (1998) Plant Physiol 117:1317
29. Shintani D, DellaPenna D (1998) Science 282:2098
30. Aratri E, Spycher SE, Breyer I, Azzi A (1999) FEBS Lett 447:91
31. Yamauchi J, Iwamoto T, Kida S, Masushige S, Yamada K, Esashi T (2001) Biochem Biophys Res Comm 285:295
32. Collakova E, DellaPenna D (2001) Plant Physiol 127:1113
33. Porfirova S, Bergmuller E, Tropf S, Lemke R, Dörmann P (2002) Proc Natl Acad Sci USA 99:12495
34. Cheng Z, Sattler S, Maeda H, Sakuragi Y, Bryant DA, DellaPenna D (2003) Plant Cell 15:2343
35. Azzi A, Gysin R, Kempná P, Munteanu A, Negis Y, Villacorta L, Visarius T, Zingg JM (2004) Ann NY Acad Sci 1031:86
36. Sattler SE, Gilliland LU, Magallanes-Lundback M, Pollard M, DellaPenna D (2004) Plant Cell 16:1419
37. Maeda H, DellaPenna D (2007) Curr Opin Plant Biol 10:260
38. Munne-Bosch S (2005a) New Phytol 166:363
39. Munne-Bosch S, Weiler EW, Alegre L, Muller M, Duchting P, Falk J (2007) Planta 225: 681
40. Soll J, Kemmerling M, Schultz G (1980) Arch Biochem Biophys 204:544
41. Soll J, Schultz G, Joyard J, Douce R, Block MA (1985) Arch Biochem Biophys 238:290
42. Rohmer M (2003) Pure Appl Chem 75:375
43. Schulz A, Ort O, Beyer P, Kleinig H (1993) FEBS Lett 318:162
44. Forbes BJR, Hamilton GA (1994) Bioorg Chem 22:343
45. Norris SR, Barrette TR, DellaPenna D (1995) Plant Cell 7:2139
46. Kleber-Janke T, Krupinska K (1997) Planta 203:332
47. Cahoon EB, Hall SE, Ripp KG, Ganzke TS, Hitz WD, Coughlan SJ (2003) Nat Biotechnol 21:1082
48. d'Harlingue A, Camara B (1985) J Biol Chem 260:15200
49. Stocker A, Ruttimann A, Woggon WD (1993) Helv Chim Acta 76:1729
50. Austin JR II, Frost E, Vidi PA, Kessler F, Staehelin LA (2006) Plant Cell 18:1693
51. Hofius D, Hajirezaei MR, Geiger M, Tschiersch H, Melzer M, Sonnewald U (2004) Plant Physiol 135:1256
52. Soll J (1987) Methods Enzymol 148:383
53. Wagner KH, Wotruba F, Elmadfa I (2001) Eur J Lipid Sci Tech 103:746
54. Arango Y, Heise K (1998) Fruits J Exp Bot 49:1259
55. Garcia I, Rodgers M, Pepin R, Hsieh TF, Matringe M (1999) Plant Physiol 119:1507
56. Dähnhardt D, Falk J, Appel J, van der Kooij TA, Schulz-Friedrich R, Krupinska K (2002) FEBS Lett 523:177
57. Falk J, Krau MN, Dáhnhardt D, Rupinska K (2002) J Plant Physiol 159:1245
58. Furuya T, Yoshikawa T, Kimura T, Kaneko H (1987) Phytochemistry 26:2741
59. Tsegaye T, Shintani DK, DellaPenna D (2002) Plant Physiol Biochem 40:913
60. Falk J, Andersen G, Kernebeck B, Krupinska K (2003) FEBS Lett 540:35
61. Dufourmantel N, Dubald M, Matringe M, Canard H, Garcon F, Job C, Kay E, Wisniewski JP, Ferullo JM, Pelissier B, Sailland A, Tissot G (2007) Plant Biotechnol J 5:118
62. Grimm B, Tanaka R (2003) US patent 564367
63. Schledz M, Seidler A, Beyer P, Neuhaus G (2001) FEBS Lett 499:15

64. Savidge B, Weiss JD, Wong YH, Lassner MW, Mitsky TA, Shewmaker CK, Post-Beittenmiller D, Valentin HE (2002) Plant Physiol 129:321
65. Keller Y, Bouvier F, d'Harlingue A, Camara B (1998) Eur J Biochem 251:413
66. Graßes T, Grimm B, Koroleva O, Jahns P (2001) Planta 213:620
67. Shpilyov AV, Zinchenko VV, Shestakov SV, Grimm B, Lokstein H (2005) Biochim Biophys Acta 1706:195
68. Karunanandaa B, Qi Q, Hao M, Baszis SR, Jensen PK, Wong YH, Jiang J, Venkatramesh M, Gruys KJ, Moshiri F, Post-Beittenmiller D, Weiss JD, Valentin HE (2005) Metab Eng 7:384
69. Norris S, Lincoln K, Abad M, Scott M, Eilers R, Hartsuyker K, Kindle K, Hirshberg J, Karunanandaa B, Moshiri F, Stein JC, Valentin HE, Venkatesh TV (2004) International patent application WO 2004013312 A2
70. Valentin HE, Lincoln K, Moshiri F, Jensen PK, Qi Q, Venkatesh TV, Karunanandaa B, Baszis SR, Norris SR, Savidge B, Gruys KJ, Last RL (2006) Plant Cell 18:212
71. Cook W, Miles D (1992) Photosynth Res 31:99
72. Demurin Y, Skoric D, Karlovic D (1996) Plant Breed 115:33
73. Shintani DK, Cheng Z, DellaPenna D (2002) FEBS Lett 511:1
74. Van Eenennaam AL, Lincoln K, Durrett TP, Valentin HE, Shewmaker CK, Thorne GM, Jiang J, Baszis SR, Levering CK, Aasen ED, Hao M, Stein JC, Norris SR, Last RL (2003b) Plant Cell 15:3007
75. Motohashi R, Ito T, Kobayashi M, Taji T, Nagata N, Asami T, Yoshida S, Yamaguchi-Shinozaki K, Shinozaki K (2003) Plant J 34:719
76. Subramaniam S, Slater S, Karberg K, Chen R, Valentin HE, Wong Y-HH (2001) International patent application WO 01/79472
77. Sattler SE, Cahoon EB, Coughlan SJ, DellaPenna D (2003) Plant Physiol 132:2184
78. Provencher LM, Miao L, Sinha N, Lucas WJ (2001) Plant Cell 13:1127
79. Stocker A, Fretz H, Frick H, Ruttimann A, Woggon WD (1996) Bioorg Med Chem 4:1129
80. Kumar R, Raclaru M, Schusseler T, Gruber J, Sadre R, Luhs W, Zarhloul KM, Friedt W, Enders D, Frentzen M, Weier D (2005) FEBS Lett 579:1357
81. Shintani DK, DellaPenna D (2003) US patent 118637
82. Gilliland LU, Magallanes-Lundback M, Hemming C, Supplee A, Koornneef M, Bentsink L, DellaPenna D (2006) PNAS 103:18834
83. Hunter SC, Cahoon EB (2007) Lipids 42:97
84. Sadre R, Gruber J, Frentzen M (2006) FEBS Lett 580:5357
85. Venkatesh TV, Karunanandaa B, Free DL, Rottnek JM, Baszis SR, Valentin HE (2006) Planta 223:1134
86. Tang S, Hass CG, Knapp SJ (2006) Theor Appl Genet 113:783
87. Bergmüller E, Porfirova S, Dörmann P (2003) Plant Mol Biol 52:1181
88. Tavva VS, Kim YH, Kagan IA, Dinkins RD, Kim KH, Collins GB (2007) Plant Cell Rep 26:61
89. Hass CG, Tang S, Leonard S, Traber MG, Miller JF, Knapp SJ (2006) Theor Appl Genet 113:767
90. Valentin HE, Lincoln K, Moshiri F, Jensen PK, Qi Q, Venkatesh TV, Karunanandaa B, Baszis SR, Norris SR, Savidge B, Gruys KJ, Last RL (2006) Plant Cell 18:212
91. Rippert P, Scimemi C, Dubald M, Matringe M (2004) Plant Physiol 134:92
92. Rocheford TR, Wong JC, Egesel CO, Lambert RJ (2002) J Am Coll Nutr 21:191
93. Collakova E, Dellapenna D (2003) Plant Physiol 131:632
94. Savidge B, Weiss JD, Wong YH, Lassner MW, Mitsky TA, Shewmaker CK, Post-Beittenmiller D, Valentin HE (2002) Plant Physiol 129:321
95. Theriault A, Chao JT, Wang Q, Gapor A, Adeli K (1999) Clin Biochem 32:309
96. Liu XL, Hua XJ, Guo J, Wang LJ, QI DM, Liu ZP, Jin ZP, Chen SY, Liu GS (2008) Biotechnol Lett 30:1275
97. Estevez JM, Cantero A, Reindl A, Reichler S, Leon P (2001) J Biol Chem 276:22901
98. Van Eenennaam A, Valentin HE, Karunanandaa B, Hao M, Aasen E, Levering C (2003a) International patent application WO 0/3016482

99. Cho EA, Lee CA, Kim YS, Baek SH, los Reyes BG, Yun SJ (2005) Mol Cells 19:16
100. Yusuf MA, Sarin NB (2007) Transgen Res 16:109
101. Teresa Capell, Paul Christou (2004) Curr Opin Biotechnol 15:148
102. Kanwischer M, Porfirova S, Bergmiiller E, Dörmann P (2005) Plant Physiol 137:713
103. Sattler SE, Cheng Z, DellaPenna D (2004) Trends Plant Sci 9:365
104. Collakova E, DellaPenna D (2003b) Plant Physiol 133:930
105. Lee BK, Kim SL, Kim KH, Yu SH, Lee SC, Zhang ZY, Kim MS, Park HM, Lee JY (2008) Plant Cell Tiss Org Cult 92:47
106. Van Eenennaam AL, Li G, Venkatramesh M, Levering C, Gong X, Jamieson AC, Rebar EJ, Shewmaker CK, Case CC (2004) Metab Eng 6:10
107. Shuangyan C, Hongjie L, Gongshe L (2006) Transgen Res 15:655
108. Broun P (2004) Curr Opin Plant Biol 7:202
109. Muir SR, Collins GJ, Robinson S, Hughes S, Bovy A, Ric De Vos CH, van Tunen AJ, Verhoeyen ME (2001) Nat Biotechnol 19:470
110. Bovy A, de Vos R, Kemper M, Schijlen E, Almenar Pertejo M, Muir S, Collins G, Robinson S, Verhoeyen M, Hughes S, Santos-Buelga C, van Tunen A (2002) Plant Cell 14:2509
111. Mendes P, Kell D (1997) Trends Biotechnol 15:6
112. Zhu XH, Galili G (2003) Plant Cell 15:845
113. Römer S, Lubeck J, Kauder F, Steiger S, Adomat C and Sandmann G (2002) Metab Eng 4:263
114. Azzi A, Zingg JM (2006) Clin Dermatol Retinoids Other Treatments 22:26

Chapter 19
Herbal Drugs of Abuse

Jochen Beyer, Olaf H. Drummer and H.C. Hans H. Maurer

Abstract Substances taken for nonmedical reasons, usually for their mind-altering effects, are called drugs of abuse. The use of psychoactive plants as drugs of abuse has had a long tradition. Most commonly abused drugs extracted from or based on natural products are illicit substances, such as cannabis products, morphine, or cocaine, but other herbal products used to produce a "high" are becoming increasingly popular drugs of abuse. Unfortunately, these "new herbal drugs" are falsely labeled as safe and legal. This chapter gives a brief historical background of herbal drugs of abuse, a description of the classical herbal drugs of abuse, and their current trends of use. The monographs of the different herbal drugs of abuse contain descriptions of the plants, their pharmacologically active compounds, and the current knowledge of their pharmacological properties.

Keywords Herbal drugs · Abuse

Abbreviations

CE	Common Era
BCE	Before Common Era
DMT	N,N-Dimethyltryptamine
LSD	Lysergic acid diethylamide
MAO	Monoamine oxidase
MMDA	3-methoxy-4,5-methylenedioxyamphetamine
NMDA	N-methyl-D-aspartic acid
RNA	Ribonucleic acid
THC	Δ9-Tehtrahydrocannabinol
TMA	3,4,5-trimethoxyamphetamine

J. Beyer (✉)
Victorian Institute of Forensic Medicine, Department of Forensic Medicine,
Monash University, 57-83 Kavanagh St, Southbank, Victoria 3006, Australia,
jochenb@vifm.org

K.G. Ramawat (ed.), *Herbal Drugs: Ethnomedicine to Modern Medicine*,
DOI 10.1007/978-3-540-79116-4_19,

19.1 Introduction

Drugs of abuse are defined as substances taken for nonmedicinal reasons, usually for their mind-altering effects. Most often these abused drugs are illicit substances such as heroin, cocaine, etc., but prescription and over-the-counter drugs are also abused. Abuse of ethical drugs (prescription or over-the-counter) occurs when people use the drugs in a manner or in quantities other than those directed, or for purposes that are not legitimate.

Many drugs abused in modern society are extracted from or based on natural products. The dangers associated with these drugs are substantially enhanced when the active substance is isolated from the plant material and used as essentially a pure drug, for example, heroin (derived from acetylation of morphine from the opium poppy) and cocaine when isolated from *Erythroxylum coca*.

Herbal products used to produce a "high" are becoming increasingly popular drugs for abuse [1]. These substances include obvious examples mentioned earlier as well as plants containing tropane alkaloids, plants containing stimulants such as caffeine, ephedrine, and khat, hallucinogenic plants, or even the alkenebenzene derivatives in nutmeg. The most common drugs of abuse of herbal origin sorted by scientific name, including the common English name, most common use, and most important active compound are summarized in Table 19.1.

The source of knowledge by users is often Internet based; "trip reports" and descriptions of the plants are shared among drug users, and often, unfortunately, the "new herbal drugs" are falsely labeled as safe and legal [2].

19.2 Historical Background of Herbal Drugs of Abuse

The use of psychoactive plants for mind-altering purposes has a long tradition. Archaeological evidence shows the use of psychoactive plants by humans for many thousands of years, often in a highly ritualized and ceremonial context [3].

The earliest archaeological evidence of a potential psychoactive plant in a cultural context is from a Neanderthal burial site in northern Iraq. Large quantities of pollen of different plants (including *Ephedra spp.*) have been found in the soil surrounding a male Neanderthal burial [4]. The pollen was dated to more than 50,000 BCE.

This finding might indicate that the body was deliberately, perhaps ritualistically, buried on a bed of plants [4].

Evidence of the use of cannabis found in Taiwan has been dated to approx. 10,000 BCE. Cord-impressed pottery with possible fiber evidence has been found in early postglacial fishing sites [3].

Preserved remains of plant and animal material including seeds of *Papaver somniferum* were found in ruins of a Stone Age settlement in Italy. The seeds were presumably cultivated for food, medical, and possible cult use and were dated to 5500 BCE [3].

Table 19.1 Herbal drugs of abuse sorted by their scientific name, including English name, common use, and most important active compound

Scientific name	Common English name	Common use	Active compound (most important)
Atropa belladonna	Deadly nightshade	Oral ingestion (infusion, berries)	Atropine, scopolamine
Banisteriopsis caapi	Ayahuasca	Oral ingestion (infusion)	Harmine, harmaline
Brugmansia spec.	Angel's trumpet	Oral ingestion (infusion), smoking (leaves and flowers)	Atropine, scopolamine
Cannabis sativa	Cannabis	Smoking, oral ingestion (cookies)	Δ9-Tetrahydrocannabinol
Datura stramonium	Thornapple	Oral ingestion (infusion), smoking (leaves and flowers)	Atropine, scopolamine
Ephedra spec.	Ephedra	Oral ingestion (infusion, herbal pills)	Ephedrine
Erythroxylum coca	Coca	Smoking, intravenous abuse, snorting (freebase)	Cocaine
Ipomea tricolor	Morning glory	Oral ingestion (infusion, seeds)	Ergine
Myristica fragrans	Nutmeg	Oral ingestion (infusion, seeds)	Myristicin, safrole
Papaver somniferum	Opium Poppy	Intravenous abuse, oral ingestion (tablets), smoking (common in Asian cultures)	Morphine, codeine
Piper methysticum	Kava	Oral ingestion (infusion, herbal pills)	Kavapyrone
Psychotria viridis	Ayahuasca	Oral ingestion (infusion)	N,N-Dimethyltryptamine
Rivea corymbosa	Morning glory	Oral ingestion (infusion, seeds)	Ergine
Salvia divinorum	Salvia	Smoking, chewing (leaves)	Salvinorin A
Tabernanthe iboga	Iboga	Oral ingestion (roots)	Ibogaine

Recently, the presence of a psychoactive compound in a 5700-year-old dried cactus "button" found in a cave in Texas has strengthened the evidence that humans recognized the psychoactive properties of plants at that time [5].

Approximately 3000 BCE *P. somniferum* was described as the "plant of happiness" on a Sumerian tablet [6]. Much evidence of the medical use and abuse of opium can be found in Egyptian, Greek, and Roman reports between 3000 and 1000 BCE [3]. Also, artifacts including vases and ornaments filled with crude opium were found and dated to this time frame [3].

In the Middle Ages, authors such as the Arabic scientist Avicenna (980–1037) and the German abbess Hildegard von Bingen (1098–1179) described numerous plants and their effects, including their psychoactivity. The description of plants including their psychoactivity was mainly continued in the 16th century by the so-called founding fathers of botany. These were the botanists Hieronymus Bock (1498–1554), Leonhart Fuchs (1501–1566), and Otto Brunfels (1488–1534). The first published systematic study of psychoactive plants was published in 1855 by Heinrich von Bibra. His book described 17 narcotic and stimulant plants including their effects on the human body [6].

The abuse of herbal drugs initiated the first treaty of international drug control. In 1912 the International Opium Convention was signed by 13 countries to provide control over the distribution of morphine and cocaine. In 1925, the convention was revised by the addition of the prohibition of hashish due to its common abuse. Today, many herbal drugs of abuse are controlled by national and international conventions and laws.

19.3 Classical Herbal Drugs of Abuse

19.3.1 Cannabis and Tetrahydrocannabinol

Cannabis is derived from *Cannabis sativa*, an herbaceous annual plant. The species *C. sativa* can be divided into the following subspecies and varieties: *C. sativa var. sativa, C. sativa var. spontanea, C. sativa spp. indica, C. sativa var. indica*, and *C. sativa var. kafiristanica* [6, 7]. The plant is highly recognizable: the leaves are typically separated into five or more serrated leaflets. The stem of the plant is extremely fibrous and is therefore a popular source of fabric fiber.

Cannabis is reported to have been first used by the Chinese in 2737 BCE. It was also known on the Indian subcontinent and the Arabian Peninsula well before the birth of Christ as an herbal remedy. The remains of cannabis seeds discovered in Germany date to 500 BCE. Cannabis was used as an antimalarial and antirheumatic drug, a narcotic, an aphrodisiac, a carminative, and as a remedy to treat nausea and headache.

Sinsemilla is a seedless, more potent form produced from the unfertilized flowering tops of female cannabis plants, often found in the USA [8]. Other forms of cannabis include Thai sticks, hashish (resin), and oil (hash oil). Modern names used to describe cannabis include "weed," "grass," "dope," "dagga," and "pot."

The most interesting compounds contained in cannabis are the cannabinoids, a group of at least 66 substances [9]. The most psychoactive cannabinoid in cannabis is Δ^9-tetrahydrocannabinol (THC). The chemical structure of THC (**1**) is shown in Fig. 19.1.

The cannabis plant is dioecious, with male and female flowers borne on separate plants. The female plant is more robust with larger flowers and, importantly, contains a higher THC content. All parts of the plant contain THC including flowers, leaves, seeds, and stalks. Oil is produced from the seeds and flower heads.

The plants prefer a warm humid climate, although it will grow in most climates. However, indoor and glasshouse cultivation are required for the cooler climates.

Cannabis is used mainly by smoking in the form of cigarettes (joints) or using cold-water pipes (bongs). It can also be made into "cookies" and eaten. Heat is essential to activate procannabinoids to THC.

THC increases the heart rate, blood pressure, and body temperature. More importantly, smoking THC produces a range of cognitive and psychomotor effects associated with a transient euphoric effect that is usually perceived as a "high." Short term memory loss can occur particularly with repeated use.

The cannabinoids are ligands of the cannabinoid G-protein coupled receptors. Evidence for the existence of a human cannabinoid receptor was found in the mid-1980s, and the CB_1 receptor was confirmed by cloning in 1990. A second cannabinoid receptor, the CB_2 receptor, was confirmed by cloning in 1993 [10]. The CB_1 receptor is predominantly found at central and peripheral nerve terminals, whereas the CB_2 receptor is mainly distributed in immune cells [9].

With the discovery of cannabinoid receptors, the question arose as to whether these receptors had endogenous ligands or whether they were targets only for THC and other plant cannabinoids. The first endogenous cannabinoid was described in 1992. The substance was identified as arachidonoyl ethanolamide and named anandamide. So far, five endogenous cannabinoid receptor agonists have been described [10].

19.3.2 Coca and Cocaine

Coca is a plant from the genus *Erythroxylum* and is native to northwestern South America. For the purposes of a plant of abuse, coca leaves are used. Cocaine is the most important psychoactive alkaloid in the plant. The chemical structure of cocaine (**2**) is shown in Fig. 19.1. The leaves contain up to 2% cocaine in a dried state [6]. Although the genus *Erythroxylum* includes several hundred species, only *E. coca* and *E. novogratense* contain significant amounts of cocaine [6]. Trace amounts of cocaine have recently been detected in 23 species of *Erythroxylum*, with content below 0.001% [11].

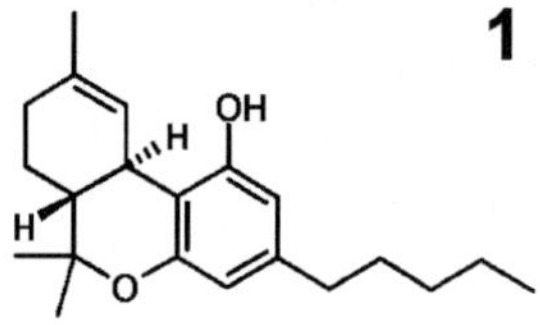

1

Δ9- Terathydrocannabinol

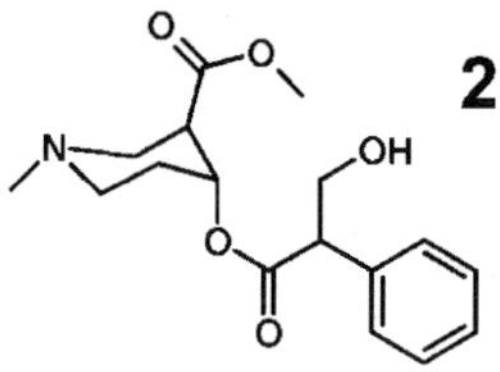

2

Cocaine

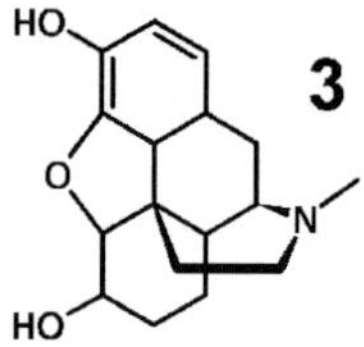

3

Morphine

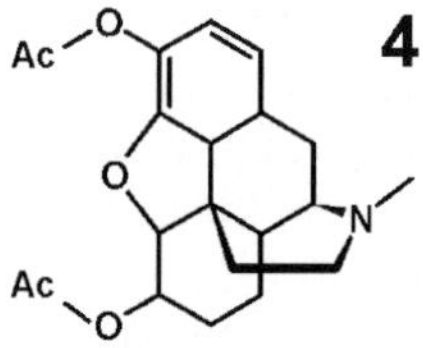

4

Diacetylmorphine

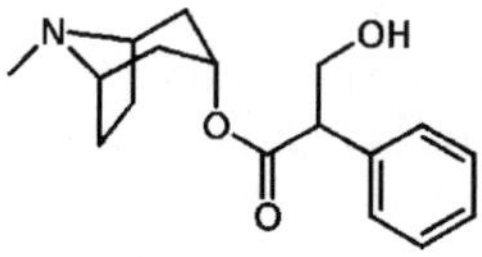

5

Atropine

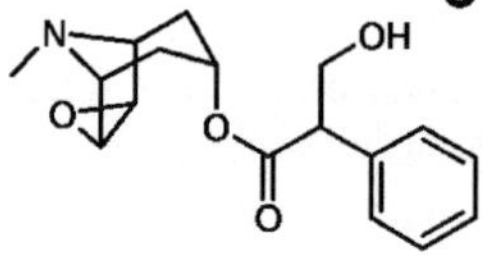

6

Scopolamine

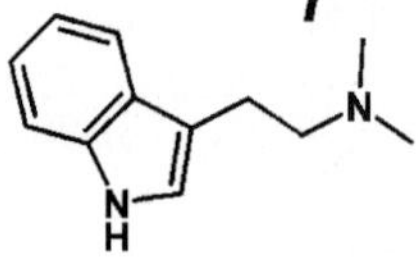

7

N,N-dimethyltryptamine

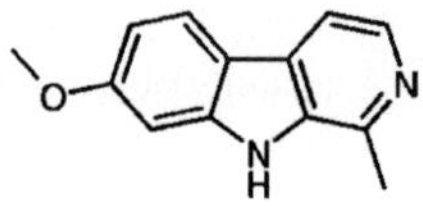

8

Harmine

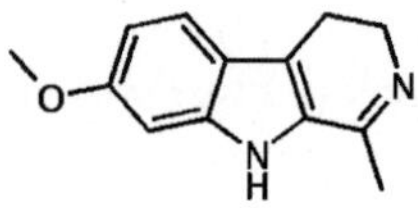

9

Harmaline

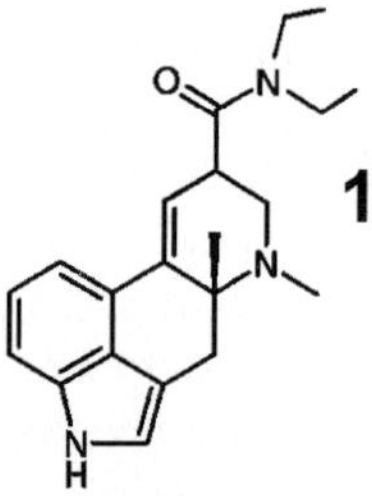

10

LSD

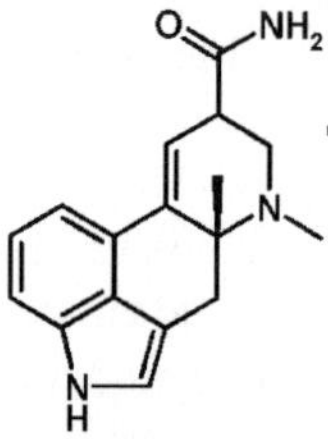

11

Ergine

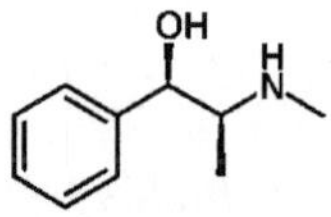

12

Ephedrine

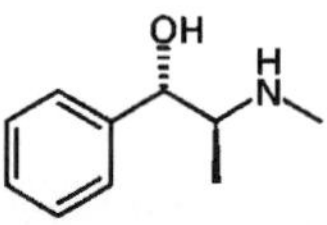

13

Pseudoephedrine

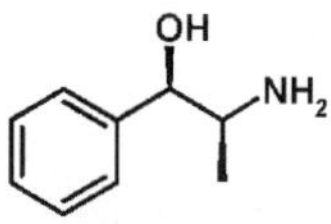

14

Norephedrine

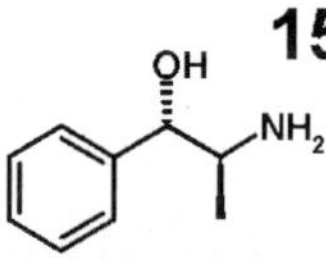

15

Norpseudoephedrine

Fig. 19.1 Chemical structures of most important active compounds of common herbal drugs of abuse

16

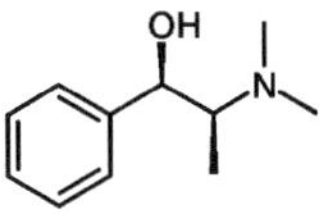

Methylephedrine

17

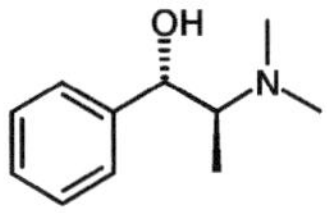

Methylpseudoephedrine

18

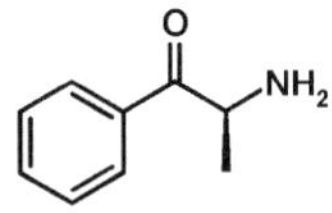

Cathinone

19

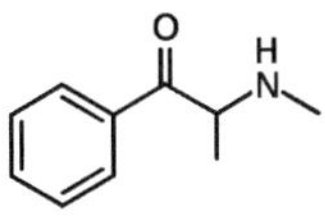

Methcathinone

20

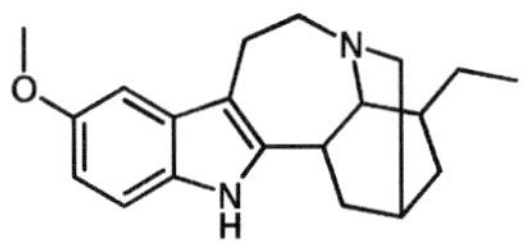

Ibogaine

21

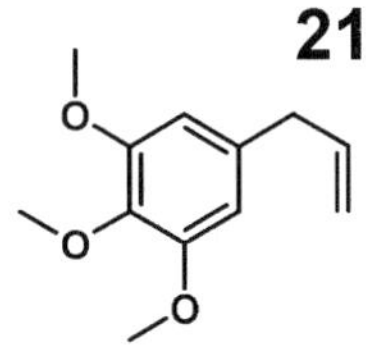

Elemicin

22

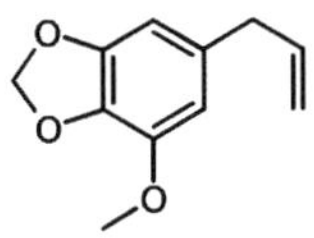

Myristicin

23

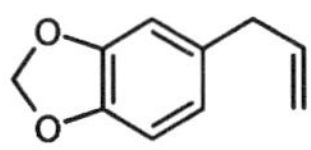

Safrole

24

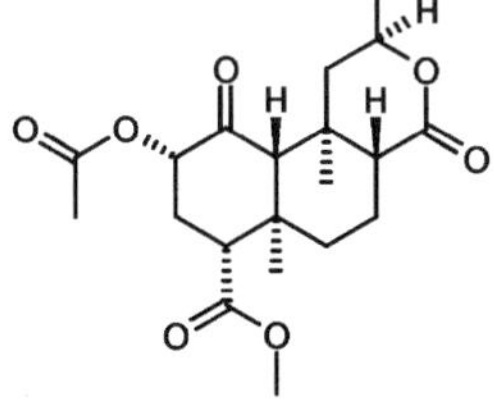

Salvinorin A

25

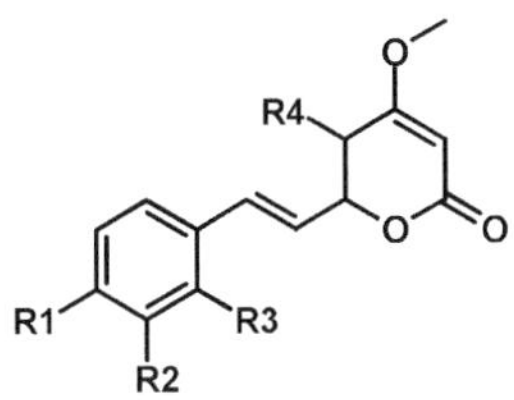

Kavapyrone

Fig. 19.1 (continued)

The use of coca has a long tradition in South America. Hair analysis of mummies dated to at least 1000 CE has proven the consumption of coca via detection of cocaine and its metabolites in hair [12]. The first European description of coca is from Amerigo Vespucci from 1499. He described how South American aboriginals would chew the leaves together with a white powder [6]. The main psychoactive compound, cocaine, was first isolated in 1859 by Albert Niemann, a German chemist [13].

In 1863, the Italian chemist Angelo Mariani introduced a coca extract in sweet wine as a tonic, the so-called Vin Mariani. This tonic was copied in 1884 in the USA by John Pemberton. One year later, Pemberton responded to the American Prohibition legislation by developing a nonalcoholic carbonated coca extract, called Coca-Cola, which contained an estimated amount of 9 mg of cocaine and on the market until 1903.

The use of coca leaves as a local anesthetic was first proposed by Samuel Percy in 1856 [13]. After the isolation of cocaine, the Austrian pharmacologist Karl Damian Ritter described narcotic effects after application on the skin. About 25 years later, cocaine was used in clinical practice as a local anesthetic in ophthalmic surgery. The advantage of cocaine as a local anesthetic is its unique combination of local anesthesia and intense vasoconstriction. Therefore, cocaine is still used for topical anesthesia by otorhinolaryngologists [13].

The anesthetic effects are pharmacologically caused by blocking of sodium channels, whereas the vasoconstriction is caused by sympathetic activation [14]. Cocaine inhibits the reuptake of catecholamines, which increases the activity of sympathetic synapses. This stimulation also occurs in the brain; therefore cocaine causes euphoria, garrulousness, and increased motor activity [14]. Cocaine also decreases fatigue and is abused as a stimulant.

In some communities coca leaves are typically mixed with an alkaline substance (such as lime) and chewed like a gum. Alternatively, coca leaves can be infused in liquid and consumed like tea such as *mate de coca*. This is a traditional method of consumption to treat altitude sickness, reduce fatigue, and increase energy and has been practiced for many centuries by some natives of South America [6].

19.3.3 Poppies and Opiates

Poppies include a number of attractive wildflower species found growing singularly or in large groups. Many species are also grown in gardens. Of all the poppy plants, the opium poppy or *Papaver somniferum* is the most important. The plant has been used for many thousands of years.

P. somniferum is an annual plant with a simple or only slightly branching stem. The leaves are ovate-oblong and of grey-blue greenish color. The flower can vary in color; pink varieties are the most common. Incisions made in the unripe capsules of *P. somniferum* release white latex that quickly hardens and turns brown after contact with oxygen. The hardened brown latex is called raw opium; each capsule releases approx. 20 to 50 mg.

The main active alkaloid in opium and other parts of *P. somniferum* is morphine, which was first isolated in 1806 by the German pharmacist Friedrich Sertuerner [15]. He called the isolated alkaloid "morphium" after the Greek god of dreams, Morpheus. It was not only the first alkaloid to be extracted from opium, but the first alkaloid ever to be isolated from any plant. With the invention of the hollow needle and the syringe, morphine was used in the treatment of postoperative and chronic pain, as well as an adjunct to general anesthetics. Unfortunately, morphine had as much potential for abuse as opium. Looking for a safer and nonaddictive opiate, Felix Hofmann synthesized in 1898 diacetylmorphine [16]. From 1898 to 1910 diacetylmorphine was marketed by Bayer under the brand name heroin. The drug was promoted as a nonaddictive morphine substitute and cough medicine for children. Bayer marketed heroin also as a substitute drug for morphine addiction before it was discovered that heroin is converted to 6-acetylamorphine and morphine by human cholineesterases [17]. Figure 19.1 shows the chemical structures of morphine **(3)** and heroin **(4)**.

The abuse of opioids was common in the late 19th century, leading to the International Opium Commission in 1909 (in Shanghai) as the first step toward international drug prohibition. Based on this meeting, the first international drug control treaty, the International Opium Convention, was signed in 1912 at The Hague. This convention went into force globally in 1919, when it was incorporated into the Treaty of Versailles [18]. Today, opium and opiates are controlled by the International Narcotics Control Board of the United Nations under Article 23 of the Single Convention on Narcotic Drugs, and subsequently under the Convention on Psychotropic Substances. Opium-producing nations are required to designate a government agency to take physical possession of illicit opium crops as soon as possible after harvest and conduct all wholesaling and exporting through that agency [19].

19.4 Current Herbal Drugs of Abuse

19.4.1 Nightshades

The botanical family of nightshades contains edible as well as poisonous plants. Well-known poisonous plants, such as *Atropa belladonna* (deadly nightshade), *Datura stramonium* (thornapple), and *Brugmansia* spp. (angel's trumpet), have been used for their psychoactive properties for hundreds of years [6]. Common names such as dwale, death's herb, or witch berry give an impression of their toxicity and use in the Middle Ages. The toxicity and pharmacological effects of deadly nightshade, for example, are also part of the etymology of the botanical name. The genus *Atropa* is named after the goddess Atropos, who is known in Greek mythology as the cutter of the life thread. The species name *belladonna* is Italian for "beautiful lady" and originates from the historical use of its berry juice by women to dilate their pupils [20].

Plants of this family are distributed generally throughout temperate and subtropical regions. Some species like *Brugmansia* sp. are often cultivated in pots as house plants.

Many plants of this family contain toxic tropane alkaloids such as (S)-(−)-hyoscyamine and (S)-(−)-scopolamine. (S)-(−)-hyoscyamine is converted during storage and/or isolation to a racemic mixture of 50% (S)-(−)-hyoscyamine and (R)-

(+)-hyoscyamine, called atropine. Atropine (**5**) and scopolamine (**6**) act pharmacologically by blocking acetylcholine receptors of the muscarine subtypes [14]. The blockage of these receptors causes symptoms such as tachycardia, dilated pupils, decreased gastrointestinal motility, dry hot skin, and dry mouth due to decreased sweat and saliva production. Apart from these peripheral effects, atropine also affects the central nervous system, causing agitation, disorientation, and hallucinations [14]. Due to the hallucinogenic properties of these alkaloids, plants are often abused. For example, this increasing misuse has prompted a prohibition by law in Florida of planting angel's trumpets.

19.4.2 Ayahuasca

Ayahuasca is the name of a psychoactive beverage that has its origin in the Amazon region. The name means "vine of the souls" in Quechua, a Native American language of South America. Traditionally, Ayahuasca is prepared by boiling or soaking the stems and barks of *Banisteriopsis caapi* together with various plants containing the psychoactive alkaloid N,N-dimethyltryptamine (DMT). The most commonly used plant containing DMT is *Psychotria viridis*. DMT is a potent sort-acting hallucinogenic agent but is not active following oral ingestion of doses up to 1000 mg. After parenteral administration of more than 25 mg, DMT shows psychoactive effects [21]. The alkaloid is metabolized via monoamine oxidase (MAO) to inactive metabolites.

The stems and bark of *B. caapi* contain β-carbolines such as harmine and harmaline. These carbolines are not psychoactive, but they are inhibitors of the enzyme MAO. The combination of these plants leads to psychoactive effects because the breakdown of the psychoactive DMT is inhibited [22]. This pharmacological "trick" is common today also using other MAO-inhibiting drugs. Such combinations are often called pharmahuasca [6]. Figure 19.1 shows the chemical structures of DMT (**7**), harmine (**8**), and harmaline (**9**).

19.4.3 Morning Glory and Lysergic Acid Amide

Morning glory is a common name for over 1000 species of plants in the family *Convolvulaceae*. If morning glory is abused, the plant is probably either *Ipomea tricolor* or *Rivea corymbosa*. Both plants are perennial twinning liana native to Central and South America. The seeds of both plants have been used by Native Americans for their hallucinogenic properties. The seeds of *I. tricolor* have been called by Aztecs *tlilitzin*, meaning "the very black," while the seeds of *R. corymbosa* were named *ololiuqui*, which is translated as "that which causes turns" [6]. Nowadays, the most commonly used term for the seeds of *I. tricolor* is badoh negro, and for the seeds of *R. corymbosa* badoh blanco [6].

The fresh or dried seeds are grounded and mixed with water and ingested orally. After ingestion, the seeds produce psychedelic effects similar to those of *Psilocybe* mushrooms or lysergic acid diethylaminde (LSD). The hallucinogenic effects of a cold water extract are not exactly the same as those of LSD, but vision of "small people" is typical [23]. Eating the seeds can induce side effects such as nausea and vomiting, probably induced by non-water-soluble alkaloids [6].

Albert Hofmann isolated in 1960 ergot alkaloids like lysergic acid amide from *R. corymbosa* [24]. The psychoactive effects of lysergic acid amide, also called ergine, were assessed by Albert Hofmann by self administration back in 1947, well before this was discovered to be a natural compound. He described a tired, dreamy state with an inability to maintain clear thoughts after intramuscular administration of 500 mg of ergine [21]. Besides lysergic acid itself, ergine is listed as a depressant in the category of Schedule III drugs in the Controlled Substances Act. This regulation probably attempts to control these substances as logical precursors of LSD. The chemical structures of LSD (**10**) and ergine (**11**) are shown in Fig. 19.1.

19.4.4 Ephedra and Kath

The genus *Ephedra* includes approx. 45 species indigenous to the temperate and subtropical regions of Asia, Europe, and America. The species *E. sinica* has been used in traditional Chinese medicine under the name Ma Huang for more than 500 years. The use of Ma Huang as a stimulant was first documented in the time of the Han Dynasty (ca. 206 BCE–220 CE) [25]. A possible translation of the Chinese name of the plant is "yellow hemp," which could explain its main use as a stimulant. Unfortunately, this assumption is not confirmed by historical sources.

The main pharmacologically active ingredients of the *Ephedra* species are the alkaloids ephedrine (**12**) and pseudoephedrine (**13**). Other pharmacologically active compounds in the plant are norephedrine (**14**), norpseudoephedrine (**15**, also called cathine), methylephedrine (**16**), and methylpseudoephedrine (**17**). These compounds are potent central nervous system stimulants and also have sympathomimetic effects on the peripheral nervous system [14]. In sport, the use of ephedrine, norpseudoephedrine, and methylephedrine has been banned by the International Olympic Committee. Because of their peripheral effects, ephedra alkaloids are often contained in cold medications. Recently, the use of pseudoephedrine in cold medications has been banned in many countries because the alkaloid is used as a precursor in the synthesis of methamphetamine [26].

Catha edulis, commonly known as khat, is not related botanically to *Ephedra*. Both plant groups are described here due to their similar active ingredients. Khat is indigenous to Ethiopia and is now cultivated in some other African countries, Arabia, and Afghanistan. The fresh leaves are chewed as soon as possible after harvesting. Khat has been used traditionally as a stimulant in Ethiopia by older men in conjunction with religious rites [27]. Today, khat leaves are chewed by men and women of all ages, mostly in countries where khat is grown. But also in other countries, emigrants try to maintain this habit. Therefore, large quantities of fresh leaves are illegally imported to other countries [27, 28]. The leaves must be consumed fresh, and they should not be older than 2 d as the psychoactive properties of the leaves decrease rapidly after harvesting [27].

The active ingredients of khat are similar to those in *Ephedra*. Besides the ephedrine alkaloids norpseudoephedrine and norephedrine, khat contains the alkaloid cathinone [27]. The pharmacological properties of cathinone are identical to those of *Ephedra* alkaloids. Due to khat abuse, the plant and its ingredients cathinone and norpseudoephedrine are controlled substances in many countries. Methcathinone is a potent derivative formed by the oxidation of pseudoephedrine [29]. Figure 19.1 shows the chemical structures of cathinone (**18**) and methcathinone (**19**).

19.4.5 Iboga

The perennial rainforest shrub *Tabernanthe iboga*, commonly known as iboga, is native to central western Africa [6]. The plant reaches about 1.5 to 2 m in height and has yellowish or pinkish flowers that turn into sweet fruits that do not contain psychoactive alkaloids. The plant is a sacrament and symbol of power in the Bwiti religion with the roots used in religious ceremonies as a "bridge to the ancestors." In small amounts (up to 5 mg/kg), the root is chewed by locals to reduce hunger and fatigue, but larger amounts (10 mg/kg or greater) will cause hallucinations and has even caused death [30, 31].

The root bark contains about 6% indole alkaloids: primarily ibogaine (12-methoxyibogamine), but also tabernanthine, ibogaline, and ibogamine.

Ibogaine (**20**) is a noncompetitive antagonist at $\alpha3\beta4$ nicotinic receptors and a sigma$_2$ receptor agonist, increases RNA expression of glial-cell-line-derived neurotrophic factor, is a weak agonist on the serotonin 5HT2A receptor, and is a weak NMDA receptor antagonist [32]. The major metabolite, noribogaine (12-hydroxyibogamine), is a potent serotonin reuptake inhibitor and is also a moderate kappa and weak mu opioid receptor full agonist. Ibogaine has also been used to treat dependency on alcohol and illicit drugs, although its efficacy is questionable [33].

It is classified as a Schedule I controlled substance in the USA and sale and possession is banned in Sweden, Denmark, Belgium, and Switzerland.

19.4.6 Nutmeg

Nutmegs are the seeds of the evergreen tropical tree *Myristica fragrans*, which is indigenous to the Spice Islands. The seeds are covered by a netlike red aril that is used to produce mace. Both seeds and mace are mainly used, but the seeds are described to be psychoactive when administered in high doses.

The psychoactive properties of nutmeg were already described by Hildegard von Bingen in the Middle Ages. Later, in 1829, the first human self-experiment describing the psychoactivity of nutmeg was provided by the physiologist Johann E. Purkyne. He described sleepiness with peaceful dreams and dizziness at higher concentrations [6]. Nutmeg intoxications due to abuse were described around the early 20th century, subsequently only a few cases were described until its resurgence in the mid-20th century. Since then, nutmeg has been discovered mainly periodically by adolescents as a natural and legal high. Widespread abuse is uncommon due to unpleasant and frightening side effects. The consumption of large amounts of nutmeg is also dangerous; some fatal intoxications and cases of nutmeg induced psychosis have been described [34].

The psychoactivity of nutmeg is believed to be caused by ingredients of the volatile oil. The main ingredients of the volatile oil of nutmeg are the alkenebenzene derivatives elemicin (**21**), myristicin (**22**), and safrole (**23**). In 1966, Shulgin [35] hypothesized that the possible psychotropic effects of myristicin may be caused by the metabolic addition of ammonia to the allyl side chain leading to the amphetamine derivative 3-methoxy-4,5-methylenedioxyamphetamine (MMDA) and that of elemicin by conversion to the designer drug 3,4,5-trimethoxyamphetamine (TMA). Although this metabolic step is unlikely, a formation of these designer drugs would explain the possible psychoactive effects of nutmeg. Although this hypothesis was widely accepted, the formation of these designer drugs in humans has never been proven. Recently, a study has shown that these proposed metabolites could not be detected in human and rat urine after ingestion of large amounts of nutmeg and the isolated ingredients itself [36]. Further studies to explain the psychoactive effects of nutmeg are needed.

19.4.7 Salvia divinorum

Salvia divinorum is a herbaceous plant native to the Mazatec region of the Sierra Madre Oriental in the Mexican state Oaxaca. This plant has been used in traditional spiritual practices by Mazatec aborigines in a manner very similar to magic mushrooms (*Psilocybe* spp.). The plant and its use were discovered by the West in 1962 by Gordon Wasson, followed by the first botanical description by Carl Epling and Carlos Játiva-M. in the same year [6]. Shortly after the first botanical description of the plant and its psychoactive properties, Albert Hofmann tried to discover the active constituents by analyzing the juice pressed from the plant. His analysis was unsuccessful, and it took until 1982 to isolate the diterpenes salvinorin A and salvinorin B. So far, 14 diterpenes have been isolated from *S. divinorum*, but salvinorin A seems to be the major active compound. Even though the plant is closely related to common sage (*S. officinalis*), neither an essential oil nor thujone, common in other *Salvia* species, has been discovered in *Salvia divinorum*.

In 1994, Daniel Siebert showed that the leaves of *S. divinorum* have psychoactive effects when the ingredients of the juice are absorbed via oral mucosa [37]. The plant material was inactive when swallowed as quickly as possible to bypass the oral mucosa. In this study the author also showed that salvinorin A produced the same psychoactive effects as the plant juice. Some years later, in 2002, salvinorin A was found to be a potent and the first nonnitrogenous agonist of the κ-opioid receptor [38], whereas salvinorin B is inactive at this receptor. A study in 2004 also reported κ-opioid receptor agonistlike discriminative effects in rhesus monkeys [39]. Since that time, the availability of the plant has increased rapidly, partially due to Internet trading. Within a few years, *S. divinorum* became a popular herbal drug of abuse.

The unique pharmacology of salvinorin A as the first nonnitrogenous opioid receptor agonist and its popularity as a drug of abuse have substantially increased the interest in it among researchers. More than 60 publications on salvinorin A have been published in the last 3 years. The chemical structure of salvinorin A (**24**) may be manipulated for the design of drugs with therapeutic potential.

19.4.8 Kava

Kava is the common English name for the western Pacific plant *Piper methysticum*. Other names for the plant are 'awa', used in Hawaii, or "yaqona," common in Fiji.

Kava is an evergreen bush growing up to 3 m tall, with heart-shaped leaves up to 20 cm in length. The plant is closely related to black pepper (*Piper nigra*) and also has a spicy taste. In 1777, the plant was first described botanically by Johann Gregor Forster, a fellow of Captain James Cook. During this trip, they also described the psychoactivity of the plant and the ceremonies by indigenous Polynesians [6]. The psychoactivity of kava is also indicated by the scientific species name *methysticum*, which is Greek for intoxicating.

Today, kava is the most important psychoactive plant in Oceania and in aboriginal cultures in the Northern Territory of Australia [20]. Traditionally Kava is prepared by grinding or chewing the rhizome, which is mixed with water or coconut milk. The effects after consumption of kava are talkative and euphoric behavior, anxiolytic effects, sense of well-being, clear thinking, and relaxed muscles.

The plant contains a mix of kavalactones and kavapyrones. Figure 19.1 shows the basic structure of kavapyrones (**25**). Kava ingredients are postulated to have pharmacological properties including blockade of voltage-gated sodium ion channels, reversible inhibition of monoamine oxidase B, and reduction of neuronal reuptake of noradrenaline. The effectiveness of kava and kava ingredients in the treatment of anxiety has been shown in clinical studies. Therefore, extracts of the plant were introduced into modern medicine as a mild anxiolytic. After the report of some deaths due to its medicinal use, kava medicines were banned. Kava containing medicine causes acute liver failure [40]. The traditional use of kava by Pacific Islanders and by some aboriginal communities is not believed to be associated with liver damage. A recent study has shown that kava feeding in rats does not cause liver damage [41]. Further investigations are necessary to demonstrate the long-term safety of kava preparations.

References

1. Richardson WH III, Slone CM, Michels JE (2007) Emerg Med Clin North Am 25:435–457
2. Dennehy CE, Tsourounis C, Miller AE (2005) Ann Pharmacother 39:1634–1639
3. Merlin MD (2003) Econom Bot 57(3):295–323
4. Solecki RS (1963) Science 139:179–193
5. El-Seedi HR, De Smet PA, Beck O, Possnert G, Bruhn JG (2005) J Ethnopharmacol 101:238–242
6. Raetsch C (2005) Enyclopedia of Psychoactive Plants, 1st edn. Park Street Press, Rochester, VT
7. Clarke RC (1981) Marijuana Botany, 1st edn. Ronin, Berkeley, CA
8. El-Sohly MA, Ross SA, Mehmedic Z, Arafat R, Yi B, Banahan BF III (2000) J Forens Sci 45:24–30
9. Pertwee RG (2006) Br J Pharmacol 147(Suppl 1):163–171
10. Howlett AC, Barth F, Bonner TI, Cabral G, Casellas P, Devane WA, Felder CC, Herkenham M, Mackie K, Martin BR, Mechoulam R, Pertwee RG (2002) Pharmacol Rev 54:161–202
11. Bieri S, Brachet A, Veuthey JL, Christen P (2006) J Ethnopharmacol 103:439–447
12. Springfield AC, Cartmell LW, Aufderheide AC, Buikstra J, Ho J (1993) Forens Sci Int 63:269–275
13. Ruetsch YA, Boni T, Borgeat A (2001) Curr Top Med Chem 1:175–182
14. Rang HP, Dale MM, Ritter JM, Flower RJ (2007) Pharmacology, 6th edn. Churchill Livingstone, New York
15. Sertürner F (1806) Journal der Pharmacie fuer Aerzte und Apotheker 14:47–93
16. Brownstein MJ (1993) Proc Natl Acad Sci USA 90:5391–5393
17. Lockridge O, Mottershaw-Jackson N, Eckerson HW, La Du BN (1980) J Pharmacol Exp Ther 215:1–8
18. Costa E, Silva JA (2004) Hum Psychopharmacol 19:215–224
19. McAllister WB (2004) Drug Alcohol Depend 76:3–8
20. Frohne D, Pfänder HJ (2005) Poisonous Plants: A Handbook for Doctors, Pharmacists, Toxicologists, Biologists and Veterinarians, 2nd edn. Manson, London
21. Shulgin A (1997) Tihkal, the Continuation, 1st edn. Transform, Berkeley, CA
22. Riba J, Valle M, Urbano G, Yritia M, Morte A, Barbanoj MJ (2003) J Pharmacol Exp Ther 306:73–83
23. Heim E, Heimann H, Lukacs G (1968) Psychopharmacologia 13:35–48
24. Hofmann A, Tscherter H (1960) Experientia 16:414
25. Chan EL, Ahmed TM, Wang M, Chan JC (1994) Am J Nephrol 14:295–301
26. Cunningham JK, Liu LM (2008) Soc Sci Med, Jan 25, epub ahead of print
27. Al-Hebshi NN, Skaug N (2005) Addict Biol 10:299–307
28. Toennes SW, Harder S, Schramm M, Niess C, Kauert GF (2003) Br J Clin Pharmacol 56:125–130
29. Belhadj-Tahar H, Sadeg N (2005) Forens Sci Int 153:99–101
30. Maas U, Strubelt S (2006) Med Hypotheses 67:960–964
31. Cheze M, Lenoan A, Deveaux M, Pepin G (2007) Forens Sci Int, Nov 14, epub ahead of print
32. Popik P, Skolnick P (1998) Alkaloids 52:197–231
33. Xu BJ, Zheng YN, Sung CK (2005) Drug Alcohol Rev 24:525–536
34. Stein U, Greyer H, Hentschel H (2001) Forens Sci Int 118:87–90
35. Shulgin AT (1966) Nature 210:380–384
36. Beyer J, Ehlers D, Maurer HH (2006) Ther Drug Monit 28:568–575
37. Siebert DJ (1994) J Ethnopharmacol 43:53–56
38. Roth BL, Baner K, Westkaemper R, Siebert D, Rice KC, Steinberg S, Ernsberger P, Rothman RB (2002) Proc Natl Acad Sci USA 99:11934–11939
39. Butelman ER, Harris TJ, Kreek MJ (2004) Psychopharmacology (Berl) 172:220–224
40. Anonymous (2002) S Afr Med J 92(12):961
41. DiSilvestro RA, Zhang W, DiSilvestro DJ (2007) Food Chem Toxicol 45:1293–1300

Chapter 20
Biological Activities of Kinetin

Kalina Duszka, Brian F.C. Clark, Frank Massino and Jan Barciszewski

Abstract Kinetin (N^6-furfuryladenine) is a cytokinin growth factor with several biological effects observed for human cells and fruit flies. It was given the name kinetin because of its ability to induce cell division. Kinetin is often used in plant cell and tissue culture for induction of callus formation (in conjunction with auxin) and to regenerate shoots from callus (with lower auxin concentration). Kinetin exists naturally in the DNA of almost all organisms tested so far, including human cells, and various plants. The mechanism of kinetin synthesis in DNA is thought to be via the production of furfural, an oxidative damage product of DNA deoxyribose, and it is quenched by the adenine base converting it into N6-furfuryladenine. Since 1994 kinetin has been thoroughly tested for its powerful antiaging effects in human skin cells and other systems. At present, kinetin is one of the most widely used components in numerous skin care cosmetics and cosmeceuticals. There are some reports published on other biological effects of kinetin in human beings, as an antiplatelet aggregation factor reducing thrombus formation, and its ability to correct genetic diseases of RNA missplicing. Kinetin was shown to be more effective in the improvement of skin texture, making it smoother and with a significant reduction in fine lines and wrinkles. It is a stable antioxidant that slows down the aging process.

Keywords cytokinins · kinetin · DNA damage · biological activity · antiaging

Abbreviations

AGEP	Advanced glycation end product
ALT	Alanine aminotranferease
AML	Acute myelogenous leukemia
AST	Aspartate aminotransferase
6-BAP	6-benzylaminopurine
BSA	Bovine serum albumin

J. Barciszewski (✉)
Institute for Bioorganic Chemistry, Polish Academy of Sciences, Noskowskiego 12/14, 61–704 Poznan, Poland
e-mail: jan.barciszewski@ibch.poznan.pl

K.G. Ramawat (ed.), *Herbal Drugs: Ethnomedicine to Modern Medicine*,
DOI 10.1007/978-3-540-79116-4_20, © Springer-Verlag Berlin Heidelberg 2009

cGMP	Cyclic guanosine-3′,5′-monophosphate
CK-MB	Muscle creatine kinase
2,4-D	2,4-dichlorophenoxyacetic acid
EEC	Esophageal epithelial cells
ERM	Ezrin/radixin/moesin
FD	Familiar dysautonomia
GDI rho	GDP dissociation inhibitor
HMDMEC	Human dermal microvascular endothelial cells
HUVEC	Human umbilical vein endothelial cells
IAA	Indole-3-acetic acid
MAPK	Mitogen-activated protein kinase
PIP_2	Phosphatidylinositol
PR-10c	Pathogenesis-related protein c
ROS	Reactive oxygen species

20.1 Introduction

Fifty years after the discovery of kinetin it remains an interesting object of scientific research and a rich source of inspiration. Long-lasting studies have deepened our knowledge about its role and significance as well as making kinetin the best known cytokinin. Recently discovered data showed us a great potential and broad field of kinetin activity as it plays a role in many processes, from regulation of growth and development of plants through antioxidative and antiaging properties to therapeutic utility.

20.2 Kinetin Occurrence, Structure, and Chemical Properties

Kinetin was isolated in 1955 from autoclaved herring sperm DNA. For many years it was considered to be a synthetic product. Now it is known that it occurs in commercially available DNA and naturally as a component of freshly extracted plant and human DNA as well as in urine and coconut water where it reaches a concentration of 0.31×10^{-3}M [1–4].

Kinetin belongs to a group of plant hormones called cytokinin. These are adenine derivatives with an additional side chain at the N6 atom. In accordance with the structure of the side chain, cytokinins are classified as isoprenoid with an isopentenyl or a hydroxyisopentenyl group or aromatic, where a benzyl group occurs. Kinetin contains a furfuryl group at N6 (Fig. 20.1) [1].

Kinetin is soluble in strong acids, alkali, and glacial acetic acid. It is slightly soluble in ethanol, butanol, acetone, and water [1]. The cytokinin shows electrochemical

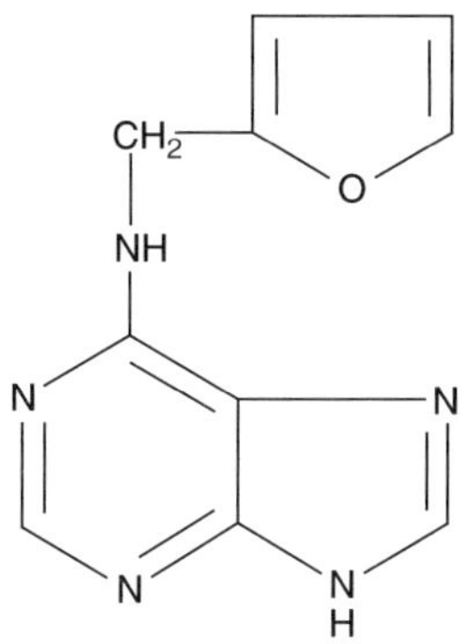

Fig. 20.1 Chemical structure of kinetin

properties that enable simple monitoring of its level. This feature is found in, among others, adducts of nucleic acids but not modified compounds of DNA [2].

20.3 Synthesis and Modification of Kinetin

Kinetin is thought to be synthesized within the cell as a result of oxidative damage processes. In such conditions DNA undergoes degradation, and as a result of oxidation of the deoxyribose 5′ carbon, furfural is formed. In fact, the main target of deoxyriboside oxidation is C1′ carbon, which results in 5-methylene-2-furanone formation. Furfural yield reaches 15% of deoxyribose C5′ oxidation products. It reacts with adenine in DNA, forms the Schiff base, and, following intramolecular rearrangement and dehydration, kinetin results [5, 6].

When kinetin occurs in DNA, the DNA polymerase incorporates thymine opposite to the modified base on the complementary strand; however, in the vicinity of kinetin misincorporation occurs. The presence of cytokinin in DNA induces conformational changes. Increased risk of mutation is lowered by activation of a repair mechanism that excises kinetin [7].

Cytokinins are present in plants as free bases, nucleosides, or nucleotide derivatives. Interconversion between these forms is maintained by enzymes involved to purine metabolism [8]. Cytokinin modifications include O-xylosylation, O-glucosylation, and N-glucosylation. All these forms are inactive. O-glucosylation provides resistance to cytokinin oxidase cleavage of the N6 side chain. The reaction is reversible and the molecule can be converted to the active form by β-glucosidases. It could serve as a storage form of inactive cytokinin. Adenine ring glucosylation could occur at N3, N7, and N9 sites. Like N9-glucoside, N7-glucoside is resistant to glucosidases since they cannot be converted to the active forms. Therefore N-glucosylation is considered to be an irreversible pathway of cytokinin inactivation [9, 11]. Cytokinins can be modified by cleavage of the side chain by cytokinin oxidase. The reaction leads to irreversible degradation [12].

20.4 Biological Activities of Kinetin

20.4.1 Influence of Kinetin on Plants

Kinetin's role is connected with the growth and development of plants. It stimulates seed germination, regulates cell division and reaction to stress, and determines apical dominance, formation, and activity of shoot meristems. Kinetin and 6-BAP counteract the inhibitory effect of darkness on epicotyl shoot formation and sprouting and raise it to the same level as in tissue incubated in light [13]. It is also implicated in the vascular development and synthesis of secondary metabolites like indol alkaloids and anthocyanins. Kinetin influences chloroplast differentiation and chlorophyll biosynthesis by stimulation of 5-aminolevulinc acid synthesis [8], [14–16]. Cytokinin inhibits lateral root initiation by blocking the pericycle founder cells' cell division cycle at the G2 to M transition phase. It stimulates lateral root elongation by stimulating the G1 to S phase transition [17–19] and strongly inhibits of RNase from the *Pisum sativum* isolated leaf, inhibits proteolysis, and maintains the integrity of cell membranes [17]. Since the majority of plant hormones bear a charge, cytokinin can be absorbed into specific membrane lipid domains [20]. It induces a rapid response at the outer face of membrane modulating ion transport [21]. The hormones affect the surface pressure decrease observed in the phospholipid monolayer relaxation. The greatest decay of surface pressure has been recorded in the presence of auxins in all systems, the lowest in the presence of cytokinins. Negatively charged hormones like IAA and 2,4-D affect monolayer disorganization to the highest degree, while the effect of positively charged compounds (kinetin, zeatin) is less pronounced. Both the cytokinins, zeatin and kinetin, display very similar effects, which suggests that the purine ring is the main steric parameter that determinines interaction with the lipid monolayer built from phospholipids as phosphatidylcholine and phosphatidic acid, having the same hydrophobic parts [22].

Exogenously applied kinetin stimulates promoter-dependent rRNA transcription of the repeated genes that encode the precursor of 17–18, 5.8, and 25–28 S ribosomal RNA in *Arabidopsis thaliana*. RNA polymerase I enhanced activity in growing cells and decreased it in quiescent cells. On average it accounts for approx. 34% of total nuclear transcription, but in cytokinin-treated cells this figure is approx. 60%. Kinetin was suggested to operate at the level of transcription initiation rather than rRNA stabilization and acts as a general regulator of protein synthesis activity and growth modulator in plant cells [23].

Kinetin influences the cell cycle. In *N. plumbaginifolia* it was shown to be indispensable in cell transition to the G2 phase and to histone kinase H1 p34cdc2 normal functioning [24]. Kinetin and 1-naphthylacetic acid (auxin) increases phosphorylation of the S6 ribosomal protein and activation of its cognate kinase, AtS6K, as well as translationaly up-regulating S6 and S18A mRNAs in *Arabidopsis* suspension cell culture [25].

Kinetin delays aging and leaf senescence and causes nutrient mobilization. It influences the activity of extracellular invertase, an enzyme responsible for nutrition remobilization from senescing leaves to other parts of the plant. The expression is inhibited by the invertase inhibitors expressed under the control of the cytokinin-inducible promoter. As a result, kinetin contributes to nutrition retention and thus inhibits senescence [26].

Kinetin interacts with the pathogenesis-related protein c (PR-10c) of the PR-10 protein family from a birch tree. PR-10c shows a low ribonuclease activity and is known to take part in response to stress. Kinetin interacts with the glycine-rich loop through its adenine moiety [27].

Kinetin serves as a chitinase inhibitor and could be an effective antifungal compound. It is considered to be the most potent of 880 druglike molecules tested. It acts as a competitive inhibitor, with its furfuran group occupying a deep pocket of the enzyme. In the pocket, oxygen atoms could interact with Tyr32, which probably determines unusually strong molecule binding [28].

It can also play a protective role. The cytokinin reduces membrane damage caused by fungal toxin fusaric acid and necrosis brought about by mercuric chloride [29].

It also combats viral infections. Kinetin treatment of *Xanthic-nc tobacco* reduces the number of lesions induced by tobacco mosaic viral infection but only slightly affects virus multiplication [30].

BT37 is a crown gall teratoma induced on tobacco by *Agrobacterium tumefaciens* containing pTi-T37, a nopaline-type Ti plasmid. The BT37 tissue treated with kinetin at 1 mg/L results in the development of relatively normal-appearing shoots. The shoots can produce viable seeds and are susceptible to *A. tumefaciens* infection. The cells lose most of the Ti plasmid sequences (T-DNA) found in BT37 DNA containing a highly conserved region of the Ti plasmid that has been found to be incorporated into all tumors. However, the cells retain sequences homologous to the ends of the T-DNA present in the teratoma tissue [31].

Kinetin brings about an increase in cytoplasmic calcium, an early event in vegetative bud formation in the moss *Physcomitrella patens* through stimulation of azidopine binding to 1,4-dihydropyridine-sensitive calcium channels [32]. The cytokinin has also been suggested to enhance phosphoinositide catabolism in soybean suspension cultures [33]. Kinetin as well as the cyclic guanosine-3′,5′-mono- phosphate (cGMP) analog 8-Br-cGMP induces a stomatal opening in *Tradescantia albiflora*. Its effect is reversibly inhibited by guanylate cyclase inhibitors [34].

It has been shown that kinetin is present in a root nodule of *Casuarina equisetifolia* modulated by *Frankia* [35]. Probably it is produced as a result of oxidative stress during infection. In that process reactive oxygen species (ROSs) are generated in high amounts in the root cells [36], which could provoke DNA damage and result in kinetin formation. In the same process the cytokinin could serve as a factor that regulates ROS amounts as its balance is necessary for infection thread progression.

20.4.2 *Influence of Kinetin on Animals*

20.4.2.1 Kinetin Antioxidative Properties

In experiments using animal cells and **other** organisms it was shown that kinetin influences many processes, regulates proliferation, and has antiaging and antioxidant properties.

Kinetin's antioxidant and scavenger activity was confirmed in vivo and in vitro. It could act in a few different ways: as a donor of hydrogen, as an enzyme, or as an activator of enzyme activity [37–40]. Because of these properties kinetin prevents damage to DNA, proteins, and other macromolecules, avoiding the accumulation of abnormal particles in organs, tissues, and cells.

Kinetin can act as a free-radical scavenger when oxygen radicals directly abstract hydrogen from the α-carbon of the amine bond of N6-furfuryladenine [37].

The kinetin-copper complex catalyzes the dismutation of superoxide into oxygen and hydrogen peroxide at the reaction rate constants 2.3×10^{-7} M^{-1} s^{-1} at pH 9.8 and 25°C [38].

Kinetin was proved to protect DNA against the formation of 8-oxodeoxyguanine, which is the result of hydrogen peroxide generation in a Fenton reaction. Inhibition of 8-oxo-dG formation was exhibited in a dose-dependent manner with a maximum efficiency of 50% at a concentration of 100 μM [41].

Kinetin protects against oxidative and glycoxidative protein damage generated in vitro by sugars and iron/ascorbate system. Glycation is a nonenzymatic reaction of binding hexose, mostly glucose to the amine group of protein or nucleotides [42]. The products of these reactions are accumulated in cells during aging [43]. Inside the cell, due to a high concentration of glucose and other reactive sugars like pentoses and α-oxoaldehydes, glycation/glycoxydation reactions progress fast [44, 45]. Kinetin at a concentration of 50 μM exhibits an 82% inhibition of bovine serum albumin (BSA)-pentosidine formation. At 200 μM the cytokinin prevents BSA aggregation during glycation and also inhibits 59 to 68% advanced glycation end product (AGEP) development [46].

20.4.2.2 Kinetin Antiaging Properties

The antiaging properties of kinetin were shown using in vitro cell cultures, in vivo on skin, and fruit flies. The fruit fly *Zaprionus*, with its diet supplemented with 125 to 625 μM kinetin, prolonged life span due to a reduction in the age-specific death rates, slowed down development, and delayed maturation of insects in the larval and pupal stages. Delayed aging is reached at the cost of decreased reproductive activity and egg-laying capacity. The molecular mechanism of kinetin activity is connected with an increase in catalase activity. The enzyme belongs to the oxydoreductase group, displays strong antioxidant activity, and catalyzes the decomposition of hydrogen peroxide into water and oxygen. A concentration of 125 μM seems to be the most effective for antiaging and life-prolonging effects. A higher concentration,

500 μM and above, gives toxic and life-shortening results. The cytokinin exerts a similar effect on human cell cultures at these doses [39, 40].

Nymphs of *Lipaphis erysimi* fed kinetin-treated *Raphanus sativus* L. showed an increased activity of catalase, glutathione peroxidase, and superoxide dismutase and a decrease in the activity of APTaze [47].

In in vitro skin cultures, mammary carcinoma and cystic disease kinetin delays the onset of several morphological and biochemical processes connected with aging. During senescence in vitro cell cultures became large and flattened, full of lysosomal residual bodies and oxidation-modified macromolecules, debris, and accumulated lipofuscin with disorganized cytoskeleton and some of them contain more than one nucleus. The addition of kinetin at 40- to 200-μM concentrations in culture media prevents fibroblasts from developing these changes. In spite of avoiding age-related degeneration, kinetin can also slightly reverse the changes. Upon removal of the cytokinin, some of the aging characteristics reverse [48]. Some properties of kinetin were proved in vivo using aged skin of hairless dogs. After 50 d of daily application of solution containing kinetin a 0.01-, 0.1-, 1-, 10-, and 96.6-mM improvement in skin texture, wrinkling, and pigmentation was observed. After 100 d rejuvenation and depigmentation became more visible. Lower concentrations of kinetin normalized hyperpigmentation and improved the aged skin structure. Throughout the treatment no adverse effect was observed, showing that kinetin is safe for long-term therapy [49].

20.4.2.3 Kinetin Influence on Animal Cell and Tissue Cultures

Kinetin influences both the epidermis and the dermis in the skin in the same way. It stimulates keratinocyte proliferation and differentiation in the epidermis, increases the amount of laminin 5 at the dermal-epidermal junction, and influences te formation of fibrillin-1 and elastin deposition as well as their organization perpendicularly to the dermal-epidermal junction in the dermis [50]. On the other hand human keratinocyte culture exhibits significant growth inhibition in media containing 40 to 200 μM kinetin concentration. At the same time it stimulates the cells to differentiate, especially strongly in the presence of calcium [51]. Kinetin retards the outgrowth of epithelium skin cultures at 1 to 0.25 mg/100 ml and increases epithelial sheet production at 0.006 to 0.015 mg/100 ml [52]. Its riboside appears to be toxic to fibroblasts, breast carcinomas, and cystic disease cells at 1 mg/100 ml and results in reduced or no outgrowth in in vitro culture. But it is not toxic at 0.1 mg/100 ml [48].

Kinetin in high concentrations (100 mg/1000 ml) acts as a toxin and triggers cytoplasm vacuolization and degenerative changes in fibroblast cell cultures. At lower doses (1 mg/1000 ml, 10 mg/1000 ml) chromatin became more sensitive to acid hydrolysis, which results in higher transcription activity. DNA amounts in the fibroblast nucleus increase after 24 and 72 h incubation with kinetin [53].

20.4.2.4 Cytokinin Influence on Cancer Cells

Cytokinins influence animal cell proliferation and differentiation, which makes them attractive as potential agents in cancer treatment. Some cytokinins, for example kinetin, isopentenyladenine, and benzyladenine, are inhibitors of ML-1, NB4, and U937 leukemia cell proliferation and stimulate the cells' mature granulocytes to differentiate as well as influence plant cancer cells [54, 55]. Also, other adenine derivatives like 6-methyladenine, 6-anilinopurine, and 2-aminopurine induce myeloid leukemia cells' HL-60 differentiation [56]. Cytokinin ribosides, including the kinetin riboside, inhibit the growth of HL-60, M4 Beu human, and B16 murine melanoma cells. Once cells are incubated with cytokinin ribosides and antioxidants, apoptosis is reduced and differentiation is increased. It triggers such a result only in the presence of antioxidants, scavengers, or the caspase inhibitor. Normally cytokinin ribosides reduce the intracellular ATP content and disturb the mitochondrial membrane potential and the accumulation of oxygen species, leading to apoptosis. This shows that cytokinin ribosides can induce differentiation but first it stimulates apoptosis [56, 57]. cAMP and ATP also stimulate the differentiation of HL-60 cells but at a much lower rate than the cytokinin derivative of adenine [55]. Diverse adenine analogs display varied influences on HL-60 growth. Methyladenine (IC_{50} 1,172–1,713 μM), adenosine (IC_{50} 685 μM), deoxyadenosine (IC_{50} 662 μM), and *trans*zeatin (IC_{50} 516 μM) appeared to be the least harmful. Other cytokinins such as kinetin (IC_{50} 48.6 μM), benzylaminopurine (IC_{50} 67.6 μM), and isopentenyloadenine (IC_{50} 47.6 μM) display very similar abilities to inhibit growth. The benzylaminopurine riboside as well as the kinetin riboside and isopentenyladenosine all show a strong inhibiting influence with its IC_{50} at 0.706 to 0.981 μM. The mechanism of cytokinin action is connected with adenine metabolism and adenine kinase activity that converts adenine to AMP. It appeared that a cytokinin could not influence cells until it was phosphorylated and converted to its nucleotide. Adenine kinase inhibition stops kinetin-triggered HL-60 differentiation, and kinetin riboside caused apoptosis [56]. The mechanism of kinetin action in HL-60 is known to abandon the P2 receptor, which is a member of the known differentiation mechanism. It involves the induction of mitogen-activated protein kinase (MAPK) and S100P [58]. S100P belongs to the S100 binding calcium protein family and it is temporarily expressed during the early stages of differentiation in esophageal epithelial cells (EEC) [56, 59, 60]. In normal neutrophil cells, S100P is strongly expressed, while in AML expression it is weak or nondetectable [61].

20.4.2.5 Therapeutic Applications of Kinetin

Familial dysautonomia (FD) is a neurodegenerative disease caused by a mutation in the IKBKAP gene coding IKAP protein. In cytosine substitution by thymine in intron 20, tissue-specific intron skipping during alternative splicing takes place [62]. IKAP participates in transcription elongation. In the case of FD, accurate splicing is particularly ineffective in the nervous system. Kinetin stimulates the inclusion of exon 20 from an endogenous gene and from a proper IKBKAP minigene. It is known that the CAA sequence at exon 5′ is necessary for kinetin to show its activity [63].

Kinetin is utilized in therapy treating Meniere's disease. A solution of 4% lidocaine and cytokinin was introduced into the tympanic cavity of patients. 87.5% of the sick reported a noticeable decrease in vestibular symptoms, and 66.7% of these patients were free of attacks for an average of 26.5 months with the same or improved hearing in 87.5% of patients [64].

20.4.2.6 Kinetin's Molecular Mechanism of Action in Animal Cells

In endothelium cells kinetin influences signaling pathways connected with the cytoskeleton. Human dermal microvascular endothelial cells (HDMEC) between the 5th and 30th passages were treated with kinetin at 50 μM. This resulted in changes in the expression of moesin, actin, and rho GDP dissociation inhibitor (GDI) and an increased activity of rho GTPase, which influences actin. Actin is a protein that forms the cytoskeleton and is connected with the rho pathway. Moesin belongs to the ERM (ezrin/radixin/moesin) protein family that connects cell membrane proteins with actin underneath the membrane. Moesin participates in signal transduction and cytoskeleton remodeling. It is modulated by phosphorylation, the phosphoinositide pathway, and is controlled mutually with rho GTPase. Rho GDI expression is elevated in aged human umbilical vein endothelial cells (HUVEC) and it suppresses rho GTPase activity. It interacts directly with ERM protein, which reduces rho GDI activity, thereby activating rho GTPase [65, 69].

Using similar pathways kinetin takes part in the regulation of cell proliferation. Cell cycle arrest in the G1 phase is connected with senescence and leads to apoptosis. This occurs when G1-specific cyclin D1 or cyclin E1, pRB, p16, p21, and p27 undergo changes. When cellular function decreases, p53 activation takes place. This suppresses cell cycle progression, stimulates a rise of p21 and p27, and induces G1 arrest. Kinetin decreases expression of p16, p27, and p53 and increases the amount of D1 cyclin. The rho pathway, as well as influencing cytoskeleton, supports cell cycle transition G1/S, and thus it promotes proliferation. The rho GTPase enhances expression of p27 and thus regulates D1 cyclin. In HDMEC treated with kinetin, rho GTPase is activated, total p16, p21, p27 is reduced, the amount of cyclin D1 is enlarged, and stimulation of G1/S transition is observed. Kinetin delays aging of endothelial cells and increases proliferation and metabolic capacity [69, 70]. Earlier experiments showed that kinetin delayed the onset of aging of fibroblasts and helped to complete cytokinesis, but it does not promote induction of the S phase. This suggests diverse activities of the cytokinin depending on the cell type [48].

There are also other cases where kinetin was proved to act through cGMP and Ca^{2+} connected pathways. At 50 to 150 μM, it inhibits platelet aggregation. It is supposed to stop Na^+/H^+ exchanger activation and phospholipase C activation and at the same time prevent phosphatidylinositol (PIP_2) metabolism and lipid signaling pathways. This results in lower intracellular alkalization and Ca^{2+} mobilization, augmented cyclic AMP synthesis, and inhibition of thromboxane A2 formation, which is known to be responsible for platelet aggregation. cAMP stops the Na^+/H^+ exchanger and leads to reduced mobilization of intracellular Ca^{2+} and phosphorylation of P47 [immunity-related GTPases (IRG) family]. At 70 μM and 150 μM kinetin decreased the amount of free radicals in collagen-activated platelet. Intravenous injection of 4 to 6 mg/kg of the cytokinin into mice prolonged bleeding time by approx. 1.9 to 2.1-fold [71, 72].

6-benzylaminopurine (6-BAP), kinetin, and zeatin induce positive inotropic effects in rat atria involving P2-purinoceptors, cGMP, and intracellular calcium release but not using pathways connected with arginine/nitric oxide, cyclooxygenase, phospholipase C, or L-type calcium channels [73].

Kinetin, as well as some other cytokinins, auxins, and gibberelinins, increases rat lung, small intestine, liver, and renal cortex guanylate cyclase activity two- to fourfold. The maximal stimulation of guanylate cyclase was observed at a 1-μM concentration of the plant hormones [74].

Cytokinins inhibits muscle creatine kinase (CK-MB), activates alanine aminotranferease (ALT) and aspartate aminotransferase (AST), and increases the level of AST, CK, and LDH, but they do not influence carbonic anhydrase and glucose-6-phosphate dehydrogenase [75, 76].

Cytokinins are also incorporated into the rRNA, tRNA, and tRNA of tobacco callus, *E. coli*, and yeast cells. Specific incorporation of kinetin into *E. coli* $tRNA^{Tyr}$ at position 37 by the putative tRNA kinetin transglycosylase takes place. The exchange reaction occurs in the presence of protein from *E. coli*, yeast, or MRC-5V2 cell extracts. Likewise, enzymes of *E. coli* or MRC-5V2 are able to catalyze incorporation of kinetin into yeast tRNA [77]. This shows a relationship between tRNA and enzymes in prokaryotic and eukaryotic organisms.

20.5 Perspectives

Pluripotency of cytokinins, especially in the context of recently discovered properties, makes it an attractive subject for experiments concerning a molecule with multiple roles and possible applications. The potential is largely connected to kinetin utility in cosmetics. As it is not only an antioxidant but also an antiaging compound that reduces wrinkles, regulates skin pigmentation, and improves overall skin appearance, kinetin has become the object of cosmetic companies' interest. Numerous reports about diverse diseases that kinetin can influence has raised its possible application in medicine. At present kinetin could be useful in FD therapy but it can also probably exert effects on other genetic disorders. Since cytokinins induce AML cell differentiation, they could possibly also affect other kinds of cancer cells.

References

1. Miller CO, Skoog F, Okumura FS, Von Saltza MH, Strong FM (1955) J Am Chem Soc 78:1375
2. Barciszewski J, Siboska GE, Pedersen BO, Clark BFC, Rattan SI (1996) FEBS Lett 393:197
3. Barciszewski J, Mielcarek M, Stobiecki M, Siborska G, Clark BFC (2000) Biochem Biophys Res Commun 179:69
4. Ge L, Yong JW, Goh NK, Chia LS, Tan SN, Ong ES (2005) J Chromatogr B Analyt Technol Biomed Life Sci 829:26
5. Barciszewski J, Siboska GE, Pedersen BO, Clark BFC, Rattan SI (1997a) Biochem Biophys Res Commun 238:317
6. Barciszewski J, Siboska GE, Pedersen BO, Clark BFC, Rattan SI (1997b) FEBS Lett 414:457
7. Wyszko E, Barciszewska MZ, Markiewicz M, Szymanski M, Markiewicz WT, Clark BFC, Barciszewski J (2003) Biochim Biophys Acta 1625:239
8. Mok DWS, Mok MC (2001) Annu Rev Plant Physiol Plant Mol Biol 52:89
9. Hou B, Lim EK, Higgins GS, Bowles DJ (2004) J Biol Chem 279:47822
10. Brzobohatý B, Moore I, Kristoffersen P, Bako L, Campos N, Schell J, Palme K. (1993) Science 262:1051
11. Letham DS, Palni LMS (1983) Annu Rev Plant Physiol 34:163
12. Armstrong DJ (1994) In: Mok DWS, Mok MC (eds) Cytokinins, Chemistry, Activity and Function. CRC Press, Boca Raton, FL, p 139
13. Molinal RV, Castelló S, García-Luis A, Guardiola JL (2007) Plant Cell Tissue Organ Cult 89:131
14. Papon N, Bremer J, Vansiri A, Andreu F, Rideau M, Creche J (2005) Planta Med 71:572
15. Pasqua G, Monacelli B, Mulinacci N, Rinaldi S, Giaccherini C, Innocenti M, Vinceri FF (2005) Plant Physiol Biochem43:293
16. Yaronskaya EB, Gritskevitch ER, Trukhanovets NL, Averina NG (2007) Russian J Plant Physiol 54:388
17. Shibaoka H. Thimann KV (1970) Plant Physiol 46:212
18. Debi RB, Taketa S, Ichii M (2005) J Plant Physiol 162:507
19. Li X, Mo X, Shou H, Wu P (2006) Plant Cell Physiol 47:1112
20. Filek M, Zę bala M, Szechyńska-Hebda M (2002) Z Naturforsch 57:696
21. Venis MA, Thomas EW, Barbier-Brygoo H, Ephritikhine G, Guen J (1990) Planta 182:232
22. Gzyl-Malchera B, Filek M, Brezesinski G, Fischer A (2007) Z Naturforsch 62:55
23. Gaudino RJ, Pikaard CS (1997) J Biol Chem 272:6799
24. Zhang K, Letham DS, John PC (1996) Planta 200:2
25. Turck F, Zilbermann F, Kozma SC, Thomas G, Nagy F (2004) Plant Physiol 134:1527
26. Balibrea LME, Garcia GMC, Fatima T, Ehness R, Lee TK, Proels R, Tanner W, Roitsch T (2004) Plant Cell 16:1276
27. Koistinen KM, Soininen P, Venäläinen TA, Häyrinen J, Laatikainen R, Peräkylä M, Tervahauta AI, Kärenlampi SO (2005) Phytochemistry 66:2524
28. Hurtado-Guerrero R, van Aalten DMF (2007) Chem Biol 14:589
29. Barna B, Fodor J, Pogány M, Király Z (2003) Pest Manag Sci 59:549
30. Balázs E, Barna B, Király Z (1976) Acta Phytopath Acad Sci Hung 11:1
31. Yang F, Simpson RB (1981) Proc Natl Acad Sci USA 78:4151
32. Schumaker KS, Gizinski MJ (1995) J Biol Chem 270:23461
33. Connett RJA, Hanke DE (1987) Planta 170:161
34. Pharmawati M, Billington T, Gehring CA (1998) Cell Mol Life Sci 54:272
35. Raman N, Elumalai S (1996) Ind J Exp Biol 34:577
36. Santos R, Hérouart D, Sigaud S, Touati D, Puppo A (2001) Mol Plant Microbe Interact 14:86
37. Rattan SI (2002) J Anti-aging Med 5:113
38. Inoue H, Hirobe M (1986) Biochem Biophys Res Commun 137:372
39. Sharma SP, Kaur P, Rattan SI (1995) Biochem Biophys Res Commun 216:1067
40. Sharma SP, Kaur J, Rattan SI (1997) Biochem Mol Biol Int 41:869

41. Olsen A, Siboska GE, Clark BFC, Rattan SI (1999) Biochem Biophys Res Commun 265:499
42. Pischetsrieder M, Seidel W, Munch G, Schinzel R (1999) Biochem Biophys Res Commun 264:544
43. Yokosuka K, Park JS, Jimbo K, Yamada K, Sato K, Tsuru M, Takeuchi M, Yamagishi S, Nagata K (2006) J Neurosurg Spine 5:324
44. Hamada Y, Araki N, Koh N, Nakamura J, Horiuchi S, Hotta N (1996) Biochem Biophys Res Commun 228:539
45. Cervantes-Laurean D, Jacobson EL, Jacobson MK (1996) J Biol Chem 271:10461
46. Verbeke P, Siboska GE, Clark BFC, Rattan SI (200) Biochem Biophys Res Commun 276:1265
47. Rup PJ, Sohal SK, Kaur H (2006) J Environ Biol 27:579
48. Orr MF, McSwain B (1960) Cancer Res 20:1362
49. Kimura T, Doi K (2004) Rejuvenat Res 7:32
50. Vicanova J, Bouez C, Lacroix S, Lindmark L, Damour O (2006) Ann NY Acad Sci 1067:337
51. Berge U, Kristensen P, Rattan SI (2006) Ann NY Acad Sci 1067:332
52. Orr MF, McSwain B (1957) Cancer 10:617
53. Kowalska E (1992) Folia Morphol 51:109
54. Honma Y, Ishii Y (2002) Leuk Lymphoma 43:1729
55. Honma Y (2003) Drug Dev Res 59:14
56. Ishii Y, Hori Y, Sakai S, Honma Y (2002) Cell Growth Differ 13:19
57. Griffaut B, Bos R, Maurizis JC, Madelmont JC, Ledoigt G (2004) Int J Biol Macromol 34:271
58. Ishii Y, Sakai S, Honma Y (2003) Biochim Biophys Acta 1643:11
59. Becker T, Gerke V, Kube E, Weber K (1992) Eur J Biochem 207:541
60. Sato N, Hitomi J (2002) Anat Rec 276:60
61. Bullinger L, Dohner K, Bair E, Frohling S, Schlenk RF, Tibshirani R, Dohner H, Pollack JR (2004) N Engl J Med 350:1605
62. Slaugenhaupt SA, Blumenfeld A, Gill SP, Leyne M, Mull J, Cuajungco MP, Liebert CB, Chadwick B, Idelson M, Reznik L, Robbins C, Makalowska I, Brownstein M, Krappmann D, Scheidereit C, Maayan C, Axelrod FB, Gusella JF (2001) Am J Hum Genet 68:598
63. Hims MM, Ibrahim el C, Leyne M, Mull J, Liu L, Lazaro C, Shetty RS, Gill S, Gusella JF, Reed R, Slaugenhaupt SA (2007) J Mol Med 85:149
64. Adunka O, Moustaklis E, Weber A, May A, von Ilberg C, Gstoettner W, Kierner AC (2003) ORL J Otorhinolaryngol Relat Spec 65:84
65. Chimini G, Chavrier P (2000) Nat Cell Biol 2:191
66. Ivetic A, Ridley AJ (2004) Immunology 112:165
67. Kamino H, Hiratsuka M, Toda T, Nishigaki R, Osaki M, Ito H, Inoue T, Oshimura M (2003) Cell Struct Funct 28:495
68. Takahashi K, Sasaki T, Mammoto A, Takaishi K, Kameyama T, Tsukita S, Takai Y (1997) J Biol Chem 272:23371
69. Lee JH, Chung KY, Bang D, Lee KH (2006) Proteomics 6:1351
70. Wagner M, Hampel B, Bernhard D, Hala M, Zwerschke W, Jansen-Durr P (2001) Exp Gerontol 36:1327
71. Sheu JR, Hsiao G, Shen MY, Chou CY, Lin CH, Chen TF, Chou DS (2003) Platelets 14:189
72. Hsiao G, Shen MY, Lin KH, Chou CY, Tzu NH, Lin CH, Chou DS, Chen TF, Sheu JR (2003) Eur J Pharmacol 465:281
73. Froldi G, Gallo U, Ragazzi E, Caparrotta L (1999) Planta Med 65:245
74. Vesely DL, Hudson JL, Pipkin JL Jr, Pack LD, Meiners SE (1985) Endocrinology 116:1887
75. Çelik I, Özbek H, Tülüce Y (2002) J Biol 26:193
76. Ciftci M, Demir Y, Ozmen I, Atici O (2003) J Enzyme Inhib Med Chem 18:71
77. Barciszewski J, Otzen D, Rattan SI, Clark BFC (1994) Biochem Int 28:805

Chapter 21
Safety Issues of Phytomedicines in Pregnancy and Paediatrics

Laura Cuzzolin and Giuseppina Benoni

Abstract In this review we discuss some aspects of herbal use among pregnant women and children, since exposure to phytomedicines is frequent in these subjects, often on a self-treatment basis.

From an analysis of the literature some interesting aspects emerge: (1) herbal consumption during pregnancy ranges from 7 to 45%, depending on the geographical area; (2) the use of herbal products during pregnancy could be due to pre-existing conditions, to symptoms/problems related to the new physiological state (nausea, vomiting, etc.), to prevent miscarriage or to prepare for childbirth; (3) the important aspect of safe use should take into account that some plants are contraindicated (oxytocic and uterine-stimulating herbs, teratogenicity, effect on neonatal birth weight), may affect concurrent conditions such as gestational diabetes and pre-eclampsia and may be dangerous for the newborn (some case reports of adverse reactions are reported); (4) herbal use in children is relatively common and on the rise; (5) few data exist about the clinical evidence for effectiveness; (6) most herbal products have not been subjected to rigorous clinical trials in children and a number of case reports are available in the literature reporting side effects due to herbal consumption by newborns (through lactation), infants and children.

Based on information summarized in this review, the use of herbs during pregnancy and in children remains inadequately studied as regards both efficacy and safety. However, data reported show that herbal products are not free of risk and suggest that their use should be contraindicated in pregnancy and a cautious behaviour adopted in children.

Keywords Herbs · Safety · Pregnancy · Children

L. Cuzzolin (✉)
Department of Medicine and Public Health-Section of Pharmacology University of Verona, Policlinico G.B. Rossi, Piazzale L.A. Scuro, 37134 Verona, Italy
e-mail: laura.cuzzolin@univr.it

K.G. Ramawat (ed.), *Herbal Drugs: Ethnomedicine to Modern Medicine*,
DOI 10.1007/978-3-540-79116-4_21,

21.1 Introduction

Phytomedicine is becoming increasingly popular for treating many different problems. The amount of information on this use is substantial and indicates that up to 50% of the general population have tried at least one herbal product [1–4]. However, phytomedicines have the potential to elicit the same types of adverse reactions as synthetic drugs, since they consist of whole extracts or, more commonly, of defined parts of plants (root, rhizome, leaf, flower head) that contain numerous active molecules [5]. Moreover, in most countries herbs are sold as unlicensed food supplements or are available to consumers as over-the-counter items in various preparations not regulated with the same scrutiny as conventional drugs, with risks of contamination or adulteration with poisonous metals, non-declared herbs or conventional medicines.

The safety of phytomedicines becomes particularly important in some subpopulations of patients such as pregnant women and children, which are more vulnerable to the effects of drugs but also of natural products for their physiological characteristics. Despite the fact that available data are insufficient to justify herbal use during pregnancy and in paediatrics, exposure to herbal products is frequent in these subjects [6, 7], often on a self-treatment basis [8, 9].

In this review we will discuss some aspects of herbal use in pregnant women and children, with the aim of alerting customers and health-care professionals to the fact that these compounds are not entirely free of risk in these subjects.

21.2 Methods

Medline and Embase searches were performed between 1990 and 2007 using the terms 'phytotherapic compounds', 'phytomedicine', 'efficacy', 'safety' and 'epidemiological data' combined with the terms 'pregnancy', 'newborns' and 'children'.

Information on the use of herbal products on the examined subpopulations was derived from medical records, prescription data, personal interviews, self-completion questionnaires or a combination of these.

21.3 Herbal Products in Pregnancy

The use of herbal products in pregnancy is common, although there is little evidence of its safety [6]. The actual incidence of this use is unknown, even if it has been quoted by some authors as varying between 7 and 45% depending on geographical area [10–14]. This frequent use is particularly worrisome because many of these compounds are taken without the physician's knowledge. In fact, while pregnant women recognize the potential risks of drug consumption, they do not realize that herbal products, if taken incorrectly, could also be toxic. This derives from the implicit belief that herbal remedies, being natural, are necessarily safe [15].

21.3.1 Epidemiological Data

There have been limited data on the extent of herbal consumption during pregnancy, but from the analysis of the available studies this consumption ranges from 7.1 to 50%.

A series of surveys conducted in Finland between 1985 and 1988 showed a rise in the use of alternative medicines (mostly herbal products) from 4 to 15% [16].

A small Australian series published in 1996 suggested that 15% of patients in an emergency department had taken herbal preparations during pregnancy [17].

Another study performed in South Africa showed that 43.7% of 577 women had used herbal products during pregnancy, mostly in the second trimester, to improve foetal conditions or to make labour easier [10].

A survey of 500 American midwives showed that approximately half were using herbs with the purpose of inducing labour [18], with evening primrose, red raspberry, castor oil and blue cohosh being the most commonly used.

A survey conducted among 1200 pregnant Nigerian women demonstrated that 12.1% consumed native herbs, with a prevalent use among nulliparous women (42%) [19].

Another survey of 200 pregnant US women demonstrated that 15% used herbal products (mostly ginger, chamomile and cola) in an attempt to relieve morning sickness [20].

Twenty-two out of 242 (9.1%) pregnant US women attending antepartum visits reported to have taken at least one herbal product, with garlic, aloe, chamomile, ginger, Echinacea and ginseng being the most frequently used [11].

Similar findings were observed in another US survey, where 20 out of 150 pregnant women (13%) had used dietary supplements, mostly Echinacea and ginger: the most commonly cited reason for starting to use an herbal product was to relieve nausea and vomiting [12].

Two hundred eleven pregnant women attending the antenatal clinic of the Women's and Children's Hospital of Adelaide (Australia) were interviewed to assess herbal use. Throughout their entire pregnancy, 71 out of 211 (33.6%), 43 out of 211 (20.4%) and 49 out of 211 (23%) of the women took at least one herbal product in the first, second and third trimester respectively. Ginger was used by about a fifth of women in the first trimester to treat morning sickness. Chamomile replaced ginger as the main herbal product taken in the second trimester. Raspberry leaf, the most frequently used herbal preparation in the third trimester, was taken mainly to prepare for labour [21].

Another survey was carried out among 305 patients presenting at an Australian antenatal clinic at 16 to 24 weeks gestation. The study showed that 37 out of 305 (12%) women used herbal products. The most commonly used herbal treatments were raspberry leaf, ginger and St John's wort to prevent miscarriage and to treat morning sickness and depression respectively [22].

Among a sample of 1203 pregnant women who answered to a questionnaire on phytotherapy consumption, 7.1% used herbal products, mostly Echinacea, St John's wort and Ephedra. This use was more common among women over 40 years old and in 46% of cases was based on a health-care provider's advice [13].

Four hundred women who gave birth at Ulleval University Hospital of Oslo (Norway) were interviewed about the use of herbal products. One hundred forty-four women (36%) had used at least one herbal product during pregnancy, in total 249 products containing 46 different herbs. The most commonly used herbs were Echinacea, iron-rich herbs, ginger, chamomile and cranberry. The proportion of use increased throughout the first, second and third trimesters of pregnancy. Approximately 40% of the women had used herbal products that were considered possibly harmful or without any information about safety in pregnancy. Only 11.8% of the women had been recommended herbal use by healthcare providers [23].

Among a sample of 500 Australian women, 36% took at least one herbal product during pregnancy. The most common herbs taken were raspberry leaf as uterine tonic, ginger to relieve nausea and chamomile for sleep or relaxation [24].

21.3.2 Herbs Commonly Used during Pregnancy

The use of herbal products during pregnancy could be due to pre-existing conditions or to symptoms/problems related to the new physiological state such as nausea, vomiting, constipation, fatigue, depression, urinary tract infections and itching.

Ginger, a well-known remedy for gestational sickness, is used for its antiemetic properties related to local gastrointestinal, anticholinergic and antihistaminic actions [25]. Some authors demonstrated significant improvements in the severity of both nausea and vomiting [26], whereas other researchers found it effective for hyperemesis [27].

A number of plants (comfrey, parsley, peppermint, evening primrose, flax seed) and fish oil, providing essential vitamins and minerals, are used during pregnancy as nutritional tonics even if their specific benefits are not clearly identifiable [28].

Women looking for natural remedies for pregnancy-associated constipation are advised to use bulk laxatives, since stimulant laxatives are contraindicated [29]. In particular, *Cascara sagrada*, senna and aloe contain anthroquinones that could cause contraction of the smooth muscle of the uterus [8].

Dandelion root is a liver tonic that gently increases bile flow: it aids digestion and waste elimination. Dandelion leaf decreases oedema and is rich in calcium and other substances that promote bone strength [30]. Scientific documentation is limited despite its long-time use [31].

Extracts from St John's wort are used by pregnant women as an antidepressant [32], since almost 10% of women experience depression during pregnancy and patients with a history of depression are at risk for puerperal worsening of mood [33].

Chamomile is traditionally used as a mild sedative and against digestive problems, but excessive use in pregnancy should be avoided [31]. Echinacea is often used prophylactically against the common cold for its antiviral and antibacterial properties [34]. The majority of the available studies report positive results, but there is not enough evidence to recommend its use during pregnancy [35]. Cranberry is used to prevent and treat urinary tract infections, even if there is no conclusive evidence to recommend its use [36]. In addition, specific herbs are traditionally used to prevent miscarriage or to prepare for childbirth at the end of pregnancy due to their potential uterine-stimulating properties. Wild yam and lobelia are reputed to prevent miscarriage [28]. Blue cohosh is thought to increase efficiency of contractions in labour [18] attributable to caulosaponin. Raspberry leaf is frequently ingested by women in the third trimester as it is thought to tone the uterus in preparation for labour, making contractions more effective and less painful [37].

21.3.3 Safety Data

Among herbs that are specifically used during pregnancy, clear evidence of negative effects exists in some cases: for other plants advisory warnings are cautious since a cause-effect relationship has not always been established beyond a reasonable doubt but data available on their safety are too limited. Moreover, in many cases toxicity related to herbal products may be due to adulteration, contamination, mislabelling or misidentification of the plant [5].

The important aspect of a safe use of herbal compounds during pregnancy may be considered from different points of view.

First, some specific plants are contraindicated for their effects on pregnancy outcomes. Oxytoxic and uterine-stimulating herbs that may potentially induce spontaneous abortion include aloe, blue and black cohosh, chamomile, golden seal, pennyroyal, parsley, rosemary, rue, tansy, chaste tree and castor. Obviously, as regards culinary herbs such as parsley, rosemary and chamomile, it is important to distinguish therapeutic doses from the lower amounts commonly used [28]. Relatively few teratogenic effects have been demonstrated for herbs or their contaminants. Based on animal studies, it is possible that certain constituents of ginger should be mutagenic in a dose-dependent manner, even if this was not demonstrated in humans [38]. Pregnant women should be advised to discontinue use of St John's wort as damage to oocytes may occur in high concentrations [39]. The use of *Ginkgo biloba* during pregnancy may cause foetal harm due to the accumulation of colchicine in the placenta [40]. Moreover, the main concern in using ginkgo leaf revolves around its antiplatelet properties that may prolong bleeding during delivery [41]. Frangulin, the major anthroquinone contained in *Cascara sagrada*, is teratogenic [28]. The potential teratogenicity of anagyrine in blue cohosh has been suggested [42], and a possible link between vascular and skeletal abnormalities and maternal consumption of this herb has been reported [43]. A recent epidemiological study underlines a possible association between hypospadia and soy milk consumption [44]. Other published data support the direct effect on neonatal birth weight by some herbs. A small dose-response effect of caffeine-containing beverages in reducing the birth weight of neonates has been reported [45]. Finnish authors tested the effects of maternal consumption of licorice on neonatal birth weight: babies with heavy exposure to licorice-containing glycyrrhizin (an inhibitor of cortisol metabolism) were not significantly lighter at birth, but they were significantly more likely to be born prematurely [46].

Secondly, certain activities of plants may affect pregnant women with concurrent conditions. Hypertensive properties of herbal products such as blue cohosh, ginger, ginseng and licorice may add to the complications of pre-eclampsia [31]. Similarly, hyperglycaemic properties of ginseng and licorice may complicate gestational diabetes, or coumarin-containing compounds (angelica, dong quai, aniseed, chamomile) could be a concern for women suffering from coagulation disorders [28].

Only a small number of clinical trials and retrospective surveys specifically report on the safety of herbal product use during pregnancy; their findings are synthesized in a recent review by Ernst [47]. Moreover, some case reports are available in the literature (Table 21.1).

Blue cohosh taken in the final month of pregnancy led to myocardial infarction, congestive heart failure and cardiovascular shock in a newborn whose mother took three times the recommended dose [42]. Another woman ingested an unspecified amount of blue and black cohosh to induce labour [48], and the newborn developed seizures, kidney damage and the need for mechanical ventilation. After maternal consumption of blue cohosh, an infant was born with severe seizures, encephalopathy and renal failure [49]. In these cases, the authors suggested the involvement of caulosaponin, a constituent that constricts coronary blood vessels and causes myocardial toxic reactions. Another case where chamomile was used to induce labour resulted in maternal anaphylaxis and foetal asphyxia [50]. Two documented cases of a malformation from herbal ingestion during pregnancy involve neonatal androgenization after maternal use of a ginseng preparation [51, 52]. A fatal neonatal liver injury was related to the mother's consumption of an herbal cough tea containing 0.6 mg pyrrolizidine alkaloids throughout pregnancy [53]. Eight newborns showed cardiorespiratory depression requiring intensive care after ingestion of infusions of *Montanoa tomentosa* by their mothers during labour: the authors hypothesized a mechanism similar to oxytocin or ergot alkaloids [54]. A case of foetal alcohol syndrome was associated with the consumption of a herbal tonic by a 29-year-old Chinese woman during her pregnancy [55]. A Japanese woman took a remedy containing *Tripterygium wilfordii* during pregnancy for rheumatoid arthritis: after an uncomplicated delivery at 38 weeks' gestation, the newborn suffered from occipital meningoencephalocele and cerebellar agenesis [56].

Table 21.1 Case reports of severe adverse drug reactions (ADRs) in newborns after maternal herbal consumption during pregnancy and/or lactation

Reference	Herb	ADRs
Jones and Lawson [42]	Blue cohosh	Myocardial infarction, congestive heart failure
Gunn and Wright [48]	Blue and black cohosh	Seizures, kidney damage, respiratory problems
Wright [49]	Blue cohosh	Seizures, encephalopathy, renal failure
Jensen-Jarolim et al. [50]	Chamomile	Asphyxia
Koren et al. [51]	Ginseng	Neonatal androgenization
Awang [52]	Ginseng	Neonatal androgenization
Roulet et al. [53]	Pyrrolizidine alkaloids	Fatal liver injury
Montoya-Cabrera et al. [54]	*Montanoa tomentosa*	Cardiorespiratory depression
Pradeepkumar et al. [55]	Herbal tonic	Alcohol syndrome
Takei et al. [56]	*Tripterygium wilfordii*	Occipital meningoencephalocele, cerebellar agenesis

21.4 Herbal Products in Paediatrics

For many families and their children the use of herbal products is an accepted adjunct or alternative to orthodox medical care. Because women make up the largest percentage of herbal users [3, 57], it is not unexpected that they also administer herbal remedies to their children. However, data about the efficacy and safety in childhood populations are scarce since most surveys have excluded children from their analyses.

21.4.1 Epidemiological Data

A number of international surveys indicate that herbal use in children is on the rise, but the majority of the available data is related to complementary/alternative medicines (CAM) in general [58, 60], and so it is difficult to extrapolate detailed information on herb consumption in children. In every case, these studies reported prevalence rates of 11 to 41% for CAM use, with the most frequently reported therapies being herbal or homeopathic medicines.

From a population-based telephone survey, 5% of the respondents reported giving their children herbal medicines in the past year [61]. In a national online survey, 41% of 520 adolescents reported to have used herbal remedies in the previous 6 months, mainly ginseng, ginger, *Ginkgo biloba*, valerian and St. John's wort [62]. The results of another study conducted in four primary paediatric care practices in Washington showed that 8% of children had been treated with herbal therapies in the prior year [58]. Other authors [63] verified herbal use among 2562 families in Kansas and Wisconsin through questionnaires: child herbal use was indicated in 35.8% of the surveys. The herbs most commonly taken by children were chamomile, *Aloe vera*, peppermint, garlic and lavender.

In other cases, information is focused on groups of patients with specific diagnoses or problems.

Some authors [64] explored the use of herbs in children with attention-deficit-hyperactivity disorder or depression: the prevalence of herbal therapy among 117 patients was 20% and, interestingly, the children's psychiatrists and paediatricians were not aware of such use. *Ginkgo biloba*, Echinacea and St John's wort were the herbs most frequently given. A survey performed in three paediatric gastroenterological centres reported that 14% of 208 children with inflammatory bowel disease used herbal medicines to treat this condition, in addition to conventional therapies [65]. In Australia, 29% of 174 asthmatic children on therapy with bronchodilators and disodium cromoglycate or inhaled steroids had used herbal preparations to ease their symptoms [66].

The results of a survey conducted in a Pittsburgh paediatric emergency department showed that approx. 6% of the parents had ever given their children herbs [67]. In another survey of 142 families in an emergency department, 45% of caregivers reported giving their child an herbal product, mostly aloe, Echinacea and sweet oil:

the most commonly reported reasons for use were colds, burns, immune stimulation and relaxation [68].

Use of herbal medicines in preoperative patients has sparked great interest because of the possibility that such therapies may alter immune response, retard wound healing or interfere with drugs or coagulation parameters: relatively little is known about the incidence of herbal use specifically in the paediatric surgical population.

Among 1021 paediatric patients surveyed through questionnaires during a preoperative visit, 12.8% used herbal remedies before the surgery: the herbs most commonly taken were echinacea, aloe and cranberry [69]. Two other surveys confirmed the use of herbal medicines also preoperatively in children. In the first study [70], among 601 children 6.4% reported to be currently taking an herbal preparation: Echinacea and arnica were the commonest used herbal remedies, but most importantly a significant number had taken herbs interacting with anaesthesia and surgery. In the other survey [71], 3.5% of 894 paediatric surgical patients had been taking herbal medications in the 2 weeks prior to surgery, mostly Echinacea. Thirty-six parents (4%) of 914 paediatric surgical patients undergoing surgical procedures at an urban children's hospital reported their child to have taken an herbal product: 15 of these children were undergoing major surgery and 13 were at risk for potential drug or anaesthetic interactions. Echinacea, chamomile and aloe were the most frequently used herbal medicines [72].

Despite the potential interference of herbal products with antitumoural chemotherapy [73], different surveys report herbal use among children with cancer as an adjunct to conventional or supportive therapy, especially to reduce the side effects of chemotherapy. An analysis of such surveys showed that the prevalence of herbal use is as high as 20% and a large proportion of users communicate this use to their physician.

Fernandez et al. [74] surveyed 583 paediatric patients diagnosed with cancer in British Columbia between 1989 and 1995. Among the 366 respondents, 42% used complementary therapies, mostly herbal agents. Among 75 children treated for cancer at the Babies and Children's Hospital of New York and interviewed on the use of unconventional therapies, 20 (27%) reported having taken at least one herbal agent, and this use was not related to the severity of chemotherapy- or radiotherapy-induced side effects, but rather to improve the general health of the child. Green tea and cat's claw were the herbs most commonly used [75]. One hundred and one paediatric cancer patients in Washington State were identified and telephone interviews were conducted: 34.7% used herbal preparations for both primary and secondary prevention, mostly to maintain good health and to avoid non-cancer illnesses: these treatments were higher in children over 5 years of age and among patients with neurological tumours [76]. Other researchers [77] interviewed by telephone parents of children with cancer in the province of Saskatchewan (Canada) to estimate the prevalence of unconventional therapies. Among families responding to the survey, 36% reported using herbal products for their child. A survey [78] was conducted among parents of children treated at the oncologic clinic of Sainte Justine Hospital in Montreal: 18 of 92 parents who completed the questionnaire (20%) reported that their child used at least one herbal product. During a 6-month period,

110 consecutive paediatric patients with cancer attending a tertiary care hospital in western Mexico were interviewed on the use of unconventional treatments: 53 patients (69%) reported to have taken herbal or plant extracts, mostly to limit the side effects of conventional therapies [79]. A detailed questionnaire was completed by 95 patients attending the Paediatric Oncology Unit of Gazi in Turkey: 25 out of 95 (26.3 %) used at least one herbal product, mostly stinging nettle. The most frequently cited reason for using an herbal product was the belief that it could help cure the cancer [80].

21.4.2 Herbs Commonly Used in Children

In paediatric patients herbal products are used to promote health, to prevent illness and to treat acute but overall chronic, recurrent or incurable conditions such as asthma, atopic dermatitis, allergic rhinitis, cystic fibrosis, inflammatory bowel disease, rheumatoid arthritis or cancer.

Regarding the important aspect of efficacy of such treatments in young patients, some studies have revealed promising results, while in other cases no clinical evidence of effectiveness has been demonstrated. In one study, herbal teas containing chamomile seemed to have a favourable effect on infantile colic [81]. Two controlled trials involving a limited number of subjects have been performed on the use of cranberry for the prevention and treatment of urinary tract infections in children with neurogenic bladder [82, 83]: both studies suggested that cranberry juice was not effective. *Echinacea* is commonly used to prevent and treat the common cold and respiratory tract infections, but its usefulness in children has yet to be clearly demonstrated. A randomized placebo-controlled trial involving 524 children 2 to 11 years old assessed the efficacy of this herbal product in reducing the duration and/or severity of upper respiratory symptoms [84]. *Echinacea* was not effective and its use was associated with an increased risk of rash. Other authors investigated the clinical efficacy of *Echinacea* in an observational study in which 1322 children were suffering from recurring upper respiratory infections. Herbal treatment was effective either in resolving symptoms or in shortening the duration of illness, but there was no placebo comparison [85]. Clinical experience suggests that the use of herbal remedies such as St John's wort in children and adolescents with affective disorders is increasing. Preliminary findings [86, 87] indicate that St John's wort may be clinically effective in the treatment of juvenile depression. Six clinical trials have assessed the efficacy of evening primrose oil used in paediatric populations to cure atopic dermatitis [88, 91] and hyperactivity disorders [92, 93]. The administration of this herbal product produced only minimal or no behavioural improvements in 49 hyperactive children, while some benefits were observed in the treatment of atopic dermatitis. Supplementation of evening primrose oil has also been evaluated in 11 diabetic children [94], but it remains unknown if this kind of therapy may have a preventive effect on diabetic vascular complications.

21.4.3 Safety Data

Most herbal products have not been subjected to rigorous clinical trials, and there remains a dearth of knowledge concerning how children are affected by these substances. Paediatric subjects are physiologically more vulnerable to certain adverse effects of herbs than adults. For example, some herbs such as senna and aloe are known cathartics and some herbal teas contain powerful diuretic compounds: these actions may cause dehydration and electrolyte disturbances quickly in an infant or young child. Moreover, some subpopulations are more susceptible to certain adverse effects of herbs than other children. Subjects with allergies may be at increased risk, since the allergic potential of some plants commonly used in children that could cause contact dermatitis, rhinitis, conjunctivitis and wheezing is well known. For example, chamomile can cause anaphylaxis and other herbs such as angelica and rue are capable of photosensitization [95]. Thus paediatricians must be cautious regarding the hazards of the long-term use of herbs, even if causality is sometimes uncertain and adverse events could have been caused by overdosing or contamination of the remedy rather than by the herbal ingredient itself. A publication from the WHO Monitoring Center included 8985 case reports of adverse events associated with herbal products observed in 55 countries during the period 1968–1997: about 100 of these events were referred to children up to 10 years of age and a further 100 were related to adolescents [96].

First of all, we must distinguish between neonates and children. In the first case, newborns could be exposed to herbal products both indirectly (during the period of lactation) and directly.

How herbs may affect lactation in breastfeeding women has not been fully explored. The excretion of herbs into breast milk is a concern, as many herbs have lipophilic chemicals that may concentrate in breast milk and be transferred to the newborn. During lactation, St John's wort should be used with caution due to potential side effects. Despite good scientific evidence that this herb does not affect maternal milk production or infant body weight [97], there is also evidence that St John's wort constituents cross into breast milk [98], and a few cases of colic, drowsiness or lethargy were reported [97]. A 32-year-old Chinese woman who took Dong quai for postpartum weakness developed acute hypertension, and her 3-week-old son's blood pressure increased to 115/59 [99]. During lactation, the safety of ginkgo leaf, used for memory boosting, varicose veins or cyclic oedema, is likewise unknown and should be avoided until high quality human studies are conducted [100]. A review article recommended that black cohosh should be avoided during lactation due to its potential hormonal effects [101, 102].

Relatively few case reports are available in the literature reporting side effects of herbal consumption by infants and children (Table 21.2).

A neonate hospitalized for suspected sepsis developed chemical burns following a topical treatment with vinegar, a common home remedy [103]. An early fatal colitis was observed in a 4-year-old boy following exposure to the alkaloids detected in *Chelidonium majus* [104]. Other case reports documented life-threatening bradycardia and respiratory depression in three small children following unintentional

Table 21.2 Case reports of adverse drug reactions (ADRs) in infants and children taking herbal products

Reference	Herb	Subject	ADR
Korkmaz et al. [103]	Vinegar	Neonate	Burns
Koopman [104]	*Chelidonium majus*	4-year-old child	Fatal colitis
Horowitz et al. [105]	Jin Bu Huan	3 small children	Bradycardia, respiratory depression
Garty [106]	Garlic	6-month-old infant	Burns
Canduela et al. [107]	Garlic	6-year-old child	Necrotic ulcers
Bakerink et al. [108]	Mint tea (pennyroyal oil)	2 infants	Multiple organ failure
Ernst [5]	Asafetida gum	5-week-old infant	Methemoglobinemia
Steenkamp et al. [109]	Pyrrolizidine alkaloids	20 children	Hepatic veno-occlusive diseases
Bagheri et al. [110]	Valerian	13-year-old child	Fulminant liver failure
Parson et al. [111]	Eucalyptus oil	149 children	Poisoning
Darben et al. [112]	Eucalyptus oil	6-year-old child	Systemic intoxication

overdosing of the Chinese remedy Jin Bu Huan containing large amounts of tetrahydropalmatine [105]. A 6-month-old infant suffered garlic burns when his father applied crushed garlic cloves to the wrists [106], while a 6-year-old child developed a necrotic ulcer on her foot after her grandmother applied crushed garlic under a bandage as a remedy for a minor sore [107]. Two cases of serious or fatal toxicity have been described in two infants who had been given 90 to 120 ml of mint tea containing pennyroyal oil for colic and minor ailments [108]. A case of methemoglobinemia in a 5-week-old infant treated with a gum asafetida preparation has been reported [5]. In a study carried out in two hospitals in South Africa, 20 children were diagnosed with hepatic veno-occlusive diseases, and pyrrolizidine alkaloid poisoning was suggested as the cause of the problem, confirmed from the presence of these alkaloids in the urine samples of four subjects [109]. The case of a 13-year-old child with fulminant seronegative liver failure requiring transplantation has been reported [110]: the most likely cause of the liver damage was *Valeriana officinalis* contained in a self-administered herbal mixture. The toxicity of eucalyptus oil in young children has been well documented and was the leading cause of hospital admission for 149 childhood poisoning in Australia [111]. In addition, systemic effects of eucalyptus oil applied topically have been reported. A 6-year-old girl was topically treated with a homemade concoction containing eucalyptus oil for pruritus: the child developed symptoms of intoxication and was admitted to hospital unconscious [112].

21.5 Conclusions

Based on the information obtained in the literature and summarized here, the following conclusions can be drawn.

Despite their widespread popularity, the use of many herbs during pregnancy and in children remains poorly documented and inadequately studied as regards both efficacy and safety. However, data reported above show that herbal products are not free of risk for pregnant women, their newborns and children, even though scientific evidence is incomplete due to a high rate of under-reporting. This under-reporting can be explained by various factors. Firstly, consumers of herbal products, comprised mostly of pregnant women, act differently than the general population in reporting adverse events to their physician [113], believing these substances to be natural and, therefore, necessarily safe [15]. Secondly, monitoring of adverse effects due to herbal consumption is inefficient, as in many countries these products are not regulated as traditional medicines. Thirdly, standard search techniques are not always able to locate certain information if the signalled adverse effect appears in uncited journals. Another important aspect is that the cause-effect relationship with these substances is not always well established due to the presence of confounding factors and to a lack of independent confirmation. In every case, until definitive data emerge, the use of herbal products should be considered contraindicated in pregnancy and a cautious behaviour should be adopted towards their consumption in children.

Finally, we would underline that physicians and other health-care professionals have an important educational role. Overall, pregnancy-care providers should be aware of the common use of herbal products by women, learn as much as possible about herbal remedies and discuss their potential risks and benefits with their patients.

References

1. Eisenberg DM, Davis RB, Ettner SL, Appel S, Wilkey S, Van Rompay M, Kessler RC (1998) J Am Med Assoc 280:1569
2. Kuo GM, Hawley ST, Weiss LT, Balkrishnan R, Volk RJ (2004) BMC Complement Altern Med 4:18
3. Gardiner P, Graham R, Legedza AT, Ahn AC, Eisenberg DM, Phillips RS (2007) Altern Ther Health Med 13:22
4. Goldstein LH, Elias M, Ron-Avraham G, Biniaurishvili BZ, Madjar M, Kamargash I, Braunstein R, Berkovitch M, Golik A (2007) Br J Clin Pharmacol 64:373
5. Ernst E (2000) Risks associated with complementary therapies. In: Dukes MNG, Aronson JK (eds) Meyler's Side-Effects of Drugs, 14th edn. Elsevier, Amsterdam, p 1649
6. Pinn G (2001) Aust Fam Phys 30:351
7. Cuzzolin L, Zaffani S, Murgia V, Gangemi M, Meneghelli G, Chiamenti GP, Benoni G (2003) Eur J Pediatr 162:820
8. Conover EA (2003) Best Pract Res Clin Endocrinol Metab 17:237
9. Sibinga E, Ottolini M, Duggan A, Wilson M (2000) Pediatr Res 47:226
10. Mabina MH, Moodley J, Pitsoe SB (1997) Trop Doct 27:84
11. Gibson PS, Powrie R, Star J (2001) Obstet Gynecol 97:S44
12. Tsui B, Dennehy CE, Tsourounis C (2001) Am J Obstet Gynecol 185:433
13. Hepner DL, Harnett M, Segal S, Camann W, Bader AM, Tsen LC (2002) Anesth Analg 94:690
14. Glover DD, Amonkar M, Rybeck BF, Tracy TS (2003) Am J Obstet Gynecol 188:1039
15. Adams C, Cannell S (2001) Menopause 8:433
16. Hemminki E, Mantyranta T, Malin M, Koponen P (1991) Scand J Soc Med 19:199
17. McLennan A, Wilson D, Taylor A (1996) Lancet 2:569
18. McFarlin BL, Gibson MH, O'Rear J, Harman P (1999) J Nurse Midwifery 44:205
19. Gharoro EP, Igbafe AA (2000) Med Sci Monit 6:84
20. Pastore L (2000) Cochrane Library 3:529
21. Maats FH, Crowther CA (2002) Aust NZ J Obstet Gynaecol 42:494
22. Pinn G, Pallett L (2002) Complement Ther Nurse Midwifery 8:77
23. Nordeng H, Havnen GC (2004) Pharmacoepidemiol Drug Safety 13:371
24. Forster DA, Denning A, Wills G, Bolger M, McCarthy E (2006) BMC Pregnancy Childbirth 6:21
25. Quian DS, Liu ZS (1992) Chung Kuo Chung His Chieh Ho Tsa Chih 12:95
26. Vutyavanich T, Kraisarin T, Ruangsri R (2001) Obstet Gynecol 97:577
27. Fischer-Rasmussen W, Kjaer SK, Dahl C, Asping U (1991) Eur J Obstet Gynecol Reprod Biol 38:19
28. Johns T, Sibeko L (2003) Birth Defects Res 68:501
29. Belew C (1999) J Nurse-Midwifery 44:231
30. Schutz K, Carle R, Schieber A (2006) J Ethnopharmacol 107:313
31. Newall CA, Anderson LA, Phillipson JD (1996) Herbal Medicines: a Guide for Health-Care Professionals. Pharmaceutical Press, London
32. Grush LR, Nierenberg A, Keefe B, Cohen LS (1998) J Am Med Assoc 280:1566

33. Altshuler LL, Cohen L, Szuba MP, Burt VK, Gitlin M, Mintz J (1996) Am J Psychiat 153:592
34. Schulten B, Bulitta M, Ballering-Bruhl B, Koster U, Schafer M (2001) Arzneimittelforschung 5197:563
35. Melchart D, Linde K, Fischer P, Kaesmayr J (2002) *Echinacea* for preventing and treating the common cold (Cochrane Review). In: The Cochrane Library, Issue 3. Update Software, Oxford
36. Jepson RG, Mihaljevic L, Craig J (2002) Cranberries for preventing urinary tract infections (Cochrane Review). In: The Cochrane Library, Issue 3. Update Software, Oxford
37. Tiran D (2003) Complement Ther Nurse Midwifery 9:176
38. Wilkinson KM (2000) Reprod Toxicol 14:507
39. Rayburn WF, Gonzalez CL, Christensen HD, Stewart JD (2001) Am J Obstet Gynecol 184:191
40. Petty HR, Fernando M, Kindzelskii AL, Zarewych BN, Ksebati MB, Hryhorczuk LM, Mobashery S (2001) Chem Res Toxicol 14:1254
41. Kudolo G (2001) Altern Ther 7:105
42. Jones TK, Lawson BM (1998) J Pediatr 132:550
43. Ortega JA, Lazerson J (1987) J Pediatr 111:87
44. North K, Golding J (2000) BJU Int 85:107
45. Bracken MB, Triche EW, Belanger K, Hellenbrand K, Leaderer BP (2003) Am J Epidemiol 157:456
46. Strandberg TE, Javenpaa AL, Vanhanen H, McKeigue PM (2001) Am J Epidemiol 153:1085
47. Ernst E (2002) Br J Obstetr Gynecol 109:227
48. Gunn TR, Wright IM (1996) NZ Med J 110:20
49. Wright IM (1999) J Pediatr 134:384
50. Jensen-Jarolim E, Reider N, Fritsch R, Breiteneder H (1998) J Allergy Clin Immunol 102:1041
51. Koren G, Randor S, Martin S, Danneman D (1990) J Am Med Assoc 264:2866
52. Awang DV (1991) J Am Med Assoc 266:363
53. Roulet M, Laurini R, Rivier L, Calame A (1988) J Pediatr 112:433
54. Montoya-Cabrera MA, Simental-Toba A, Sanchez-Rodriguez S, Escalante-Galindo P, Aguilar-Conteras A (1998) Gac Med Mex 134:611
55. Pradeepkumar VK, Tan KW, Ivy NG (1996) Aust NZ J Obstet Gynaecol 36:420
56. Takei A, Nagashima G, Suzuki R, Hokaku H, Takahashi M, Miyo T, Asai J, Sanada Y, Fujimoto T (1997) Pediatr Neurosurg 27:45
57. Tindle HA, Davis RB, Phillips RS, Eisenberg DM (2005) Altern Ther Health Med 11:42
58. Ottolini MC, Hamburger EK, Loprieato JO, Coleman RH, Sachs HC, Madden R, Brasseux C (2001) Ambul Pediatr 1:122
59. Sawni-Sikand A, Schubiner H, Thomas RL (2002) Ambul Pediatr 2:99
60. Fong DP, Fong LK (2002) Austral Fam Phys 31:388
61. Wheaton AG, Blanck HM, Gizlice Z, Reyes M (2005) Ann Epidemiol 15:678
62. Wilson KM (2003) J Adolesc Health 32:127
63. Lohse B, Stotts JL, Priebe JR (2006) J Am Diet Assoc 106:227
64. Cala S, Crismon ML, Baumgartner J (2003) Pharmacotherapy 23:222
65. Heuschkel R, Afzal N, Wuerth A, Zurakowski D, Leichtner A, Kemper K, Tolia V, Bousvaros A (2002) Am J Gastroenterol 97:382
66. Shenfield G, Lim E, Allen H (2002) J Pediatr Child Health 38:252
67. Pitetti R, Singh S, Hornyak D, Garcia SE, Herr S (2001) Pediatr Emerg Care 17:165
68. Lanski SL, Greenwald M, Perkins A, Simon HK (2003) Pediatrics 111:981
69. Lin YC, Bioteau AB, Ferrari LR, Berde CB (2004) J Clin Anesth 16:4
70. Crowe S, Lyons B (2004) Pediatr Anesth 14:916
71. Everett LL, Birmingham PK, Williams GD, Brenn BR, Shapiro JH (2005) Pediatr Anesth 15:455
72. Noonan K, Arensman RM, Hoover JD (2004) J Pediatr Surg 39:500
73. Orbach O (1999) N Engl J Med 340:569

74. Fernandez C, Stutzer CA, MacWilliam L, Fryer C (1998) J Clin Oncol 16:1279
75. Kelly KM, Jacobson JS, Kennedy DD, Braudt SM, Mallick M, Weiner MA (2000) J Pediatr Hematol/Oncol 22:412
76. Neuhouser ML, Patterson RE, Schwartz SM, Hedderson MM, Bowen DJ, Standish LJ (2001) Prev Med 33:347
77. Bold J, Leis A (2001) J Pediatr Oncol Nurs 18:16
78. Martel D, Bussieres J-F, Theoret Y, Lebel D, Kish S, Moghrabi A, Laurier C (2005) Pediatr Blood Cancer 44:660
79. Gomez-Martinez R, Tlacullo-Parra A, Garibaldi-Covarrubias R (2007) Pediatr Blood Cancer 49:820
80. Karadeniz C, Pinarli FG, Oguz A, Gursel T, Canter B (2007) Pediatr Blood Cancer 48:540
81. Weizman Z, Alkrinawi S, Goldfarb D, Bitran C (1993) J Pediatr 122:650
82. Foda M, Middlebrook P, Gatfield CT, Potvin G, Wells G, Schillinger JF (1995) Can J Urol 2:98
83. Schlager TA, Anderson S, Trudell J, Hendley JO (1999) J Pediatr 135:698
84. Taylor JA, Weber W, Standish L, Quinn H, McGann M, Calabrese C (2003) J Am Med Assoc 290:2824
85. Götte K, Roschke I (2001) Prakt Pädiatrie 111:95
86. Findling RL, McNamara NK, O'Riordan MA, Reed MD, Demeter CA, Branicky LA, Blumer JL (2003) J Am Acad Child Adolesc Psych 42:908
87. Simeon J, Nixon MK, Milin R, Jovanovic R, Walker S (2005) J Child Adolesc Psychopharmacol 15:293
88. Bordoni A, Biagi PL, Masi M, Ricci G, Fanelli C, Patrizi A (1987) Drugs Exp Clin Res 14:291
89. Biagi PL, Bordoni A, Masi M, Ricci GP, Fanelli C, Patrizi A, Ceccolini E (1988) Drugs Exp Clin Res 14:285
90. Biagi PL, Bordoni A, Hrelia S, Celadon M, Ricci GP, Canella V, Patrizi A, Specchia F, Masi M (1994) Drugs Exp Clin Res 20:77
91. Hederos CA, Berg A (1996) Arch Dis Child 75:494
92. Aman MG, Mitchell EA, Turbott SH (1987) J Abnorm Child Psychol 15:75
93. Arnold LE, Kleykamp D, Votolato NA (1989) Biol Psychiat 25:222
94. Arisaka M, Arisaka O, Yamashiro Y (1991) Prostagl Leukot Essent Fatty Acids 43:197
95. Woolf AD (2003) Pediatrics 112:240
96. Farah MH, Edwards R (2000) Pharmacoepidemiol Drug Safety 9:105
97. Lee A, Minhas R, Matsuda N, Lam M, Ito S (2003) J Clin Psychiat 64:966
98. Klier CM, Schafer MR, Schmid-Siegel B, Lenz G, Mannel M (2002) Pharmacopsychiatry 35:29
99. Nambiar S, Schwartz RH, Constantino A (1999) West J Med 171:152
100. Sierpina VS, Wollschlaeger B, Blumenthal M (2003) Am Fam Physic 68:923
101. Mahady GB, Fabricant D, Chadwick LR, Dietz B (2002) Nutr Clin Care 5:283
102. Dugoua JJ, Seely D, Perri D, Koren G, Mills E (2006) Can J Clin Pharmacol 13:e257
103. Korkmaz A, Sahiner U, Yurdakok M (2000) Pediatr Dermatol 17:34
104. Koopman H (1937) Vergiftungsf lle 8:93
105. Horowitz RS, Gomez H, Moore LL, Fulton B, Feldhaus K, Brent J, Stermitz FR, Beck JJ, Alessi JR, De Smet PAGM (1993) Mor Mortal Wkly Rep 42:633
106. Garty B (1993) Pediatrics 91:658
107. Canduela V, Mongil I, Carrascosa M, Docio S, Cagisgas P (1995) Br J Dermatol 132:161
108. Bakerink JA, Gospe SM, Dimand RJ, Eldridge MW (1996) Pediatrics 98:944
109. Steenkamp V, Stewart MJ, Zuckerman M (2000) Ther Drug Monit 22:302
110. Bagheri H, Brouè P, Lacroix I, Larrey D, Olives JP, Vaysse P, Ghisolfi J, Montastruc JL (1998) Therapie 53:77
111. Parson BJ, Dobbin M, Tibbals J (1997) Aust N Z J Publ Health 21:297
112. Darben T, Cominos B, Lee CT (1998) Australasian J Dermatol 39:265
113. Barnes J, Mills SY, Abbot NC, Willoughby M, Ernst E (1998) Br J Clin Pharmacol 45:496

Index